The Biology of
Acinetobacter

Taxonomy, Clinical Importance,
Molecular Biology, Physiology,
Industrial Relevance

FEDERATION OF EUROPEAN MICROBIOLOGICAL SOCIETIES SYMPOSIUM SERIES

A Continuation Order Plan is available for this series. A continuation order will bring delivery of each new volume immediately upon publication. Volumes are billed only upon actual shipment. For further information please contact the publisher.

The Biology of
Acinetobacter

Taxonomy, Clinical Importance, Molecular Biology, Physiology, Industrial Relevance

Edited by

K. J. Towner

Department of Microbiology and PHLS Laboratory
University Hospital
Nottingham, United Kingdom

E. Bergogne-Bérézin

Département de Microbiologie, CHU Bichat
Paris, France

and

C. A. Fewson

Department of Biochemistry
University of Glasgow
Glasgow, United Kingdom

PLENUM PRESS • NEW YORK AND LONDON

Library of Congress Cataloging in Publication Data

The Biology of *Acinetobacter*: Taxonomy, Clinical Importance, Molecular Biology,
Physiology, Industrial Relevance / edited by K. J. Towner, E. Bergogne-Bérézin, and
C. A. Fewson.
    p.   cm. — (FEMS symposium; no. 57)
  Based on the proceedings of the 2nd International Workshop on Acinetobacter held
under the auspices of the Federation of European Microbiological Societies, on Sept.
6–7, 1990 in Paris, France.
  Includes bibliographical references and index.
  ISBN 0-306-43902-6
  1. Acinetobacter — Congresses. I. Towner, K. J. II. Bergogne-Bérézin, E. III. Fewson,
C. A. IV Federation of European Microbiological Societies. V. International Workshop
on Acinetobacter (2nd: 1990: Paris, France) VI. Series.
QR82.N4B56 1991   91-12826
589.9'5 — dc20   CIP

Based on the proceedings of a symposium held under the auspices of the
Federation of European Microbiological Societies,
on September 6–7, 1990, in Paris, France

ISBN 0-306-43902-6

© 1991 Plenum Press, New York
A Division of Plenum Publishing Corporation
233 Spring Street, New York, N.Y. 10013

PREFACE

The 1st International Workshop on *Acinetobacter* was held on 6th September, 1986, in Manchester, England, in association with the 14th International Congress of Microbiology. That occasion was so well attended and productive that there were soon discussions about how, when and where the next meeting should be held. This time, however, there was sufficient confidence to think of a more substantial meeting and to plan for the proceedings to be published. It emerged that there was wide agreement that the time was ripe to take stock of the entire biology of *Acinetobacter*: its occurrence and taxonomy; its molecular biology, biochemistry and physiology; its clinical importance and its industrial and commercial applications. The 2nd International Workshop on *Acinetobacter* took place from 6th to 7th September, 1990, at the Institut Pasteur, Paris, and was sponsored by the Federation of European Microbiological Societies. There were about 100 participants from 19 countries. The backbone of the meeting consisted of 23 plenary lectures. There were 28 posters and the meeting closed with a general discussion which went on long after the official finishing time despite all the counter-attractions of a sunny Parisian Friday afternoon. Indeed discussions continued while cruising along the Seine and while dining at the top of the Tour Montparnasse. However, the vitality and usefulness of even the most successful meeting is difficult to transmit by the printed word. The editors decided, therefore, not simply to publish the proceedings of the Workshop, but rather to commission a set of review articles with the aim of giving an up-to-date and rounded picture of our current knowledge of the genus *Acinetobacter*. These reviews, together with a number of articles that expand on research talks presented at the Workshop, form this book, the timely production of which owes much to the expert assistance of Linda Bowering, Gerry Holmes and Gillian Johnston, to whom the editors express their grateful thanks.

In addition to sponsorship by the Federation of European Microbiological Societies, the Workshop was supported by financial contributions from Beecham, Boots, Bristol-Meyers, Glaxo, ICI, Lilly, Merck Sharpe & Dohme, Pharmuka, Roussel and Unilever, to all of whom the organisers and participants are greatly indebted.

K.J. Towner
E. Bergogne-Bérézin
C.A. Fewson

CONTENTS

ACINETOBACTER: PORTRAIT OF A GENUS

K.J.Towner[a], E.Bergogne-Bérézin[b] and C.A.Fewson[c]

[a]Department of Microbiology & PHLS Laboratory
University Hospital
Nottingham NG7 2UH, UK

[b]Laboratoire de Microbiologie
CHU Bichat
75877 Paris Cedex 18, France

[c]Department of Biochemistry
University of Glasgow
Glasgow G12 8QQ, UK

INTRODUCTION

The genus *Acinetobacter* originally proposed by Brisou and Prévot (1954) comprised a heterogenous collection of non-motile Gram-negative organisms which could be distinguished from other similar organisms by their lack of pigmentation (Ingram and Shewan, 1960). Now classified in the family Neisseriaceae, the present generic description (Juni, 1984) allows unambiguous identification of strains to the genus level. A difficulty arises in that many current members of the genus have been classified previously under a variety of different names (reviewed by Henriksen, 1973), and much of the early literature concerning this group of organisms is, therefore, difficult to interpret owing to confusion over nomenclature and the lack of a widely-accepted classification scheme. Delineation of species within the genus is still the subject of much research. The latest developments are described in detail elsewhere (Grimont and Bouvet, this volume), but the reader should bear in mind that many of the *Acinetobacter* strains described in this volume still occupy a somewhat nebulous taxonomic position within the genus.

Why the interest in this genus? Acinetobacters are ubiquitous organisms in the environment and have formed the subject of several

The Biology of Acinetobacter, Edited by K. J. Towner *et al.*
Plenum Press, New York, 1991

previous reviews (e.g. Henriksen, 1973, 1976; Juni, 1978). They have been implicated in a variety of food spoilage and human disease processes. They are often resistant to commonly used antibiotics and may form a reservoir of antibiotic resistance genes, particularly in hospital environments. They are easy to isolate and can be grown in the laboratory in a simple mineral medium containing a single carbon and energy source; indeed the vast majority of isolates resemble saprophytic pseudomonads in being able to use any one of a large number of organic compounds as a carbon and energy source. Although the utilisation of carbohydrates is relatively uncommon, a major biochemical feature of the genus is the range of compounds that can be metabolised. This range encompasses aliphatic alcohols (including 2,3-butanediol), amino acids, dicarboxylic and fatty acids, *n*-alkanes, alicyclic compounds, and many aromatic compounds. Members of the genus are, therefore, particularly suitable organisms for studying unusual peripheral biochemical pathways, and may also have an important role to play in the biodegradation of a variety of hazardous and unpleasant aromatic pollutants and industrial residues. All three of the major modes of gene transfer are known to occur in *Acinetobacter*, and it is already clear from detailed studies of gene structure, organisation and genetic regulation that *Acinetobacter* has some unusual and exciting features that may help in gaining an understanding of important processes that occur during prokaryotic evolution in general. The aim of this introductory article is to present a broad overview of the genus *Acinetobacter* and to highlight the most important features, many of which are described in detail in the remainder of this volume.

IDENTIFICATION AND TAXONOMY

Morphological and Biochemical Identification

Members of the genus *Acinetobacter* are short, plump, Gram-negative rods that often become more coccoid in the stationary phase. Cells commonly occur in pairs or chains of variable length that are occasionally difficult to destain. No spores are formed. Flagella are absent but, although generally considered to be non-motile, "twitching" and "gliding" motility has been reported, particularly on semi-solid media (Barker and Maxsted, 1975; Henrichsen, 1975; Mukherji and Bhopale, 1983). Many strains are encapsulated. All strains of *Acinetobacter* are non-fastidious strict aerobes that are catalase-positive and oxidase-negative. Most strains can grow in a simple minimal medium containing a single carbon and energy source at a temperature between 20 and 30°C. The GC content is 38-45 mol %. There is no single biochemical test that enables differentiation of this genus, but they can be readily distinguished from similar bacteria by means of a combination of standard nutritional tests for non-fastidious, non-fermentative organisms. Unambiguous attribution of strains to the genus *Acinetobacter* is achieved by means of the DNA transformation test described by Juni (1972).

Members of the genus *Acinetobacter* have suffered a long history of extensive taxonomic changes which has inhibited a proper comparative study of their properties. At least 15 different "generic" names have been used to describe these organisms (Baumann et al., 1968), the most common of which are *Bacterium anitratum* (Shaub and Hauber, 1948), *Herellea vaginicola* and *Mima polymorpha* (Debord, 1939), *Micrococcus calcoaceticus* and "B5W" (Juni, 1978), and *Moraxella glucidolyţica* and *Moraxella lwoffii* (Piéchaud et al., 1956; Brisou, 1957). The genus *Acinetobacter*, as conceived by Brisou (1957) and Piéchaud et al. (1956), included both oxidase-positive (*Moraxella*) and oxidase-negative strains. In 1971 the Subcommittee on the Taxonomy of *Moraxella* and Allied Bacteria proposed that the genus *Acinetobacter* should include only the oxidase-negative strains. Classified in the family Neisseriaceae, the genus *Acinetobacter* originally comprised only one species, *A. calcoaceticus*, with, according to the Approved Lists of Bacterial Names (Skermann et al., 1980), two varieties: var. *anitratus* (formerly *Herellea vaginicola*) and var. *lwoffii* (formerly *Mima polymorpha*). The latest definition of *Acinetobacter* (Bouvet and Grimont, 1987) was arrived at by using a combination of phenotypic properties (including carbon source utilisation tests) and identification of genotypic species by means of modern taxonomic methods (genetic transformation, DNA hybridisation and RNA sequence comparison). This new definition of *Acinetobacter* has resulted in the identification of 17 genomic species and 19 biotypes. The major genomic species comprise *A. baumannii* (with nine biotypes), *Acinetobacter* sp. 3 (six biotypes), *A. haemolyticus, A. johnsonii, A. junii, A. lwoffii, A. calcoaceticus, Acinetobacter* sp. 6 and *Acinetobacter* sp. 11 (for further details see Grimont and Bouvet, this volume). Simultaneously, independent studies performed by Tjernberg and Ursing (1989) and Nishimura et al. (1988b) have provided data compatible with the groupings outlined above. Phenotypic group 5 has been named *A. radioresistens* (Nishimura et al. 1988a; Ino and Nishimura, 1990). Thus a satisfactory classification of the genus *Acinetobacter* is now at the disposal of microbiologists, and *Acinetobacter* is recognised as a group clearly distinct from *Moraxella* and *Neisseria*.

OCCURRENCE AND ISOLATION

Strains of *Acinetobacter* can be obtained easily from soil, water and sewage with appropriate enrichment techniques (Baumann, 1968), and various methods have been developed for their enumeration (LaCroix and Cabelli, 1982; Bifulco et al., 1989). It has been estimated that acinetobacters may constitute 0.001% of the total heterotrophic aerobic population of soil and water (Baumann, 1968). They have been found to be common in American rural drinking water supplies (Bifulco et al., 1989) and to be present at densities exceeding 10^4 organisms /100 ml in freshwater ecosystems and 10^6 organisms /100 ml in raw sewage (LaCroix and Cabelli,

1982). Although they can be isolated from heavily polluted water, they are found more frequently near the surface of freshwater and where freshwater flows into the sea (Droop and Jannasch, 1977). Acinetobacters are also found in a variety of foodstuffs, including milk products, fresh meat, eviscerated chicken carcasses and other poultry meats (Koburger, 1964; Barnes and Thornley, 1966; Eribo and Jay, 1985). Members of the genus are responsible for spoilage of a number of foods, including chickens, eggs, bacon and fish, even when stored under refrigerated conditions (Shewan et al., 1960; Thornley et al., 1960; Gardner, 1971; Jay, 1982) or following irradiation (Firstenberg-Eden et al., 1980). Acinetobacters are normal inhabitants of human skin and have consequently been implicated as presumed causal or contributory agents in a variety of infectious disease processes (see below). They also have some veterinary significance (e.g. Dickie, 1978; Gibson and Eaves, 1981). Acinetobacters have also been isolated from a bizarre range of other sources, including laboratory dental pumice (Williams et al., 1983), eye shadows (Dawson and Reinhardt, 1981) and stored buffalo semen (Ramachandra et al., 1981).

Isolation of acinetobacters can be achieved on standard laboratory media such as nutrient agar or trypticase soy agar. Most strains have an optimum growth temperature of 30-35°C, and grow well at 37°C, but some isolates are mesophilic and may be unable to grow at 37°C (Breuil and Kushner, 1975). Baumann (1968) described an enrichment culture procedure for isolating members of the genus from soil and water which involves the use of a simple mineral medium with acetate as the sole carbon and energy source, and Holton (1983) described a selective differential medium for the isolation of acinetobacters from clinical specimens. Acinetobacters have a slightly acid pH optimum for growth, and vigorous aeration at a pH of 5.5-6.0 favours their enrichment.

CLINICAL IMPORTANCE OF *ACINETOBACTER* SPP.

Extent of the Problem

Strains of *Acinetobacter* are distributed widely in the hospital environment and have been isolated from various inanimate sources such as ventilatory equipment, blood collection tubes, mattresses and plastic urinals. Nevertheless, it seems that the skin of patients and staff is the likely source for most outbreaks of nosocomial *Acinetobacter* infection (Bergogne-Bérézin et al., 1987; Noble, this volume). It has been shown that 29% of nurses who had contact with infected patients carried epidemic strains (French et al., 1980), but *Acinetobacter* spp. may become established in hospitals as part of the resident skin flora in addition to being simply transient contaminants on the hands of staff and patients. *A. baumannii* is the main species associated with nosocomial infection (Bouvet and Grimont, 1987), whereas other species such as *A. haemolyticus, A. junii, A.*

johnsonii or *A. lwoffii* seem to be isolated primarily from environmental sources and only occasionally from clinical samples.

The pathogenic role of *Acinetobacter* is limited to nosocomial infections, but a dramatic increase in the incidence of this organism has been noted in many reports during the last few years, particularly when compared with other common organisms causing nosocomial infection. Joly-Guillou et al. (1990b) reported that *Acinetobacter* strains comprised 1-9% of all bacterial species isolated from clinical specimens between 1984 and 1988. The major sites of infection do not differ from those of other nosocomial Gram-negative pathogens, with the lower respiratory tract and the urinary tract accounting for 15-28% of total *Acinetobacter* infections (Retailliau et al., 1979; Joly-Guillou et al., 1990b). Nosocomial pneumonia caused by *Acinetobacter* spp. is a particular problem in intensive care units (Cunha et al., 1980; Vieu et al., 1980; Muller-Serieys et al., 1989), and often involves patients with severe underlying disease requiring assisted mechanical ventilation with equipment that is capable of aerosolising the organism. Some isolates of *Acinetobacter* may reflect colonisation rather than infection (French et al., 1980), but the presence of pneumonia with a lobar distribution, purulent secretions, and isolation of the organism as the sole or predominant pathogen from bronchial brushings confirms the pathogenic role of *Acinetobacter*. Nosocomial urinary tract infections caused by *Acinetobacter* spp. occur at rates varying from 2-61%, depending on local outbreaks (Hoffmann et al., 1982; Mayer and Zinner, 1985). Other nosocomial infections occasionally reported in the medical literature include meningitis, skin and wound infections, burn wound superinfections and bacteraemia (Bergogne-Bérézin et al., 1987).

The severity and difficulties associated with *Acinetobacter* nosocomial infection are partly related to the high rates of resistance to most major antibiotics found in *Acinetobacter*. The evolution of antibiotic resistance in *Acinetobacter* spp. over the last 20 years has resulted in a progressive decrease in susceptibility to commonly used antimicrobial drugs, to the extent that acinetobacters now exhibit one of the most impressive patterns of antibiotic resistance found in nosocomial bacteria. Acinetobacters are frequently resistant to most available antibacterial drugs of the ß-lactam and aminoglycoside classes, with increasing proportions of clinical isolates becoming resistant to the initially effective fluoroquinolones. Even resistance to imipenem, which was for several years the most effective drug in treating *Acinetobacter* nosocomial infections, has now been described in several reports. Detailed analysis of the unusual large molecular mass chromosomal ß-lactamases produced by members of the genus has raised the exciting hypothesis that these enzymes could represent an intermediate evolutionary step between "normal" ß-lactamases and penicillin binding proteins (Hood and Amyes, this volume). A full description of the different antibiotic resistance mechanisms in *Acinetobacter* can be found in the review by Bergogne-Bérézin and Joly-Guillou (this volume).

Pathogenicity and Virulence Factors

The virulence of *Acinetobacter* has been studied primarily in animal models involving mice (Lutz et al., 1956; Daly et al., 1962; Obana et al., 1985; Obana, 1986; reviewed by Avril, this volume) and rats (Loeffelholz and Modrzakowski, 1988). Strains of *A. baumannii* have limited virulence in mice (LD_{50} values of 10^6-10^8 cfu/mouse) when inoculated intraperitoneally, even in neutropenic mice treated with cyclophosphamide. In guinea pigs, the virulence of *Acinetobacter* strains has been shown to be even more limited following sub-cutaneous inoculation (Lutz et al., 1956; Von Graevenitz, 1983). Analysis of possible antigenic determinants has shown that *Acinetobacter* produces a capsular polysaccharide formed of L-rhamnose and D-glucose, or galactose (Juni, 1978), but apart from its possible role in mediating adhesion to cells (Rosenberg and Rosenberg, 1981), the capsule does not seem to play an important virulence role. In contrast, the production of "slime", a mucous extravascular substance, does seem to result in enhanced virulence of *Acinetobacter* (Obana, 1986; Avril, this volume). Lipopolysaccharide from *Acinetobacter* has certain unusual structural features (Borneleit and Kleber, this volume) and may also have pathogenic properties.

Typing and Epidemiology

Analysis of the cell structure of acinetobacters has also provided the basis for certain typing systems. Hospital outbreaks of *Acinetobacter* infection require suitable typing methods for epidemiological studies. Novel systems that have been investigated in recent years include fluorescent antibody serotyping to detect capsular antigens, phage-typing (Bergogne-Bérézin et al., 1987), bacteriocin-typing, cell envelope protein patterns (Dijkshoorn et al., 1987) and plasmid analysis (Gerner-Smidt, 1989; Harstein et al., 1990; Joly-Guillou et al., 1990a; Vila et al., this volume). Determination of antibiograms and biotypes has also proved to be useful, but most investigators have concluded that, at present, a single reliable typing system for *Acinetobacter* does not exist, and that epidemiological studies require a combination of several different approaches. The current situation regarding typing systems is reviewed by Bouvet (this volume), and examples of the application of some of these typing systems to outbreaks of nosocomial *Acinetobacter* infection are described by Joly-Guillou et al. (this volume) and Vila et al. (this volume).

GENETICS AND MOLECULAR BIOLOGY

Plasmid and Transposon Behaviour

There is still a relative paucity of information about the detailed role of plasmids and transposons in the overall biology of *Acinetobacter*. The clinical importance of antibiotic resistance in nosocomial isolates of *Acinetobacter* has already been alluded to above. A variety of plasmids and transposons conferring antibiotic resistance can be transferred to *Acinetobacter* (Towner, this volume), where they may form an important hospital reservoir of antibiotic resistance genes. Although much is known about the biochemical mechanisms of antibiotic resistance in *Acinetobacter* (Bergogne-Bérézin and Joly-Guillou, this volume), little is known about the structure and regulation of these genes in *Acinetobacter* at the molecular level. The complex nature of the problem and the efforts needed to unravel it are well-illustrated by Elisha and Stern (this volume). Plasmids also seem to be involved in determining a range of other phenotypic properties in *Acinetobacter* (Towner, this volume). Again, there is a relative lack of information, but it seems clear that while some plasmids can transfer readily between *Acinetobacter* and other genera, there is also a pool of plasmid-mediated genetic information which is confined largely to *Acinetobacter* and not readily mobilisable to other organisms. This seems to result, however, from a barrier to transfer rather than expression (Towner, this volume). Further information on plasmid stability and expression of plasmid-encoded genes in *Acinetobacter* is provided by Winstanley et al. (this volume), with particular reference to the release of genetically engineered microorganisms (GEMs) into the natural environment.

Chromosome Organisation and Gene Regulation

Acinetobacter has been shown to have a circular chromosomal linkage map (Towner, 1978), and a total of 29 genetic loci have now been mapped (Vivian, this volume). It is not possible, as yet, to discern any overall features in the genome arrangement, but all three major modes of gene transfer (transformation, conjugation and transduction) have been demonstrated to occur in *Acinetobacter* (Vivian, this volume), and so the genetic tools for future work aimed at defining the gross topological structure of the chromosome are readily applicable. Methods for random insertional mutagenesis, marker rescue and the cloning of *Acinetobacter* genes in appropriate vectors are also available (Goosen et al., 1987; Hunger et al., 1990; articles by Gutnick et al.; Ornston and Neidle; Vosman et al., this volume).

So far as fine structure of the chromosome is concerned, particular attention has focused on the organisation and regulation of genes concerned with tryptophan biosynthesis (Haspel et al., this volume) and the ß-ketoadipate degradative pathway (Ornston and Neidle, this volume). With regard to tryptophan biosynthesis, striking similarities exist with respect to the organisation and regulation of the *trp* genes from *A. calcoaceticus* and *Pseudomonas putida*, but the complete regulatory circuits for this pathway have yet to be fully resolved in *Acinetobacter* (Haspel et al., this volume). Studies of *Acinetobacter* genes encoding the ß-ketoadipate pathway have indicated homologies between the *Acinetobacter* and *Pseudomonas* systems, but with different forms of transcriptional regulation. Although the GC content of the structural genes for the various enzymes differs by about 20%, there is a striking level of amino acid sequence identity. The difference in GC content is reflected in distinctive patterns of codon usage (articles by

Haspel et al.; Ornston and Neidle; White et al., this volume). Studies by L.N.Ornston and colleagues have demonstrated that considerable gene rearrangements must have occurred as the ß-ketoadipate pathway diverged in *Acinetobacter* and *Pseudomonas*, and their results have generated the exciting hypothesis that sequence exchange between DNA slippage structures causes mutations leading to genetic divergence, while repair events involving sequence exchange between slipped DNA strands may contribute to conservation of the evolved sequences. The experiments and discoveries leading to these stimulating concepts are described in detail by Ornston and Neidle (this volume).

BIOCHEMISTRY AND PHYSIOLOGY

Energetics of Growth, Electron Transfer and Oxidative Phosphorylation, and Transport Mechanisms

There have been systematic examinations of the effects of variables such as substrate, temperature, oxygen status, growth rate and nutrient limitation on the yields and composition of several strains of *Acinetobacter* (Abbott et al., 1974; Kleber et al., 1975; Du Preez and Toerien, 1978; Du Preez et al., 1981, 1984; Fewson, 1985; Hardy and Dawes, 1985), and in some cases substrate inhibition has been found to be important (Abbott, 1973; Jones et al., 1973; Fewson, 1985). Growth yields and respiratory efficiencies of *A. calcoaceticus* NCIB8250 have been measured in the course of batch and continuous culture experiments (Fewson, 1985; Hardy and Dawes, 1985). Growth yields were generally found to be rather low, probably because the effective P/O ratio is only about one under most conditions, even though there may be two or more sites of oxidative phosphorylation (Fewson, 1985). Failure to make use of all the potential synthesis of ATP is consistent with the conclusions that the redox chain of *A. calcoaceticus* HO1-N is divided into two equivalent sites of energy conservation (Ensley et al., 1981) and that proton translocating loops 0, 1 and 2 are active in *A. lwoffii* 4B (Meyer and Jones, 1973; Jones et al., 1977a). Results of experiments to determine $H+/O$ quotients all seem to suggest that there are at least two sites of oxidative phosphorylation (Meyer and Jones, 1973; Jones et al., 1977a; Hardy and Dawes, 1985). There is evidence that some strains of *Acinetobacter* which cannot grow on glucose as sole source of carbon and energy can nevertheless use the dehydrogenation of glucose to energise active transport systems and to provide ATP for biosynthesis (Kitagawa et al., 1986a,b). However, the observation that glucose can increase the molar growth yield on acetate (Müller and Babel, 1986; van Schie et al., 1987) may have to be subjected to more detailed interpretation in order fully to understand its quantitative significance (Duine, this volume).

The nature of the electron transfer chain in acinetobacters is not entirely clear despite many year's work (e.g. Whittaker, 1971; Meyer and Jones, 1973; Jones et al., 1977a; Ensley and Finnerty, 1980; Ensley et al.,

1981; Hardy and Dawes, 1985; Asperger and Kleber, this volume). In general, under conditions of high aeration a cytochrome o-containing oxidase and a cytochrome b_{554} are the predominant species, whereas under oxygen-limited conditions a cytochrome d-containing oxidase is predominant (Ensley and Finnerty, 1980; Hardy and Dawes, 1985; Geerlof et al., 1989). In recent work with *A. calcoaceticus* strain LMD 79.41, cytochrome o oxidase, cytochrome d oxidase and cytochrome b_{554} have all been partially purified and characterised (Geerlof et al., 1989). Cytochrome b_{562} is involved in linking glucose dehydrogenase to the main electron transfer chain (Beardmore-Gray and Anthony, 1986; Dokter et al., 1988; Geerlof et al., 1989; Duine, this volume). There is no evidence that *Acinetobacter* spp. contain a cytochrome c, and this accords with their status as oxidase-negative organisms. Several strains contain cytochrome P450 and this is reviewed by Asperger and Kleber (this volume). An NADH dehydrogenase connected with the respiratory chain has been solubilised, purified and partially characterised (Borneleit and Kleber, 1983). Ubiquinones, but not menaquinone, have been found in acinetobacters, including the neotype strain ATCC 23055. The number of isoprene units attached to the benzoquinone nucleus is typically nine, but small amounts of the next higher and lower isoprenologues are usually found (Whittaker, 1971; Collins and Jones, 1981; Nishimura et al., 1986, 1988a; L.M. Fixter, unpublished results); however, some *Acinetobacter* strains appear to contain either approximately equal amounts of CoQ_8 and CoQ_9, or just CoQ_8 (Denis et al., 1975).

It is clear that acinetobacters contain active transport systems for several substrates, and that in some strains these can be energised by glucose dehydrogenation (van Schie et al., 1985; Kitagawa et al., 1986 a,b). There is also evidence that some aromatic compounds are taken up by specific transport systems (Fewson, this volume). The cytoplasmic membrane of *A. calcoaceticus* NCIB8250 appears to be impermeable to monosaccharides (Cook and Fewson, 1973), but Midgley et al. (1986a,b) have presented evidence that the apparent impermeability of strain NCIB8250 to low concentrations of methyltriphenylphosphonium ion is partly the consequence of an efflux system. Some strains of *Acinetobacter* produce siderophores and iron-repressible outer membrane proteins (Smith et al., 1990; P.Williams and K.J.Towner, unpublished results), and these observations could have implications for explaining the organism's virulence in relation to the availability of iron. The entire question of uptake of substances into acinetobacters, and efflux systems from them, requires a great deal more effort if we are to understand their physiology and pathogenicity adequately.

Metabolism and Enzymology

Early work on the metabolism of *Acinetobacter* was fully reviewed by Juni (1978); since then there has been a steady flow of papers describing enzymes, co-factors, metabolic pathways and products, as well as other aspects of the biochemistry and physiology of the genus. In general, these have confirmed the view that *Acinetobacter* is typical of other Gram-negative eubacteria, but that it has distinctive metabolic features which are well-adapted to its lifestyle and support its nutritional versatility.

Metabolism centres on a Krebs tricarboxylic acid cycle which has a rather unusual pattern of regulation. The activities of pyruvate dehydrogenase, citrate synthase, isocitrate dehydrogenase and 2-oxoglutarate dehydrogenase are all modulated by AMP, and this may act in a concerted manner to direct metabolite flux through the cycle. This distinctive multi-point control of the cycle may be a significant feature in the regulation of energy metabolism in *Acinetobacter* (Weitzman, this volume). There is a repressible glyoxylate bypass (references in Juni, 1978; Asperger and Kleber, this volume). Phosphoenolpyruvate carboxylase serves an anaplerotic function under tight control, especially by nucleotides. Again, AMP appears to play an important role. The activation of phosphoenolpyruvate carboxylase by AMP may be physiologically advantageous because, under such conditions, there may be a high demand for oxaloacetate by the tricarboxylic acid cycle (Richter and Kleber, 1980). Oxaloacetate can be degraded by a coupled reaction involving malic enzyme and NAD-linked malate dehydrogenase (Juni, 1978). Many acinetobacters can grow in butane 2,3-diol and convert it into acetate by a cyclic mechanism (Juni, 1978). A plethora of complex catabolic pathways for the degradation of aromatic and alicyclic compounds, as well as alkanes, amino acids and related compounds, feed into the tricarboxylic acid cycle (articles by Asperger and Kleber; Fewson; Ornston and Neidle; Trudgill, this volume). Biosynthetic pathways, such as those involved in gluconeogenesis and amino acid synthesis, are subtly influenced by the need to integrate with the overall metabolism of the organism. This requirement shows up with respect to the details of regulation of gene expression as well as the regulation of enzyme activity by allosteric and other mechanisms (e.g. Mukkada and Bell, 1971; Juni, 1978; Haspel et al., this volume). There is no good evidence for the operation in *Acinetobacter* spp. of global metabolic regulatory systems, such as those involving cyclic AMP, but presumably they exist, and it has been suggested that there is a Pho regulon which is controlled by a single protein analogous to the product of the *phoB* gene of *E. coli* (Yashphe et al., 1990).

Many strains of *Acinetobacter* accumulate poly-ß-hydroxybutyrate and wax esters as energy reserves, and some strains accumulate polyphosphate, either as an energy reserve or as a phosphate reserve (Fixter and Sherwani, this volume). The mechanisms involved in polyphosphate metabolism provide a particular challenge because our present knowledge of the enzymes that might be involved is insufficient to provide a coherent explanation for the synthesis and degradation of polyphosphate in the genus as a whole.

Amongst the more special features of the genus is the ability of at least one strain to grow on malonate, apparently by converting it into malonyl-CoA followed by decarboxylation to acetyl-CoA (Kim and Kim 1985; Bahk and Kim, 1987; Lim and Kim, 1987). The acetate kinase of this strain appears to be quite different from the equivalent enzyme in *E. coli* (Kim and Park, 1988). Strain JC1 is a carboxydobacterium, and its inducible carbon monoxide dehydrogenase has been isolated and partially characterised (Kim and Kim, 1989; Kim et al., 1989). There has also been a report that some acinetobacters, isolated from the surfaces of soybean nodules (where there may be appreciable concentrations of hydrogen because of the action of nitrogenase), can oxidise hydrogen (Wong et al., 1986).

Despite the name '*anitratus*', many strains of *Acinetobacter* can assimilate nitrate, presumably without appreciable accumulation of nitrite (Villalobo et al., 1977; Juni, 1978). Ammonia appears to be assimilated by an NADP-dependent glutamate dehydrogenase, at least in strain NCIB8250. When glutamate is used as the nitrogen source for growth, the initial step involves transamination to form aspartate; if, in addition, the glutamate is serving as the source of carbon and energy, then aspartase is induced and releases surplus ammonia (A.C. Borthwick, L.M. Fixter and C.A. Fewson, unpublished results). A sulphur transferase and a rhodanase have both been isolated from acinetobacters (Vandenbergh and Berk, 1980; Layh et al., 1982; Aird et al., 1987). *Acinetobacter lwoffii* GB1 can volatilise mercury from media containing $HgCl_2$ (Bhattacharyya and Mandal, 1987), and enzymic reduction of $HgCl_2$ has been demonstrated in cell extracts (Chandhuri et al., 1990).

The enzymes of nucleic acid metabolism in *Acinetobacter* have not been much studied at the molecular level, although the overall pattern was established some time ago (Juni, 1978). *A. calcoaceticus* has at least three restriction endonucleases, *Acc*I, *Acc*II and *Acc*III, whose recognition sequences and cleavage sites make them useful for genetic recombination work (Kita et al., 1985).

Acinetobacters seem to be quite adept at producing extracellular enzymes (Juni, 1978); examples include lipases and esterases (Breuil and Kushner, 1975), amylase (Onishi and Hidaka, 1978), collagenase (Monboisse et al., 1979), endo-ß-1,6-glucanase (Kotohda et al., 1979) and ß-lactamase (Blechschmidt et al., 1989). There are several interesting periplasmic enzymes (Borneleit and Kleber, this volume), including an insulin-cleaving metalloprotease (Fricke and Aurich, 1989), and membrane-bound enzymes in abundance, including an NAD-independent malate dehydrogenase (Jones et al., 1977b), proteases (Fricke et al., 1987), leucine aminopeptidase (Ludewig et al., 1987), alanine aminopeptidase (Jahreis et al., 1989) and NAD-independent aldehyde, alcohol, lactate and mandelate dehydrogenases (references in Asperger and Kleber, this volume; Fewson, this volume).

So far as co-factors are concerned, acinetobacters have played a major role in the elucidation of the biological role of pyrroloquinoline quinone (PQQ), particularly in respect of its function in glucose dehydrogenase (Jongejan and Duine, 1989; Duine, this volume).

INDUSTRIAL, COMMERCIAL, CLINICAL AND
OTHER APPLICATIONS

A quite astonishing range of uses for *Acinetobacter* and its products can be gleaned from the scientific and patent literature (Table 1), although for one reason or another, not all of them may have proved to be successful. Around 200 processes involving *Acinetobacter* had been patented, or patents had been applied for, by mid-1990. In some cases the patents gave almost world-wide coverage, such as those for extracellular emulsifiers of oils and hydrocarbons in the European Community, the United States, Canada, Japan, Norway and Denmark. Perhaps the most novel and surprising of all the suggestions for using *Acinetobacter* is that for the enhancement of the clarity of fingerprints. The organisms proliferate on the material present in the deposit of barely detectable prints; they then multiply and form colonies on the ridges and, as a result, the characteristic fingerprint pattern becomes visible (Harper et al., 1987).

Products

The biochemistry, molecular biology and applications of *Acineto-bacter* bioemulsifying agents are extensively reviewed by Gutnick et al. (this volume), and these compounds appear likely to bring the organism to the attention of a new audience. Exploitation of the ability of acinetobacters to accumulate simple wax esters (Neidleman and Geigert, 1984; Fixter and Sherwani, this volume) will largely depend on the economics of the process, particularly the cost of feedstocks. These waxes might perhaps be used for the production of scarce alcohols. It is clear that the composition of the waxes, in terms of chain length and degree of unsaturation, can be strongly influenced by the conditions under which the organism is grown (M.J. McAvoy, L.M. Fixter and C.A. Fewson, unpublished results), and these observations might be worthy of industrial application.

Pollution Control

The involvement of *Acinetobacter* strains in the removal of phosphate from waste-water is reviewed by Gutnick et al. (this volume), and this process, which is important in minimising eutrophication, depends on the ability of the organisms to accumulate polyphosphate, which is in turn discussed by Fixter and Sherwani (this volume). The degradation of pollutant aromatic and alicyclic compounds and hydrocarbons which enable *Acinetobacter* to be used for effluent and other waste treatment has also been reviewed (articles by Asperger and Kleber; Fewson, this volume), and there have been attempts to increase the degradative capability of acinetobacters by genetic engineering (e.g. Liaw and Srinivasan, 1990).

Biotransformations

The general versatility of acinetobacters with regard to their ability to metabolise aromatic, alicyclic and a great range of other compounds underlies some of the possible commercial biotransformation processes (articles by Asperger and Kleber; Fewson; Trudgill, this volume), including the production of L-carnitine (Kleber, this volume). Production of gluconic acid depends on the possession of glucose dehydrogenase, a fascinating

Table 1. Industrial, commercial, clinical and other applications of *Acinetobacter*

Production of emulsifying agents, e.g. Emulsan, Biodispersan
  -to aid oil biodegradation
  -to stabilise heavy oil in water emulsions and coal in water slurries (in order to facilitate transportation by reduction in viscosity and to increase the efficiency of combustion)
  -to help prevent dental plaque
  -to facilitate grinding of limestone for use in paper making
Waste and effluent treatment
  -phosphate removal from waste-water
  -effluent purification in the petrochemical industry
  -degradation of cyclohexane, acrylonitrile, polyacrylate, acrylamide, polyethylene, poly(vinyl alcohol) resins, linear alkylbenzene sulphonates, aromatic hydrocarbons, cresols, phthalic acid esters
Anti-tumour and anti-leukaemia agents, e.g. asparaginase-glutaminase
Production of optically-active carboxylic acids and their enantiomeric esters, gluconic acid, racemic thiol-containing esters, amino acids and related compounds, L-carnitine, unsaturated organic acids
Components of cosmetics, detergents and shampoos, e.g. fragrant compositions, bioemulsifiers, moisturisers
Wax esters, e.g. for the production of medium-chain alcohols
Enzymes, e.g. *Acc*I restriction endonuclease, mutarotase, lipases, glucose dehydrogenase, aspartylglycosylamine amidohydrolase
Coal and petroleum desulphurisation
Manganese leaching from ores
Production of immune adjuvants
Decaffeination processes and debittering of orange juice
Production of protein-rich biomass
Development of latent fingerprints

enzyme in its own right (Duine, this volume) and one which may have applications in the construction of glucose sensors (D'Costa et al., 1986) and in the determination of PQQ (B.W. Groen and J.A. Duine, unpublished results).

Biomass Production

Production of single cell protein for animal feed is often limited by costs of possible feedstocks. Economically-viable processes must be based on raw materials that are cheap, plentiful and in constant supply. Two possible applications of *Acinetobacter* for single cell protein production involve growth on palm oil (Koh et al., 1985, 1987) or the effluent from the industrial production of oil from coal (Du Preez et al., 1985). In addition, ethanol (Abbott et al., 1973) or alkanes (Asperger and Kleber, this volume) might, at some time in certain parts of the world, become cheap enough to support protein production in the form of *Acinetobacter* biomass.

Glutaminase-Asparaginase

The possible use of glutaminase-asparaginase as an anti-cancer agent has led to this being perhaps the best-studied of all *Acinetobacter* enzymes. The logic of the approach is that cancerous cells may lack the enzymic capability to synthesise adequate amounts of certain non-essential amino acids, and if the circulatory level of these amino acids is reduced the cancerous cells will be selectively killed. Parenteral administration of L-asparaginase has caused regression of lymphomas and leukaemias in experimental animals and humans. The *E. coli* enzyme has been used in the treatment of acute lymphoblastic leukaemia for many years, and other amidohydrolases have been used in clinical trials (Ammon et al., 1988). glutaminase-asparaginase has been purified from *A. calcoaceticus* (Joner et al., 1973; Joner, 1976) and from *A. "glutaminasificans"* (Holcenberg, 1985). The latter enzyme uses both L-glutamine and L-asparagine as substrates and is a tetramer of identical subunits (M_r = 35,500; 331 residues) (Tanaka et al., 1988). Detailed studies of its kinetic properties and substrate binding site have been undertaken (Holcenberg et al., 1978; Steckel et al., 1983; Holcenberg, 1985). It has been crystallised (Wlodawer et al., 1977; Ammon et al., 1983), and a preliminary quaternary structure has been reported showing that each subunit folds into two domains: the amino terminal domain contains a five-stranded ß-sheet surrounded by five α-helices, while the carboxyl terminal domain contains three α-helices (Ammon et al., 1988). Various studies have been aimed at increasing the circulatory life of the enzyme and at reducing immunological complications; these have included chemical modifications such as succinylation and glycosylation (e.g. Holcenberg et al., 1975, 1979; Blazek and Benbough, 1981), and immobilisation in microspheres (e.g. Edman et al., 1987). There is an exhaustive literature on the possible application of *Acinetobacter* glutaminase-asparaginase in cancer therapy (e.g. Kien et al., 1985; Ammon et al., 1988).

ENVOI

Acinetobacter is a genus that has, for many years, been somewhat ignored by the main body of microbiologists. However, rapidly dawning interest in the genus resulted in the meeting which formed the basis for this book, and it is hoped that this brief introductory article will encourage the casual reader to delve more deeply into the wealth of scientific material which follows. Members of the genus *Acinetobacter* have a plethora of interesting facets to their lifestyles, some of which are unique, while others provide us with exciting insights into prokaryotic behaviour in general. We firmly believe that scientists, of whatever microbiological speciality, cannot fail to be stimulated by a closer acquaintance with this fascinating group of organisms.

REFERENCES

Abbott, B.J., 1973, Ethanol inhibition of a bacterium (*Acinetobacter calcoaceticus*) in chemostat cultures, *J. Gen. Microbiol.*, 75:383.

Abbott, B.J., Laskin, A.I., and McCoy, C.J., 1973, Growth of *Acinetobacter calcoaceticus* on ethanol, *Appl. Microbiol.*, 25:787.

Abbott, B.J., Laskin, A.J., and McCoy, C.J., 1974, Effect of growth rate and nutrient limitation on the composition and biomass yield of *Acinetobacter calcoaceticus, Appl. Microbiol.*, 28:58.

Aird, B.A., Heinrikson, R.L., and Westley, J., 1987, Isolation and characterization of a prokaryotic sulfurtransferase, *J. Biol. Chem.*, 262:17327.

Ammon, H.L., Murphy, K.C., Sjolin, L., Wlodawer, A., Holcenberg, J.S., and Roberts, J., 1983, The molecular symmetry of glutaminase-asparaginases: rotation function studies of the *Pseudomonas* 7A and *Acinetobacter* enzymes, *Acta Cystallog.*, B39:250.

Ammon, H.L., Weber, I.T., Wlodawer, A., Harrison, R.W., Gilliland, G.L., Murphy, K.C., Sjolin, L., and Roberts, J., 1988, Preliminary crystal studies of *Acinetobacter glutaminasificans* glutaminase-asparaginase, *J. Biol. Chem.*, 263:150.

Bahk, Y.Y., and Kim, Y.S., 1987, Acidic isocitrate lyase from *Acinetobacter calcoaceticus* grown on malonate, *Han'guk Saenghwa Hakhoechi*, 20:191.

Barker, J., and Maxsted, H., 1975, Observations on the growth and movement of *Acinetobacter* on semisolid media, *J. Med. Microbiol.*, 8:443.

Barnes, E.M., and Thornley, M.J., 1966, The spoilage of eviscerated chickens stored at different temperatures, *J. Food Technol.*, 1:113.

Baumann, P., 1968, Isolation of *Acinetobacter* from soil and water, *J. Bacteriol.*, 96:39.

Baumann, P., Doudoroff, M., and Stanier, R.Y., 1968, A study of the *Moraxella* group II. Oxidase negative species (genus *Acinetobacter*), *J. Bacteriol.*, 95:1520.

Beardmore-Gray, M., and Anthony, C., 1986, The oxidation of glucose by *Acinetobacter calcoaceticus*: interaction of the quinoprotein glucose dehydrogenase with the electron transport chain, *J. Gen. Microbiol.*, 132:1257.

Bergogne-Bérézin, E., Joly-Guillou, M.L., and Vieu, J.F., 1987, Epidemiology of nosocomial infections due to *Acinetobacter calcoaceticus, J.Hosp. Infect.*, 10:105.

Bhattacharyya, G., and Mandel, A., 1987, Volatization of mercury by a soil bacterium *Acinetobacter lwoffi* GB1, *Ind. J. Exp. Biol.*, 25:347.

Bifulco, J.M., Shirey, J.J., and Bissonnette, G.K., 1989, Detection of *Acinetobacter* spp. in rural drinking water supplies, *Appl. Environ. Microbiol.*, 55:2214.

Blazek, R., and Benbough, J.E., 1981, Improvement in the persistence of microbial asparaginase and glutaminase in the circulation of the rat by chemical modifications, *Biochim. Biophys. Acta*, 677:220.

Blechschmidt, B., Borneleit, P., Emanoulidis, I., Lehman, W., and Kleber, H.-P., 1989, Extracellular location of a ß-lactamase produced by *Acinetobacter calcoaceticus, Appl. Microbiol. Biotech.*, 32:85.

Borneleit, P., and Kleber, H.-P., 1983, Purification and properties of the membrane-bound NADH dehydrogenase from hydrocarbon-grown *Acinetobacter calcoaceticus, Biochim. Biophys. Acta*, 722:94.

Bouvet, P.J.M., and Grimont, P.A.D., 1987, Identification and biotyping of clinical isolates of *Acinetobacter, Ann. Inst. Past. Microbiol.*, 138:569.

Breuil, C., and Kushner, D.J., 1975, Lipase and esterase formation by psychrophilic and mesophilic *Acinetobacter* species, *Can. J. Microbiol.*, 21:423.

Brisou, J., 1957, Contribution à l'étude des *Pseudomonadaceae*. Précisions taxonomiques sur le genre *Acinetobacter, Ann. Inst. Past.*, 92:134.

Brisou, J., and Prévot, A.R., 1954, Études de systématique bacterienne. X. Révision des èspeces réunies dans le genre *Achromobacter, Ann. Inst. Past.*, 86:722.

Chandhuri, J., Bhattacharyya, G., Ghosh, D.K., and Mandal, A., 1990, Enzymatic reduction of mercuric chloride by cell-free extract of a mercury-resistant strain of *Acinetobacter lwoffi, Ind. J. Exp. Biol.*, 28:138.

Collins, M.D., and Jones, D., 1981, Distribution of isoprenoid quinone structural types in bacteria and their taxonomic implications, *Microbiol. Rev.*, 45:316.

Cook, A.M., and Fewson, C.A., 1973, Role of carbohydrates in the metabolism of *Acinetobacter calcoaceticus, Biochim. Biophys. Acta*, 320:214.

Cunha, B.A., Klimek, J.J., Gracewski, J., McLaughlin, J.C., and Quintiliani, R., 1980, A common source outbreak of *Acinetobacter* pulmonary infections traced to Wright respirometers, *Postgrad. Med. J.*, 56:169.

Daly, A.K., Postic, B., and Kass, E.H., 1962, Infections due to organisms of the genus *Herellea*, *Arch. Intern. Med.*, 110:580.

Dawson, N.L., and Reinhardt, D.J., 1981, Microbial flora of in-use, display eye shadow testers and bacterial challenges of unused eye shadows, *Appl. Environ. Microbiol.*, 42:297.

D'Costa, E.J., Higgins, I.J., and Turner, A.P.F., 1986, Quinoprotein glucose dehydrogenase and its application in an amperometric glucose sensor, *Biosensors*, 2:71.

Debord, G.G., 1939, Organisms invalidating the diagnosis of gonorrhea by the smear method, *J. Bacteriol.*, 38:119.

Denis, F.A., D'Oultremont, P.A., Debacq, J.J., Cherel, J.M., and Brisou, J., 1975, Distributions des ubiquinones (coenzyme Q) chez les bacilles a Gram negatif, *Compt. Rend. Seanc. Soc. Biol.*, 169:380.

Dickie, C.W., 1978, Equine myositis and septicemia caused by *Acinetobacter calcoaceticus* infection, *J. Amer. Vet. Med. Assoc.*, 57:530.

Dijkshoorn, L., van Vianen, W., Degener, J.E., and Michel, M.F., 1987, Typing of *Acinetobacter calcoaceticus* strains isolated from hospital patients by cell envelope protein profiles, *Epidemiol. Infect.*, 99:659.

Dokter, P., van Wielink, J.E., van Kleef, M.A.G., and Duine, J.A., 1988, Cytochrome *b*-562 from *Acinetobacter calcoaceticus* LMD 79.41. Its characteristics and role as electron acceptor for quinoprotein genome dehydrogenase, *Biochem. J.*, 254:131.

Droop, M.R., and Jannasch, H.W., 1977, Bacterial indication of water pollution, *Adv. Aquat. Microbiol.*, 1:346.

Du Preez, J.C., and Toerien, D.F., 1978, The effect of temperature on the growth of *Acinetobacter calcoaceticus*, *Water SA*, 4:10.

Du Preez, J.C., Toerien, D.F., and Lategan, P.M., 1981, Growth parameters of *Acinetobacter calcoaceticus* on acetate and ethanol, *Eur. J. Appl. Microbiol. Biotech.*, 13:45.

Du Preez, J.C., Lategan, P.M., and Toerien, D.F., 1984, Influence of the growth rate on the macromolecular composition of *Acinetobacter calcoaceticus* in carbon-limited chemostat culture, *FEMS Microbiol. Lett.*, 23:71.

Du Preez, J.C., Lategan, P.M., and Toerien, D.F., 1985, Utilization of short chain monocarboxylic acids in an effluent of a petrochemical industry by *Acinetobacter calcoaceticus*, *Biotech. Bioeng.*, 27:128.

Edman, P., Artursson, P., Bjork, E., and Davidson, B., 1987, Immobilized L-asparaginase-L-glutaminase from *Acinetobacter glutamin-asificans* in microspheres: some properties in vivo and in an extracorporeal system, *Int. J. Pharmaceut.*, 34:225.

Ensley, B.D., and Finnerty, W.R., 1980, Influences of growth substrates and oxygen on the electron transport system of *Acinetobacter* sp. HO1-N, *J. Bacteriol.*, 142:859.

Ensley, B.D., Irwin, R.M., Carreira, L.A., Hoffman, P.S., Morgan, P.S., Morgan, T.V., and Finnerty, W.R., 1981, Effects of growth substrates and respiratory chain composition on bioenergetics in *Acinetobacter* sp. strain HO1-N, *J. Bacteriol.*, 148:508.

Eribo, B.E., and Jay, J.M., 1985, Incidence of *Acinetobacter* species and other Gram-negative, oxidase-negative bacteria in fresh and spoiled ground beef, *Appl. Environ. Microbiol.*, 49:256.

Fewson, C.A., 1985, Growth yields and respiratory efficiency of *Acinetobacter calcoaceticus*, *J. Gen. Microbiol.*, 131:865.

Firstenberg-Eden, R., Rowley, D.B., and Shattuck, G.E., 1980, Factors affecting inactivation of *Moraxella - Acinetobacter* cells in an irradiation process, *Appl. Environ. Microbiol.*, 40:480.

French, G.L., Casewell, M.W., Roncoroni, A.J., Knight, S., and Phillips, I., 1980, A hospital outbreak of antibiotic-resistant *Acinetobacter anitratus*, epidemiology and control, *J. Hosp. Infect.*, 1:125.

Fricke, B., and Aurich, H., 1989, Characterization of a periplasmic insulin-cleaving metalloproteinase from *Acinetobacter calcoaceticus*, *Biomed. Biochim. Acta*, 9:661.

Fricke, B., Jahreis, G., Sorger, H., and Aurich, H., 1987, Proteases in different membrane fractions of *Acinetobacter calcoaceticus*, *J. Basic Microbiol.*, 27:75.

Gardner, G.A., 1971, Microbiological and chemical changes in lean Wiltshire bacon during aerobic storage, *J. Appl. Bacteriol.*, 34:465.

Geerlof, A., Dokter, P., van Wielink, J.E., and Duine, J.A., 1989, Haem-containing protein complexes of *Acinetobacter calcoaceticus* as secondary electron acceptors for quinoprotein glucose dehydrogenase, *Ant. v. Leeuw.*, 56:81.

Gerner-Smidt, P., 1989, Frequency of plasmids in strains of *Acinetobacter calcoaceticus*, *J. Hosp. Infect.*, 14:23.

Gibson, J.A., and Eaves, L.E., 1981, Isolation of *Acinetobacter calcoaceticus* from an aborted equine foetus, *Aust. Vet. J.*, 57:530.

Goosen, N., Vermaas, D.A., and van de Putte, P., 1987, Cloning of the genes involved in synthesis of coenzyme pyrroloquinolone-quinone from *Acinetobacter calcoaceticus*, *J. Bacteriol.*, 169:303.

Hardy, G.A., and Dawes, E.A., 1985, Effect of oxygen concentration on the growth and respiratory efficiency of *Acinetobacter calcoaceticus*, *J. Gen. Microbiol.*, 131:855.

Harper, D.R., Clare, C.M., Heaps, C.D., Brennan, J., and Hussain, J., 1987, A bacteriological technique for the development of latent fingerprints, *Forens. Sci. Internat.*, 33:209.

Harstein, A.I., Morthland, V.H., Rourke, J.W., Freeman, J., Garber, S., Sykes, R., and Rashad, A.L., 1990, Plasmid DNA fingerprinting of *Acinetobacter calcoaceticus* subspecies *anitratus* from intubated and mechanically ventilated patients, *Infect. Control Hosp. Epidemiol.*, 11:531.

Henrichsen, J., 1975, The occurrence of twitching motility among Gram-negative bacteria, *Acta Path. Microbiol. Scand.*, B83:171.

Henriksen, S.D., 1973, *Moraxella, Acinetobacter* and the *Mimiae*, *Bacteriol. Rev.*, 37:522.

Henriksen, S.D., 1976, *Moraxella, Neisseria, Branhamella* and *Acinetobacter, Ann. Rev. Microbiol.*, 30:63.

Hoffman, S., Mabeck, C.E., and Vejsgaard, R., 1982, Bacteriuria caused by *Acinetobacter calcoaceticus* biovars in a normal population and in general practice, *J. Clin. Microbiol.*, 16:443.

Holcenberg, J.S., 1985, Glutaminase from *Acinetobacter glutaminasificans, Meth. Enzymol.*, 113:257.

Holcenberg, J.S., Schmer, G., Teller, D.C., and Roberts, J., 1975, Biologic and physical properties of succinylated and glycosylated *Acinetobacter* glutaminase-asparaginase, *J. Biol. Chem.*, 250:4165.

Holcenberg, J.S., Ericsson, L., and Roberts, J., 1978, Amino acid sequence of the diazooxonorleucine binding site of *Acinetobacter* and *Pseudomonas* 7A glutaminase-asparaginase enzymes, *Biochemistry*, 17:411.

Holcenberg, J.S., Camitta, B.M., Borella, L.D., and Ring, B.J., 1979, Phase 1 study of succinylated *Acinetobacter* L-glutaminase-L-asparaginase, *Cancer Treatment Reports*, 63:1025.

Holton, J., 1983, A note on the preparation and use of a selective differential medium for the isolation of *Acinetobacter* spp. from clinical sources, *J. Appl. Bacteriol.*, 54:141.

Hunger, M., Schmucker, R., Kishan, V., and Hillen, W., 1990, Analysis and nucleotide sequence of an origin of DNA replication in *Acinetobacter calcoaceticus* and its use for *Escherichia coli* shuttle vectors, *Gene*, 87:45.

Ingram, M., and Shewan, J.W., 1960, Introductory reflections on the *Pseudomonas-Achromobacter* group, *J. Appl. Bacteriol.*, 23:373.

Ino, T., and Nishimura, Y., 1990, Radiation sensitivity of *Acinetobacter* species, *J. Gen. Appl. Microbiol.*, 36:55.

Jahreis, G., Sorger, H., and Aurich, H., 1989, Eine membrangebundene Alaninaminopeptidase aus *Acinetobacter calcoaceticus*. I. Isolierung und Reinigung des Enzymes, *Biomed. Biochim. Acta*, 48:617.

Jay, J.M., 1982, Modern Food Microbiology, 3rd edn., Van Nostrand, New York.

Joly-Guillou, M.L., Bergogne-Bérézin, E., and Vieu, J.F., 1990a, A study of the relationships between antibiotic resistance phenotypes, phage-typing and biotyping of 117 clinical isolates of *Acinetobacter* spp., *J. Hosp. Infect.*, 16:49.

Joly-Guillou, M.L., Bergogne-Bérézin, E., and Vieu, J.F., 1990b, Epidémiologie et résistance aux antibiotiques des *Acinetobacter* en milieu hospitalier. Bilan de 5 annees, *Press. Med.*, 19:357.

Joner, P.E., 1976, Purification and properties of L-asparaginase B from *Acinetobacter calcoaceticus, Biochim. Biophys. Acta*, 438:287.

Joner, P.E., Kristiansen, T., and Einarsson, M., 1973, Purification and properties of L-asparaginase A from *Acinetobacter calcoaceticus, Biochim. Biophys. Acta*, 327:146.

Jones, G.L., Jansen, F., and McKay, A.J., 1973, Substrate inhibition of the growth of bacterium NCIB8250 by phenol, *J. Gen. Microbiol.*, 74:139.

Jones, C.W., Brice, J.M., and Edwards, C., 1977a, The effect of respiratory chain composition on the growth efficiencies of aerobic bacteria, *Arch. Microbiol.*, 115:85.

Jones, M., Fernandes, R., and King, H.K., 1977b, Solubilization studies and properties of particulate NAD-independent malate dehydrogenase of *Acinetobacter calcoaceticus* (NCIB8250), *Biochem. Soc. Trans.*, 5:258.

Jongejan, J.A., and Duine, J.A., eds., 1989, PQQ and Quinoproteins, Kluger, Dordrecht.

Juni, E., 1972, Interspecies transformation of *Acinetobacter*: genetic evidence for a ubiquitous genus, *J. Bacteriol.*, 112:917.

Juni, E., 1978, Genetics and physiology of *Acinetobacter*, *Ann. Rev. Microbiol.*, 32:349.

Juni, E., 1984, Genus III. Acinetobacter. Brisou and Prevot 1954, *in* "Bergey's Manual of Systematic Bacteriology, vol. 1," p.303, N.R.Grieg and J.G.Holt, eds., Williams and Wilkins, Baltimore.

Kien, C.L., Anderson, A.J., and Holcenberg, J.S., 1985, Tissue nitrogen-sparing effect of high protein diet in mice with or without ascites tumour treated with *Acinetobacter* glutaminase-asparaginase, *Cancer Res.*, 45:4876.

Kim, S.J., and Kim, Y.S., 1985, Isolation of a malonate-utilizing *Acinetobacter calcoaceticus* from soil, *Misaengmul Hakhoechi*, 23:230.

Kim, Y.J., and Kim, Y.M., 1989, Induction of carbon monoxide dehydrogenase during heterotrophic growth of *Acinetobacter* sp. strain JC1 DSM3803 in the presence of carbon monoxide, *FEMS Microbiol. Lett.*, 59:207.

Kim, Y.S., and Park, C., 1988, Inactivation of *Acinetobacter calcoaceticus* acetate kinase by diethylpyrocarbonate, *Biochim. Biophys. Acta*, 956:103.

Kim, K.S., Ro, Y.T., and Kim, Y.M., 1989, Purification and some properties of carbon monoxide dehydrogenase from *Acinetobacter* sp. strain JC1 DSM3803, *J. Bacteriol.*, 171:958.

Kita, K., Hiraoka, N., Oshima, A., Kadonishi, S., and Obayashi, A., 1985, *Acc*II, a new restriction endonuclease from *Acinetobacter calcoaceticus*, *Nucl. Acids Res.*, 13:8685.

Kitagawa, K., Tateishi, A., Nakano, F., Matumoto, I., Morohoshi, T., Tanino, T., and Osui, T., 1986a, Generation of energy coupled with membrane-bound glucose dehydrogenase in *Acinetobacter calcoaceticus*, *Agric. Biol. Chem.*, 50:1453.

Kitagawa, K., Tateishi, A., Nakano, F., Matsumoto, T., Morohoshi, T., Tanino, T., and Osui, T., 1986b, Sources of energy and energy coupling reactions of the active transport systems in *Acinetobacter calcoaceticus*, *Agric. Biol. Chem.*, 50:2939.

Kleber, H.-P., Frikke, B., Klossek, P., and Aurich, H., 1975, Effect of growing conditions on protein content and amino acid composition of *Acinetobacter calcoaceticus*, *Ukran. Biochem. J.*, 47:422.

Koburger, S.A., 1964, Isolation of *Mima polymorpha* from dairy products, *J. Dairy Sci.*, 47:646.

Koh, J-S., Yamakawa, T., Kodama, T., and Minoda, Y., 1985, Rapid and dense culture of *Acinetobacter calcoaceticus* on palm oil, *Agric. Biol. Chem.*, 49:1411.

Koh, J-S., Yamakawa, T., Kodama, T., and Minoda, Y., 1987, Cell mass production of *Acinetobacter calcoaceticus* in palm oil-limited chemostat culture, *J. Ferment. Tech.*, 65:229.

Kotohda, S., Suzuki, F., Katsuki, S., and Sato, T., 1979, Purification and some properties of endo-ß-1,6-gluconase from *Acinetobacter* sp., *Agric. Biol. Chem.*, 43:2029.

LaCroix, S.J., and Cabelli, V.J., 1982, Membrane filter method for enumeration of *Acinetobacter calcoaceticus* from environmental waters, *Appl. Environ. Microbiol.*, 43:90.

Layh, G., Eberspaecher, J., and Lingens, F., 1982, Rhodanase in chloridazon-degrading bacteria, *FEMS Microbiol. Lett.*, 15:23.

Liaw, H.J., and Srinivasan, V.R., 1990, Expression of an *Erwinia* sp. gene encoding diphenyl ether cleavage in *Escherichia coli* and an isolated *Acinetobacter* strain PE 7, *Appl. Microbiol. Biotech.*, 32:686.

Lim, K.J., and Kim, Y.S., 1987, Malonate uptake by *Acinetobacter calcoaceticus*, *Han'guk Saenghwa Hakhoechi*, 20:286.

Loeffelholz, M.J., and Modrzakowski, M.C., 1988, Antimicrobial mechanisms against *Acinetobacter calcoaceticus* of rat polymorphonuclear leukocyte granule extracts, *Infect. Immun.*, 56:552.

Ludewig, M., Fricke, B., and Aurich, H., 1987, Leucine aminopeptidase in intracytoplasmic membranes of *Acinetobacter calcoaceticus*, *J. Basic Microbiol.*, 27:557.

Lutz, A., Grooten, O., Velu, H., and Velu, M., 1956, Role pathogene et frequence des bacteries du type *Bacterium anitratum*, *Ann. Inst. Past.*, 91:110.

Mayer, K.H., and Zinner, S.H., 1985, Bacterial pathogens of increasing significance in hospital-acquired infections, *Rev. Infect., Dis.*, 7(suppl. 3):S371.

Meyer, D.J., and Jones, C.W., 1973, Distribution of cytochromes in bacteria: relationship to general physiology, *Int. J. Syst. Bacteriol.*, 23:459.

Midgley, M., Iscandar, N.S., and Dawes, E.A., 1986a, The interaction of phosphonium ions with *Acinetobacter calcoaceticus*: evidence for the operation of an efflux system, *Biochim. Biophys. Acta*, 856:45.

Midgley, M., Iscandar, N.S., Parish, M., and Dawes, E.A., 1986b, A fluorescent probe for a phosphonium ion efflux system in bacteria, *FEMS Microbiol. Lett.*, 34:187.

Monboisse, J-C., Labadie, J., and Gouet, P., 1979, Induction et repression de la synthase de collagenase chez *Acinetobacter* sp., *Can. J. Microbiol.*, 25:611.

Mukherji, S., and Bhopale, N., 1983, Gliding motility of *Acinetobacter anitratus*, *J. Clin. Pathol.*, 36:484.

Mukkada, A.J., and Bell, E.J., 1971, Partial purification and properties of the fructose-1,6-diphosphatase from *Acinetobacter lwoffi, Arch. Biochem. Biophys.*, 142:22.

Müller, R.H., and Babel, W., 1986, Glucose as an energy donor in acetate growing *Acinetobacter calcoaceticus, Arch. Microbiol.*, 144:62.

Muller-Serieys, C., Lesquoy, J.B., Perez, E., Fichelle, A., Boujeois, B., Joly-Guillou, M.L., and Bergogne-Bérézin, E., 1989, Infections nosocomiales à *Acinetobacter, Press. Med.*, 18:107.

Neidleman, S.L., and Geigert, J., 1984, Biotechnology and oleochemicals: changing patterns, *J. Amer. Oil Chem. Soc.*, 61:290.

Nishimura, Y., Kanbe, K., and Iizuka, H., 1986, Taxonomic studies of aerobic coccobacilli from seawater, *J. Gen. Appl. Microbiol.*, 32:1.

Nishimura, Y., Ino, T., and Iizuka, H., 1988a, *Acinetobacter radioresistans* sp. nov. isolated from cotton and soil, *Int. J. Syst. Bacteriol.*, 38:209.

Nishimura, Y., Kanzaki, H., and Iizuka, H., 1988b, Taxonomic studies of *Acinetobacter* species based on the electrophoretic analysis of enzymes, *J. Basic Microbiol.*, 6:363.

Obana, Y., 1986, Pathogenic significance of *A. calcoaceticus*. Analysis of experimental infection in mice, *Microbiol. immunol.*, 30:645.

Obana, Y., Nishino, T., and Tanino, T., 1985, *In vitro* and *in vivo* activities of antimicrobial agents against *Acinetobacter calcoaceticus, J. Antimicrob. Chemother.*, 15:441.

Onishi, H., and Hidaka, O., 1978, Purification and properties of amylase produced by a moderately halophilic *Acinetobacter* sp., *Can. J. Microbiol.*, 24:1017.

Piéchaud, D., Piéchaud, M., and Second, L., 1956, Variétés protéolytiques de *Moraxella lwoffi* et de *Moraxella glucidolytica, Ann. Inst. Past.*, 90:517.

Ramachandra, R.N., Rao, M.S., Raghavan, R., and Keshavamurthy, B.S., 1981, Isolation of *Acinetobacter calcoaceticus* from extended frozen buffalo semen used for artificial insemination, *Curr. Sci.*, 50:1066.

Retailliau, H.F., Hightower, W.A., Dixon, R.E., and Allen, J.R., 1979, *Acinetobacter calcoaceticus*: a nosocomial pathogen with an unusual seasonal pattern, *J. Infect. Dis.*, 139:371.

Richter, K., and Kleber, H.-P., 1980, Regulation of phosphoenolpyruvate carboxylase from *Acinetobacter calcoaceticus* by nucleotides, *FEBS Lett.*, 109:202.

Rosenberg, M., and Rosenberg, E., 1981, Role of adherence in growth of *Acinetobacter calcoaceticus* RAG1 on hexadecane, *J. Bacteriol.*, 148:51.

Shaub, I.G., and Hauber, F.D., 1948, A biochemical and serological study of a group of identical unidentifiable Gram-negative bacilli in human sources, *J. Bacteriol.*, 56:379.

Shewan, J.W., Hobbs, G., and Hodgkiss, W., 1960, The *Pseudomonas* and *Achromobacter* groups of bacteria in the spoilage of marine white fish, *J. Appl. Bacteriol.*, 23:468.

Skerman, V.B.D., McGowan, V., and Sneath, P.A., eds., 1980, Approved lists of bacterial names, *Int. J. Syst. Bacteriol.*, 30:225.

Smith, A.W., Freeman, S., Minett, W.G., and Lambert, P.A., 1990, Characterization of a siderophore from *Acinetobacter calcoaceticus*, *FEMS Microbiol. Lett.*, 70:29.

Steckel, J., Roberts, J., Phillips, F.S., and Chou, T-C., 1983, Kinetic properties and inhibition of *Acinetobacter* glutaminase-asparaginase, *Biochem. Pharmacol.*, 32:971.

Tanaka, S., Robinson, E.A., Appella, E., Miller, M., Ammon, H.L., Roberts, J., Weber, I.T., and Wlodawer, A., 1988, Structures of amidohydrolases. Amino acid sequence of a glutaminase-asparaginase from *Acinetobacter glutaminasificans* and preliminary crystallographic data for an asparaginase from *Erwinia chrysanthemi*, *J. Biol. Chem.*, 263:8583.

Thornley, M.J., Ingram, M., and Barnes, E.M., 1960, The effects of antibiotics and irradiation on the *Pseudomonas-Achromobacter* flora of chilled poultry, *J. Appl. Bacteriol.*, 23:487.

Tjernberg, I., and Ursing, J., 1989, Clinical strains of *Acinetobacter* classified by DNA-DNA hybridization, *APMIS*, 1987:595.

Towner, K.J., 1978, Chromosome mapping in *Acinetobacter calcoaceticus*, *J. Gen. Microbiol.*, 104:175.

van Schie, B.J., Hellingwerf, K.J., van Dijken, J.P., Elferinck, M.G.L., van Dijl, J.M., Kuenen, J.G., and Konings, W.N., 1985, Energy transduction by electron transfer via a pyrrolo-quinoline quinone dependent glucose dehydrogenase in *Escherichia coli*, *Pseudomonas aeruginosa* and *Acinetobacter calcoaceticus* (var. *lwoffi*), *J. Bacteriol.*, 163:493.

van Schie, B.J., Rouwenhorst, R.J., de Bont, J.A.M., van Dijken, J.P., and Kuenen, J.G., 1987, An in vivo analysis of the energetics of aldose oxidation by *Acinetobacter calcoaceticus*, *Appl. Microbiol. Biotech.*, 26:560.

Vandenbergh, P.A., and Berk, R.S., 1980, Purification and characterization of rhodenase from *Acinetobacter calcoaceticus*, *Can. J. Microbiol.*, 26:281.

Vieu, J.F., Bergogne-Bérézin, E., Joly, M.L., Berthelot, G., and Fichelle, A., 1980, Epidémiologie d'*Acinetobacter calcoaceticus*, *Press. Med.*, 9:3551.

Villalobo, A., Roldan, J.M., Rivas, J., and Cardenas, J., 1977, Assimilatory nitrate reductase from *Acinetobacter calcoaceticus*, *Arch. Microbiol.*, 112:127.

Von Graevenitz, A., 1983, *Acinetobacter calcoaceticus* var. *anitratus*: okologie, isolierung und Pathogenitat, *in* "Neus Von Alten Erregen und nene Erreges," p.112, C.Kraseman, ed., de Gruyter, Berlin.

Whittaker, P.A., 1971, Terminal respiration in *Moraxella lwoffi* (NCIB8250), *Microbios*, 4:65.

Williams, H.N., Falkler, W.A., and Hasler, J.F., 1983, *Acinetobacter* contamination of laboratory dental pumice, *J. Dent. Res.*, 62:1073.

Wlodawer, A., Roberts, J., and Holcenberg, J.S., 1977, Characterization of crystals of L-glutaminase-asparaginase from *Acinetobacter glutaminasificans* and *Pseudomonas* 7A, *J. Mol. Biol.*, 112:515.

Wong, T.Y., Graham, L., O'Hara, E., and Maier, R.J., 1986, Enrichment for hydrogen-oxidising *Acinetobacter* spp. in the rhizosphere of hydrogen-evolving soybean root nodules, *Appl. Environ. Microbiol.*, 52:1008.

Yashphe, J., Chikarmane, H., Iranzo, M., and Halvorson, H.O., 1990, Phosphatases of *Acinetobacter lwoffi*. Localization and regulation of synthesis by orthophosphate, *Curr. Microbiol.*, 20:273.

TAXONOMY OF *ACINETOBACTER*

P.A.D. Grimont and P.J.M. Bouvet

Unité des Entérobactéries
INSERM Unité 199
Institut Pasteur
75724 Paris Cedex 15
France

INTRODUCTION

The history of the oxidase-negative, non-motile, Gram-negative diplobacilli now constituting the genus *Acinetobacter* has been confused for many years (Henriksen, 1973, 1976). These bacteria were originally classified into various genera including *"Bacterium"*, *Neisseria*, *Alcaligenes*, *"Mima"*, *"Herellea"*, *"Achromobacter"* and *Moraxella*. For some time this group of bacteria was referred to as the oxidase-negative *Moraxella*. The application of modern methods of taxonomy, including carbon source utilisation tests, genetic transformation, DNA hybridisation, and rRNA sequence comparison (by hybridisation or direct sequencing), has now resolved the earlier confusion. Progress in *Acinetobacter* taxonomy has now provided a new insight into *Acinetobacter* ecology and epidemiology.

DELINEATION OF THE GENUS *ACINETOBACTER*

The nutritional studies of Baumann et al. (1968) clearly showed that oxidase-negative strains differed from oxidase-positive strains of the genus *Moraxella*. Baumann et al. (1968), in search of a generic name, exhumed the name *Acinetobacter*, originally coined by Brisou and Prévot (1954), to accommodate non-motile *"Achromobacter"* strains. DNA from oxidase-negative strains of *Moraxella* (*Acinetobacter* spp.) was found to transform a competent strain (strain BD413*trpE27*) of that group, whereas DNA from oxidase-positive strains failed to transform this strain (Juni, 1972). *Acinetobacter* strains were therefore recognised as a natural group

The Biology of Acinetobacter, Edited by K. J. Towner *et al.*
Plenum Press, New York, 1991

distinct from the genera *Moraxella* and *Neisseria*. The current definition of the genus *Acinetobacter* is as follows (adapted from Juni, 1984): strictly aerobic, non-motile, oxidase-negative coccobacilli; Gram-negative, but sometimes difficult to destain; grows well on complex media between 20 and 30°C without growth factor requirements; nitrates are rarely reduced; extracted DNA is able to transform strain BD413*trpE27*; the G + C content of the DNA is between 39 and 47 mol %.

The genus *Acinetobacter* constitutes a discrete phylogenetic branch in superfamily II (Van Landschoot et al., 1986) of the Proteobacteria. The closest genera are *Moraxella, Pseudomonas* (fluorescent group), and *Xanthomonas* (Woese et al., 1985; Van Landschoot et al., 1986).

DELINEATION OF SPECIES IN THE GENUS *ACINETOBACTER*

Only two species, *A. calcoaceticus* and *A. lwoffii* were listed in the Approved Lists of bacterial names (Skerman, 1980), and only one species was given in Bergey's Manual of Systematic Bacteriology (Juni, 1984), although early DNA hybridisation experiments with a nitrocellulose filter method had shown the genus *Acinetobacter* to be heterogeneous (Johnson et al., 1970). Using a cut-off value of 50% relatedness to delineate DNA groups, Johnson et al. (1970) found five such groups in the collection of strains they studied.

There is presently a general agreement to define a genomic species as a group of strains which are at least 70% related by DNA hybridisation with divergence values (ΔT_m) below 5°C (Wayne et al., 1987). Genomic species which can be identified with phenotypic tests can be given a formal species name (Wayne et al., 1987). Using the above criteria to interpret the results of DNA relatedness data, Bouvet and Grimont (1986) delineated 12 genomic species (numbered 1 to 12) among 85 strains, although the separation between species 8 and 9 was uncertain. A total of ten genomic species could be identified by phenotypic tests. The type strains of *A. calcoaceticus* and *A. lwoffii* fell in genomic species 1 and 8 respectively. Since a reference strain of *"Achromobacter haemolyticus"* (Stenzel and Mannheim, 1963) fell in genomic species 4, the epithet was revived in a new combination, *Acinetobacter haemolyticus*. Genomic species 2, 5 and 7 were named *A. baumannii, A. junii* and *A. johnsonii* respectively. The other genomic species, which contained less than ten strains or which could not be identified unambiguously using phenotypic tests, were not named. This work was later expanded to include 27 proteolytic strains which did not fit into any of the 12 genomic species described previously (Bouvet and Jeanjean, 1989). Five additional genomic species were delineated and numbered 13 to 17. In spite of these additional efforts, several strains remained ungrouped.

Independent work, carried out by Tjernberg and Ursing (1989) (abbreviated as "T&U" study) on 168 consecutive clinical isolates and 30 reference strains, showed a good correlation with the results of Bouvet and Grimont (1986) (abbreviated as "B&G" study). Ten genomic species were common to both studies. Three additional genomic species (numbered 13 to 15) were described in the T&U study. Genomic species 13 (T&U) was represented by a single strain in the B&G study. Group 14 (T&U) contained a reference strain of "*Achromobacter conjunctivae*" (Stenzel and Mannheim, 1963) which was ungrouped in the B&G study. Genomic species 14 (T&U) corresponded to group 13 of Bouvet and Jeanjean (1989). Genomic species 15 (T&U) was not represented in the studies of Bouvet and Grimont (1986) and Bouvet and Jeanjean (1989). Thus, the data obtained in two different laboratories were fully compatible.

Nishimura et al. (1987) delineated five DNA relatedness groups among 31 strains. The reference strains included in the study of Nishimura et al. (1987) allowed them to identify groups 2 and 3 as *A. lwoffii* and *A. johnsonii*, respectively. Group 4 corresponded to both *A. haemolyticus* and genomic species 6 (B&G). However, a careful examination of the data from Nishimura et al. (1987) shows that the reference strain of genomic species 6 was only 65% related by the filter hybridisation method to the reference strain of *A. haemolyticus*. It has been shown (Bouvet and Grimont, 1986) that a DNA relatedness value of 65% obtained by the filter method corresponds to about 44% relatedness as determined by the S1 nuclease method (i.e. the method used by Bouvet and Grimont, 1986). The data obtained in Tokyo and Paris are thus not contradictory and only the interpretation differs. A more serious discrepancy occurred with group 1 (Nishimura et al., 1987), which corresponded to both *A. calcoaceticus* and *A. baumannii*. The type strains of both species showed 98% relatedness (Nishimura et al., 1987). When strains from both laboratories were exchanged, the strain used as the type strain of *A. calcoaceticus* in Tokyo (IAM 12087) was found to be a typical *A. baumannii*, both by phenotypic tests and DNA relatedness. Group 5 contained radiation-resistant *Acinetobacter* strains isolated from soil and cotton samples (Nishimura et al., 1987). This group was later named *A. radioresistens* (Nishimura et al., 1988a). Tjernberg and Ursing (1989) showed that *A. radioresistens* corresponded with B&G genomic species 12. Although this species was poorly represented in the B&G study, it represented the second largest group (31 isolates from pathological samples) in the T&U study. Radiation resistance should be studied in clinical isolates contained in this group to determine whether this property can define the species.

Table 1 compares and summarises the species designations and DNA groups delineated in the different laboratories.

CHARACTERISATION OF *ACINETOBACTER* SPECIES

Protein Electrophoresis Patterns

Polyacrylamide gel electrophoresis (PAGE) of whole cell proteins is increasingly used for the characterisation of bacteria. Whole cell protein electrophoresis of *Acinetobacter* strains yields a number of patterns

Table 1. A comparison of the DNA groups delineated in different laboratories.

Reference strain	DNA relatedness group according to			Present nomenclature
	Bouvet and Grimont,1986; Bouvet and Jeanjean,1989	Tjernberg and Ursing, 1989	Nishimura et al., 1987, 1988a	
ATCC 23055^t=IAM12087^t	1	1	1	A.calcoaceticus
CIP70.34^t=IAM12088^t	2	2	1	A.baumannii
CIP70.29=ATCC19004	3	3	nt	Not named
CIP70.11=ATCC17903	ug	13	nt	Not named
CIP64.3^t=ATCC17906^t	4	4	4	A.haemolyticus
CIP64.5^t=ATCC17908^t	5	5	nt	A.junii
CIPA165=ATCC17979	6	6	4	Not named
CIP64.6^t=ATCC17909^t	7	7	3	A.johnsonii
ATCC15309^t=CIP64.10^t	8	8	2	A.lwoffii
CIP70.31=ATCC9957	9	8	nt	Not named
CIP70.12=ATCC17924	10	10	ug	Not named
CIP63.46=NCIB8250 =ATCC11171	11	11	ug	Not named
strain FO-1^t=IAM13186^t	(12)^a	12	5	A.radioresistens
CIP64.2=ATCC17905	13	14	nt	Not named
Bouvet 382	14	nt	nt	Not named
Bouvet 240	15	nt	nt	Not named
CIP70.18=ATCC17988	16	ug	nt	Not named
Bouvet 942	17	nt	nt	Not named
Tjernberg 151a	nt	15	nt	Not named

[a]unpublished result; [t]type strain.
nt, not tested; ug, ungrouped.

(Alexander et al., 1984, 1988). Unfortunately these data have not yet been integrated in the newer taxonomic scheme outlined above and it is not yet known whether whole cell protein patterns can differentiate the species delineated by DNA hybridisation.

Electrophoresis of outer membrane proteins was applied to the strains previously characterised with DNA hybridisation by Ino and Nishimura (1989). Patterns generally contained 20-25 distinct protein bands including one or two major proteins with an apparent molecular size of 20,000-50,000. Correspondence between outer membrane protein patterns and some DNA relatedness groups was fairly good (Ino and Nishimura, 1989).

Cell envelope protein patterns of a number of strains were determined by Dijkshoorn et al. (1987a,b). These strains were later identified using the latest taxonomic scheme (Bouvet et al., 1990). With the exception of *A. haemolyticus* (E patterns) and *Acinetobacter* sp. 3 (D patterns), in which strains showed related patterns within a species, all other studied species showed a broad range of patterns. Cell envelope protein patterns may thus be more useful for typing isolates than for species identification.

Enzyme Electrophoresis

Nishimura et al. (1988b) used PAGE to examine 22 strains from culture collections and 14 isolates from soil and cotton samples for the presence of twelve enzymes. The strains divided into four clusters (Z-1, Z-2, Z-3 and Z-4). Cluster Z-1 further divided into three subclusters (Z-1a, Z-1b, Z-1c). Cluster Z-1 included four different *Acinetobacter* spp. (*A. calcoaceticus*, *A. baumannii*, *A. haemolyticus*, and *Acinetobacter* sp. 6); cluster Z-2 included two species (*A. johnsonii* and *Acinetobacter* sp. 11); cluster Z-3 corresponded to *A. radioresistens* (equivalent to B&G *Acinetobacter* sp. 12); cluster Z-4 corresponded to *A. lwoffii*.

Picard et al. (1989) examined the electrophoretic polymorphism of malate and glutamate dehydrogenase and diverse esterases produced by *Acinetobacter* strains belonging to known genomic species. Following treatment of the data by correspondence analysis (a type of factor analysis), the results were found to correlate with the distribution of strains into genomic species.

Serological Cross-Reactions

Traub (1989) has developed a taxonomic scheme for typing *A. baumannii*, the most prevalent species in hospitals. Twenty serovars have been delineated. With one exception, the sera directed against the 20 serovars of *A. baumannii* did not cross-react with seven serovars of *Acinetobacter* sp. 3. This supports the separation of sp. 3 from *A. baumannii*, which otherwise can only be separated by the upper limit to its growth temperature (see below).

When total bacterial DNA is cleaved by a restriction endonuclease and the DNA fragments separated by agarose gel electrophoresis, a large number of DNA fragments can be visualised after staining with ethidium bromide. Although restriction patterns obtained from two strains can be directly compared, the number of fragments is too large for a pattern to be recorded for later comparison. If the DNA fragments in the agarose gel are transferred to a nylon membrane and then hybridised with a universal probe containing labelled *Escherichia coli* 16 + 23S rRNA, a simpler pattern of hybridised fragments (rRNA gene restriction pattern) can be visualised (Grimont and Grimont, 1986). Depending upon the bacterial group studied and, to a lesser extent, upon the restriction endonuclease chosen, the level of identification provided by the method is at or below the species level. This method is currently being applied to *Acinetobacter* strains (F. Grimont, S. Jeanjean, P. Bouvet, and P.A.D. Grimont, manuscript in preparation). It has been found that rRNA gene restriction patterns generated by endonuclease *Cla*I contain between four and 14 fragments. A single pattern was observed with four strains of genomic species 3. A total of 13 different patterns were given by 20 *A. baumannii* strains. In other genomic species it was found that almost every strain gave a unique pattern.

Physiological and Biochemical Characterisation

Optimal temperature for growth is around 30°C; however, the upper growth temperature differs between species. *A. johnsonii* strains cannot grow at 37°C, whereas only strains of *A. baumannii* can grow at 44°C (Table 2).

The ability to produce acid from glucose has long been used to differentiate *A. calcoaceticus sensu lato* from *A. lwoffii sensu lato*. Acid production with D-glucose results from the oxidation of glucose to gluconic acid (Aubert et al., 1952). The glucose dehydrogenase involved is actually a non-specific aldose dehydrogenase that can also oxidise a number of different aldoses (Hauge, 1960). The phenons delineated by Baumann et al. (1968) often contained both glucose-oxidisers and non-oxidisers. Duine et al. (1979) demonstrated that glucose dehydrogenase was a quinoprotein. Glucose dehydrogenase is an inactive apoenzyme and pyrroloquinoline quinone is a co-factor required for enzymic activity. Strains producing the apoenzyme, but not the co-factor, will not oxidise glucose unless the co-factor is added to the medium (Duine, this volume). Bouvet and Bouvet (1989) found that the following percentages of strains (in parentheses) produced acid from glucose without added pyrroloquinoline quinone: *A. calcoaceticus* (100), *A. baumannii* (99), sp. 3 (100), *A. haemolyticus* (63), sp. 6 (50), *A. lwoffii* (5), sp. 10 (100), and sp. 12 (40). No strains of *A. junii, A. johnsonii* or sp. 11 produced acid from glucose under these conditions. When pyrroloquinoline quinone was added, all genomic species contained strains producing acid from glucose (percentages of strains in parentheses): *A. baumannii* (100), *A. haemolyticus* (69), *A. junii* (72),

Table 2. Current identification scheme for *Acinetobacter* genomic species

DNA relatedness group[a]

	1	2	3	4	5	6	7	8/9	10	11	12	13	14	15	16	17
Growth at: 44°C	-	+	-	-	-	-	-	-	-	-	-	-	-	-	-	-
41°C	-	+	+	-	90	-	-	-	-	-	-	-	-	-	-	-
37°C	+	+	+	+	-	+	-	+	+	+	-	-	-	+	+	75
Acid from D-glucose	+	95	+	60	-	50	-	6	+	-	40	89	+	+	+	-
Gelatin hydrolysis	-	-	-	96	-	+	-	-	-	-	-	+	+	+	+	+
Utilisation of:																
DL-Lactate	+	+	+	-	+	-	+	99	+	+	+	+	+	+	+	+
DL-4-Aminobutyrate	+	+	+	+	90	-	35	40	+	+	+	11	+	-	25	+
trans-Aconitate	+	99	+	52	-	-	-	-	-	-	-	11	67	-	-	50
Citrate	+	+	+	90	82	+	98	-	+	+	-	+	+	+	+	+
Glutarate	+	+	+	-	-	-	-	-	+	+	+	-	+	-	-	+
Aspartate	+	+	+	64	40	66	61	-	+	+	+	-	-	-	-	-
Azelate	+	90	+	-	-	-	-	+	50	25	+	-	+	-	-	-
ß-Alanine	+	95	95	-	+	-	-	-	+	+	-	-	+	+	75	+
L-Histidine	+	98	94	96	+	+	-	-	+	+	-	+	+	+	+	+
D-Malate	-	98	+	96	-	66	20	60	+	+	-	+	+	+	+	+
Malonate	+	98	85	-	-	-	15	-	-	-	+	11	+	-	50	50
Histamine	-	-	-	-	-	-	-	-	75	+	-	-	-	-	-	-
L-Phenylalanine	+	87	66	-	-	-	-	-	-	-	+	+	+	+	+	+
Phenylacetate	+	87	66	-	-	-	-	94	25	50	+	+	+	+	+	+

a numbering according to Bouvet and Grimont (1986); Bouvet and Jeanjean (1989).   Named species are: 1, *A. calcoaceticus*; 2, *A. baumannii*; 4, *A. haemolyticus*; 5, *A. junii*; 7, *A. johnsonii*; 8/9, *A. lwoffii*; 12, *A. radioresistens*.

+ = all strains positive; - = all strains negative.   Numbers are percentages of strains positive for a particular character.

sp. 6 (50), *A. johnsonii* (97), *A. lwoffii* (90), sp. 11 (100), and sp. 12 (100) (Bouvet and Bouvet, 1989). Since glucose does not enter *Acinetobacter* cells, only the rare strains which can both utilise gluconate and oxidise glucose can utilise glucose in a minimal medium (Baumann et al., 1968; Bouvet and Grimont, 1986). The taxonomic significance of acid production from carbohydrates in *Acinetobacter* taxonomy is thus of less importance than was once thought.

The nutritional abilities (growth in a defined medium with ammonia as nitrogen source and a given carbon source) of strains are essential properties to consider in studying the taxonomy of the genus *Acinetobacter* (Baumann et al. 1968; Bouvet and Grimont, 1986). A simplified identification scheme for named and unnamed *Acinetobacter* species has been published (Bouvet and Grimont, 1987) and is updated in Table 2.

CLINICAL AND ECOLOGICAL SIGNIFICANCE OF NEW TAXONOMIC CHANGES

Progress in taxonomy does not always meet with warm acceptance from those who use identification schemes or simply refer to bacteria by scientific names in their routine work. It is thus important for the taxonomist to indicate how a new taxonomic scheme can generate new concepts in the users' fields. It seems that different *Acinetobacter* species may have different habitats, and these are outlined below. More complete accounts of the occurrence of the different *Acinetobacter* species in clinical situations have been given elsewhere (Bergogne-Bérézin and Joly-Guillou, this volume) and the interested reader should refer to this review for full details.

The habitat of *A. calcoaceticus sensu stricto* is where Beijerinck originally found this bacterium, i.e. in the soil (Beijerinck, 1911). This species is very easy to isolate from soil by enrichment using a minimal liquid medium with ammonium ions as nitrogen source, acetate as carbon source, and aerobic conditions (Baumann et al., 1968). *A. calcoaceticus* is seldom associated with human infection; indeed, only three cases, from wounds or sputum, are known to us (Tjernberg and Ursing, 1989), and it is noteworthy that these three isolates were somewhat divergent from the type strain, showing T_m values ranging from 3.0 to 5.5°C (rounded to the nearest 0.5°C) in hybridisation experiments (Tjernberg and Ursing, 1989); such a level of divergence can indicate a relationship at the sub-specific level (Grimont, 1988).

A. baumannii has, to date, only been isolated from humans. It is the species of *Acinetobacter* most commonly associated with nosocomial infections (Bergogne-Bérézin and Joly-Guillou, this volume). Of 291 isolates recovered from 264 patients, this species comprised 244 isolates

(84%) (Bouvet and Grimont, 1987). No habitat, other than man, is known for this species. Thus, its presence on inanimate objects in the hospital environment (Bouvet and Grimont, 1987) can be interpreted as contamination from an infected patient. Since *A. calcoaceticus* and *A. baumannii* are similar in many metabolic properties, biotechnological applications should use *A. calcoaceticus* rather than *A. baumannii* wherever possible.

Acinetobacter sp. 3 has been found in soil and clinical samples. Although infrequent in French studies (Bouvet and Grimont, 1987), this species constituted the largest group of clinical isolates studied by Tjernberg and Ursing (1989).

A. haemolyticus has been isolated occasionally from patients (about 3% of total *Acinetobacter* isolates), the hospital environment, and activated sludge.

A. junii has been isolated both from clinical samples and the environment.

A. johnsonii has been isolated from the environment (soil, activated sludge, river sediment), animals or animal products (chicken, raw milk), the skin of healthy people, and rarely from clinical samples (cerebrospinal fluid). This was the most frequent *Acinetobacter* species isolated from activated sludge in Australia (Duncan et al., 1988). It seems to be a common contaminant of refrigerated eviscerated chicken (Bouvet and Grimont, 1987). This species is commonly found on the skin (hands) of nurses, but is also found on the skin of people not working in a hospital environment (Bouvet and Grimont, 1987). Combined with the fact that this species cannot grow at 37°C, knowledge of this habitat requires that the clinical significance of an isolate (especially from cerebrospinal fluid) should be critically evaluated, with a strong suspicion of hand-borne contamination. In an epidemiological investigation, the presence of *Acinetobacter* strains on the hands of personnel is meaningless unless the identification procedure differentiates *A. baumannii* from *A. johnsonii* and other species.

A. lwoffii has been isolated from human patients (about 3 to 6% of total *Acinetobacter* isolates), hands of uninfected personnel, eviscerated chickens and activated sludge (Bouvet and Grimont, 1987; Duncan et al., 1988).

A. radioresistens, originally found in association with cotton, has also been isolated from patients (Tjernberg and Ursing, 1989).

The habitats of the other genomic species are, as yet, poorly understood.

CONCLUSIONS

Accurate identification to the species level of a clinical isolate belonging to the genus *Acinetobacter* is essential since: (i) identification is a primary epidemiological marker; (ii) habitats may differ between species; (iii) the antibiotic susceptibility of *A. baumannii* compared with other species is quite different (Freney et al., 1989). Further work is required to better delineate, characterise, and name unnamed species, and to make such identification readily available to more laboratories. More precise identification should improve our knowledge of the ecology of the different *Acinetobacter* species. The numerous biochemical studies performed with *Acinetobacter* strains will have a significantly greater scientific impact when the strains used in such studies have been precisely identified.

REFERENCES

Alexander, M., Ismail, F., Jackman, P.J.H., and Nobel, W.C., 1984, Fingerprinting *Acinetobacter* strains from clinical sources by numerical analysis of electrophoretic protein patterns, *J. Med. Microbiol.*, 18:55.

Alexander, M., Rahman, M., Taylor, M., and Noble, W.C., 1988, A study of the value of electrophoretic and other techniques for typing *Acinetobacter calcoaceticus, J. Hosp. Infect.*, 12:273.

Aubert, J.P., Milhaud, G., and Gavard, R., 1952, Étude preliminaire d'un systeme enzymatique bactérien oxydant le glucose en acide gluconique, *Compt. Rend. Séances Heb. Acad. Sci.*, 235:1165.

Baumann, P., Doudoroff, M., and Stanier, R.Y., 1968, A study of the *Moraxella* group. II. Oxidative-negative species (genus *Acinetobacter*), *J. Bacteriol.*, 95:1520.

Beijerinck, M.W., 1911, Uber Pigmentbildung bei Essigbakterien, *Proc. Royal Acad. Sci. (Amsterdam)*, 13:1066.

Bouvet, P.J.M., and Bouvet, O.M.M., 1989, Glucose dehydrogenase activity in *Acinetobacter* species, *Res. Microbiol.*, 140:531.

Bouvet, P.J.M., and Grimont, P.A.D., 1986, Taxonomy of the genus *Acinetobacter* with the recognition of *Acinetobacter baumannii* sp. nov., *Acinetobacter haemolyticus* sp. nov., *Acinetobacter johnsonii* sp. nov., and *Acinetobacter junii* sp. nov., and emended descriptions of *Acinetobacter calcoaceticus* and *Acinetobacter lwoffii, Int. J. Syst. Bacteriol.*, 36:228.

Bouvet, P.J.M., and Grimont, P.A.D., 1987, Identification and biotyping of clinical isolates of *Acinetobacter, Ann. Inst. Past. Microbiol.*, 138:569.

Bouvet, P.J.M., and Jeanjean, S., 1989, Delineation of new proteolytic genomic species in the genus *Acinetobacter, Res. Microbiol.*, 140:291.

Bouvet, P.J.M., Jeanjean, S., Vieu, J.F., and Dijkshoorn, L., 1990, Species, biotype, and bacteriophage type determinations compared with cell envelope protein profiles for typing *Acinetobacter* strains, *J. Clin. Microbiol.*, 28:170.

Brisou, J., and Prévot, A.R., 1954, Études de systématique bactérienne. X. Révision des espèces réunies dans le genre *Achromobacter*, *Ann. Inst. Past.*, 86:722.

Dijkshoorn, L., Michel, M.F., and Degener, J.E., 1987a, Cell envelope protein profiles of *Acinetobacter calcoaceticus* strains isolated in hospitals, *J. Med. Microbiol.*, 23:313.

Dijkshoorn, L., van Vianen, W., Degener, J.E., and Michel, M.F., 1987b, Typing of *Acinetobacter calcoaceticus* strains isolated from hospital patients by cell envelope protein profiles, *Epidemiol. Infect.*, 99:659.

Duine, J.A., Frank, J., and van Zeeland, K., 1979, Glucose dehydrogenase from *Acinetobacter calcoaceticus:* a "quinoprotein", *FEBS Lett.*, 108:443.

Duncan, A., Vasiliadis, G.E., Bayly, R.C., and May, J.W., 1988, Genospecies of *Acinetobacter* isolated from activated sludge showing enhanced removal of phosphate during pilot-scale treatment of sewage, *Biotech. Lett.*, 10:831.

Freney, J., Bouvet, P.J.M., and Tixier, C., 1989, Identification et détermination de la sensibilité aux antibiotiques de 31 souches cliniques d'*Acinetobacter* autres que *A. baumannii*, *Ann. Biol. Clin.*, 47:41.

Grimont, P.A.D., 1988, Use of DNA reassociation in bacterial classification, *Can. J. Microbiol.*, 34:541.

Grimont, F., and Grimont, P.A.D., 1986, Ribosomal ribonucleic acid gene restriction patterns as potential taxonomic tools, *Ann. Inst. Past. Microbiol.*, 137B:165.

Hauge, J.G., 1960, Kinetics and specificity of glucose dehydrogenase from *Bacterium anitratum*, *Biochim. Biophys. Acta*, 45:263.

Henriksen, S.D., 1973, *Moraxella, Acinetobacter* and the *Mimae, Bacteriol. Rev.*, 37:522.

Henriksen, S.D., 1976, *Moraxella, Neisseria, Branhamella* and *Acinetobacter, Ann. Rev. Microbiol.*, 30:63.

Ino, T., and Nishimura, Y., 1989, Taxonomic studies of *Acinetobacter* species based on outer membrane protein patterns, *J. Gen. Appl. Microbiol.*, 35:213.

Johnson, J.L., Anderson, R.S., and Ordal, E.J., 1970, Nucleic acid homologies among oxidase-negative *Moraxella* species, *J. Bacteriol.*, 101:568.

Juni, E., 1972, Interspecies transformation of *Acinetobacter*: genetic evidence for a ubiquitous genus, *J. Bacteriol.*, 112:917.

Juni, E., 1984, Genus III. *Acinetobacter* Brisou et Prévot 1954, *in* "Bergey's Manual of Systematic Bacteriology, vol. 1," p.303, N.R. Grieg and J.G. Holt, eds. Williams and Wilkins, Baltimore.

Nishimura, Y., Kano, M., Ino, T., Iizuka, H., Kosako, Y., and Kaneko,T., 1987, Deoxyribonucleic acid relationship among the radiation-resistant *Acinetobacter* and other *Acinetobacter*, *J. Gen. Appl. Microbiol.*, 33:371.

Nishimura, Y., Ino, T., and Iizuka, H., 1988a, *Acinetobacter radioresistens* sp. nov. isolated from cotton and soil, *Int. J. Syst. Bacteriol.*, 38:209.

Nishimura, Y., Kanzaki, H., and Iizuka, H., 1988b, Taxonomic studies of *Acinetobacter* species based on the electrophoretic analysis of enzymes, *J. Basic Microbiol.*, 6:363.

Picard, B., Goullet, P., Bouvet, P.J.M., Decoux, G., and Denis, J.B., 1989, Characterization of bacterial genospecies by computer-assisted statistical analysis of enzyme electrophoretic data, *Electrophoresis*, 10:680.

Skerman, V.B.D., McGowan, V., and Sneath, P.H.A., eds., 1980, Approved lists of bacterial names, *Int. J. Syst. Bacteriol.*, 30:225.

Stenzel, W., and Mannheim, W., 1963, On the classification and nomenclature of some nonmotile and coccoid diplobacteria, exhibiting the properties of Achromobacteriaceae, *Int. Bull. Bacteriol. Nomen. Taxon.*, 13:195.

Tjernberg, I., and Ursing, J., 1989, Clinical strains of *Acinetobacter* classified by DNA-DNA hybridization, *APMIS* 97:595.

Traub, W.H., 1989, *Acinetobacter baumannii* serotyping for delineation of outbreaks of nosocomial cross-infection, *J. Clin. Microbiol.*, 27:2713.

Van Landschoot, A., Rossau, R., and De Ley, J., 1986, Intra- and intergeneric similarities of the ribosomal ribonucleic acid cistrons of *Acinetobacter*, *Int. J. Syst. Bacteriol.*, 36:150.

Wayne, L.G., Brenner, D.J., Colwell, R.R., Grimont, P.A.D., Kandler, O., Krichevsky, M.I., Moore, L.H., Moore, W.E.C., Murray, R.G.E., Stackebrandt, E., Starr, M.P., and Truper, H.G., 1987, Report of the ad hoc committee on reconciliation of approaches to bacterial systematics, *Int. J. Syst. Bacteriol.*, 37:463.

Woese, C.R., Weisburg, W.G., Hahn, C.M., Paster, B.J., Zablen, L.B., Lewis, B.J., Macke, T.J., Ludwig, W., and Stackebrandt, E., 1985, The phylogeny of purple bacteria: the gamma subdivision, *Syst. Appl. Microbiol.*, 6:25.

TYPING OF *ACINETOBACTER*

P.J.M. Bouvet

Unité des Entérobactéries
INSERM Unité 199
Institut Pasteur
75724 Paris Cedex 15
France

INTRODUCTION

Need for a Typing System in the Genus Acinetobacter

In the last 25 years, outbreaks of *Acinetobacter* infection have been a growing concern in hospitals (Bergogne-Bérézin et al., 1987, this volume; Noble, this volume). These infections can be difficult to treat since strains are commonly resistant to multiple antimicrobial agents. In 1986 the taxonomy of the genus *Acinetobacter* was changed extensively. At least 19 DNA hybridisation groups have now been delineated (Bouvet and Grimont, 1986; Nishimura et al., 1988; Bouvet and Jeanjean, 1989; Tjernberg and Ursing, 1989) and five new species have been described: *A. baumannii, A. haemolyticus, A. junii, A. johnsonii* and *A. radioresistens.* In addition, the original descriptions of *A. calcoaceticus* and *A. lwoffii* have been emended (Grimont, this volume).

Preliminary results published in 1987 (Bouvet and Grimont, 1987) showed that *A. baumannii* was the most prevalent *Acinetobacter* species isolated from clinical specimens, with 253 of 299 (84.6%) nosocomial isolates being identified as *A. baumannii.* Additional supporting data are now available (Freney et al., 1987; Giammanco et al., 1989; Hercouet et al., 1989; Traub, 1989). In contrast, Tjernberg and Ursing (1989) studied 168 strains isolated consecutively during one year from hospitals and outpatient clinics in the city of Malmo, Sweden. Only six strains (3.6%) were identified as *A. baumannii.* The most prevalent *Acinetobacter* species in this study were *A. lwoffii* (22.6%), *A.radioresistens* (17.9%), *Acinetobacter* sp. 3 (16.7%) and *A. junii* (10.7%). Although the ratio of hospital to outpatient isolates was not disclosed, this statistic may strongly influence the proportion of isolates obtained. A previous study (Gerner-Smidt et al., 1985) showed that most *Acinetobacter* isolates (153 of 185) obtained in general practice were "*A. calcoaceticus* biovar *lwoffii*" (possibly *A. lwoffii, A. junii, A. johnsonii,* or *Acinetobacter* spp. 11, 12...etc.), whereas 93 of 143 strains (65%) isolated in hospitals were of "biovar *calcoaceticus*" (most probably *A. baumannii*). Thus, there is an acute need for a typing system allowing differentiation of *A. baumannii* strains.

Several methods, such as bacteriophage typing or cell envelope protein profiles, which were originally developed for epidemiological purposes before 1986, should be reassessed in light of the new taxonomic knowledge. Since 1986, biotyping (typing by biochemical characteristics) and serotyping methods have been developed for *Acinetobacter* strains identified to the species level according to the new classification. Molecular methods such as low-frequency-cleavage restriction endonuclease digest analysis, enzyme electrophoresis and rRNA gene restriction patterns are also potential tools for studying epidemiology. A few examples of potential typing methods will be reviewed below.

BIOTYPING

Early attempts at biotyping *Acinetobacter* isolates relied on gelatinase production, haemolysis, and acid production from glucose (Gilardi, 1978). Within the newly described *Acinetobacter* species (Bouvet and Grimont, 1986; Bouvet and Jeanjean, 1989), gelatinase production is restricted to very few species, i.e. *A. haemolyticus*, *Acinetobacter* sp. 6 and *Acinetobacter* DNA groups 13 to 17. These species have never been implicated in outbreaks of *Acinetobacter* infection.

For many decades D-glucose oxidation has been used to distinguish *Acinetobacter* strains or "species" (*calcoaceticus* and *lwoffii*). We have recently published a study of D-glucose oxidation by 1134 strains of *Acinetobacter* (Bouvet and Bouvet, 1989). Four species, i.e. *A. calcoaceticus sensu stricto, A. baumannii* (with a few exceptions), *Acinetobacter* sp. 3 and *Acinetobacter* sp. 10, were represented solely by glucose-oxidising strains. *A. lwoffii* (with a few exceptions), *A. junii, A. johnsonii* and *Acinetobacter* sp. 11 were represented solely by glucose non-oxidising strains. Other species, i.e. *A. haemolyticus*, *Acinetobacter* sp. 6 and *Acinetobacter* sp. 12, contained both glucose-oxidising and glucose non-oxidising strains. These results confirmed that the former taxonomic scheme for the genus *Acinetobacter* (two species or biovars differing only by glucose oxidation) was untenable and that identification of *Acinetobacter* strains to the species level should be performed using better methods such as carbon source utilisation tests.

Table 1. Scheme for identifying biotypes of *Acinetobacter baumannii*

	Biotype																		
	1	2	3	4	5	6	7	8	9	10	11	12	13	14	15	16	17	18	19
Utilisation of:																			
levulinate	+	+	+	+	-	-	-	-	-	-	-	-	-	+	-	+	+	-	+
citraconate	-	-	-	-	+	+	+	-	-	-	-	-	+	-	-	+	-	-	+
L-phenylalanine and phenylacetate	+	+	+	-	+	+	+	+	+	+	-	-	-	-	-	+	+	-	+
4-hydroxybenzoate	+	+	-	-	+	+	-	+	+	-	+	-	+	+	-	+	-	+	+
L-tartrate	+	-	-	-	+	-	-	+	-	-	-	-	-	-	+	+	+	+	-

+ = growth after 2 days;   - = no growth after 2 days

Further attempts at biotyping were published by Towner and Chopade (1987) using a commercially available identification system.   This 20-test system (mainly carbon source utilisation tests) allowed the differentiation of 209 possible biotypes. A total of 31 different biotypes were identified among clinical isolates from Nottingham hospitals. This method, if used alone, still does not allow the identification of *Acinetobacter* strains to the species level.

More recently, a biotyping system has been devised to type isolates of *A. baumannii,* the species usually implicated in nosocomial outbreaks of infection. This system was based on the utilisation of six carbon sources (levulinate, citraconate, L-phenylalanine, phenylacetate, 4-hydroxybenzoate and L-tartrate), and allowed the recognition of 19 biotypes (Table 1), of which biotypes 1, 2, 6 and 9 were most frequently encountered.   This biotyping system has, in some instances, been valuable in typing nosocomial strains (Giammanco et al., 1989; Hercouet et al., 1989; Buisson et al., 1990). The following sections will compare biotyping data with results obtained by other typing methods such as bacteriophage typing or cell envelope protein profiles.

PHAGE TYPING

A phage typing system developed ten years ago at the Institut Pasteur (Vieu et al., 1979; Santos-Ferreira et al., 1984) has been found to be useful for the study of hospital outbreaks (Vieu et al., 1980).   Unfortunately, this phage typing system was developed in the absence of a well-defined taxonomic structure and has only been used in a single laboratory in the world. Briefly, the bacteriophage typing system is based on the use of two complementary sets of bacteriophages isolated from sewage (Santos-Ferreira et al., 1984). The first set included 21 bacteriophages and distinguished 116 phage types (Vieu et al., 1979). A further nine new phage types have been added subsequently. The proportion of untypable strains, due to their insensitivity to this set of bacteriophages, was 30% (Santos-Ferreira et al., 1984). A second set of 14 phages was introduced to sub-divide the untypable strains. A total of 21 phage types (called sub-types) was recognised by this second set of phages, and the overall proportion of untypable strains was reduced to about 20% (Santos-Ferreira et al., 1984). Typing of strains isolated from countries other than France has sometimes proved to be difficult. In the study of Giammanco et al. (1989), performed in Palermo, Italy, only 19 of 54 *A. baumannii* strains examined (35%) were correctly typed, while six strains had a new phage type and 25 strains were resistant to all phages tested.

Data on species, biotypes and bacteriophage types of 528 hospital isolates from published and unpublished results are available for comparison. In all, 355 isolates were from 16 French hospitals, 68 strains were from an Italian hospital, and 105 strains were from 11 hospitals in the Netherlands. Data from 444 *A. baumannii* strains were examined in order to correlate biotypes and phage types. A total of 380 strains (85.5%) were typable, of which 331 strains belonging to 11 biotypes were typable using the first phage set, while 49 strains belonging to three biotypes were typable using the second set of phages. The results presented in Tables 2 and 3 list the number of typable strains for each biotype that were susceptible to the different phages of the first phage set (Table 2) or the second phage set (Table 3). In the first system, phages 2, 5, 6, 11, 14-18, 20 and 21 were nearly always inactive on the 331 strains tested. Most biotype 1 and biotype 2 strains were susceptible to phages 9 and 12. Most biotype 1 strains were susceptible to phage 10, whereas biotype 2 strains were not. Most strains (76%) of biotype 2 were susceptible to phage 7. Phages 1 and 13 seemed to be specific for biotypes 6 and 9. All strains of biotypes 6 and 9 were typable with the first system, compared with 55% and 84% of the strains in biotypes 1 and 2 respectively. *Acinetobacter* sp. 3 strains, which are phenotypically and genetically related to *A. baumannii*, were also typable (25 of 29 strains studied; data not shown). Among strains belonging to species other than *A. baumannii* or *Acinetobacter* sp. 3, only seven of 59 strains (12%) were typable. The typable strains belonged to *A. haemolyticus* (two strains), *A. johnsonii* (three strains), *A. lwoffii* (one strain), and *Acinetobacter* sp. 10 (one strain).

Evidence from some studies has cast doubt upon the reproducibility of this method. In the study of Buisson et al. (1990), of 24 *A. baumannii* biotype 6 strains studied, 23 belonged to phage type 17 and one strain to phage type 16. Phage type 17 (susceptibility to phages 1, 12 and 13) differed from phage type 16 (susceptibility to phages 1, 12 and 13) by susceptibility to

Table 2. Susceptibility to the 21 phages of the first phage typing system of 331 typable A. *baumannii* strains

Biotype No.	No. strains typable by this set	Total strains typable	Number of strains susceptible to phage no.																				
			1	2	3	4	5	6	7	8	9	10	11	12	13	14	15	16	17	18	19	20	21
1	36	66	1	2	7	6	3	3	6	8	23	23	0	22	0	0	0	0	0	0	14	0	0
2	88	105	0	0	3	3	0	0	67	44	56	0	0	62	0	0	0	0	0	0	2	0	2
6	136	136	134	1	6	9	1	1	39	0	2	33	0	5	132	4	3	1	0	0	0	0	0
9	54*	54	40	4	7	7	2	0	6	2	12	13	0	16	38	0	2	0	0	0	0	0	1
8	4	4	0	0	0	0	0	0	1	0	0	2	0	2	1	0	0	0	1	0	1	0	0
10	1	1	0	0	1	1	0	0	0	0	0	0	0	0	0	0	0	0	0	0	0	0	0
11	6	6	1	4	5	2	3	1	2	0	0	4	0	0	1	0	0	0	0	0	0	0	0
14	2	4	1	0	1	1	0	0	1	0	0	2	2	0	0	0	0	0	0	0	0	0	0
17	2	2	0	1	1	2	1	0	0	0	0	1	0	0	0	0	0	0	0	0	0	0	0
18	1	1	0	0	0	0	0	0	0	0	0	0	0	0	0	0	0	0	0	0	0	0	0
19	1	1	0	1	1	1	1	1	0	0	1	0	0	0	0	0	0	0	0	0	0	0	0

*One strain was typable by both phage typing systems.

41

Table 3. Susceptibility to the 14 phages of the second phage typing system of 49 typable *A. baumannii* strains

Biotype No.	No. strains typable by this set	Total strains typable	Number of strains susceptible to phage no.													
			1	2	3	4	5	6	7	8	9	10	11	12	13	14
1	30	66	1	0	30	30	29	4	3	1	1	0	0	0	0	13
2	17	105	1	0	17	17	17	4	0	0	1	0	0	0	0	3
14	2	4	0	0	2	2	2	2	0	1	0	0	0	0	0	2

a single phage. In the same study, ten biotype 1 strains belonged to phage type 86 (susceptibility to phage 9), phage type 88 (susceptibility to phages 9, 10 and 12) and phage type 89 (susceptibility to phages 10 and 12). It therefore seems that the phage typing system is of limited value in its present form and should be extended with other phages in order to increase the percentage of typable strains. Further studies should be done with strains that are well-defined epidemiologically in order to evaluate the reproducibility and reliability of this phage typing method.

SEROTYPING

There have been numerous attempts to type *Acinetobacter* strains by serological reactions (see the review by Henriksen, 1973; Adam, 1979; Das and Ayliffe, 1984). Limited success has been obtained and most schemes have been rendered obsolete by taxonomic developments. Recently, Traub (1989) studied 152 clinical isolates of *A. baumannii* from 152 patients and developed a serotyping system for this species using polyclonal rabbit immune sera. A total of 20 serovars were delineated. The sera were fairly specific for *A. baumannii*. Most nosocomially significant *A. baumannii* isolates belonged to serovar 4. Several outbreaks of nosocomial cross-infection caused by serovars 4 and 10 were identified (Traub, 1989).

CELL ENVELOPE PROTEIN PATTERNS

Cell envelope protein profiles have been shown to be a valuable tool in typing *Acinetobacter* strains (Alexander et al., 1984; Dijkshoorn et al., 1987a,b, 1989; Crombach et al., 1989). Cell envelope protein patterns allowed differentiation among strains associated with epidemiological outbreaks (Dijkshoorn et al., 1987a) or cross-infection (Dijkshoorn et al., 1987b). Bouvet et al. (1990) subsequently identified, biotyped, and phage-typed 120 strains for which cell envelope protein patterns had been determined by Dijkshoorn et al. (1987a,b) These strains could be classified into 15 protein patterns (designated A to H and J to P). Patterns that were similar, but not identical, were designated with a capital letter followed by a number (B1-4, D1-3, E1-4). The correlation between identification, biotype determination and cell envelope protein profile is shown in Table 4. A total of 90 strains belonging to *A. baumannii* (nine biotypes), *Acinetobacter* sp. 3 (six biotypes) and *A. calcoaceticus* could be assigned to 16 different protein patterns A, B1, B2, B4, C, D1, D2, D3, G, H, J, K, L, M, N. Among these strains, 17 had unique protein profiles.

A. baumannii biotypes 1 and 2 were indistinguishable by protein profile (protein pattern A), but could of course be differentiated by biotyping. Most biotype 6 strains had related type B protein patterns, but two strains had pattern J, one strain pattern M, and one strain a unique pattern. *A. haemolyticus* strains had related type E profiles. Most strains belonging to other species had unrelated unique patterns. Multiple repeat

Table 4. Correlation between identification, biotyping and cell envelope protein patterns[a] (data from Bouvet et al.,1990)

Species	A	B1	B2	B3	B4	C	D1	D2	D3	E	F	G	H	J	K	L	M	N	O	P	Unique
A. baumannii																					
biotype 1	11	0	0	0	0	0	0	0	0	0	0	0	0	0	0	0	0	0	0	0	0
biotype 2	17	0	0	0	0	0	0	0	0	0	0	0	0	0	0	0	0	0	0	0	2
biotype 5	0	0	0	0	0	0	0	0	0	0	0	0	0	0	0	0	0	0	0	0	1
biotype 6	0	7	0	1	0	0	0	0	0	0	0	0	0	2	0	0	1	0	0	0	1
biotype 8	0	0	5	0	0	0	1	1	0	0	0	0	0	0	0	0	0	0	0	0	2
biotype 9	0	0	0	0	0	2	0	0	0	0	0	4	0	0	0	0	0	0	0	0	5
biotype 12	0	0	0	0	0	0	0	0	0	0	0	0	0	0	0	2	0	0	0	0	0
biotype 16	0	0	0	0	0	0	0	0	0	0	0	0	5	0	0	0	0	0	0	0	5
biotype 19	0	0	0	0	1	0	0	0	0	0	0	0	0	0	0	0	0	0	0	0	0
Acinetobacter sp. 3																					
biotype 1	0	0	0	0	0	0	0	0	0	0	0	0	0	0	2	0	0	0	0	0	0
biotype 8	0	0	0	0	0	0	3	0	0	0	0	0	0	0	0	0	0	0	0	0	2
biotype 9	0	0	0	0	0	0	1	0	0	0	0	0	5	0	0	2	0	1	0	0	1
biotype 11	0	0	0	0	0	0	0	0	4	0	0	0	0	0	0	0	0	0	0	0	0
biotype 12	0	0	0	0	0	0	0	0	1	0	0	0	0	0	0	0	0	0	0	0	0
biotype 18	0	0	0	0	0	0	0	0	0	0	0	0	0	0	0	0	0	0	0	0	0
A. haemolyticus	0	0	0	0	0	0	0	0	0	6	0	0	0	0	0	0	0	0	0	0	1
A. johnsonii	0	0	0	0	0	0	0	0	0	0	0	0	0	0	0	0	0	0	0	1	1
A. junii	0	0	0	0	0	0	0	0	0	0	0	0	0	0	0	0	0	0	0	0	2
A. lwoffii	0	0	0	0	0	0	1	0	0	0	2	0	0	0	0	0	0	0	0	0	10
A. calcoaceticus	0	0	0	0	0	0	0	0	0	0	0	0	0	0	0	0	0	0	0	0	3
Acinetobacter sp.6	0	0	0	0	0	0	0	0	0	0	0	0	0	0	0	0	0	0	0	0	1
Acinetobacter sp.11	0	0	0	0	0	0	0	0	0	0	0	0	0	0	0	0	0	0	0	0	1
Ungrouped strains	0	0	0	0	0	0	0	0	0	0	0	0	0	0	0	0	0	0	1	0	3

[a] a total of 118 different strains were studied.

isolates from the same patient were indistinguishable by either typing method. Strains with profiles G or L (Dijkshoorn et al., 1987b; Bouvet et al., 1990) which were demonstrated to be responsible for cross-infection could be easily detected by using identification and biotyping and, to a lesser extent, phage typing. However, when strains from other origins were examined, this correlation could not be reproduced; indeed a study by Giammanco et al. (1989) led to an opposite conclusion. Analysis by polyacrylamide gel electrophoresis of the cell envelope protein failed to show any diversity, not only within, but also between some of the biotypes of *A. baumannii*. Alexander et al. (1988), in a study comparing the value of electrophoretic and other techniques (plasmid analysis, antibiograms and biochemical tests) for typing *Acinetobacter*, also pointed out the reproductibility problems associated with this method.

PLASMID PROFILE TYPING AND PLASMID FINGERPRINTING

With other bacterial species, determination of plasmid profiles has proved to be an inexpensive and easily performed epidemiological tool with good discriminatory power, especially when used in conjunction with other typing methods (Mayer, 1988). Few studies have been performed with the genus *Acinetobacter*, although Gerner-Smidt (1989) investigated the plasmid content of 93 clinical isolates of *Acinetobacter*. Unfortunately, strain identification followed the older taxonomic schemes for the genus. Plasmids were found in 75 of 93 (81%) strains studied, including 56 of 71 epidemiologically unrelated strains. The number of plasmids found varied between strains (range 0-20 with a median of two). A single plasmid was found in 13% of strains. Among the glucose-oxidising strains, 77% had plasmids, compared with 82.8% of glucose non-oxidising strains. Three strains from the Neurosurgical Department and six strains from the Intensive Care Unit were identified by Gerner-Smidt (1989) as atypical *A. baumannii* biotype 9 (malonate not utilised) and all of these strains contained a plasmid of the same size. Two additional strains from the Intensive Care Unit were found to contain two plasmids and were also identified as *A. baumannii* biotype 9. The finding of only one or two plasmids in these strains illustrates the major limitation in the use of plasmid profiles for epidemiological purposes. It is well-known that microorganisms may gain or lose plasmids in hospitals and that repeated culturing and storing may result in loss of plasmids. A separate study by Vila et al. (1989) used plasmid profiles and restriction endonuclease digestion (plasmid fingerprinting) to examine 25 clinical glucose-oxidising strains (one strain from each patient). The strains could be divided into two groups on the basis of assimilation of two substrates (phenylacetate and adipate), the number of plasmid bands (one or two), and restriction endonuclease digestion profiles (identical for strains within a group and different between groups.) Determination of plasmid profiles seems to be a simple and reproducible tool for use with *Acinetobacter* strains. The method should, however, be used in conjunction with other typing methods because of the risk of plasmid instability and because 20% of strains do not contain detectable plasmid

DNA. Plasmid profiles have been used to either verify the identity of bacterial isolates or show that they are different. It should, however, be noted that while in some cases plasmid profile analysis seems to be the most discriminating typing method, in other cases plasmid profiles have been of no value. Restriction endonuclease digestion of plasmid DNA may be of additional value when only a single plasmid can be identified or when similar plasmid profiles occur in strains which cannot be differentiated by other methods.

ANTIBIOTIC RESISTANCE PATTERNS

Antibiotic resistance patterns have been used to differentiate strains involved in outbreaks of infection, especially when a new resistance phenotype emerges (Castle et al., 1978; French et al., 1980; Devaud et al., 1982; Holton, 1982; Gerner-Smidt et al., 1985; Allen and Green, 1987; Crombach et al., 1989). Since October 1984, outbreaks of infections caused by strains resistant to kanamycin and structurally related aminoglycosides, including amikacin, have occurred in France. Resistance to aminoglycosides was demonstrated to result from synthesis of a new type of 3'-aminoglycoside phosphotransferase [APH(3')] (Lambert et al., 1988). This resistance was carried in the particular *A. baumannii* strain investigated by a 63 kb plasmid, self-transferable to other *Acinetobacter* species such as *A. haemolyticus* or *A. lwoffii*, but not to *Escherichia coli*. This new kanamycin resistance gene (*aphA-6*) has been sequenced by Martin et al. (1988). Buisson et al. (1990) studied amikacin-resistant, tobramycin-sensitive *Acinetobacter* strains isolated in a hospital and demonstrated, using dot blot hybridisation with an *aphA-6*-intragenic probe, that the gene coding for amikacin resistance had disseminated among different *Acinetobacter* species (*A. baumanni, A. haemolyticus, A. junii, A. johnsonii, A. lwoffii, Acinetobacter* sp. 3 and sp. 11) and different biotypes of *A. baumannii*. In addition, this phenotype was encountered in saprophytic *A. johnsonii* isolates recovered from the hands of 11 healthy workers among the medical staff. The finding of this type of resistance in *A. johnsonii*, which is a normal saprophyte of the human skin, and its isolation from the hands of medical staff, suggests that this species may have played a role in the transmission and dissemination by hand carriage of the amikacin resistance gene to more pathogenic species. Typing of *Acinetobacter* strains by antimicrobial resistance pattern must, therefore, be interpreted with care and only after identification of the strain to the species level.

LOW-FREQUENCY-CLEAVAGE RESTRICTION ENDONUCLEASE ANALYSIS

Analysis of bacterial restriction endonuclease digests provides a direct and sensitive means of detecting minor genomic differences between microorganisms. Restriction endonucleases specifically cleave DNA into different lengths, depending on the number and position of the individual recognition sequences, provided that they have not been modified in any way. A DNA polymorphism involves a change in the size of a restriction fragment. If a change occurs in a genomic DNA sequence, this can delete or create a new recognition site and result in the generation of a restriction fragment length polymorphism (RFLP). The DNA fragments generated by restriction enzyme digestion are separated according to size by electrophoresis in agarose gels to give a pattern of bands. The usefulness of such patterns as diagnostic tools is limited by their complexity; indeed they may comprise >50 bands of various sizes, depending on the cutting frequency of the restriction endonuclease used and the size of the organism's genome. Large DNA molecules can now be separated by pulsed-field gel electrophoresis, making it possible to carry out DNA analysis with low-frequency-cleavage restriction endonucleases. These produce fewer fragments, usually <40, and electrophoretic patterns can be compared more precisely and easily. This technique was used by Allardet-Servent et al. (1989) to investigate an outbreak of *A. calcoaceticus* infection (although the species was not precisely determined by these authors) in a urologic department. Two series of strains were studied. The first (eight strains) consisted of isolates obtained during three weeks from blood or urine cultures of patients admitted to the urologic department. The second group (five strains) comprised occasional isolates from different departments of the same hospital. The eight strains of the first group were of the same phage type and all but three had different antibiotypes. Following digestion with *Sma*I and separation of the resulting DNA fragments with pulsed-field electrophoresis (over a 48 h period), all these isolates presented the same pattern of about 30 distinct bands. The patterns of the strains belonging to the second group were all different and distinct from that of the former group. The authors concluded that low-frequency-cleavage restriction endonuclease analysis of genomic DNA is a versatile and precise method for epidemiological investigation.

ELECTROPHORETIC ANALYSIS OF ISOENZYMES

Electrophoretic analysis of isoenzymes has been suggested as a method for studying the genetic diversity of bacteria (Selander et al., 1986, 1987) and has been used for the phenotypic characterisation of bacterial species defined by DNA-DNA hybridisation (Goullet and Picard, 1986). Electrophoretic analysis of esterases has also been used as an epidemiological typing method (Picard and Goullet, 1987). Electrophoretic analysis (Picard et al., 1989) of 27 esterases and two dehydrogenases from 81 genetically characterised *Acinetobacter* strains (Bouvet and Grimont, 1986) enabled clear computerised differentiation of the 12 genospecies. The number of electrophoretic variants for each enzyme among the *Acinetobacter* strains varied between two and 17 (with a mean of 6.5 for each enzyme). Preliminary analysis of 40 unrelated *A. baumannii* strains belonging to 14 biotypes demonstrated that all patterns were dissimilar. Further studies on epidemiologically related strains should be carried out in order to properly assess the utility of this method for typing *Acinetobacter* isolates.

CONCLUSIONS

As an initial step in typing, it is necessary to establish which particular *Acinetobacter* species are of clinical and epidemiological relevance. Once this is known, new isolates should first be identified to the species level in order to trace epidemic strains.

Bacteriophage typing of *Acinetobacter* is used in only one laboratory in the world. Furthermore, its apparent lack of reproducibility and the relatively high percentage of untypable strains (particularly when strains isolated from countries other than France are investigated) makes its use as a typing method difficult. However, when strains are typable, the method can successfully sub-divide strains within a biotype or protein profile.

The recently developed serotyping system for *A. baumannii* and *Acinetobacter* sp. 3, should become a rapid and accurate tool in the surveillance of nosocomial outbreaks. Comparison of cell envelope protein profiles also seems to be a precise method for the relative classification of *Acinetobacter*. Its feasibility depends mainly on the epidemiological context. If clinical isolates from patients in a ward or small area are to be typed, the method allows particular strains to be identified. In contrast, if large numbers of isolates from the hospital environment or the skin of healthy people need to be examined, comparison of protein profiles may be difficult and laborious and will not identify species.

Identification to the species level and biotyping using physiological and nutritional properties are easy and accurate methods for screening strains. When used together, identification/biotyping and cell envelope protein profile typing has enabled all the strains so far investigated to be typed or fingerprinted. At the present time it seems that a combination of these two methods may be the best method for performing epidemiological studies. Newer methods, such as analysis following treatment with low-frequency-cleavage restriction endonucleases, electrophoretic analysis of isoenzymes, or restriction endonuclease analysis of rRNA genes, still require further evaluation using strains that are well-defined both taxonomically and epidemiologically.

REFERENCES

Adam, M.M., 1979, Classification antigenique des "Acinetobacter", *Ann. Microbiol. (Inst.Past.)*, 130A:404.

Alexander, M., Ismail, F., Jackman, P.J.H., and Noble, W.C., 1984, Fingerprinting *Acinetobacter* strains from clinical sources by numerical analysis of electrophoretic protein patterns, *J. Med. Microbiol.*, 18:55.

Alexander, M., Rahman, M., Taylor, M., and Noble, W.C., 1988, A study of the value of electrophoretic and other techniques for typing *Acinetobacter calcoaceticus*, *J. Hosp. Infect.*, 12:273.

Allardet-Servent, A., Bouziges, N., Carles-Nurit, M.J., Bourg, G., Gouby, A., and Ramuz, M., 1989, Use of low-frequency-cleavage restriction endonucleases for DNA analysis in epidemiological investigations of nosocomial bacterial infections, *J. Clin. Microbiol.*, 27:2057.

Allen, K.D., and Green, H.T., 1987, Hospital outbreak of multi-resistant *Acinetobacter anitratus*: an airborne mode of spread, *J. Hosp. Infect.*, 9:110.

Bergogne-Bérézin, E., Joly-Guillou, M.L., and Vieu, J.F., 1987, Epidemiology of nosocomial infections due to *Acinetobacter*, *J. Hosp. Infect.*, 10:105.

Bouvet, P.J.M., and Bouvet, O.M.M., 1989, Glucose dehydrogenase activity in *Acinetobacter* species, *Res. Microbiol.*, 140:531.

Bouvet, P.J.M., and Grimont, P.A.D., 1986, Taxonomy of the genus *Acinetobacter* with the recognition of *Acinetobacter baumannii* sp. nov., *Acinetobacter haemolyticus* sp. nov., *Acinetobacter johnsonii* sp. nov., and *Acinetobacter junii* sp. nov., and emended descriptions of *Acinetobacter calcoaceticus* and *Acinetobacter lwoffii*, *Int. J. Syst. Bacteriol.*, 36:228.

Bouvet, P.J.M., and Grimont, P.A.D., 1987, Identification and biotyping of clinical isolates of *Acinetobacter*, *Ann. Inst. Past. Microbiol.*, 138:569.

Bouvet, P.J.M., and Jeanjean, S., 1989, Delineation of new proteolytic genomic species in the genus *Acinetobacter*, *Res. Microbiol.*, 140:291.

Bouvet, P.J.M., Jeanjean, S., Vieu, J.F., and Dijkshoorn, L., 1990, Species, biotype, and bacteriophage type determinations compared with cell envelope protein profiles for typing *Acinetobacter* strains, *J. Clin. Microbiol.*, 28:170.

Buisson, Y., Tran Van Nhieu, G., Ginot, L., Bouvet, P.J.M., Schill, H., Driot, L., and Meyran, M., 1990, Nosocomial outbreaks due to amikacin-resistant tobramycin-sensitive *Acinetobacter* species: correlation with amikacin usage, *J. Hosp. Infect.*, 15:83.

Castle, M., Tenney, J.H., Weinstein, M.P., and Eickhoff, T.C., 1978, Outbreak of a multiply-resistant *Acinetobacter* in a surgical intensive-care unit: epidemiology and control, *Heart Lung*, 65:641.

Crombach, W.H.J., Dijkshoorn, L., van Noort-Klaassen, M., Niessen, J., and van Knippenberg-Gordebeke, G., 1989, Control of an epidemic spread of a multi-resistant strain of *Acinetobacter calcoaceticus* in a hospital, *Intens. Care Med.*, 15:166.

Das, B.C., and Ayliffe, G.A.J., 1984, Serotyping of *Acinetobacter calcoaceticus*, *J. Clin. Pathol.*, 37:1388.

Devaud, M., Kayser, F.H., and Bachi, B., 1982, Transposon-mediated multiple antibiotic resistance in *Acinetobacter* strains, *Antimicrob. Agents. Chemother.*, 22:323.

Dijkshoorn, L., Michel, M.F., and Degener, J.E., 1987a, Cell envelope protein profiles of *Acinetobacter calcoaceticus* strains isolated in hospitals, *J. Med. Microbiol.*, 23:313.

Dijkshoorn, L., Van Vianen, W., Degener, J.E., and Michel, M.F., 1987b, Typing of *Acinetobacter calcoaceticus* strains isolated from hospital patients by cell envelope protein profiles, *Epidemiol. Infect.*, 99:659.

Dijkshoorn, L., Wubbels, J.L., Beunders, A.J., Degener, J.E., Boks, A.L., and Michel, M.F., 1989, Use of protein profiles to identify *Acinetobacter calcoaceticus* in a respiratory care unit, *J. Clin. Pathol.*, 42:853.

French, G.L., Casewell, M.W., Roncoroni, A.J., Knight, S., and Phillips, I., 1980, A hospital outbreak of antibiotic-resistant *Acinetobacter anitratus*: epidemiology and control, *J. Hosp. Infect.*, 1:125.

Freney, J., Bouvet, P.J.M., and Tixier, C., 1989, Identification et détermination de la sensibilité aux antibiotiques de 31 souches cliniques d'*Acinetobacter* autres que *A. baumannii*, *Ann. Biol. Clin.*, 47:41.

Gerner-Smidt, P., 1989, Frequency of plasmids in strains of *Acinetobacter calcoaceticus*, *J. Hosp. Infect.*, 14:23.

Gerner-Smidt, P., Hansen, L., Knudsen, A., Siboni, K., and Sogaard, I., 1985, Epidemic spread of *Acinetobacter calcoaceticus* in a neurosurgical department analyzed by electronic data processing, *J. Hosp. Infect.*, 6:166.

Giammanco, A., Vieu, J.F., Bouvet, P.J.M., Sarzana, A., and Sinatra, A., 1989, A comparative assay of epidemiological markers for *Acinetobacter* strains isolated in a hospital, *Zbl. Bakt.* 272:231.

Gilardi, G.L., 1978, Identification of miscellaneous glucose non-fermenting Gram-negative bacteria, *in* "Glucose Nonfermenting Gram-negative Bacteria in Clinical Microbiology," p.45, G.L. Gilardi, ed., CRC Press, West Palm Beach.

Goullet, P., and Picard, B., 1986, Comparative esterase electrophoretic polymorphism of *Escherichia coli* isolates obtained from animal and human sources, *J. Gen. Microbiol.*, 132:1843.

Henriksen, S.D., 1973, *Moraxella*, *Acinetobacter* and the *Mimae*, *Bacteriol. Rev.*, 37:522.

Hercouet, H., Bousser, J., Donnio, P.Y., and Avril, J.L., 1989, Activité *in vitro* des antibiotiques sur les souches hospitalières de *Acinetobacter baumannii*, *Pathol. Biol.*, 37:612.

Holton, J., 1982, A report of a further hospital outbreak caused by a multi-resistant *Acinetobacter anitratus*, *J. Hosp. Infect.*, 3:305.

Lambert, T., Gerbaud, G., and Courvalin, P., 1988, Transferable amikacin resistance in *Acinetobacter* spp. due to a new type of 3'-aminoglycoside phosphotransferase, *Antimicrob. Agents Chemother.*, 32:15.

Martin, P., Jullien, E., and Courvalin, P., 1988, Nucleotide sequence of *Acinetobacter baumanni aphA-6* gene: evolutionary and functional implications of sequence homologies with nucleotide-binding proteins, kinases, and other aminoglycoside-modifying enzymes, *Mol. Microbiol.*, 2:615.

Mayer, L.W., 1988, Use of plasmid profiles in epidemiological surveillance of disease outbreaks and in tracing the transmission of antibiotic resistance, *Clin. Microbiol. Rev.*, 1:228.

Nishimura, Y., Ino, T., and Iizuka, H., 1988, *Acinetobacter radioresistens* sp. nov. isolated from cotton and soil, *Int. J. Syst. Bacteriol.*, 38:209.

Picard, B., and Goullet, P., 1987, Epidemiological complexity of hospital aeromonas infections revealed by electrophoretic typing of esterases, *Epidemiol. Infect.*, 98:5.

Picard, B., Goullet, P., Bouvet, P.J.M., Decoux, G., and Denis, J.B., 1989, Characterisation of bacterial genospecies by computer-assisted statistical analysis of enzyme electrophoretic data, *Electrophoresis*, 10:680.

Santos-Ferreira, M.O., Vieu, J.F., and Klein, B., 1984, Phage-types and susceptibility to 26 antibiotics of nosocomial strains of *Acinetobacter* isolated in Portugal, *J. Internat. Med.Res.*, 12:364.

Selander, R.K., Caugant, D.A., Ochman, H., Musser, J.M., Gilmour, M.N., and Whittman, T.S., 1986, Methods of multilocus enzyme electrophoresis for bacterial population genetics and systematics, *Appl. Environ. Microbiol.*, 51:873.

Selander, R.K., Musser, J.M., Caugant, D.A., Gilmour, M.N., and Whittam, T.S., 1987, Population genetics of pathogenic bacteria, *Microb. Pathogen.*, 3:1.

Tjernberg, I., and Ursing, J., 1989, Clinical strains of *Acinetobacter* classified by DNA-DNA hybridization, *APMIS*, 97:595.

Towner, K.J., and Chopade, B.A., 1987, Biotyping of *Acinetobacter calcoaceticus* using the API 2ONE system, *J. Hosp. Infect.*, 10:145.

Traub, W.H., 1989, *Acinetobacter baumannii* serotyping for delineation of outbreaks of nosocomial cross-infection, *J. Clin. Microbiol.*, 27:2713.

Vieu, J.F., Minck, R., and Bergogne-Bérézin, E., 1979, Bacteriophages et lysotypie de *Acinetobacter*, *Ann. Microbiol. (Inst. Past.)*, 130A:405.

Vieu, J.F., Bergogne-Bérézin, E., Joly, M.L., Berthelot, G., and Fichelle, A., 1980, Epidémiologie d'*Acinetobacter calcoaceticus*, *Press. Med.*, 9:3351.

Vila, J., Almela, M., and Jimenez de Anta, M.T., 1989, Laboratory investigation of hospital outbreak caused by two different multiresistant *Acinetobacter calcoaceticus* subsp. *anitratus* strains, *J. Clin. Microbiol.*, 27:1086.

HOSPITAL EPIDEMIOLOGY OF *ACINETOBACTER* INFECTION

W.C. Noble

Department of Microbial Diseases
Institute of Dermatology
United Medical and Dental Schools
St Thomas' Hospital
Lambeth Palace Road
London SE1 7EH
UK

INTRODUCTION

Any consideration of hospital infection must take account of the background, or community, level of infection on which hospital experience is superimposed. It must also take account of the availability of typing schemes, since these determine the level of discrimination possible; these are discussed elsewhere in this volume. Hospital isolates of *Acinetobacter* are generally resistant to many antibiotics (see for example Garcia et al., 1983; Obana et al., 1985), although this is more often a characteristic of *A.calcoaceticus* var.*anitratus* than of var.*lwoffii*. It may be that the predominance of var.*anitratus* in hospitals is a result of this resistance, rather than a greater virulence, and also that it is the resistance which draws attention to the spread of the strain. Gerner-Smidt et al. (1985) successfully used antibiotic resistance alone to detect outbreaks of infection, but over-reliance on antibiotic resistance may be misleading. Carlquist et al. (1982), in a study of infection in 37 patients over five months in an intensive care unit (ICU), found a gradual shift in zone size to three aminoglycosides and carbenicillin. They interpreted this as changes in the resistance of a single strain, although no other typing techniques were used to confirm this. No changes in resistance were seen in *Pseudomonas aeruginosa* isolates from the ICU in this period, nor in other non-ICU acinetobacters in the same hospital. However, an increase in resistance was also recorded by Stone and Das (1985) in a single strain additionally characterised by serotype and bacteriocin type. The most acceptable approach in hospital epidemiology would therefore seem to be the use of several typing methods, obtaining a concensus on identity with data from several sources (Alexander et al., 1988). In this way simultaneous outbreaks of infection by two strains of *A.calcoaceticus* var.*anitratus* have been detected (Vila et al., 1989). Various aspects of *Acinetobacter* infection have been previously reviewed by Vivian et al. (1981), Bergogne-Bérézin and Joly-Guillou (1985) and Bergogne-Bérézin et al. (1987).

NORMAL OCCURRENCE OF *ACINETOBACTER*

Both *A.calcoaceticus* var.*lwoffii* and *A.calcoaceticus* var.*anitratus* are normal inhabitants of human skin and are found in at least the axillae, groin, toewebs and antecubital fossa of about 25% of the population (Taplin et al., 1963; Somerville and Noble, 1970). Patients with skin disease, such as eczema, frequently carry *Acinetobacter* on their lesions, although primary infection of the skin is rare (Glew et al., 1977; Gaughan et al., 1979). As a result of observed colonisation, skin has been suggested as a source for infection/contamination of the blood stream (Al-Khoja and Darrell, 1979; Hoppe et al., 1983). Al Khoja and Darrell (1979) reported that up to half of patients with kidney disease might be colonised and that persistent carriage could last for up to at least 9 weeks. In countries where there is a well-defined summer and winter with a marked temperature swing, acinetobacters are more commonly recovered from the skin in the summer months (Kloos and Musselwhite, 1975). This is presumably the result of greater sweating in the summer, especially in males. Higher skin colonisation rates in summer are reflected in higher prevalence rates of endemic infection with a marked excess of young males. A survey of infection in 81 hospitals in the USA (Retailliau et al., 1979) found a pronounced summer peak of infection for *A. calcoaceticus* var.*anitratus*. Patients aged 21-30 years were in marked excess, although traces of this could also be seen in infection with other organisms (Table 1). Seasonal effects have also been recorded in single hospitals (Holton, 1982; Smego, 1985).

Buisson et al. (1990) have observed amikacin-resistant, tobramycin-sensitive strains of *A.baumannii* infecting patients together with *A.johnsonii* strains with the same resistance pattern on the skin. They speculated that exchange of resistance plasmids between strains of *Acinetobacter* may occur on the skin, but were unable to substantiate this, although such mechanisms are well established for other members of the skin flora such as the staphylococci.

ENDEMIC INFECTION

The endemic rate of infection is hard to determine since it often fails to arouse interest. Occasional infections are reported from almost all sites on the body, but it is rare for the number of patients at risk or other denominator to be given. Galvao et al.

Table 1. Age distribution of endemic infection due to *Acinetobacter*,
Pseudomonas and other organisms

| Age | Percentage Distribution | | | | | |
| | *Acinetobacter* | | *Pseudomonas* | | Other | |
(Years)	Male	Female	Male	Female	Male	Female
<1	1.9	4.1	2.8	2.2	8.5	4.4
1-10	4.9	2.2	2.1	2.0	2.3	1.4
11-20	9.6	7.9	6.4	4.0	5.0	7.1
21-30	15.2	10.7	10.4	6.8	7.5	13.1
31-40	6.3	9.4	6.6	6.5	6.1	9.3
41-50	9.3	13.5	7.4	8.0	8.3	10.1
51-60	13.8	14.2	13.4	13.2	15.2	12.7
61-70	17.8	13.8	22.0	20.1	21.1	16.2
71-80	15.9	15.1	20.2	21.2	18.0	15.6
>80	5.4	9.1	8.7	16.0	8.0	10.3
Total	100.1	100	100	100	100	100.2

Based on Retailliau et al. (1979)

denominator to be given. Galvao et al. (1989) observed 23 episodes of infection in 450 patient years of chronic ambulatory peritoneal dialysis (CAPD), making acinetobacters the second most frequent Gram-negative bacillary infection in CAPD. There were eight var.*anitratus* and three var.*lwoffii* in the strains speciated. Endemic infection in a respiratory care unit has been reported by Dijkshoorn et al. (1989), who used protein gel electrophoresis patterns to identify and observe the spread of several different strains; perhaps significantly, environmental strains sometimes differed from those recovered from the patient. A similar study over 3.5 years has been reported, based on antibiotic resistance patterns (Sakata et al., 1989), again with a number of strains evident. In a study in a neurosurgical unit, also over a period of 3.5 years, Gerner-Smidt et al. (1985) found that isolations from sputum were more frequent than from any other sites. Although the isolates were sub-divided on the sole basis of antibiotic sensitivity patterns, it was clear that small epidemics, each of a few cases, were superimposed on the endemic pattern. A similar mixed pattern was reported by Larson (1984) in her 10 year study of infection in a single hospital.

EPIDEMIC SPREAD AND SOURCES OF INFECTION

Contaminated Apparatus

Most reports of epidemic spread seem to have been associated with the use of contaminated apparatus. Thus, over a period of six months, contaminated resuscitators were associated with infection due to a single *Acinetobacter* type in nine of ten babies in a special care baby unit (Stone and Das, 1985). Similarly, contaminated Wright respirometers resulted in eight patients developing pneumonia and a further nine being colonised in the upper respiratory tract over a period of about 30 days (Cunha et al., 1980). A single serotype and antibiogram again accounted for all infections/colonisations and no other source was found. Cold water room humidifiers were reported by Smith and Massonari (1977) as the source of infection for 24 patients, 21 of whom had positive blood cultures and fever. Large numbers of *Acinetobacter* were easily demonstrated up to 2 m from the humidifiers and it seems probable that skin contamination by the airborne route subsequently resulted in infection via the indwelling venous catheters possessed by 22 of the 24 patients. In a similar incident (Snydman et al., 1977), contamination of the water reservoir in mist tents resulted in false-positive blood cultures in 11 children. Since the nose and skin of the children was heavily contaminated with *Acinetobacter*, it seems probable that contamination of the needle occurred during vene-puncture, resulting in a contaminated sample. Indeed, over a period of several years in a single hospital, Hoppe et al. (1983) were able to distinguish both genuine blood stream infection and contamination which they attributed to skin as a source.

Other Types of Environmental Contamination

Two other outbreaks, associated with contamination of mattresses and peritoneal dialysis bags respectively, have been reported. In the first, an outbreak of infection in burns patients lasted 21 months, during which time 63 of 103 patients were colonised by *Acinetobacter* (Sheretz and Sullivan, 1985). Infections occurred in 43 of the 63 patients, with the burn itself accounting for 34, the urinary tract 16, the blood stream seven and the lower respiratory tract five patients. Wet mattresses were found to harbour *Acinetobacter* and the epidemic ended when mattresses were removed as the patient was discharged. In the second outbreak, Abrutyn et al. (1978) found that the water used to warm peritoneal dialysis fluid was contaminated with *Acinetobacter*, as a result of which 13 patients suffered peritonitis. Contamination of the rubber bung is thought to have resulted in contamination of the dialysis fluid when organisms were carried into the bag on the prong of the administration set.

In a further (hopefully unique) incident, contamination of intrathecal methotrexate for leukemia patients resulted in meningitis caused by *A. calcoaceticus* var. *anitratus* in eight patients and death in three (Kelkar et al., 1989). The outbreak was associated with the re-use of needles, which no doubt resulted in contamination of the drug during reconstitution. An unusual outbreak of infection due to *A. calcoaceticus* var.*lwoffii* has also been associated with contaminated parenteral nutrition fluid. Although this source was not proven, nine of 28 babies in an intensive care unit were receiving parenteral nutrition and seven became infected with *Acinetobacter*, with signs of severe clinical infection and septic shock (Ng et al., 1989).

Skin Carriage

Many authors comment on the contamination or colonisation of patients' or staff skin and the spread of *Acinetobacter* which can occur as a result. For example, infection by a single biotype and antibiogram of *Acinetobacter* in intensive care patients was found to be strongly associated with intubation and skin carriage amongst the hospital staff (Buxton et al., 1978). One therapist with chronic dermatitis was found to be a semi-permanent carrier and to have contaminated the apparatus used for the intubation. Similarly, Allen and Green (1987) reported wide spread of a multi-resistant *A. calcoaceticus* var.*anitratus* strain which colonised 10 and infected 18 of 38 patients, probably by a mixture of hand contact and airborne spread. Staff members also become colonised as a result of working in wards during an epidemic. French et al. (1980) reported that 11 of 38 staff in a urological ward became carriers of a gentamicin-resistant strain during an outbreak, as did two microbiologists investigating the outbreak. *A. calcoaceticus* var.*anitratus* survives better on the skin than does var.*lwoffii* (Musa et al., 1990), another factor which may bias the distribution of the species in hospitals.

Airborne Spread

Dissemination of acinetobacters via the air is repeatedly mentioned in association with spread from environmental sources and as one mode of contamination of the skin. Airborne dispersal of Gram-negative bacilli is relatively rare (Ayliffe and Lowbury, 1982); however, acinetobacters are known to survive drying better than most Gram-negative bacteria and will thus tend to survive aerosolisation. Once dried, *Acinetobacter* may persist for at least a week on dried cloth (Buxton et al., 1978) or at least five days on formica surfaces (Musa et al., 1990).

Cordes et al. (1981) reported community-acquired airborne infection in foundry workers. Three men developed pneumonia due to *A. calcoaceticus* var.*anitratus* serotype 7J. Two of the men died and, at post-mortem examination, were found to have pneumoconiosis and iron particles in the lung. *Acinetobacter*, of unknown origin, were recovered from the air of the foundry and 15% of the 167 workers there, but only 2% of 93 local community residents had significantly high serum titres to serotype 7J. Contamination of the skin was found in only one of 107 samples and from none of 106 throat swabs in men at the foundry.

Table 2. Percentage distribution of infections due to
Acinetobacter calcoaceticus

	(1)		(2)	(3)
	all isolates	*Acinetobacter* only		
Respiratory tract	29	22	42	15
Urinary tract	27	36	10	43
Surgical wound	21.5	12	9	27
Blood stream	10.5	20	18	15
Peritoneum	1	1	17	0
Other	11	8	5	0
Total patients with Acinetobacter	1372	612	85	33

Data taken from (1) Retailliau et al., 1979; (2) Larson, 1984;
(3) Muller-Serieys et al., 1989.

Urinary Tract Infection

The second, and about equally numerous, group of infections is
urinary tract infection. Hoffman et al. (1982) studied urinary tract infection
in the community and reported 14 isolations of *A.calcoaceticus*
var.*anitratus* and 93 of var.*lwoffii* from the 97 persons participating in the
study. Normal as well as particularly susceptible individuals were at risk of
infection. Similarly, French et al. (1980) studied an epidemic of infection in
a teaching hospital, caused by a single strain of gentamicin-resistant
A.calcoaceticus var.*anitratus*, and found that the index case and 25 of the
40 infected patients were from a urological ward. Twenty-five of the 31
patients with infected urinary tracts had been catheterised.

RATES OF INFECTION

It is difficult to determine rates of infection since these clearly differ
according to local circumstances; differences in reporting strategies also
obscure any comparison. Retailliau et al. (1979) reported that

Acinetobacter infection occurred in 3.11 per 10,000 patients discharged, whilst Larson (1984) gave *Acinetobacter* as representing 1.4% of 6115 hospital infections, with no increase over a ten year period. According to Hoffman et al. (1982), *Acinetobacter* cultivars formed 4% of community-acquired urinary tract infection, of which only 13% were *A. calcoaceticus* var.*anitratus*. In contrast, community-acquired *Acinetobacter* pneumonia is a rare event with only about 30 reported cases (Gottlieb and Barnes, 1989). Table 2 shows the distribution of sites of infection in three published series. In a briefly reported outbreak (LeMaistre et al., 1985) in a surgical intensive care unit, 37 patients were recorded as colonised, of whom 16 had overt clinical infection with a highly-resistant *A. calcoaceticus* var.*anitratus* strain. The distribution of sites colonised/infected in this one outbreak closely followed the general trend in consisting of 22 respiratory tract, 13 urinary tract, seven blood stream, six surgical wounds and three other infections. A generally similar distribution was also found during an outbreak in a neurosurgical ward (Allen and Green, 1987), although the most common infections with clinical significance were those of the respiratory tract.

CONCLUSIONS

Acinetobacter, usually *A. calcoaceticus* var.*anitratus*, is not a frequent cause of hospital infection, but may be intractable because of multiple antibiotic resistance. Infection may occur at any site, but is most frequent in the respiratory tract, the urinary tract and surgical wounds. Both varieties of *Acinetobacter* are found on human skin as normal residents; however, especially when exposed to aerosols of *Acinetobacter*, skin may become heavily colonised and subsequently act as the major source for contamination or clinically evident infection of deeper tissue.

REFERENCES

Abrutyn, E., Goodhart, G.L., Roos, K., Anderson R., and Buxton A., 1978, *Acinetobacter calcoaceticus* outbreak associated with peritoneal dialysis, *Amer. J. Epidemiol.*, 107:328.

Alexander, M., Rahman, M., Taylor, M., and Noble W.C., 1988, A study of the value of electrophoretic and other techniques for typing *Acinetobacter calcoaceticus*, *J. Hosp. Infect.*, 12:273.

Al-Khoja, M.S., and Darrell, J.H., 1979, The skin as a source of Acinetobacter and Moraxella species occurring in blood cultures, *J. Clin. Pathol.*, 32:497.

Allen, K.D., and Green, H.T., 1987, Hospital outbreak of multiresistant *Acinetobacter anitratus*, an airborne mode of spread?, *J. Hosp. Infect.*, 9:110.

Ayliffe, G.A.J., and Lowbury, E.J.L., 1982, Airborne infection in hospital, *J. Hosp. Infect.*, 3:217.

Bergogne-Bérézin, E., and Joly-Guillou, M.L., 1985, An underestimated nosocomial pathogen *Acinetobacter calcoaceticus*, *J. Antimicrob. Chemother.*, 16:535.

Bergogne-Bérézin, E., Joly-Guillou, M.L., and Vieu, J.F. 1987, Epidemiology of nosocomial infections due to *Acinetobacter calcoaceticus*, *J. Hosp. Infect.*, 10:105.

Buisson, Y., Tran van Nhien, G., Ginot, L., Bouvet B., Schill, H., Driot, L., and Meyran, M., 1990, Nosocomial outbreaks due to amikacin-resistant tobramycin-sensitive Acinetobacter species: correlation with amikacin usage, *J. Hosp. Infect.*, 15:88.

Buxton, A.E., Anderson, R.L., Werdegar, D., and Atlas, E., 1978, Nosocomial respiratory tract infection and colonization with *Acinetobacter calcoaceticus*. Epidemiologic characteristics, *Amer. J. Med.*, 65:507.

Carlquist, J.F., Conti, M., and Burke J.P., 1982, Progressive resistance in a single strain of *Acinetobacter calcoaceticus* recovered during a nosocomial outbreak, *Amer. J. Infect. Control,* 10:43.

Cordes, L.G., Brink, E.W., Checko, P.J., Lentnek, A., Lyons, R.W., Hayes, P.S., Wu, T.C., Tharr D.G., and Fraser, D.W., 1981, A cluster of Acinetobacter pneumonia in foundry workers, *Ann. Int. Med.,* 95:688.

Cunha, B.A., Klimek, J.J., Gracewski, J., McLaughlin, J.C., and Quintiliani, R., 1980, A common source outbreak of Acinetobacter pulmonary infection traced to Wright respirometers, *Postgrad. Med. J.,* 56:169.

Dijkshoorn, L., Wubbels, J.L., Beunders, A.J., Degener, J.E., Boks, A.L., and Michel, M.F., 1989, Use of protein profiles to identify *Acinetobacter calcoaceticus* in a respiratory unit, *J. Clin. Pathol.,* 42:853.

French, G.L., Casewell, M.W., Roncoroni, A.J., Knight, S., and Phillips, I., 1980, A hospital outbreak of antibiotic-resistant *Acinetobacter anitratus*: epidemiology and control, *J. Hosp. Infect.,* 1:125.

Galvao, C., Swartz, R., Rocher, L., Reynolds, J., Starmann, B., and Wilson, D., 1989, Acinetobacter peritonitis during chronic peritoneal dialysis, *Amer. J. Kidney Dis.,* 14:101.

Garcia, I., Fainstein, V., Le Blanc, B., and Bodey G.P., 1983, In vivo activities of new B-lactam antibiotics against *Acinetobacter* spp., *Antimicrob. Agents Chemother.*, 24:297.

Gaughan, M., White, P.M., and Noble, W.C., 1979, Skin as a source of Acinetobacter/Moraxella species, *J. Clin. Pathol.,* 32:1193.

Gerner-Smidt, P., Hansen, L., Knudsen, A., Siboni, K., and Sogaard, I., 1985, Epidemic spread of *Acinetobacter calcoaceticus* in a neurological department analysed by electronic data processing, *J. Hosp. Infect.*, 6:166.

Glew, R.H., Moellering, R.C., and Kunz, L.J., 1977, Infection with *Acinetobacter calcoaceticus (Herellea vaginicola)*: clinical and laboratory studies, *Medicine,* 56:79.

Gottlieb, T., and Barnes, D.J., 1989, Community acquired acinetobacter pneumonia, *Aust. NZ. J. Med.,* 19:259.

Hoffman, S., Mabeck, C.E., and Vejlsgaard, R., 1982, Bactiuria caused by *Acinetobacter calcoaceticus* biovars in a normal population and in general practice, *J. Clin. Microbiol.*, 16:443.

Holton, J., 1982, A report of a further hospital outbreak caused by a multi resistant *Acinetobacter anitratus*, *J. Hosp. Infect.*, 3:305.

Hoppe, M., Potel, J., and Malottke, R., 1983, Clinical importance of *A.calcoaceticus* isolations from blood cultures and vein catheters, *Zent. Bakt. Hyg. A.*, 256:80.

Kelkar, R., Gordon, S.M., Giri, N., Rao, K., Ramakrishnan, G., Saikia, T., Nair, C.N., Kurkure, R.A., Pai, S.K., Jarvis, W.R., and Advani, S.H., 1989, Epidemic iatrogenic *Acinetobacter* spp. meningitis following administration of intrathecal methotrexate, *J. Hosp. Infect.*, 14:233.

Kloos, W.E., and Musselwhite, M.S., 1975, Distribution and persistence of *Staphylocccus* and *Micrococcus* species and other aerobic bacteria on human skin, *Appl. Microbiol.*, 30:381.

Larson, E., 1984, A decade of nosocomial acinetobacter, *Amer. J. Infect. Control*, 12:14.

LeMaistre, A., Mortensen, J., La Rocco, M., and Robinson, A., 1985, An outbreak of multiply resistant *Acinetobacter anitratus* in an intensive care unit, *Clin. Microbiol. Newslett.*, 7:102.

Muller-Serieys, C., Lesquoy, J.B., Perez, E., Fichelle, A., Boujeois, B., Joly-Guillou, M.L., and Bergogne-Bérézin, E., 1989, Nosocomial Acinetobacter infections. Epidemiology and therapeutic problems, *Press. Med.*, 18:107.

Musa, E.K., Desai, N., and Casewell, M.W., 1990, The survival of *Acinetobacter calcoaceticus* inoculated on fingertips and on formica, *J. Hosp. Infect.*, 15:219.

Ng, P.C., Herrington, R.A., Beane, C.A., Ghonheim, A.T.M., and Dear, P.R.F., 1989, An outbreak of acinetobacter septicaemia in a neonatal intensive care unit, *J. Hosp. Infect.*, 14:363.

Obana, Y., Nishino, T., and Tanino T., 1985, *In vitro* and *in vivo* activities of antimicrobial agents against *Acinetobacter calcoaceticus*, *J. Antimicrob. Chemother.*, 15:441.

Retailliau, H.F., Hightower, A.W., Dixon, R.E., and Allen, J.R., 1979, *Acinetobacter calcoaceticus*: a nosocomial pathogen with an unusual seasonal pattern, *J. Infect. Dis.*, 139:371.

Sakata, H., Fujita, K., Maruyama, S., Kakehashi, H., Mori, Y., and Yoshioka, H., 1989, *Acinetobacter calcoaceticus* biovar *anitratus* septicaemia in a neonatal intensive care unit: epidemiology and control, *J. Hosp. Infect.*, 14:15.

Sheretz, R.J., and Sullivan, M.L., 1985, An outbreak of infection in burn patients: contamination of patients' mattresses, *J. Infect. Dis.*, 151:252.

Smego, R.A., 1985, Endemic nosocomial *Acinetobacter calcoaceticus* bacteraemia. Clinical significance, treatment and prognosis, *Arch. Int. Med.*, 145:2174.

Smith, P.W., and Massonari, R.M., 1977, Room humidifiers as the source of Acinetobacter infections, *J. Amer. Med. Assoc.*, 237:795.

Snydman, D.R., Maloy, M.F., Brock, S.M., Lyons, R.W., and Rubin, S.J., 1977, Pseudobacteraemia: false-positive blood cultures from mist tent contamination, *Amer. J. Epidemiol.*, 106:154.

Somerville, D.A., and Noble, W.C., 1970, A note on the Gram-negative bacilli of human skin, *Eur. J. Clin. Biol. Res.*, 40:669.

Stone, J.W., and Das, B.C., 1985, Investigation of an outbreak of infection with *Acinetobacter calcoaceticus* in a special care baby unit, *J. Hosp. Infect.*, 6:42.

Taplin, D., Rebell, G., and Zaias, N., 1963, The human skin as a source of Mima-Herella infections, *J. Amer. Med. Assoc.*, 186:952.

Vila, J., Almela, M., and Jiminez de Anta, M.I., 1989, Laboratory investigation of hospital outbreak caused by two different multiresistant *Acinetobacter calcoaceticus* subsp.*anitratus* strains, *J. Clin. Microbiol.*, 27:1086.

Vivian, A., Hinchcliffe, E., and Fewson, C.A., 1981, *Acinetobacter calcoaceticus*: some approaches to a problem, *J. Hosp. Infect.*, 2:199.

EPIDEMIOLOGY OF *ACINETOBACTER* STRAINS ISOLATED

FROM NOSOCOMIAL INFECTIONS IN FRANCE

M.L. Joly-Guillou[a], E. Bergogne-Bérézin[a] and J.F. Vieu[b]

[a]Laboratoire de Microbiologie, CHU Bichat
46 rue Huchard, 75877 Paris Cedex 18, France

[b]Institut Pasteur, rue du Dr.Roux
75724 Paris Cedex 15, France

INTRODUCTION

Acinetobacter has emerged as an important nosocomial pathogen during the last 20 years and, as described by Noble (this volume), is now frequently involved in outbreaks of hospital infections or colonisations. The increasing importance of infections caused by *Acinetobacter*, and the multi-resistance of the strains involved (Bergogne-Bérézin and Joly-Guillou, this volume), suggests that studies of the epidemiology and resistance phenotypes will be required for the prevention of further infections with this organism. Nosocomial infections with *Acinetobacter* have been a particular problem in France. This article reviews our experience to date with this organism.

EPIDEMIOLOGY OF HOSPITAL INFECTIONS CAUSED BY *ACINETOBACTER*

Taxonomy of Acinetobacter Isolates from French Hospitals

The current taxonomic situation of the genus *Acinetobacter* has been reviewed by Grimont (this volume). Using the taxonomic scheme proposed by Bouvet and Grimont (1986), 84-90% of *Acinetobacter* isolates from patients in French hospitals can currently be identified as *A.baumannii*. Remaining isolates consist primarily of *A.haemolyticus*, *A.lwoffii*, *A.johnsonii*, *A.junii*, or unnamed species, especially genospecies 3 (Bouvet and Grimont, 1987; Joly-Guillou and Bergogne-Bérézin, 1990a).

Strains of *A.baumannii* are widespread in nature. They can be isolated from soil and water samples (Baumann, 1968), and also form part of the bacterial flora found on the skin of healthy adults; indeed, up to 25% of normal adults may carry *Acinetobacter* spp. on their skins (Rosenthal and Tager, 1975). The human skin is thus an important reservoir in hospitals for *Acinetobacter* and has been implicated in several outbreaks of hospital infection (French et al., 1980; Holton, 1982). Other outbreaks of nosocomial infection (reviewed by Noble, this volume) have often occurred in connection with the use of respiratory equipment. *Acinetobacter* has frequently been associated with nosocomial respiratory tract infection in France and represented 18% of the organisms isolated following fibre-optic bronchoscopy in French hospitals (unpublished data). Several studies (Buxton et al., 1973; Cunha et al., 1980; Hartstein et al., 1988) and our own data (Vieu et al., 1980) have also implicated ventilatory equipment as an environmental reservoir. Thus in 1980, six patients in a surgical intensive care unit at the Bichat Hospital (Paris) acquired acinetobacter pneumonia following contamination of a valve respirator.

Acinetobacter spp. other than *A.baumannii* are normally not responsible for infections in patients, but are frequently found in the hospital environment. They are often found on human skin and are frequently carried by staff. These species are generally more susceptible to antibiotics (Freney et al., 1989).

Acinetobacter Infections

Although most isolates of *Acinetobacter* from clinical specimens reflect colonisation rather than infection (Rosenthal, 1974), this bacterium has been isolated from various types of opportunistic infection (reviewed by Noble, this volume). Various factors predisposing to severe infections, such as underlying disease, malignancy, burns or major surgery, have been identified and are common to many other pathogens. In a French hospital (Bichat Hospital, Paris), acinetobacters represented 1.85% of bacteria isolated from hospitalised patients. The distribution by site of *Acinetobacter* infections did not differ from other nosocomial Gram-negative bacteria. Thus, among 1,800 acinetobacters isolated between 1980 and 1989 in France, 63% were from intensive care units (surgical, medical and burn units), while 32% came from wounds, 30% from urines, 18% from respiratory tract infections and 13% from blood cultures (Joly-Guillou and Bergogne-Bérézin, 1985, 1990a).

Nosocomial respiratory tract infection with *Acinetobacter* spp. seems to be one of the most frequent clinical manifestations of *Acinetobacter* infection in France, particularly in medical intensive care units. In a study of nosocomial pneumonia in patients receiving continuous mechanical ventilation in a medical ICU of Bichat Hospital, *Acinetobacter* and *Pseudomonas aeruginosa* were the most frequently isolated Gram-negative bacilli (15 and 31% respectively) from protected pulmonary brushes (Fagon et al., 1989).

TYPING OF *ACINETOBACTER* IN FRENCH HOSPITALS

The various typing methods available for *Acinetobacter* have been reviewed by Bouvet (this volume). This section describes the application of some of these methods to isolates of *Acinetobacter* from French hospitals.

Antibiogram Typing

Antibiogram typing has proved useful in differentiating outbreak strains when there are marked similarities in resistance patterns (Bartoli et al., 1986). A recent phenotypic analysis of the distribution of ß-lactamases in *Acinetobacter* strains (Joly-Guillou et al., 1988) has been applied as an epidemiological marker to define outbreaks of nosocomial cross infections. However, the main limitation of this method as an epidemiological tool is its lack of specificity, as several strains with the same resistance pattern appear to be different. A progressive increase in resistance to commonly used antimicrobial drugs has been observed during the last few years in French hospitals (Godineau-Gauthey et al., 1988; Joly-Guillou et al., 1990a), and has also been reported in other countries (Castle et al., 1978; Murray and Moellering, 1979; Vila et al., 1989). In a recent study, 34% of the strains tested were resistant to all ß-lactams, with the exception of imipenem, together with the aminoglycoside and quinolone antibiotics commonly used in hospitals. Thus, while antibiogram typing is important to detect new resistance phenotypes, it is not a reliable epidemiological marker and utilisation of phenotypic analysis requires a complementary typing system.

Biotyping

The phage-typing system developed at the Pasteur Institute by Vieu et al. (1970) has been modified so that it currently recognises 125 phage-types and 25 sub-types. Outbreaks of *Acinetobacter* infection seem to be associated with a limited number of phage-types, whereas in the environment a large number of phage-types are present. In a recent investigation of 117 acinetobacters isolated from outbreaks of infection in intensive care units of a French Hospital (Bichat Hospital, Paris) during 1987 and 1988, phage-types 17 and 124 seemed to be particularly important. Although these phage-types were not distributed as a function of geographical origin or site of infection, a significant difference in their distribution between 1987 and 1988 was observed. Phage-types such as phage-type 17 seem to be ubiquitous and are found in various countries (Santos-Ferreira et al., 1984). Other phage-types seem to be limited to a single region or hospital, as seems to be the case for phage-type 124 at Bichat Hospital. Using a combination of biotyping and phage-typing, we have been able to establish the importance of three sets of markers: biotype 9 / phage-type 124 (30%); biotype 6 / phage-type 17 (26%); biotype 6 / phage-type 124 (15%) (Joly-Guillou et al., 1990b).

These sets of markers were, generally, also associated with multi-resistant phenotypes. The combination of these typing systems has demonstrated that patients can be colonised with several different strains, but that only a few strains are endemic and regularly isolated from infections.

Other Typing Systems

Plasmid analysis. This method has, to date, not been widely used in French hospitals. A study in progress (M.L. Joly-Guillou, unpublished results) has demonstrated the presence of six different plasmid profiles among 15 *Acinetobacter* isolates, with plasmids between 15 and 80 Md in size, but further evaluation of this technique will be required.

Bacteriocin typing. Andrews (1986) has described a technique for typing acinetobacters by bacteriocin production. An investigation of 138 *Acinetobacter* strains isolated during 1987 at Bichat Hospital, Paris (A. Bauernfeind, unpublished results), provided rather inconclusive results. Of the nine bacteriocin types investigated, seven (numbers, 1,3,4,5,6,7 and 8) were produced by 15-46% of the strains, while the remaining two types were produced by 2% and 3% of the strains respectively. 6.5% of the strains were untypable. The strains produced between one and eight bacteriocin types, and a total of 31 patterns were identified. Several predominant patterns (4, 7, 4 + 6, 3 + 4 + 6 and 1 + 5 + 8) represented 60% of the strains.

Other suggested typing systems (reviewed by Bouvet, this volume) have, to our knowledge, yet to be evaluated in French hospitals.

CONCLUSIONS

A number of epidemic outbreaks of infection caused by *Acinetobacter* spp. have occurred during recent years in French hospitals, particularly among immunocompromised patients (e.g. those suffering from burns) and in intensive care units. Epidemiological investigations have been hampered by the lack of a convenient typing scheme. From our experience it seems that a combination of the various typing methods described above will be required for epidemiological purposes. Biotyping and antibiogram typing are methods that can easily be employed in routine microbiological laboratories. Other possible typing systems (plasmid profiles, envelope proteins, bacteriocins, serotypes etc.) do not seem, at the present time, to be suitable for routine use in hospital laboratories.

REFERENCES

Andrews, H.J., 1986, *Acinetobacter* bacteriocin typing, *J. Hosp. Infect.*, 7:169.
Bartoli, M., Leguenedal, R., Hengy, C., Courrier, P., Thonnon, J., and Courtois, D., 1986, Recherche de l'identité des souches

d'*Acinetobacter calcoaceticus* responsables d'infections hospitalières par antibiotypie á l'aide du traitement informatique, *Med. Mal. Infect.*, 1:46.

Baumann, P., 1968, Isolation of *Acinetobacter* from soil and water, *J. Bacteriol.*, 96:39.

Bouvet, P.J.M., and Grimont, P.A.D., 1986, Taxonomy of the genus *Acinetobacter* with the recognition of *Acinetobacter baumannii* sp. nov., *Acinetobacter haemolyticus* sp. nov., *Acinetobacter johnsonii* sp. nov., and *Acinetobacter junii* sp. nov., and emended descriptions of *Acinetobacter calcoaceticus* and *Acinetobacter lwoffii*, *Int. J. Syst. Bacteriol.*, 36:228.

Bouvet, P.J.M., and Grimont, P.A.D., 1987, Identification and biotyping of clinical isolates of *Acinetobacter*, *Ann. Inst. Past. Microbiol.*, 138:569.

Buxton, A.E., Anderson, R.L., Werdegar, D. and Atlas, E., 1973, Nosocomial respiratory tract infection and colonization with *Acinetobacter calcoaceticus*. Epidemiologic characteristics, *Amer. J. Med.*, 65:507.

Castle, M., Tenney, J.H., Weinstein, M.P., and Eickhoff, T.C., 1978, Outbreak of a multiply-resistant *Acinetobacter* in a surgical intensive-care unit: Epidemiology and control, *Heart Lung*, 7:641.

Cunha, B.A., Klimeck, J.J., Gracewski, J., McLaughlin, J.C., and Quintiliani, R., 1980, A common source outbreak of *Acinetobacter* pulmonary infections traced to Wright respirometers, *Postgrad. Med. J.*, 56:169.

Fagon, J.Y., Chastre, J., Domart, Y., Trouillet, J.L., Pierre, J., Darne, C., and Gibert, C., 1989, Nosocomial pneumonia in patients receiving continuous mechanical ventilation, *Amer. Rev. Resp. Dis.*, 139:877.

French, G.L., Casewell, M.W., Roncoroni, A.J., Knight, S. and Phillips, I., 1980, A hospital outbreak of antibiotic resistant *Acinetobacter anitratus*: epidemiology and control, *J. Hosp. Infect.*, 1:125.

Freney, J., Bouvet, P.J.M., and Tixier, C., 1989, Identification et determination de la sensibilité aux antibiotiques de 31 souches cliniques d'*Acinetobacter* autres que *A. baumannii*, *Ann. Biol. Clin.*, 47:41.

Godineau-Gauthey, N., Lesage, D., Tessier, F., Kollia, D., and Daguet, G.L., 1988, *Acinetobacter calcoaceticus* variété *anitratus* ou *Acinetobacter baumannii*, étude de la sensibilité aux antibiotiques de 65 souches hospitalières, *Med. Mal. Infect.*, 2 bis:124.

Hartstein, A.I., Rashad, A.L., Liebler, J.M., Actis, L.A., Freeman, J., Rourke, J.W., Stibolt, T.B., Tomalsky, M.E., Ellis, G.R., and Crosa, J.H., 1988, Multiple intensive care unit outbreak of *Acinetobacter calcoaceticus* subspecies *anitratus* respiratory infection with contaminated, reusable ventilator circuits and resuscitation bags, *Amer. J. Med.*, 85:624.

Holton, J., 1982, A report of a further hospital outbreak caused by a multiresistant *Acinetobacter anitratus, J. Hosp. Infect.*, 3:305.

Joly-Guillou, M.L., and Bergogne-Bérézin, E., 1985, Evolution d'*Acinetobacter calcoaceticus* en milieu hospitalier de 1971 à 1984,*Press. Méd.*, 14:2331.

Joly-Guillou, M.L., Vallée, E., Bergogne-Bérézin, E., and Philippon, A., 1988, Distribution of betalactamases and phenotype analysis in clinical strains of *Acinetobacter, J. Antimicrob. Chemother.*, 22:597.

Joly-Guillou, M.L., Bergogne-Bérézin, E., and Vieu, J.F., 1990a, Epidémiologie et résistance des *Acinetobacter* en milieu hospitalier. Bilan de 5 années, *Press.Méd.*, 8:358.

Joly-Guillou, M.L., Bergogne-Bérézin, E., and Vieu, J.F. 1990b, A study of relationships between antibiotic resistance phenotypes, phage-typing and biotyping of 117 clinical isolates of *Acinetobacter* spp., *J. Hosp. Infect.*, in press.

Murray, B.E., and Moellering, R.C., 1979, Aminoglycoside-modifying enzymes among clinical isolates of *Acinetobacter calcoaceticus* subsp. *anitratus (Herellea vaginicola)*: explanation for high level aminoglycoside resistance, *Antimicrob. Agents Chemother.*, 15:190.

Rosenthal, S., 1974, Sources of *Pseudomonas* and *Acinetobacter* species found in human culture materials, *Amer. J. Clin. Path.*, 62:807.

Rosenthal, S., and Tager, I.B., 1975, Prevalence of Gram-negative rods in the normal pharyngeal flora, *Ann. Int. Med.*, 83:355.

Santos-Ferreira, M.O., Vieu, J.F., and Klein, B., 1984, Phage-types and susceptibility to 26 antibiotics of nosocomial strains of *Acinetobacter* isolated in Portugal, *J. Intern. Med. Res.*, 12:364.

Vieu, J.F., Minck, R., and Bergogne-Bérézin, E., 1970, Bacteriophages et lysotypie d'*Acinetobacter, Ann. Microbiol. Inst. Past.*, 130A:403.

Vieu, J.F., Bergogne-Berezin, E., Joly, M.L., Berthelot, G., and Fichelle, A., 1980, Epidémiologie d'*Acinetobacter calcoaceticus, Nouv. Press. Med.*, 46:3551.

Vila, J., Almela, M., and Jimenez de Anta, T., 1989, Laboratory investigation of hospital outbreak caused by two different multiresistant *Acinetobacter calcoaceticus* subsp. *anitratus* strains, *J. Clin. Microbiol.*, 27:1086.

MOLECULAR EPIDEMIOLOGICAL ANALYSIS OF

NOSOCOMIAL *ACINETOBACTER BAUMANNII* ISOLATES

J. Vila[a], M.A. Canales[a], M.A. Marcos[a], R. Gomez-Lus[b]
and M.T. Jimenez de Anta[a]

[a]Departmento de Microbiología
Hospital Clinic de Barcelona
Universidad de Barcelona
Barcelona, Spain

[b]Departmento de Microbiología
Universidad de Zaragoza
Zaragoza, Spain

INTRODUCTION

The incidence of nosocomial infections caused by *Acinetobacter calcoaceticus* var. *anitratus*, now designated as *A. baumannii*, has been steadily rising in recent years (Mayer and Zinner, 1985; Bergogne-Bérézin et al., 1987). Several outbreaks of nosocomial infections caused by multiply-resistant strains of *A. baumannii* have been documented (e.g. Castle et al., 1978; Holton, 1982; Vila et al., 1989). Many of these outbreaks have occurred in intensive care units where extensive use of antibiotics can select for the emergence of multiply-resistant strains (Bergogne-Bérézin and Joly-Guillou, this volume; Noble, this volume). The various methods which are available for typing acinetobacters have been reviewed elsewhere (Bouvet, this volume). The purpose of this short report is to describe and compare the results obtained with some of these methods when used to determine the molecular epidemiology of *A. baumannii* isolates endemic in a Spanish hospital.

BIOTYPING

The 49 strains included in this study were isolated in our laboratory during the period September 1988 to August 1989 from a wide variety of clinical specimens obtained from patients admitted to the hospital. Repeat isolates from the same patient were excluded.

The Biology of Acinetobacter, Edited by K. J. Towner *et al.*
Plenum Press, New York, 1991

Table 1. Biochemical characteristics of the two *A. baumannii* groups

Test	Group One	Group Two
Morphology	CC	CC
Growth at:		
44°C	+	+
41°C	+	+
37°C	+	+
Glucose fermentation	-	-
Glucose oxidation	+	+
Haemolysis	-	-
Nitrate reduction	-	-
Indole	-	-
Arginine dihydrolase	-	-
Urea, Christensen	-	-
Esculin hydrolysis	-	-
Gelatin hydrolysis	-	-
ß-galactosidase	-	-
Assimilation of:		
Glucose	-	-
Arabinose	+	+
Mannose	-	-
Mannitol	-	-
N-Acetylglucosamine	-	-
Maltose	-	-
Gluconate	-	-
Caprate	+	+
Adipate	-	+
Malate	+	+
Citrate	+	+
Phenylacetate	-	+
Catalase	+	+
Oxidase	-	-

CC, coccobacillus; -, negative reaction; +, positive reaction

Presumptive identification of the genus *Acinetobacter* was on the basis of the following characteristics: Gram-negative coccobacilli; strictly aerobic; non-motile; catalase-positive; oxidase-negative. The commercial API 2ONE microsystem was used for further identification. Supplementary tests to identify *A. baumannii* were as follows (Bouvet and Grimont, 1986): growth at 37ºC, 41ºC and 44ºC; lack of haemolysis.

The API 2ONE system identified all 49 isolates as *A. calcoaceticus* var. *anitratus*, with two different analytical profile index numbers: 0001051 and 0001073. Further investigations (Table 1) showed that all the strains were able to grow at 41° and 44°C, demonstrating that both groups were *A. baumannii*. Thus two biotypes (Groups One and Two) could be defined on the basis of these biochemical studies.

SDS-PAGE PROTEIN PATTERNS

A loopful of overnight growth from a MacConkey agar plate was suspended in 10 ml of brain heart infusion broth and incubated at 37°C for 18 h on an orbital shaker. Cells were harvested by centrifugation and processed with a slight modification of the procedure described by Alexander et al. (1988). SDS-PAGE was performed by the method described by Laemmli (1970), with a concentration of 12% acrylamide in the running gel. Boiled samples (5 μl) were applied to the gel and electrophoresed at a constant current of 30 mA in a Mini-Protean gel apparatus (Bio-Rad). Gels were stained, either with Coomassie brilliant blue or by the silver stain method. One-dimensional SDS-PAGE of whole-cell protein extracts from the 49 *A. baumannii* strains produced patterns containing 30-40 discrete bands with molecular sizes varying from 15 to 100 kd. Proteins < 15 kd in size were not resolved by the electrophoretic conditions used in this study. Fig. 1 illustrates the profiles obtained with 11 of the 49 strains. The total protein profiles were similar when the gel was stained with Coomassie brilliant blue, whereas a protein band with an apparent molecular size of 70 kd could be seen in all the strains belonging to Group Two when the gel was stained by the silver stain method. Similar results were obtained with the remaining 38 strains not shown in Fig. 1. Thus PAGE revealed two distinct protein profiles that were related to the groups defined by biotyping.

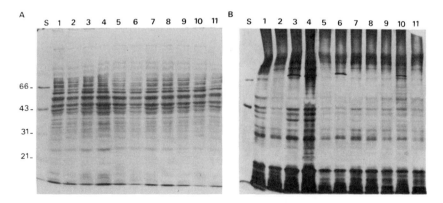

Fig. 1. Electrophoretic protein patterns. Lanes 1,2,5,7,8,9 and 11 represent strains of *A. baumannii* belonging to Group One; lanes 3,4,6 and 10 represent strains of *A. baumannii* belonging to Group Two. Gels were stained either with Coomassie brilliant blue (A) or by the silver stain method (B). Molecular size standards are shown on lane S.

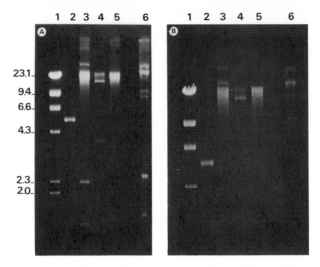

Fig. 2. Analysis of plasmid DNA. Plasmid DNA was separated by electrophoresis in a 0.7% agarose gel at 30 V for 18 h (A) or 30 h (B). Lanes: 1, HindIII-digested lambda DNA molecular size markers (kb); 2, pBR322; 3, example of a strain from Group One (plasmid isolated using large scale preparation method); 4, A. baumannii strain isolated in another hospital; 5, example of a strain from Group Two; 6, example of a strain from Group One (plasmid isolated using small scale preparation method.

PLASMID PROFILES AND RESTRICTION DIGESTS OF TOTAL DNA

Plasmid DNA or total cellular DNA was isolated essentially as described by Birnboim and Doly (1979) or Roussel and Chabbert (1978) respectively. Plasmid profile analysis revealed two groups of A. baumannii, with the patterns shown in Fig. 2. Four major plasmid bands were identified in Group One strains when the plasmids were isolated by a large scale preparation method (500 ml), while six major plasmid bands were observed following isolation with a mini-scale preparation (10 ml). No plasmid bands were observed in Group Two strains by either method.

Restriction endonuclease digestion (HindIII, EcoRV or BamHI) of genomic A. baumannii DNA yielded complex patterns that required careful analysis. While similarity could be detected between strains of the same group, the fingerprints were clearly different between the A. baumannii groups defined previously by biotyping, PAGE/SDS whole-cell protein patterns and plasmid profile analysis.

ANTIMICROBIAL SUSCEPTIBILITIES

Susceptibility to antimicrobial agents was determined by two methods: (i) the disc diffusion method with Mueller-Hinton medium, as advocated by the National Committee for Clinical Laboratory Standards (1984); (ii) MICs

Table 2.   Antimicrobial susceptibility patterns of the two *A. baumannii* groups

Antimicrobial agent	MICs (mg/l)	
	Group One	Group Two
Ampicillin	>32(R)	>32(R)
Mezlocillin	>256(R)	64(MS)
Piperacillin	256(R)	64(MS)
Ticarcillin	>256(R)	<16(S)
Azlocillin	>256(R)	64(MS)
Amoxicillin/Clavulanic acid	>16/8(R)	16/8(R)
Cephalothin	>32(R)	>32(R)
Cefamandole	>32(R)	>32(R)
Cefuroxime	>32(R)	>32(R)
Cefoxitin	>32(R)	>32(R)
Moxalactam	>32(R)	>32(R)
Cefotaxime	>32(R)	<8(S)
Cefoperazone	128(R)	16(S)
Ceftazidime	32(R)	<8(S)
Cefsulodin	>32(R)	32(R)
Ceftizoxime	64(R)	<8(S)
Ceftriaxone	64(R)	<8(S)
Chloramphenicol	>16(R)	>16(R)
Gentamicin	>16(R)	8(MS)
Amikacin	64(R)	<16(S)
Tobramycin	>16(R)	<4(S)
Netilmicin	<1(S)	>16(R)
Trimethoprim-sulphamethoxazole	>16/304(R)	>16/304(R)
Imipenem	<0.5(S)	<0.5(S)
Colistin	<2(S)	<2(S)
Ciprofloxacin	<0.5(S)	<0.5(S)
Ofloxacin	1(S)	<0.5(S)

R, resistance;   S, susceptibility;   MS, moderate susceptibility

were determined with the Sceptor system (BBL Microbiology Systems, Cockeysville, Md.) as described previously (Vila et al., 1989). The results (Table 2) allowed the 49 strains to be clearly separated into two groups. The more resistant group coincided with Group One established by the molecular epidemiological markers described above.   In this group, of the 27 antibiotics tested, only colistin, netilmicin, imipenem and the new fluoroquinolones had some inhibitory activity.   The antibiotic susceptibility pattern of the Group Two isolates consisted of resistance to ampicillin, cephalothin, cefamandole, cefuroxime, cefsulodin, chloramphenicol, netilmicin and trimethoprim-sulphamethoxazole; moderate susceptibility to mezlocillin, piperacillin and gentamicin, and susceptibility to the other antibiotics tested.

DISCUSSION

The data generated with the different techniques for epidemiological typing allowed us to sort the strains included in this study into two distinct groups. Firstly, when the API 20NE system was used as a biotyping method, two distinct biotypes were found. Growth at 41°C and 44°C showed that both biotypes were *A. baumannii*. A biotyping system described by Bouvet and Grimont (1987) has differentiated 17 biotypes for this species and we are now trying to correlate these biotypes with Groups One and Two.

The use of SDS-PAGE to analyse either whole cell proteins or cell envelope proteins of *A. calcoaceticus* has been described previously (Alexander et al., 1984; Dijkshoorn et al., 1987; Alexander et al., 1988). Whereas biotyping was able to sub-divide the isolates into two groups, no substantial difference in protein patterns was revealed when gels were stained with Coomassie brilliant blue. However, two groups could be differentiated when the silver stain method was used. Both groups coincided with those previously identified by the API 2ONE system. The same two groups were also identified by plasmid profile analysis and antibiotic susceptibility profiles. Finally, Allardet-Servent et al. (1989) have described previously how fragments of DNA generated by restriction endonucleases which cleave infrequently can be used for epidemiological investigations of *A. calcoaceticus* outbreaks following separation by pulsed field gel electrophoresis. In the present study, cleavage with three restriction endonucleases, followed by separation of the DNA fragments on conventional agarose gels was used. The fingerprints obtained were clearly different between the two groups.

Despite its low virulence, *A. calcoaceticus* has often been described as a cause of hospital infections (Noble, this volume). We have described previously (Vila et al., 1989) an outbreak in our hospital caused by two different multi-resistant *A. calcoaceticus* var. *anitratus* strains, which diminished in importance during the period that the respiratory intensive care unit was closed for repairs. The number of isolates of *A. calcoaceticus* var. *anitratus* subsequently increased in number and stimulated the investigations outlined in this study. The Group One strains coincided with one of the groups described in our original study, whereas the Group Two strains were totally different in terms of genetic and phenotypic markers (data not shown). Similar isolates were found throughout the hospital, and attempts are currently being made to identify the nosocomial reservoirs.

There is no universally accepted method for the epidemiological typing of *A. baumannii* outbreaks. A serotyping method has recently been used for delineation of outbreaks caused by *A. baumannii* (Traub, 1989), but serotyping demands a large number of antisera which are difficult and expensive to prepare and standardise. In our study we have tried to perform an epidemiological analysis which avoids cumbersome procedures and does not require sophisticated equipment (e.g. an ultracentrifuge or pulsed field gel electrophoresis apparatus). We have found that the combination of genotypic and phenotypic methods described in this report is useful for epidemiological studies of *A. baumannii* and believe that the methods can be readily applied in any routine clinical microbiology laboratory.

ACKNOWLEDGEMENT

This research was supported by grant 88/0206 from DGCYT-Spain.

REFERENCES

Alexander, M., Ismail, F., Jackman, P.J.H., and Noble, W.C. 1984, Fingerprinting *Acinetobacter* strains from clinical sources by numerical analysis of electrophoretic protein patterns, *J. Med. Microbiol.*, 18:55.

Alexander, M., Rahman, M., Taylor, M., and Noble, W.C., 1988, A study of the value of electrophoretic and other techniques for typing *Acinetobacter calcoaceticus, J. Hosp. Infect.*, 12:273.

Allardet-Servent, A., Bouziges, N., Carles-Nurit, M.J., Bourg, G., Gouby, A., and Ramuz, M., 1989, Use of low frequency cleavage restriction endonucleases for DNA analysis in epidemiological investigations of nosocomial bacterial infections, *J. Clin. Microbiol.*, 27:2057.

Bergogne-Bérézin, E., Joly-Guillou, M.L., and Vieu, J.F., 1987, Epidemiology of nosocomial infections due to *Acinetobacter calcoaceticus, J. Hosp. Infect.*, 10:105.

Birnboim, H.C., and Doly, J., 1979, A rapid alkaline extraction procedure for screening recombinant plasmid DNA, *Nucl. Acid Res.*, 7:1513.

Bouvet, P.J.M., and Grimont, P.A.D., 1986, Taxonomy of the genus *Acinetobacter* with the recognition of *Acinetobacter baumannii sp. nov., Acinetobacter haemolyticus sp. nov., Acinetobacter johnsonii sp. nov., and Acinetobacter junii sp. nov.* and emended description of *Acinetobacter calcoaceticus and Acinetobacter lwoffii, Int. J. Syst. Bact.*, 36:228.

Bouvet, P.J.M., and Grimont, P.A.D., 1987, Identification and biotyping of clinical isolates of *Acinetobacter, Ann. Inst. Past. Microbiol.*, 138:569.

Castle, M., Tenney, J.M., Weinstein, M.P.. and Eickhoff, T.C., 1978, Outbreak of a multiply resistant *Acinetobacter* in a surgical intensive care unit. Epidemiology and control, *Heart Lung*, 7:641.

Dijkshoorn, L., Michel, M.F., and Degener, J.E., 1987, Cell envelope protein profiles of *Acinetobacter calcoaceticus* isolated in hospitals, *J. Med. Microbiol.*, 23:313.

Holton, J., 1982, A report of a further hospital outbreak caused by a multi-resistant *Acinetobacter anitratus*, *J. Hosp. Infect.*, 3:305.

Laemmli, V.K., 1970, Cleavage of structural proteins during the assembly of the head of bacteriophage T4, *Nature*, 227:680.

Mayer, K.H., and Zinner, J.H., 1985, Bacterial pathogens of increasing significance in hospital-acquired infections, *Rev. Infect. Dis.*, 7(Suppl.3):5371.

National Committee for Clinical Laboratory Standards, 1984, "Performance Standards for Antimicrobial Disk Susceptibility Test, 3rd edn.," National Committee for Clinical Laboratory Standards, Villanova, Pa.

Roussel, A.F., and Chabbert, Y.A., 1978, Taxonomy and epidemiology of Gram-negative bacterial plasmids studied by DNA-DNA filter hybridisation in formamide, *J. Gen. Microbiol.*, 104:269.

Traub, W.H., 1989, *Acinetobacter baumannii* serotyping for delineation of outbreaks of nosocomial cross-infection, *J. Clin. Microbiol.*, 27:2713.

Vila, J., Almela, M., and Jimenez de Anta, M.T., 1989, Laboratory investigation of hospital outbreak caused by two different multiresistant *Acinetobacter calcoaceticus subsp. anitratus* strains, *J. Clin. Microbiol.*, 27:1086.

FACTORS INFLUENCING THE VIRULENCE OF *ACINETOBACTER*

J.-L. Avril and R. Mesnard

Laboratoire de Bactériologie-Virologie
Faculté de Médecine
Université de Rennes
Avenue du Prof.L.Bernard
35043 Rennes Cedex
France

INTRODUCTION

Members of the genus *Acinetobacter* can frequently be isolated from healthy people and are also commonly present in soil and water as free-living saprophytes.   In spite of an increasing number of reports that *Acinetobacter* induces respiratory and urinary tract infections and septicaemia (Bergogne-Bérézin et al., 1987), there have been few investigations of the nature of its pathogenicity.   Although only one species, *A. calcoaceticus*, is described in the latest edition of Bergey's Manual of Systematic Bacteriology (Juni, 1984) the taxonomy of *Acinetobacter* has recently been extensively revised (Grimont and Bouvet, this volume).   *A. baumannii* is now the most prevalent *Acinetobacter* species isolated from clinical specimens, but for historical reasons this review will refer to *Acinetobacter* without detailed consideration of individual species.

Virulent organisms exhibit pathogenicity when introduced into the host in very small numbers.   This property can be subdivided into invasiveness, including the ability to enter, multiply and spread within the host tissues, and toxigenicity, including the ability to produce toxic substances.

This article considers the possible factors which may contribute to the virulence of *Acinetobacter*.

INVASIVENESS

Polysaccharide Capsule

Invasiveness of bacteria may be associated with surface components that protect the bacteria from phagocytosis. By comparing two sets of mutant *Acinetobacter* strains with their corresponding parent strains, an enzymic decapsulation study has demonstrated that polysaccharide capsules can decrease the ability of *Acinetobacter* to adhere to hydrocarbon (Rosenberg et al., 1983a,b). This was presumably the result of making the outer bacterial surface more hydrophilic. Similarly, the surface hydrophobicity of human isolates of *Acinetobacter* also appears to play an important role in their attachment properties. A recent study has shown that hydrophobicity is higher in *A. baumannii* strains isolated from infected catheters and tracheal devices than in environmental isolates (Boujaafar et al., 1990).

Adherence to Epithelial Cells and to Hydrocarbons

Acinetobacter RAG-1, a strain isolated following growth selection on hydrocarbon, adhered to human epithelial cells when grown under conditions which promoted its adherence to hexadecane (Rosenberg and Rosenberg, 1981; Rosenberg et al., 1981), but showed only poor adherence to epithelial cells when grown under conditions which resulted in low affinity towards hexadecane. A mutant derived from RAG-1 that was deficient in its ability to adhere to hydrocarbon was similarly unable to adhere to epithelial cells.

The production of fimbriae by strains of *Acinetobacter* was reported by Henricksen and Blom (1975). These authors found two major types of fimbriae: "thin" fimbriae, ca. 3 nm in diameter, and "thick" fimbriae, ca. 5 nm in diameter. Presence of the thick fimbriae in *Acinetobacter* strains was shown to correlate with twitching motility, but no physiological role was proposed for the thin fimbriae. *Acinetobacter* RAG-1, a hydrocarbon-degrading strain which adheres avidly to hydrocarbons, possesses numerous thin fimbriae, ca. 3.5 nm in diameter, on the cell surface. A non-adherent mutant of this strain lacks these fimbriae (Rosenberg et al., 1982). Adherence of wild-type RAG-1 cells to hexadecane was considerably reduced after shearing treatment. In addition, RAG-1 cells were agglutinated in the presence of specific antibody, whereas the non-adherent cells were not.

Enzymic Activities

Invasiveness may be aided by enzymes that favour the spread of bacteria. Poh and Loh (1985) determined the enzymic profiles of clinical isolates of *Acinetobacter* by conventional plate tests and the rapid API-ZYM system. No activity could be detected for DNase, elastase, haemolysin, protease or gelatinase in the majority of strains tested. These enzymes are therefore unlikely to be involved in the virulence of *Acinetobacter*. In contrast, significant levels of butyrate esterase (C-4), caprylate esterase (C-8) and leucine arylamidase were detected in all isolates tested. These enzymes hydrolyse short chain fatty acids and may therefore be involved in causing damage to tissue lipids. No trypsin, chymotrypsin, alkaline-phosphatase or glucosidase activities were present.

Iron Uptake

The ability of bacteria to assimilate iron appears to be related to invasiveness. Some virulent strains of Enterobacteriaceae are known to synthesise an hydroxamate type of iron chelator termed aerobactin. However, in a study of 50 *Acinetobacter* clinical isolates (Martinez et al., 1987), no aerobactin-producing strains were detected.

TOXIGENICITY

The virulence of clinical isolates of *Acinetobacter* has been studied in mice by Obana et al. (1985). Acute systemic infection was induced in mice by intraperitoneal injection of *Acinetobacter*. The virulence of 40 *Acinetobacter* clinical isolates was expressed in terms of the number of viable cells of the test strains required to kill 50% of the mice challenged. Following intraperitoneal challenge, the LD_{50} values ranged from 10^3 to $>10^6$ viable cells/mouse, but for most of the strains were $>10^6$ viable cells/mouse. However, the virulence of a few *Acinetobacter* strains was found to be equivalent to the virulence of *Escherichia coli, Serratia marcescens* and *Pseudomonas aeruginosa*. A particularly interesting observation was that a considerable enhancement of virulence was obtained if strains were injected as suspensions in 3% hog gastric mucin.

Slime

The virulence of five slime-producing strains was compared with the virulence of three non-slime-producing strains in mice by Obana (1986). However, no difference was observed between the two groups (LD_{50}s of 10^7-10^8 viable cells/mouse) and no correlation could be made between virulence and the amount of slime produced.

Disease is as dependent on the host as on the infecting microorganism. Infections caused by *Acinetobacter* are often nosocomial. The experiments described above were therefore also performed with mice that had been immunosuppressed with cyclophosphamide, but no significant increase in virulence was observed.

Obana (1986) also studied the mortality of mice resulting from mixed infection. Mortality was higher in the case of mixed infection with *P. aeruginosa* and a slime-producing strain of *Acinetobacter* than in the case of single infection with a variety of inoculum sizes. Even in a mixture with a non-slime-producing strain, the virulence of *P. aeruginosa* was greater than in single infection. Similar results were observed for mixed infections with either *E. coli* or *S. marcescens*.

The chemical composition of the slime produced by *Acinetobacter* was similar for two strains tested. In each case there was a high sugar content and a substance giving a positive orcinol reaction was also present. About 30% of the total composition remained to be determined, but the DNA content was approximately 3%, a value which is significantly different from the slime of *P. aeruginosa* in which the main component is DNA. Nevertheless, it is known that *Acinetobacter* slime exhibits lethal activity in mice, although intraperitoneal inoculation of mice with slime obtained from *Acinetobacter* TMS 262 and *Acinetobacter* TMS 266 resulted in LD_{50} values of 1.8 and 0.09 mg/mouse respectively, thereby showing a considerable difference between the two strains.

To study the effect of *Acinetobacter* slime on the virulence of various other Gram-negative bacilli in mice, concurrent intraperitoneal inoculations of bacteria and increasing doses of slime have been carried out. Enhancement of virulence was proportional to the dose of slime inoculated, with the slime of *Acinetobacter* causing a ten-fold or greater enhancement in the lethal activity of *E. coli, S. marcescens* or *P. aeruginosa*.

The virulence-enhancing effect of slime has also been demonstrated by intraperitoneal inoculation into mice of a mixture of a sub-lethal dose of *Acinetobacter* slime and the Gram-negative species mentioned above. The bacterial counts in the peritoneal exudate showed a gradual decrease in the absence of slime, but increased in the presence of slime with subsequent death of the mice. This enhancement of virulence apparently resulted from the cytoxicity of slime for mouse neutrophils and the inhibition of the cellular migration of neutrophils into the peritoneal exudate of mice.

Toxins

Virtually all species of Gram-negative bacteria produce lipopolysaccharide (LPS) components of their cell walls that can function as endotoxins, of which lipid A is the active principle. Brade and Galanos (1983) compared the biological activities of the LPS and lipid A from *Acinetobacter* with those from *Salmonella*. The results showed that the LPS and lipid A from *Acinetobacter* had toxic and other biological properties similar to those found with LPS and lipid A from Enterobacteriaceae. These properties included lethal toxicity in mice, pyrogenicity in rabbits, complement inactivation *in vitro*, a local Shwartzmann reaction, mitogenicity for mouse-spleen B lymphocytes and a positive limulus amoebocytes lysate test. It therefore seems that the in-vivo production of endotoxin is probably responsible for the characteristic signs of disease and death observed following *Acinetobacter* septicaemia. No protein toxins have so far been described in *Acinetobacter*.

ROLE OF HOST ANTIMICROBIAL SYSTEMS

During phagocytosis, ingested bacterial cells are exposed to high concentrations of various bactericidal proteins released from polymorphonuclear leukocyte (PMN) granules into phagolysosomes. Studies of acid extracts from PMN granules, including fractionation of such extracts by gel permeation chromatography, indicate that the bactericidal activity of the various fractions depends on the particular organism tested. The acid extract fraction from PMN granules with the lowest molecular size (peak D) is particularly active against *Acinetobacter* (Greenwald and Ganz, 1987). This fraction consisted of a mixture of three human defensin peptides: HNP1, HNP2 and HNP3. Purified defensins reproduced the bactericidal activity, suggesting that defensins could play a major role in the killing of *Acinetobacter* by human neutrophils.

Because of its readily alterable membrane, *Acinetobacter* has been used in studies designed to elucidate the role of bacterial cell envelope structure in interactions with PMNs (Loeffelholz and Modrzakowski, 1987). An alkane-induced increase in the outer membrane permeability of *Acinetobacter* was concomitant with an enhanced susceptibility to the oxygen-independent antimicrobial activity of extracted contents from rat PMN granules. The change in outer membrane permeability was not associated with an alteration to LPS O-antigen. A further study by Loeffelholz and Modrzakowski (1988) demonstrated that rat PMN granule extract reduced the viability of *Acinetobacter* by inhibiting the transport and incorporation of macromolecule precursors. The granule extract activity that increased outer membrane permeability did not appear to be directly responsible for the observed decrease in viability.

CONCLUSIONS

Knowledge of the factors influencing the virulence of *Acinetobacter* is still at an elementary stage. Several of the characteristics considered in this review seem to play a role in enhancing the virulence of particular strains. However, on the basis of the available evidence, it can be suggested that the observed pathogenicity of *Acinetobacter* does not normally result from the simple presence of the organism, but is often related to the clinical status of the infected patient. Thus virulence is greatly enhanced in patients who are either compromised hosts or who have polymicrobial infections.

REFERENCES

Bergogne-Bérézin, E., Joly-Guillou, M.L., and Vieu, J.F., 1987, Epidemiology of nosocomial infections due to *Acinetobacter calcoaceticus*, *J. Hosp. Infect.*, 10:105.

Boujaafar, N., Freney, J., Bouvet, P.J.M., and Jeddi, M., 1990, Cell surface hydrophobicity of 88 clinical strains of *Acinetobacter baumannii*, *Res. Microbiol.*, 141:477

Brade, H., and Galanos, C., 1983, Biological activities of the lipopoly-saccharide and lipid A from *Acinetobacter calcoaceticus*, *J. Med. Microbiol.*, 16:211.

Greenwald, G.I., and Ganz, T., 1987, Defensins mediate the microbiocidal activity of human neutrophil granule extract against *Acinetobacter calcoaceticus*, *Infect. Immun.*, 55:1365.

Henricksen, J., and Blom, J., 1975, Correlation between twitching motility and possession of polar fimbriae in *Acinetobacter calcoaceticus*, *Acta Path. Microbiol. Scand.*, B83:103.

Juni, E., 1984, Genus III. *Acinetobacter* Brisou et Prévot 1954, *in* "Bergey's Manual of Systematic Bacteriology, vol.1," p.303, N.R. Grieg and J.G. Holt, eds., Williams and Wilkins, Baltimore.

Loeffelholz, M.J., and Modrzakowski, M.C., 1987, Outer membrane permeability of *Acinetobacter calcoaceticus* mediates susceptibility to rat polymorphonuclear leukocyte granule contents, *Infect. Immun.*, 55:2296.

Loeffelholz, M.J., and Modrzakowski, M.C., 1988, Antimicrobial mechanisms against *Acinetobacter calcoaceticus* of rat polymorphonuclear leukocyte granule extract, *Infect. Immun.*, 56:522.

Martinez, J.L., Cercenado, E., Baquero, F., Perez-Diaz, J.C., and Delgado-Tribarren, A., 1987, Incidence of aerobactin production in Gram-negative hospital isolates, *FEMS Microbiol. Lett.*, 43:351.

Obana, Y., 1986, Pathogenic significance of *Acinetobacter calcoaceticus*: analysis of experimental infection in mice, *Microbiol. Immunol.*, 30:645.

Obana, Y., Nishino, T., and Tanino, T., 1985, In vitro and in vivo activities of antimicrobial agents against *Acinetobacter calcoaceticus*, *J. Antimicrob. Chemother.*, 15:441.

Poh, C.L., and Loh, G.K., 1985, Enzymatic profile of clinical isolates of *Acinetobacter calcoaceticus*, *Med. Microbiol. Immunol.*, 174:29.

Rosenberg, M., and Rosenberg, E., 1981, Role of adherence in growth of *Acinetobacter calcoaceticus* RAG-1 on hexadecane, *J. Bacteriol.*, 148:51.

Rosenberg, M., Perry, A., Bayer, E.A., Gutnick, D.L., Rosenberg, E., and Ofek, T., 1981, Adherence of *Acinetobacter calcoaceticus* RAG-1 to human epithelial cells and to hexadecane, *Infect. Immun.*, 33:29.

Rosenberg, M., Bayer, E.A., Delarea, J., and Rosenberg, E., 1982, Role of thin fimbriae in adherence and growth of *Acinetobacter calcoaceticus* RAG-1 on hexadecane, *Appl. Environ. Microbiol.*, 44:929.

Rosenberg, E., Gottlieb, A., and Rosenberg, M., 1983a, Inhibition of bacterial adherence to hydrocarbons and epithelial cells by emulsan, *Infect. Immun.*, 39:1024.

Rosenberg, E., Kaplan, N., Pines, O., Rosenberg, M., and Gutnick, D., 1983b, Capsular polysaccharides interfere with adherence of *Acinetobacter calcoaceticus* to hydrocarbon, *FEMS Microbiol. Lett.*, 17:157.

ANTIBIOTIC RESISTANCE MECHANISMS IN *ACINETOBACTER*

E. Bergogne-Bérézin and M.L. Joly-Guillou

Laboratoire de Microbiologie
CHU Bichat
46 rue Huchard
75877 Paris Cedex 18
France

INTRODUCTION

Antibiotic therapy for infectious diseases is now over 40 years old. In the early years of the antibiotic era, *Staphylococcus aureus*, *Streptococcus pneumoniae* and ß-haemolytic streptococci were the predominant organisms involved in hospital-acquired infections. Since 1960 the continuous development and therapeutic introduction of new potent antimicrobial agents has resulted in significant changes in the character of nosocomial infections (Finland, 1979). In the early 1960s staphylococci subsided in importance as the major cause of nosocomial infection and were replaced by enteric Gram-negative bacilli, enterococci and fungi (Mayer and Zinner, 1985). Since 1980, increased numbers of aerobic Gram-negative bacilli , including *Acinetobacter* spp., have been noted as major bacterial agents of hospital infections (Glew et al., 1977; Retailliau et al., 1979; Bergogne-Bérézin and Joly-Guillou, 1985). The pathogenic role of *Acinetobacter* is limited to nosocomial infections (Glew et al., 1977; Retailliau et al., 1979; Holton, 1982; Joly-Guillou and Bergogne-Bérézin, 1985; Bergogne-Bérézin et al., 1987) and its role in community-acquired infection has only seldom been mentioned in the medical literature (Rudin et al., 1979). The changing profile of nosocomial infections is not only related to the sequential introduction of new antimicrobial agents, but also reflects various ecological factors, especially major changes in medical and surgical technologies in intensive care units. In this context, one of the most difficult problems confronting the clinician who deals with nosocomial infections is that of microbial resistance to antibiotics (Finland, 1979; McGowan, 1983). *Acinetobacter* is a genus that has developed one of the most impressive patterns of antibiotic resistance during the last 20 years

The Biology of Acinetobacter, Edited by K. J. Towner *et al.*
Plenum Press, New York, 1991

(Bergogne-Bérézin et al., 1971; Pinter and Kantor, 1971; Devaud et al., 1982; Holton, 1982; Sanders, 1983; Joly-Guillou and Bergogne-Bérézin, 1985b; Shannon et al., 1986; Joly-Guillou et al., 1987; Godineau-Gauthey et al., 1988; Freney et al., 1989). Nosocomial isolates of *Acinetobacter* currently exhibit antibiotic resistance profiles which result in infections that are extremely difficult to treat. An increasing number of reports in the medical literature document the high rates of resistance in *Acinetobacter*, although there are few studies of the precise mechanisms of resistance to ß-lactams, aminoglycosides, fluoroquinolones and the other major antibiotic classes. Several studies of the genetic determinants of resistance in *Acinetobacter* have been published, but current knowledge of transferable resistance plasmids in this genus is still at an early stage. This article deals with antibiotic resistance in *Acinetobacter* spp.. A review of past and current resistance profiles is followed by a description of known mechanisms of resistance and current information on the genetic determinants of resistance in *Acinetobacter*.

EVOLUTION OF RESISTANCE TO ANTIBIOTICS IN *ACINETOBACTER* SPP

Overall Susceptibilities and Resistance in Acinetobacter spp

In the historical development of drug resistance among Gram-negative bacilli, there has been a progressive decrease in susceptibility to commonly used antimicrobial drugs among strains of Enterobacteriaceae and *Pseudomonas* (McGowan, 1983; Mayer and Zinner, 1985); this has also occurred with *Acinetobacter* (Bergogne-Bérézin and Joly-Guillou, 1985). In the early 1970s, *Acinetobacter* infections were treated successfully with gentamicin, minocycline, nalidixic acid, ampicillin or carbenicillin, either as single agents or in combined therapy. In early in-vitro studies, the majority of *Acinetobacter* clinical isolates were susceptible to ampicillin (61-70%), gentamicin (92.5%), chloramphenicol (57%), or nalidixic acid (97.8%) (Bergogne-Bérézin et al., 1971). Of 86 *Acinetobacter* strains isolated in 1969, only 3.4% were resistant to carbenicillin, which in those days was a new ß-lactam under investigation (Bergogne-Bérézin et al., 1971). In the same study, a comparison between strains isolated from medical wards (120 isolates) and those from surgery or intensive care units (120 isolates) showed a significant difference in susceptibilities; the former showed higher proportions of isolates susceptible to ampicillin (80% vs. 41.6%), cephaloridine (65.8% vs. 29.1%) and streptomycin (84.1% vs. 42.5%). The difference in gentamicin susceptibility between the two groups of strains was more limited; indeed the vast majority of *Acinetobacter* isolates were susceptible to gentamicin (95.8% of strains from medical wards; 87.5% of those from intensive care units). Strains of *A.calcoaceticus* var. *anitratus*, as identified before the current taxonomy became available (Bergogne-Bérézin et al., 1987; Freney et al., 1989), were significantly more resistant than var. *lwoffii* strains in early studies (Bergogne-Bérézin et al., 1971; Pinter and Kantor, 1971), as well as in more recent investigations

(Bergogne-Bérézin and Joly-Guillou, 1985, 1989; Joly-Guillou and Bergogne-Bérézin, 1985b; Godineau-Gauthey et al., 1988; Freney et al., 1989). A change in the isolation frequency of var. *anitratus (A. baumannii* according to the new *Acinetobacter* species definition) (Freney et al., 1989) has been noted, rising from 77.5% of the strains isolated in 1971-1980 to 94.5% in 1984-85 (Joly-Guillou and Bergogne-Bérézin, 1985b). Among various factors associated with increasing resistance of *Acinetobacter* to antibiotics, the increased incidence of the more resistant var. *anitratus* may also have changed the reported overall antibiotic resistance of clinical isolates of *Acinetobacter*. All other named or unnamed *Acinetobacter* spp. (*A. lwoffii, A. johnsonii, A. haemolyticus, A. junii*) (Freney et al., 1989), which may be isolated from the hospital environment, but are seldom involved in nosocomial infections, are much more susceptible to most antibacterial agents (Joly-Guillou and Bergogne-Bérézin, 1985b; Bergogne-Bérézin et al., 1987; Freney et al., 1989). Thus most studies focus on the susceptibility or resistance of *A. calcoaceticus* var. *anitratus (A. baumannii)* strains that are responsible for the majority of outbreaks of nosocomial infections associated with this genus.

Current Resistance to Antibiotics in Acinetobacter spp

The association of increased antibiotic usage in hospitals with generalised increases in antibiotic resistance is an accepted "fact" of the current therapeutic era (McGowan, 1983; Mayer and Zinner, 1985). Successive increases in the resistance of clinical strains of *Acinetobacter* have been reported periodically (Bergogne-Bérézin et al., 1971; Dowding, 1979; Joly-Guillou and Bergogne-Bérézin, 1985b; Godineau-Gauthey et al., 1988). High proportions of strains have become resistant to older antibiotics and, at the present time,certain *Acinetobacter* strains are resistant to most commonly used antibacterial drugs, including aminopenicillins, ureidopenicillins, cephalosporins of the first (cephalothin) and second (cefamandole) generation (Morohoshi and Saito, 1977; Joly-Guillou and Bergogne-Bérézin, 1985b), cephamycins such as cefoxitin (Garcia et al., 1983; Joly-Guillou and Bergogne-Bérézin, 1985b), most aminoglycosides-aminocyclitols (Dowding, 1979; Devaud et al., 1982; Goldstein et al., 1983; Medeiros, 1984; Joly-Guillou and Bergogne-Bérézin, 1985b; Joly-Guillou et al., 1987) chloramphenicol and tetracyclines. Contrasting data are sparsely distributed in the literature for some antibiotics: minocycline and doxycycline were cited in one study as the most active of 21 antibiotics tested, including ten ß-lactams, three aminoglycosides and nalidixic acid (Obana et al., 1985), while piperacillin and aminoglycosides were effective in a study of the in-vitro activity of 25 antibacterial agents (Garcia et al., 1983). These differences in the susceptibilities of *Acinetobacter* isolates might be attributable to environmental factors such as the different patterns of antibiotic usage that exist in Japan (Obana et al., 1985), the USA (Garcia et al., 1983) or France (Joly-Guillou and Bergogne-Bérézin, 1985b). Table 1 summarises the susceptibility of *A. calcoaceticus* var. *anitratus (A. baumannii)* to

Table 1. Susceptibility of *A.calcoaceticus* var. *anitratus* to antimicrobial agents (MICs in mg/l)

Antibiotic	No of isolates	MIC range	MIC_{50}	MIC_{90}	Year	Ref.
Cephalothin	51	1.56->100	50	>100	1983	(1)
	100	>128	>128	>128	1988	(2)
Cefoxitin	51	6.25->100	100	>100	1983	(1)
	100	0.12->128	128	>128	1988	(2)
Cefuroxime	51	3.12->100	25	50	1983	(1)
Cefamandole	51	3.12->100	100	>100	1983	(1)
Cefoperazone	51	6.25->100	100	>100	1983	(1)
	100	4->128	44	>128	1988	(2)
Cefotaxime	40	0.12-128	1.0	16	1981	(3)
	116	0.12-128	7.9	22	1986	(4)
	100	0.12->128	16	100	1988	(2)
Moxalactam	40	0.5-64	4.0	32	1981	(3)
	51	1.56-100	50	100	1983	(1)
	100	2.0->128	50	120	1988	(2)
Ceftazidime	40	0.25-32	1.0	8.0	1981	(3)
	51	1.56-25	6.25	12.5	1983	(1)
	116	0.12-32	4.4	10	1986	(4)
	100	0.12-128	6.5	22	1988	(2)
Aztreonam	51	6.25->100	100	>100	1983	(1)
	100	0.12->128	18	87	1988	(2)
Temocillin	51	3.12->100	>100	>100	1983	(1)
	100	>128	>128	>128	1988	(2)
Ticarcillin	51	1.56->100	25	50	1983	(1)
	100	1.0->256	>256	>256	1988	(2)
Piperacillin	51	3.12->100	12.5	50	1983	(1)
	100	1.0->128	100	>128	1988	(2)

continued

Table 1.(continued)

Mezlocillin	51	3.12->100	25	100	1983	(1)
	100	1.0->128	>128	>128	1988	(2)
Imipenem	40	0.01-0.25	0.12	0.25	1981	(3)
	51	0.05-12.5	0.20	0.39	1983	(1)
	134	0.01-4.0	0.28	0.61	1986	(4)
	100	0.12->128	0.37	0.60	1988	(2)
	54	0.5-4.0	2.0	2.0	1988	(5)
Erythromycin	51	12.5->100	>100	>100	1983	(1)
	84[a]	---	6.25	12.5[b]	1985	(6)
Tetracycline	51	0.78->100	25	100	1983	(1)
	84[a]	---	3.13	6.25[b]	1985	(6)
Doxycycline	84[a]	---	<0.19	0.39[b]	1985	(6)
Minocycline	84[a]	---	<0.1	<0.19[b]	1985	(6)
Gentamicin	51	0.10-50	0.39	3.12	1983	(1)
	54	1.0-128	64	128	1988	(5)
Tobramycin	51	0.10-32	0.39	0.78	1983	(1)
	96	0.12->128	3.2	44	1986	(7)
	54	1.0-32	16	32	1988	(5)
Amikacin	51	0.20-12.5	1.56	3.12	1983	(1)
	96	0.25->128	4.6	74	1986	(7)
	54	1.0-64	32	64	1988	(5)

[a] strains classified as "A.calcoaceticus"

[b] MIC_{80} = concentration required to inhibit 80% of strains

References: (1) Garcia et al., 1983;   (2) Joly-Guillou et al., 1988; (3) Phillips et al., 1981   (4) Bergogne-Bérézin and Joly-Guillou, 1986; (5) Chow et al., 1988;   (6) Obana et al., 1985; (7) personal unpublished data.

various antimicrobial agents.   Combined data for each compound confirm the general trend towards decreased susceptibility of strains as a function of time.   With some major antibiotics, such as the third generation cephalosporins (e.g. cefotaxime or ceftazidime), imipenem, tobramycin and amikacin, the data show clearly that the minimal inhibitory concentrations (MICs) have increased between 1981 and 1988.   These data will be analysed in a subsequent section on mechanisms of resistance.

Multiresistance has been documented since the first studies of outbreaks of *Acinetobacter* infections. In our earliest experience (Bergogne-Bérézin et al., 1971), several multiresistant strains were isolated in a surgical ward prior to any initiation of antibiotic therapy. One strain was resistant to streptomycin and tetracycline, while a second isolate was resistant to ampicillin, cephaloridine and chloramphenicol. Other reports of multiresistant strains can be found in the medical literature. Retailliau et al. (1979) described nosocomial *Acinetobacter* strains that were resistant to ampicillin, cephalosporins and chloramphenicol. French et al. (1980) described an outbreak caused by a strain of *A. anitratus* in which the isolate was resistant to 18 antibiotics. Devaud et al. (1982) characterised multiresistant isolates of *A. calcoaceticus* which were resistant simultaneously to ampicillin, gentamicin, kanamycin, streptomycin, tetracycline, chloramphenicol and sulphamethoxazole, and were able to subsequently identify the genetic markers and transfer the resistance determinants to *Escherichia coli* K12. Similarly, in a strain of *A. calcoaceticus* resistant to ampicillin, aminoglycosides, chloramphenicol, sulphonamides and trimethoprim, the resistance genes could be transferred to *E. coli* by conjugation (Goldstein et al., 1983). More recently, multiresistant *A. calcoaceticus* var. *anitratus* strains isolated from a hospital outbreak of infection were resistant to 14 antibiotics and were inhibited only by netilmicin and imipenem (Vila et al., 1989). In our most recent study (Muller-Serieys et al., 1989), the predominant phenotypes of multiresistant *Acinetobacter* strains included resistance to penicillins, cephalosporins, gentamicin, amikacin (but not tobramycin), nalidixic acid, trimethoprim and sulphamethoxazole. Strains with such a resistance pattern were isolated in intensive care units, from respiratory nosocomial infections, urinary tract infections and septicaemia. More details of the mechanisms of resistance in these multiresistant strains will be provided in a subsequent section.

MECHANISMS OF RESISTANCE TO ß-LACTAMS IN *ACINETOBACTER* SPP

After the first recognised outbreaks of *Acinetobacter* infections (Bergogne-Bérézin et al., 1971; Pinter and Kantor, 1971), it was noted that isolates became rapidly resistant to aminopenicillins and cephalosporins of the first (cephalothin, cephaloridine) and second (cefuroxime, cefoxitin, cefamandole) generations (Retailliau et al., 1979; Garcia et al., 1983; Bergogne-Bérézin and Joly-Guillou, 1985). Indeed, it has been stated that *Acinetobacter* is "intrinsically insensitive" (Goldstein et al., 1983) to most ß-lactams, especially ampicillin and cephalosporins. However, as mentioned above, early studies with ampicillin and carbenicillin (Bergogne-Bérézin et al., 1971), and later with ureidopenicillins such as piperacillin (Philippon et al., 1980; Devaud et al., 1982; Joly-Guillou and Bergogne-Bérézin, 1985b), showed that these compounds were active against 70-80% of *Acinetobacter* strains. At a later stage, when *Acinetobacter* became resistant to early ß-lactams, the third generation cephalosporins were initially active, particularly cefotaxime (Phillips et al., 1981) and ceftazidime (Philipps et al., 1981; Bergogne-Bérézin and Joly-Guillou, 1986). However, within a few years, a rapid increase in resistance to carboxypenicillins and, to a lesser degree, ureidopenicillins and third generation cephalosporins, led several authors to investigate the possible production of inactivating enzymes by *Acinetobacter* strains.

Penicillinases in Acinetobacter spp

The commonest mechanism of resistance to penicillins is inactivation of these agents by ß-lactamases encoded either by the chromosome or by plasmids (Medeiros, 1984; Bauernfeind, 1986). ß-Lactamases have been detected in almost all Gram-positive and Gram-negative bacteria, including *Acinetobacter* spp. Resistance of *Acinetobacter* to ampicillin, carboxypenicillins and ureidopenicillins has been attributed to the presence of a penicillinase. Both TEM-1 (Philippon et al., 1980; Goldstein et al., 1983) and TEM-2 (Devaud et al., 1982) ß-lactamases have been found in representative isolates of epidemic strains that also produced aminoglycoside-modifying enzymes. The genes encoding resistance to penicillins (as well as those encoding resistance to aminoglycosides, tetracyclines, chloramphenicol and sulphonamides) have been identified on plasmids (see subsequent section on genetic determinants). In our own study (Joly-Guillou et al., 1988), 100 *Acinetobacter* strains selected from clinical strains isolated in our hospital between 1980 and 1986 (including five strains of var. *lwoffii* and three ATCC strains) were analysed for their ß-lactamase activities. ß-Lactamase activity was detected in crude extracts by the gel iodometric method (Labia and Barthelemy, 1979) in the presence of substrate inhibitors. Two ß-lactamase inhibitors were used to determine the inhibitory profiles of ß-lactamase specific activities (Philippon et al., 1980): clavulanic acid, which specifically inhibits penicillinases of the TEM type, and cloxacillin, which inhibits cephalosporinases. Isoelectric focusing of crude enzyme extracts was performed according to a standard procedure (Matthew and Harris, 1976): analytical isoelectric focusing was carried out on sheets of polyacrylamide gel at 200-400 V (for about 18 h), with 30 μl samples (extracts) applied near the anode on blotting paper discs (for more details see Matthew and Harris, 1976; Joly-Guillou et al., 1988). The enzymes were detected with benzylpenicillin as a substrate by the iodometric overlay method; their isoelectric point (pI) was determined by comparison with enzymes of known pI, as described previously (Joly-Guillou et al., 1988). The results showed that the 100 *Acinetobacter* strains divided into two populations with different ticarcillin susceptibilities: 24 susceptible strains (geometric mean MIC <32 mg/l) and 76 resistant strains (geometric mean MIC >256 mg/l). Analysis of ß-lactamase activities showed the presence of a penicillinase activity in 41% of resistant strains (Table 2). A TEM-1 type enzyme (pI 5.4) was identified in 34% of the strains, while 7% of the strains produced a ß-lactamase of pI 6.3, characterised subsequently as a CARB-type enzyme, which inactivated ampicillin and carbenicillin, but not

Table 2. ß-lactamase distribution in clinical strains of *A.calcoaceticus*

Enzyme	Inhibited by			% strains
	clavulanic acid	cloxacillin	sulbactam	
None detected	-	-	-	18
Penicillinase				41
TEM type (pI 5.4)	+	-	+	(34)
CARB type (pI 6.3)	+	-	+	(7)
Cephalosporinase (pI >8)	-	+	+	9
Penicillinase + cephalosporinase				32
TEM type + C[a]	+	+	+	(30)
CARB type + C[a]	+	+	+	(2)

[a] high pI (>8.0) enzyme assumed to be a chromosomal cephalosporinase
data taken from Joly-Guillou et al. (1988)

methicillin or cloxacillin, and had a relatively low activity against cephalosporins. Subsequent analysis of the characteristics of this enzyme was carried out with either cell-free extracts of a selected strain that was highly resistant to ticarcillin or with purified enzyme obtained from a sonic lysate of cells from a 4 l culture of the selected strain (Paul et al., 1989). The substrate profile of the purified enzyme was characterised by preferential hydrolysis of penicillins (ampicillin, carbenicillin, azlocillin), while cloxacillin, cephalothin and cephaloridine were very poorly hydrolysed. The enzyme was inhibited by *p*-chloromercuribenzoate (PCMB), which inhibits many enzymes containing cysteine residues (Bauernfeind, 1986), and also by cloxacillin and the ß-lactamase inhibitors clavulanic acid and sulbactam, but not by chloride ions (100 mM). Immunological studies were performed with several antisera prepared against purified enzymes (CARB-3, TEM-1, OXA-2, OXA-4 and OXA-6), but only antiserum to CARB-3 neutralised the enzyme activity and no cross-neutralisation was obtained with other antisera. Thus this new enzyme appeared to be a CARB-type enzyme, but since it did not co-focus with CARB-3 (pI = 5.7), it was designated CARB-5. Its molecular mass was estimated from gel filtration (on Ultrogel AcA54) to be 28,000, similar to other CARB enzymes described previously (Philippon et al., 1980; Medeiros, 1984). This was the first demonstration of a CARB-type enzyme in strains of *A. calcoaceticus* var. *anitratus*. Substrate and inhibition profiles of this novel ß-lactamase from *A. calcoaceticus* are given in Table 3. This novel enzyme belongs to the carbenicillinase group of transferable ß-lactamases. This is a homogenous group of enzymes which is also characterised by a high level of enzyme production in the host strains. The penicillinases described above (TEM-1, TEM-2 and CARB-5) belong to the group of constitutive plasmid-determined ß-lactamases described by Bauernfeind (1986) and Medeiros (1984).

Cephalosporinases in Acinetobacter spp

The older cephalosporins, such as cephalothin, cephaloridine or cefazoline, can also be inactivated by plasmid-mediated ß-lactamases. The third generation cephalosporins (e.g. cefotaxime, ceftazidime, ceftriaxone, or moxalactam), characterised by two substituents on the C-7 side chain (methoxyimino and aminothiazole groups), exhibit a particularly high ß-lactamase stability (Knothe and Dette, 1983), but most of the chromosomally-determined ß-lactamases preferentially hydrolyse cephalosporins, including many of the newer ß-lactams which resist hydrolysis by the plasmid-determined ß-lactamases (Medeiros, 1984). As described earlier (Table 1), less than 50% of *Acinetobacter* strains are inhibited at acceptable concentrations of cefotaxime, moxalactam, or even ceftazidime, which was initially the most active drug (Phillips et al., 1981; Bergogne-Bérézin and Joly-Guillou, 1986). Almost all Gram-negative bacilli produce a chromosomally-determined ß-lactamase that is species-specific (Medeiros, 1984). The rapid increase of resistance to third generation cephalosporins (Joly-Guillou et al., 1987) has led several authors to investigate the possible presence of an inducible cephalosporinase in *Acinetobacter* spp. Before the third generation of cephalosporins became

Table 3. Substrate and inhibition profiles of the novel ß-lactamase from *A. calcoaceticus* strains A85-135   (Paul et al., 1989)

	Cell free extract	Purified fraction
Substrate profile[a]		
Ampicillin	86	80
Carbenicillin	76	61
Azlocillin	57	61
Methicillin	1	16
Oxacillin	1	3
Cloxacillin	1	2
Cephalothin	10	4
Cephaloridine	19	8
Cefotaxime	<1	<0.5
Cefsulodin	<1	<0.5
Imipenem	<1	<0.5
Inhibition profile[b]		
Cl$^-$ (100 mM)	0	0
CLAV (1 μM)	58	94.5
SUL (1 μM)	-	91.3
Cloxacillin (1 mM)	-	>90
pCMB (0.5 mM)	-	>90
Antiserum:		
anti-CARB-3	>80	
anti-TEM-1	50	
anti-OXA-2	<30	
anti-OXA-4	<30	
anti-OXA-6	<30	

[a] V_{max} values, expressed relative to the V_{max} for benzyl-penicillin (100%).
[b] percentage of neutralised activity at the indicated inhibitor concentration in the presence of benzylpenicillin.   The inhibitory effects of clavulanic acid (CLAV) and sulbactam (SUL) were determined after preincubation with the enzyme for 10 min at 37°C.

available in 1977, an initial study (Morohoshi and Saito, 1977) had demonstrated production of an inducible class I type ß-lactamase (cephalosporinase) in *Acinetobacter*. The enzyme closely resembled the enzymes produced by *Proteus, Citrobacter* and *Pseudomonas* spp. in terms of molecular mass (30,000), sensitivity to inhibitors and inducibility. Further studies of resistance mechanisms in *Acinetobacter* spp. (Devaud et al., 1982; Goldstein et al., 1983) cited the presence of a cephalosporinase, but this was not clearly demonstrated since the reported experiments focused on plasmid-mediated enzymes. In describing the two major groups of cephalosporinases outlined by Sawai et al., (1982), Medeiros (1984) described the enzyme produced by *Acinetobacter* strains as belonging to the "typical cephalosporinase group", which includes inducible enzymes such as those produced by *Enterobacter aerogenes, Ent. cloacae, Serratia marcescens and Ps. aeruginosa*. One of the main characteristics of "typical cephalosporinases" is their narrow spectrum of substrate specificity; they have little or no activity against benzylpenicillin, ampicillin and carbenicillin, but render bacteria resistant to various cephalosporins. The contribution of such cephalosporinases to ß-lactam resistance can be based on at least two major mechanisms (Piddock and Wise, 1985): (i) hydrolysis of older cephalosporins, which requires high enzyme affinity (K_m) for the antibiotic (Medeiros, 1984) - this is the case with the *Acinetobacter* cephalosporinase; indeed the initial resistance of *Acinetobacter* to most first and second generation cephalosporins, as described above, was associated with enzymic hydrolysis of antibiotics (Joly-Guillou and Bergogne-Bérézin, 1985b); (ii) although third generation cephalosporins are normally resistant to hydrolysis by ß-lactamases, the amount of ß-lactamase (cephalosporinase) produced in the periplasmic space of the cell increases with exposure to ß-lactams such as cefoxitin or ceftazidime (induction), and it has been postulated (Sanders, 1983) that the ß-lactamase protects the bacterial cell by binding the antibiotic molecules rather than splitting them - thus the enzyme acts as a periplasmic "sponge" ("sponge effect" or "trapping") which keeps the antibiotic from reaching its target site on the cytoplasmic membrane. Such a mechanism may occur in *Enterobacter* spp., *S. marcescens* and *Ps. aeruginosa*, and has also been suggested for *A. anitratus* (Sanders, 1983; Medeiros, 1984). However, since the report in 1977 of an inducible cephalosporinase in *Acinetobacter* spp., no clear demonstration of such an enzyme can be found in the literature; the hypotheses of cephalosporinase activities mentioned above were based mostly upon clinical experience and indirect evidence of cephalosporinase production rather than in-vitro experiments (Sanders, 1983). Joly-Guillou et al. (1988) also looked for cephalosporinase production in 100 *Acinetobacter* clinical strains by using the inhibitory activity of cloxacillin and isoelectric focusing of crude enzyme extracts (Matthew and Harris, 1976; Joly-Guillou et al., 1988). The presence of cephalosporinase activity correlated with variable bands of ß-lactamase activity located at pI >8. The results of this study showed that 32% of the strains possessed a cephalosporinase plus a penicillinase, while only 9% exhibited cephalosporinase activity alone. Among the strains with evidence of cephalosporinase activity, a group of nine strains showed low-level resistance to third generation cephalosporins (geometric mean MICs were 14.7 mg cefotaxime /l and 4.3 mg ceftazidime /l); a second group of nine strains exhibited high-level resistance to all cephalosporins, with geometric mean MICs of 54.8 mg cefotaxime /l and 17.2 mg ceftazidime /l. Analysis of the results suggested that ceftazidime might be less susceptible to the cephalosporinase of *Acinetobacter* than was cefotaxime. Otherwise, ß-lactamase inducibility was not clearly demonstrated in these strains of *Acinetobacter*, although there was evidence that the second group of strains, showing a high level of resistance to third generation cephalosporins, were producing an increased amount of cephalosporinase constitutively (Bauernfeind, 1986), resulting in a certain amount of cefotaxime hydrolysis and a lesser, but detectable, degree of ceftazidime hydrolysis. As described elsewhere (Piddock and Wise, 1985), these compounds are hydrolysed, but at a slower rate than in conventional enzymic hydrolysis of first generation cephalosporins. More recently, a further investigation (Hood and Amyes, 1989) of chromosomally mediated ß-lactamases from the genus *Acinetobacter* used a novel isoelectric focusing method to identify, in eight strains, four different ß-lactamases with pIs ranging from 7.3 to 8.8. Additional minor bands at pI 9.8 and 10.1 were also revealed for one enzyme by this modification of conventional isoelectric focusing techniques. Further details of these enzymes are given elsewhere (Hood and Amyes, this volume).

New Enzymes in Acinetobacter spp

The study of Hood and Amyes (1989) provided information on enzymes previously undescribed in *Acinetobacter* spp. and illustrated the apparently limitless power of bacteria to adapt themselves to new antibiotics and to develop new mechanisms of resistance (Wiedeman et al., 1989). This has happened recently with *Klebsiella pneumoniae* (Sirot et al., 1988) and other Enterobacteriaceae, particularly *E. coli*. In epidemic strains of *K. pneumoniae*, plasmid-mediated broad-spectrum ß-lactamases have been characterised. Clinical isolates encoding these novel ß-lactamases appeared in France in 1984-85 and showed decreased susceptibilities to aminopenicillins, carboxypenicillins, ureidopenicillins, most cephalosporins, including third generation compounds (cefotaxime and ceftazidime), and a monobactam (aztreonam). These strains were still, however, susceptible to imipenem, cephamycins (cefoxitin, cefotetan) and moxalactam. In late 1989 a strain of *A. baumannii* isolated in our hospital from a urine sample exhibited an unusual susceptibility profile (Joly-Guillou and Bergogne-Bérézin, 1990), with resistance to ticarcillin and piperacillin (MICs >256 mg/l), cefoxitin (64 mg/l), cefotaxime (32 mg/l) and ceftazidime (16 mg/l). In addition, synergy was observed between the inhibition zones obtained in disc sensitivity tests with amoxicillin/clavulanic acid and cefotaxime or ceftazidime. This was an unusual pattern of response to cephalosporins - most strains resistant to cefotaxime and ceftazidime that produce cephalosporinase (Joly-Guillou et al., 1988) exhibit antagonism between the above mentioned zones, due to the induction of cephalosporinase by the potent cephalosporinase inducer clavulanic acid (Sanders, 1983). Analysis of crude extracts by the gel iodometric method in the presence of substrate inhibitors (Labia and Barthelemy, 1979; Joly-Guillou and Bergogne-Bérézin, 1988), showed that: (i) the strain was a ß-lactamase producer; (ii) ß-lactamase activity was inhibited by clavulanic acid, which specifically inhibits penicillinase activity; (iii) ß-lactamase activity was not inhibited by cloxacillin, which specifically inhibits cephalosporinase activity. Analytical isoelectric focusing failed to show the presence of either a TEM-type or a CARB-5 enzyme, and no cephalosporinase activity was detected, although some enzymic activity was observed with a pI of 7.7. It can be postulated that this particular *Acinetobacter* isolate may produce one of the novel broad spectrum ß-lactamases described recently by Sirot et al. (1988) and designated "cefotaximases" (e.g. CTX-1 of pI 6.3) or "ceftazidimases" (CAZ-1 to CAZ-5), the latter including enzymes of pI 7.7 which otherwise belong to the SHV ß-lactamase family. A possible role for *A. calcoaceticus* as a reservoir or vehicle of transmissible antibiotic resistance in the hospital environment has been suggested by Chopade et al. (1985); indeed, many clinical isolates contain a variety of different plasmids (as described in a subsequent section). Trimethoprim R plasmids have been shown to be transmissible from *E. coli* to *A. calcoaceticus* (Chopade et al., 1985), which suggests the potential for interchange of resistance markers with *E. coli* and other Enterobacteriaceae in close contact with *Acinetobacter* spp. in the clinical environment. If the presence of plasmid-mediated broad-spectrum ß-lactamases is confirmed in *Acinetobacter* strains, it will be the first description of this new class of enzyme in aerobic Gram-negative bacilli and will further confirm the epidemiological role of *Acinetobacter* strains in the spread of resistance in hospitals.

ß-Lactamase Inhibitors

In order to overcome ß-lactamase production and achieve effective antibiotic therapy of infections due to ß-lactamase producers, several ß-lactamase inhibitors have been designed (Wise, 1982) and their in-vitro activities assayed in combination with various penicillins and cephalosporins, especially against ß-lactamase-producing aerobic Gram-negative bacilli (Philippon et al., 1980; Jacobs et al., 1986). Among the well-known ß-lactamase inhibitors, clavulanic acid and sulbactam have been introduced into medical practice; more recently, the newly developed compound tazobactam (YTR 830) has also been investigated. Although ß-lactamase inhibitors have poor intrinsic antibacterial activity (Wiedeman et al., 1989), sulbactam is surprisingly effective against *Acinetobacter* spp. (Kitzis et al., 1983). Similarly, in our own study (E. Vallée, M.L. Joly-Guillou and E. Bergogne-Bérézin, unpublished data), sulbactam was less active (Table 4) than tazobactam against ticarcillin-resistant strains (range MICs: 1-32 mg/l), but both were active by themselves against ticarcillin-susceptible strains, with geometric mean MICs of sulbactam and tazobactam of 2.1 and 6.5 mg/l respectively. Whether or not these figures are clinically relevant is questionable. It was of more interest to examine the inhibitory activities of these ß-lactamase inhibitors against the ß-lactamases produced by *Acinetobacter* strains. MICs of ticarcillin, cefotaxime and ceftazidime

Table 4. Susceptibility of 46 clinical strains[a] of *A.calcoaceticus* to ticarcillin in the absence and presence of ß-lactamase inhibitors

Drugs		MICs (mg/1)			
		Range	50% of strains	90% of strains	Geometric mean
Ticarcillin	Tic^S	2 - 32	10	25.2	11.5
	Tic^R	2048 - >2048	>2048	>2048	>2048
Sulbactam	Tic^S	1 - 8	1.5	3.5	2.1
	Tic^R	16 - 32	16	22.7	18
Tazobactam	Tic^S	1 - 16	6	13.3	6.5
	Tic^R	1 - 32	5.6	14.8	7.1
Ticarcillin + clavulanic acid (2 mg/1)	Tic^S	0.5 - 32	9.7	22.9	10.6
	Tic^R	32 - >2048	112	>2048	-
Ticarcillin + sulbactam	Tic^S	0.5 - 4	1.12	2.9	1.55
	Tic^R	8.0 - 64	16.0	30.08	15.50
Ticarcillin + tazobactam	Tic^S	1 - 16	3.75	11.6	5.48
	Tic^R	1 - 32	23.7	99.55	33.02

[a]17 strains were ticarcillin-susceptible or moderately resistant; 29 strains were ticarcillin-resistant (MIC >64 mg/1)

were therefore determined for these drugs alone and in combination with ß-lactamase inhibitors at a fixed concentration ratio 1:1. Synergy was defined as a significant decrease in the geometric mean MICs of the antibiotic/inhibitor compared with those for antibiotic alone. Simultaneously, in order to determine the inhibition profiles of ß-lactamases, clavulanic acid, cloxacillin, sulbactam and tazobactam were added in gel, made from the classical iodine-iodide starch system, at a fixed concentration of 0.5 mg/l, to crude extracts of bacterial cells in order to specifically inhibit penicillinase and/or cephalosporinase activities. Clavulanic acid caused a significant decrease in the MICs for the ticarcillin-resistant strains, ($MIC_{50} = 112$ mg/l), but not a true reversion to susceptibility, (Table 4). The MICs of ticarcillin, either singly or combined with tazobactam, were identical for ticarcillin-susceptible strains, but were notably diminished for ticarcillin-resistant strains, which returned to susceptible levels (geometric mean MIC of 33.02 mg/l). Similarly, sulbactam in combination with ticarcillin significantly reduced the MICs of ticarcillin (geometric mean MIC of 15.50 mg/l). The geometric mean MICs of cefotaxime and ceftazidime are shown in Table 5 in relation to the enzymic profiles of the tested strains. Combinations of cefotaxime and inhibitors produced a significant reduction in the geometric mean MICs, when compared with those of cefotaxime alone, for strains that only produced cephalosporinase (geometric mean MIC of cefotaxime was 41.8 mg/l, compared with 2.2 and 7.2 mg/l when combined with sulbactam or tazobactam, respectively). This observation was not made for ceftazidime, with which the geometric mean MICs were identical or only slightly diminished when tested in combination with inhibitors. Using crude bacterial extracts, it was found that tazobactam and sulbactam inhibited both cephalosporinase and penicillinase activities. These data require further investigation before any clinical application can be proposed.

Imipenem

Imipenem is a new carbapenem antibiotic with the broadest antibacterial spectrum of any ß-lactam antibiotic currently available. It is active against many multiresistant pathogens, including strains of *Enterobacter, Serratia* and *Ps. aeruginosa*. No cross-resistance has been observed between imipenem and other ß-lactam antibiotics; indeed, imipenem is theoretically not inactivated by ß-lactamases, and "trapping" of the antibiotic by high concentrations of ß-lactamase in the periplasmic space (as described above for third generation cephalosporins) does not seem to be the mechanism by which imipenem is inactivated. With the earliest preparation of this drug (*N*-formimidoyl-thienamycin), and before the antibiotic was clinically available, we started a survey of the imipenem susceptibility of *Acinetobacter* strains isolated since 1981 (Bergogne-Bérézin and Joly-Guillou, 1986). All strains were initially very susceptible to imipenem and no changes in susceptibility occurred for six years (Fig. 1). The MICs of imipenem were in the range 0.016-4 mg/l for *A. anitratus* and 0.016-1 mg/l for *A. lwoffii*, with geometric mean MICs of 0.33 and 0.20 mg/l respectively. This survey, which is still in progress, has now shown the

Table 5. In-vitro activity of sulbactam and tazobactam in combination with cefotaxime and ceftazidime against *A.calcoaceticus*

Enzymes produced	Geometric mean MICs (mg/l)								
	CTZ	CTZ + SUL	SUL	CTX + SUL	CTX	CTX + TZB	TZB	CTZ + TZB	CTZ
None detected	2.8	1	1.4	1	5.6	1.7	2.3	1.4	2.8
Penicillinase	3.2	2.5	16	6.3	6.3	6.3	8.0	4.0	3.2
Cephalosporinase	8.0	1.7	2.3	2.2	41.8	7.2	8.9	5.2	8.0
Penicillinase + cephalosporinase	6.5	6.8	18.3	11.9	20.9	5.5	6.8	3.6	6.5

CTZ: ceftazidime; SUL: sulbactam; CTX: cefotaxime; TZB: tazobactam

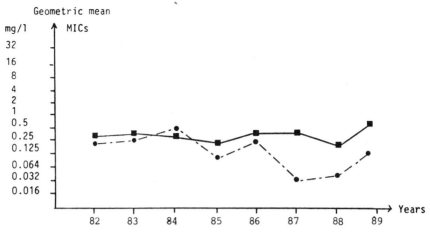

Fig. 1. Imipenem susceptibility of *Acinetobacter* spp.

emergence of a few resistant strains: two resistant strains of *A. baumannii* belonging to two different phage types were isolated in 1987 from a single patient who had been treated with imipenem (Bergogne-Bérézin et al., 1987), and, within the last two years, 20 strains resistant to imipenem have been isolated in our hospital. The MICs of imipenem for these isolates ranged from 4 to 8 mg/l, which actually indicates a diminished susceptibility rather than high-level resistance. However, the minimal bactericidal concentrations (MBCs) of imipenem for these strains ranged from 32 to 128 mg/l. Study of their resistance phenotypes has shown that these strains were resistant to ticarcillin (with no significant effect of clavulanic acid), piperacillin, moxalactam and cefoxitin; they were variably resistant to cefotaxime and ceftazidime. Analysis of their enzyme content has shown the variable presence of a TEM-1 (pI 5.4) or TEM-2 (pI 5.6) enzyme, a cephalosporinase, and a new enzyme (pI 7.6) found in two strains which were resistant to all ß-lactams. A few strains, resistant to imipenem only, did not exhibit any ß-lactamase activity. Although ß-lactamases capable of inactivating imipenem, produced by *Ps. maltophilia, Flavobacterium odoratum, Bacteroides fragilis* and *Aeromonas hydrophila,* have been reported occasionally in the literature (Medeiros, 1984; Shannon et al., 1986; Quinn et al., 1988; Wiedeman et al., 1989), most studies of imipenem resistance refer to alterations in bacterial outer membrane permeability. Studies on the penetration of ß-lactams through porin channels have shown that the rate of penetration appears to be determined by physico-chemical properties of the drug, such as molecular mass, hydrophobicity and electrical charge (Hancock, 1984; Piddock and Wise, 1985; Nayler, 1987; Quinn et al., 1988). Imipenem crosses the outer membrane of *E. coli* K12 extremely rapidly, probably as a result of its compact structure, low molecular mass and its zwitterionic charge (Hancock, 1984; Quinn et al., 1988). The mechanism of resistance to imipenem in most resistant bacteria appears to be related to a selective decrease in permeability across the bacterial outer membrane, usually associated with discrete alterations in the electrophoretic profiles of outer membrane proteins, resulting in a decreased rate of antibiotic uptake.

This has been analysed primarily in *Ps. aeruginosa* (Quinn et al., 1988), in which impaired penetration of imipenem through the outer membrane is related to the lack of an outer membrane protein that is present in susceptible parent strains (Quinn et al., 1988). It has also been established that alteration of the lipopolysaccharide in the outer membrane may contribute substantially to diminished permeability and thus to ß-lactam resistance (Piddock and Wise, 1985; Nayler, 1987). So far as *Acinetobacter* spp. are concerned, it has been shown that a selective alteration in outer membrane permeability affects the susceptibility of the organism to the oxygen-independent antimicrobial activity of rat polymorphonuclear leucocyte granule contents (Leoffelholz and Modrzakowski, 1987). This alteration in outer membrane permeability, related to lipopolysaccharide structure, was shown to be associated with the presence of plasmid RP1 (Loeffelholz and Modrzakowski, 1986). RP1, which confers resistance to ampicillin, carbenicillin, kanamycin, neomycin and tetracycline, physically alters the outer membrane structure of the bacteria. Whether these RP1-mediated outer membrane alterations could play a role in resistance to imipenem and other antibiotics in *A. calcoaceticus* remains to be elucidated. Similarly, the question as to whether the impaired penetration of imipenem through the outer membrane observed in *Ps. aeruginosa* can be extrapolated to *Acinetobacter* remains to be determined. Further investigations of possible enzymic inactivation of imipenem in resistant *Acinetobacter* strains also remain to be carried out.

MECHANISMS OF RESISTANCE TO AMINOGLYCOSIDES IN *ACINETOBACTER* SPP

Aminoglycosides have been used since the mid-1940s. The emergence of resistance as a result of the interplay between use of aminoglycosides and selective pressure (McGowan, 1983) was observed during the early phase of aminoglycoside therapy when streptomycin and kanamycin were the major available antibiotics. There are three major mechanisms of resistance: (i) alteration of the target site on the ribosomes; (ii) reduced aminoglycoside permeability; (iii) inactivation of the aminoglycoside by enzymic modification. The latter mechanism is the most common type of aminoglycoside resistance found among clinical isolates of staphylococci and Gram-negative bacilli. Aminoglycosides are modified either by acetylation of an amino group (acetyltransferases), by phosphorylation (phosphotransferases), or by adenylation of hydroxyl groups (adenyltransferases). The distribution of aminoglycoside-modifying enzymes in various bacterial species is listed in Table 6 (Bauernfeind et al., 1986). A number of studies since 1975, have shown the presence of aminoglycoside modifying enzymes in clinical isolates of *Acinetobacter* (Uwaydah and Taql-Eddin, 1976; Dowding, 1979; Bergogne-Bérézin et al., 1980; Gomez-Lus et al., 1980; Murray and Moellering, 1980; Devaud et al., 1982; Goldstein et al., 1983). Resistance of *Acinetobacter* to tobramycin,

Table 6. Substrate profiles of aminoglycoside modifying enzymes (based on Bauernfeind et al., 1986)

Enzyme	Modification of						Organisms
	Genta-micin C	Netil-micin	Tobra-mycin	Amika-cin	Kana-mycin A	Strepto mycin	

Phosphorylating enzymes

APH(3')I APH(3')II	-	-	-	-/+	+		Enterobacter-iaceae, *Pseudomonas*, *Acinetobacter*, *Staphylococcus*, *S. faecalis*
APH(2")	+	-	+	-	+		*Staphylococcus*, *S.faecalis*
APH(3")	-	-	-	-	-	+	Enterobacter-iaceae, *Pseudomonas*, *Acinetobacter*, Gram-positive bacteria

Adenylating enzymes

AAD(2")	+	+	+	-	+		Enterobacter-iaceae, *Pseudomonas*, *Acinetobacter*
AAD(4')	-	-	+	+	+		*Staphylococcus*
AAD(3")	-	-	-	-	-	+	Enterobacter-iaceae, *Pseudomonas*, *Acinetobacter*
AAD(6)	-	-	-	-	-	+	*Staphylococcus*

continued

Table 6.(Continued)

Acetylating enzymes

AAC(6')	$-/+^a$	+	+	+	+	Enterobacter-iaceae, *Pseudomonas*, *Acinetobacter*, *Staphylococcus*, *S.faecalis*
AAC(3)I	+	-	±	-	-	Enterobacter-iaceae, *Pseudomonas*, *Acinetobacter*
AAC(2')	+	+	+	-	-	*Prov.stuartii* *P.rettgeri*, *Acinetobacter*
AAC(3)III	+	+	+	-	+	*Pseudomonas*
AAC(3) II	+	+	+	-	±	Enterobacter-iaceae

+, modified; -, not significantly modified
[a]gentamicin C_1: -ve; C_{1A} and C_2: +ve

dibekacin and sisomycin, plus moderate resistance to kanamycin and amikacin, were related to the presence of an aminoglycoside 6'-*N*-acetyltransferase (AAC(6)) (Bergogne-Bérézin et al., 1980; Gomez-Lus et al., 1980; Murray and Moellering, 1980). Other studies have characterised acetylating enzymes of the AAC(2') (Dowding, 1979) and AAC(3)I types (Gomez-Lus et al., 1980), which modify gentamicin and tobramycin, phosphotransferases of the APH(3')I (Gomez-Lus et al., 1980) and APH(3')II types (Murray and Moellering, 1980), which confer resistance to kanamycin, and adenyltransferases of the AAD(2") (Murray and Moellering, 1980) and AAD(3") types (Devaud et al., 1982; Goldstein et al., 1983), which modify streptomycin and spectinomycin. Thus the presence of modifying enzymes belonging to the three main classes is responsible for the resistance of *Acinetobacter* strains to a great number of aminoglycosides. In an extensive study (Bergogne-Bérézin et al., 1980), we investigated the in-vitro activity of seven aminoglycosides against 215 Acinetobacter strains isolated from patients between 1971 and 1980. The strains isolated between 1971 and 1975 were initially susceptible to the aminoglycosides gentamicin, sisomycin and tobramycin, with only 10 to 20% resistant strains isolated in 1971. A similar investigation (Uwaydah and Taql-Eddin, 1976) of the susceptibility of 15 *Acinetobacter* strains to gentamicin and tobramycin found MICs ranging from 0.12 to 0.9 mg/l. In our study, a sudden rise in resistance to gentamicin, kanamycin, sisomycin and tobramycin occurred in 1975, correlating with the appearance of aminoglycoside-modifying enzymes. An acetylating enzyme of the AAC(3)I type and a phosphorylating enzyme of the APH(3′)I type were found to be present in cell-free extracts (Bergogne-Bérézin et al., 1980). More recently, a new phenotype has appeared in France, with resistance to gentamicin (GEN^r), sisomycin (SIS^r) and amikacin (AMK^r), but not to tobramycin (TOB^s). Twelve *Acinetobacter* strains with this phenotype (GEN^r SIS^r AMK^r TOB^s) were analysed in late 1986; the results indicated the presence of AAC(3)I and APH(3′)I enzymes that inactivate lividomycin, but not amikacin. A possible new phosphotransferase was also found in four of the 12 *Acinetobacter* strains. Enzymic phosphorylation of amikacin, which has been reported previously in *E. coli* and *Ps. aeruginosa* (Joly-Guillou et al., 1987), was found in *A. calcoaceticus* var. *anitratus*. This new enzyme also inactivated kanamycin A,B and C, neomycin, neamine, paromomycin, seldomycin, butirosin, lividomycin, sisomycin and tobramycin. Thus the presence of an AAC(3)I and the new phosphorylating enzyme, provisionally designated APH(3′)III, modifying butirosin, lividomycin and amikacin, may account for the new *Acinetobacter* phenotype described above and provides a new mechanism of amikacin resistance. These observations have been supported by a further study (Lambert et al., 1988) which demonstrated that resistance to amikacin was related to the synthesis of a new type of transferable APH(3′) in a clinical strain of *Acinetobacter*. The authors of this study showed that the gene conferring resistance to kanamycin/amikacin was carried by a 63 kb plasmid, pIP1841, self-transferable to *A. baumannii, A. haemolyticus* and *A. lwoffii*, but not to *E. coli*. Of the APH(3′) types detected in clinical isolates, only this new type of enzyme, subsequently designated APH(3′)VI (Lambert et al., 1990), modifies amikacin in the manner described above (Joly-Guillou et al., 1987; Lambert et al., 1988). Alternative mechanisms of resistance to aminoglycosides, such as diminished permeability or alteration of the ribosomal target sites, have not yet been investigated in *Acinetobacter*.

RESISTANCE OF *ACINETOBACTER* TO FLUOROQUINOLONES

During the last few years, the availability of an increasing number of new fluorinated quinolones has given rise to numerous studies on their comparative in-vitro activities, mostly against various multiply-resistant Gram-negative bacilli (King and Philips, 1986; Rolston et al., 1988). Additional clinical evaluations of the efficacy of the new quinolones have included their use in treating severe nosocomial infections involving Gram-negative bacilli. *A. baumannii* was not tested extensively in the series of strains included in the original in-vitro studies (Bergogne-Bérézin and Joly-Guillou, 1985; Godineau-Gauthey et al., 1988; Rolston et al., 1988) and thus only limited data are available on the initial susceptibility of *Acinetobacter*. clinical strains to new quinolones. Even so, the extremely interesting data generated by these early in-vitro studies (Bergogne-Bérézin and Joly-Guillou, 1985; Joly-Guillou and Bergogne-Bérézin, 1985a; King and Phillips, 1986) showed enhanced activity against *A. baumannii* of new quinolones such as pefloxacin, ciprofloxacin, norfloxacin and ofloxacin, compared with the parent compound, nalidixic acid. In 1985-86, a study with a limited number of strains (King and Phillips, 1986) and our own studies of 130 clinical *Acinetobacter* strains (Bergogne-Bérézin and Joly-Guillou, 1985; Joly-Guillou and Bergogne-Bérézin, 1985a), compared susceptibility to the new fluorinated quinolones with susceptibilities to other antibiotics. Very promising results were obtained, indicating that ciprofloxacin, ofloxacin, and pefloxacin were substantially more active *in vitro* than either nalidixic acid, third generation cephalosporins or aminoglycosides. The greatest activity was exhibited by ofloxacin, with an MIC_{50} of 0.6 mg/l. The MIC_{50} of ciprofloxacin was 0.7 mg/l, with mean MICs of 0.98, 1.03 and 1.64 mg/l for ciprofloxacin, ofloxacin and pefloxacin respectively. Similarly, King and Phillips (1986) reported that the MICs of ciprofloxacin for 35 strains of *Acinetobacter* ranged from 0.004 to 1 mg/l, with geometric mean MICs of 0.053, 0.067 and 0.158 mg/l for ciprofloxacin, ofloxacin and pefloxacin respectively. Since 1985, the new quinolones, especially those compounds which can be administered parenterally, have been used increasingly in the treatment of severe nosocomial infections caused by multiply-resistant bacteria.

It has been suggested that the isolation frequency of quinolone-resistant variants is higher in non-fermenting organisms than in enterobacteria; thus it might be expected that emergence of resistant mutants to the newer quinolones, which occurred rapidly in *Ps. aeruginosa* (Neu, 1988), might also occur in *Acinetobacter*. A few years after addition of the new quinolones to the antibacterial armamentarium, we carried out a new in-vitro investigation using *Acinetobacter* strains collected in 1988-89. We had noted in our routine work that *Acinetobacter* strains were becoming increasingly resistant to pefloxacin. Table 7 compares our latest data (June 1989) on the susceptibility of 100 recent isolates to new quinolones with our previous results (Joly-Guillou and Bergogne-Bérézin, 1985a). The results showed that the 1989 MIC values (MIC_{50}, MIC_{90} and geometric mean MICs) had increased five to ten-fold for pefloxacin and ciprofloxacin, but not for ofloxacin. In addition, the MIC ranges were extended, e.g. from 0.016-32 mg/l to 0.03-128 mg/l for ciprofloxacin and from 0.064-16 mg/l to 0.03-32 mg/l for ofloxacin. The data obtained with newer compounds (lomefloxacin, temafloxacin, difloxacin) did not indicate any significant improvement in activity compared to previous compounds. As noted in our earlier studies, the most active fluorinated quinolone against *Acinetobacter* clinical strains was ofloxacin, while difloxacin exhibited a very similar activity to that of ciprofloxacin. When the distribution of the strains as a function of MIC values was examined (Fig. 2), a bimodal distribution corresponding to two populations (susceptible and resistant) was apparent. Using the breakpoints recommended in France of 1 and 4 mg/l, the proportions of fully susceptible strains were only 27, 31 and 35% for pefloxacin, ciprofloxacin and ofloxacin respectively. Such a threatening

Table 7. MICs of fluoroquinolones (mg/l) for strains of *A. baumannii* isolated during 1985 (130 strains) and 1989 (100 strains)

Compounds	Range	MIC_{50}	MIC_{90}	Geometric Mean
Pefloxacin				
1985	0.25-16	1.05	6.71	1.64
1989	0.125-128	14.15	74.66	10.26
Ciprofloxacin				
1985	0.016-32	0.72	4.64	0.98
1989	<0.03-128	7.3	60.44	5.76
Ofloxacin				
1985	0.064-16	0.69	3.73	1.03
1989	<0.03-32	3.37	14.6	2.03
Lomefloxacin				
1989	0.06-256	10.18	85.33	7.29
Temafloxacin(A.52254)				
1989	<0.03-64	2.8	41.14	2.83
Difloxacin(A.56619)				
1989	<0.03-128	7.73	88	5.32

situation might have resulted from the early and large usage of pefloxacin in France; indeed these data can be compared with other studies (e.g. Chow et al., 1988) showing a more limited MIC range for ciprofloxacin of 0.5-4 mg/l and MIC_{50} and MIC_{90} values of 2 and 4 mg/l respectively for a low inoculum of 10^4 cfu. It should, however, be noted that the latter values were notably increased by a larger inoculum (10^6 cfu), reaching 8 and 16 mg/l, which suggests, as in our study, a large proportion of resistant strains (MICs >4 mg/l)

A further in-vitro study with our susceptible strains was carried out in a microdilution system which permitted us to evaluate the minimal bactericidal concentrations (MBCs) for each strain (Table 8). The results of this study indicated a very close relationship between the MICs and MBCs (inoculum: 10^5 cfu) for all new quinolones tested. The ratios of geometric mean MBCs/MICs ranged from 1.11 to 1.21, which confirmed the excellent bactericidal activity of this class of antibiotics against susceptible *Acinetobacter* strains. Very few studies of antibiotic combinations have been carried out, although the available in-vitro data (Chow et al., 1988)

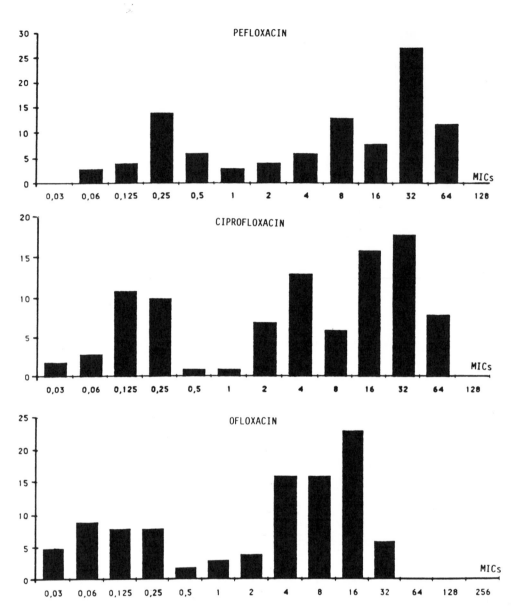

Fig. 2. MICs of pefloxacin, ciprofloxacin and ofloxacin for 100 *A. baumannii* strains

suggest that synergistic interactions with ciprofloxacin rarely occur; most combinations exhibit at least additive effects at clinically achievable serum concentrations. Multiply-resistant *Acinetobacter* strains involved in nosocomial infections should respond to combinations of ciprofloxacin with amikacin, tobramycin, or ceftazidime.

Even though the frequency of mutational resistance was as low as 10^{-9} in early studies, the emergence of resistant variants in clinical *Acinetobacter* strains is a serious threat which may result in the failure of

Table 8. MICs and MBCs (mg/l) of fluoroquinolones for susceptible strains of *A. baumannii* (number of strains examined in parentheses)

| | | MIC_{50} | MBC_{50} | MIC_{90} | MBC_{90} | Geometric Mean | | |
						MIC	MBC	MBC/MIC
Pefloxacin	(34)	0.31	0.32	1.16	1.8	0.45	0.52	1.15
Ciprofloxacin	(29)	0.2	0.24	0.47	0.85	0.31	0.37	1.19
Ofloxacin	(35)	0.24	0.29	1.47	1.8	0.38	0.46	1.21
Lomefloxacin	(34)	0.33	0.38	0.9	0.92	0.45	0.51	1.13
Temafloxacin	(34)	<0.12	<0.12	0.48	0.485	0.18	0.2	1.11
Difloxacin	(35)	<0.12	<0.12	0.9	0.95	0.197	0.24	1.21

antibiotic therapy. Cross-resistance may develop by mutations that result in an altered DNA gyrase, reduced permeability, or both. Precise mechanisms of resistance in *Acinetobacter* spp. remain to be analysed, although DNA gyrase has been shown to exist in all bacterial species examined to date (Piddock and Wise, 1989). Mutations occur in *Ps. aeruginosa, H. influenzae* and *C. freundii* which give rise to a DNA gyrase A sub-unit that is less susceptible to inhibition by quinolones. The *gyrA* mutations result in decreased susceptibility to all quinolones, but not cross-resistance to chemically unrelated antibiotics. There have also been several reports of decreased uptake of quinolones causing reduced susceptibility, and permeability mutants have been described that show cross-resistance to ß-lactams, e.g. *norB* mutations in *E. coli* and *Ps. aeruginosa* with presumed *nalB* mutations (Piddock and Wise, 1989). It is possible to select for cross-resistance to all fluoroquinolones by in-vitro serial passage of organisms in the presence of increasing concentrations of the drugs (Neu, 1988). Repeated transfers may lead to increases in MICs from 0.12 to 16 mg/l for *Ps. aeruginosa* and several Enterobacteriaceae, with concomitant increases in the MICs of ß-lactams. This suggests that resistance is due to alteration of the bacterial outer membrane and to permeability factors. In our experience with *Acinetobacter*, combined resistance to all ß-lactams (83%), all aminoglycosides (42%) and quinolones (30%) is occurring with an increasing incidence (35% in 1989) (Muller-Serieys et al., 1989). Although not yet proven, it is a reasonable assumption that diminished permeability in *Acinetobacter*, as described for imipenem resistance, might also be an acceptable explanation for certain types of multiple resistance, including resistance to fluoroquinolones, in some *Acinetobacter* strains. Moreover, among these strains were isolates with high levels of resistance to penicillins and cephalosporins, but without any evidence for ß-lactamase activity, which again suggests the involvement of diminished outer membrane permeability (Joly-Guillou et al., 1988). Simultaneous resistance to ß-lactams and quinolones in these strains also suggests the possible involvement of a *gyrB* mutation (Piddock and Wise, 1989). This hypothesis remains to be investigated.

GENETIC DETERMINANTS OF RESISTANCE IN *ACINETOBACTER* SPP

Available information on plasmids and transposons in *Acinetobacter* spp. has been reviewed in detail elsewhere (Towner, this volume). So far as the genetic determinants of resistance are concerned, a limited number of investigations have provided information on plasmid-mediated resistance. Although *Acinetobacter* spp. can easily acquire plasmids from *E. coli* (Towner and Vivian, 1976; Chopade et al., 1985), the opposite transfer seems to be a rare event (Gomez-Lus et al., 1980; Goldstein et al., 1983). Plasmid patterns of nosocomial *Acinetobacter* strains were first analysed by Murray and Moellering (1980) who demonstrated, in a study of aminoglycoside-modifying enzymes, that loss of resistance in the strains

Table 9. Plasmids conferring resistance in *Acinetobacter* spp.

Plasmid content of *Acinetobacter* strains	Molecular size	Resistance pattern	Ref.
pUZ12[a]	-	Ap, Km, Sm, Su.	(1)
pUZ13[a]	-	Ap, Tl, Gm, Si, Cm, Sm, Su.	(1)
not named	5 Md	Gm, Km, Sm, Ne, Sp.	(2)
not named	16 Md	Gm, Sm, Su, Cm, Tc.	(3)
pIP1031	167 kb	Ap, Cm, Km, Sm, Su, Tp.	(4)
pIP1841	63 kb	Ne, Km, Ak.	(5)

Ap, ampicillin; Km, kanamycin; Sm, streptomycin; Su, sulphonamides; Tl, ticarcillin; Gm, gentamicin; Si, sisomycin; Cm, chloramphenicol; Ne, neomycin; Sp, spectinomycin; Ak, amikacin; Tc, tetracycline; Tp, trimethoprim.

[a]member of the P incompatibility group

References: (1) Gomez-Lus et al., 1980; (2) Murray and Moellering, 1980; (3) Devaud et al., 1982; (4) Goldstein et al., 1983; (5) Lambert et al., 1988.

studied was associated with the physical loss of a plasmid; however, no transfer of resistance, either by filter matings or transformation, was achieved, and it was not shown conclusively that the plasmid carried the genes encoding resistance.   Nevertheless, despite the inability to demonstrate transfer of any of these markers, loss of resistance to kanamycin, gentamicin, sulphadiazine, tetracycline and carbenicillin, suggested a plasmid location for this genetic information. In another study, a large self transferable plasmid (pIP1031), belonging to incompatibility group C(6) and encoding a TEM-1 ß-lactamase and two aminoglycoside modifying enzymes, (AAD(3")(9)) and APH(3')(5")-I), was transferred from a clinical *Acinetobacter* strain to *E. coli* by conjugation at an extremely low transfer frequency (5 x 10^{-10} - Goldstein et al., 1983). The multiresistance of an epidemic strain described by Devaud et al. (1982) could not be transferred to any recipient by various mating procedures, but mobilisation of resistance to various drugs could be achieved after the conjugative plasmid RP4 was transferred into the epidemic strain. These authors also showed that mobilisation resulted from transposition of a 16 Md DNA sequence from the *Acinetobacter* chromosome on to plasmid RP4. It was suggested that a plasmid conferring resistance to multiple antibiotics (gentamicin, sisomicin, kanamycin, neomycin, streptomycin, chloramphenicol, tetracycline, ampicillin and cephalothin) had originally been transferred from the hospital flora into *Acinetobacter* spp., and that the DNA sequence conferring multiple resistance had transposed and stably integrated into the *Acinetobacter* chromosome. More recently, in a study investigating the mechanism of amikacin resistance in *Acinetobacter* spp. (Lambert et al., 1988), the gene conferring resistance to kanamycin and amikacin in a clinical strain isolated from a urine sample was found to be encoded by a 63 kb plasmid, pIP1841. This plasmid was self-transferable to other *Acinetobacter* strains (*A. baumannii, A. haemolyticus* and *A. lwoffii*), but not to *E. coli*. The authors suggested that the transfer barrier resulted from a defect in conjugation, plasmid replication, or both, in the recipient bacteria (*E. coli*). The same authors (Lambert et al., 1990) showed subsequently that resistance to gentamicin, netilmicin, streptomycin, sulphonamide, ticarcillin and tobramycin could also be transferred to strains of *A. haemolyticus* and *A. lwoffii*. Hybridisation experiments using an *aphA*-6 probe (the structural gene for the APH(3′)VI enzyme - Martin et al., 1988) and 14 selected *Acinetobacter* isolates demonstrated that various size fragments hybridised with the probe, suggesting that the resistance determinants were located in different genomic environments. Thus the authors demonstrated that the *aphA*-6 gene was involved in the dissemination of amikacin resistance in *Acinetobacter* spp. responsible for nosocomial infections, and that spread of this resistance was due to different plasmids carrying this epidemic gene. Various other plasmids conferring resistance or other phenotypic properties have also been studied in *Acinetobacter* (Towner and Vivian, 1976; Towner, 1983; Chopade et al., 1985); their transfer properties and behaviour in *A. calcoaceticus* are described elsewhere (Towner, this volume). Data on naturally-occurring plasmids conferring resistance in *Acinetobacter* spp. are gathered together in Table 9.

CONCLUSIONS

Although a certain amount of work has been done on resistance mechanisms in *Acinetobacter* spp., many investigations remain to be carried out, especially with regard to new antimicrobial agents and the emergence of resistance to imipenem and the new fluoroquinolones. Problems of outer membrane permeability are suspected, but not yet demonstrated. The evolutionary processes which result in the emergence of new mechanisms of resistance following the therapeutic introduction of new antibiotics render the task increasingly complicated. This article has tried to summarise current knowledge on resistance patterns and mechanisms of resistance in *Acinetobacter* spp. It is clear that epidemic spread of genes conferring antimicrobial resistance in hospitals will require careful analysis of plasmid and transposon behaviour in *Acinetobacter* spp. in order to determine the possible role of this organism as a reservoir or vehicle of transmissible drug resistance in the hospital environment.

ACKNOWLEDGEMENTS

We are grateful to Eric Vallée for his contribution to the study of ß-lactamase inhibitors, to K.J. Towner for communication of his publications and data on plasmids in *Acinetobacter* spp., to T. Lambert and colleagues for communication of their most recent data on aminoglycoside resistance, and to Marie-Jeanne Julliard for careful preparation of the manuscript.

REFERENCES

Bauernfeind, A., 1986, Classification of beta-lactamases, *Rev. Infect. Dis.*, 8 (Suppl. 5):5470.

Bauernfeind, A., Hörl, G., and Mönch, V., 1986, Changes in microbial ecology by therapeutic use of aminoglycosides, *Scand. J. Infect. Dis.*, Suppl.49:106.

Bergogne-Bérézin, E., and Joly-Guillou, M.L., 1985, An underestimated nosocomial pathogen, *Acinetobacter calcoaceticus, J. Antimicrob. Chemother.*, 16:535.

Bergogne-Bérézin, E., and Joly-Guillou, M.L., 1986, Comparative activity of imipenem, ceftazidime and cefotaxime against *Acinetobacter calcoaceticus, J. Antimicrob. Chemother.*, 18(Suppl.E):35.

Bergogne-Bérézin, E., and Joly-Guillou, M.L., 1989, Activity of new quinolones against Acinetobacter from nosocomial infections, *Quinolones Bull.*, 5:23.

Bergogne-Bérézin, E., Zechovsky, N., Piechaud, M., Vieu, J.F., and Bordini, A., 1971, Sensibilité aux antibiotiques de 240 souches de *"Moraxella"* oxydase-négative (Acinetobacter) isolées d'infections hospitalières. Evolution sur 3 ans, *Path. Biol.*, 19:981.

Bergogne-Bérézin E., Joly, M.L., Moreau, N., and Le Goffic, F., 1980, Aminoglycoside modifying-enzymes in clinical isolates of *Acinetobacter calcoaceticus, Curr. Microbiol.*, 4:361.

Bergogne-Bérézin, E., Joly-Guillou, M.L., and Vieu, J.F., 1987, Epidemiology of nosocomial infections due to *Acinetobacter calcoaceticus, J. Hosp. Infect.*, 10:105.

Chopade, B.A., Wise, P.J., and Towner, K.J., 1985, Plasmid transfer and behaviour in *Acinetobacter calcoaceticus* EBF65/65, *J. Gen. Microbiol.*, 131:2805.

Chow, A.W., Wong, J., and Bartlett, K.H., 1988, Synergistic interactions of ciprofloxacin and extended spectrum beta-lactams or aminoglycosides against *Acinetobacter calcoaceticus* ss. *anitratus*, *Diagn. Microbiol. Infect. Dis.*, 9:213.

Devaud, M., Kayser, F.H., and Bachi, B., 1982, Transposon-mediated multiple antibiotic resistance in *Acinetobacter* strains, *Antimicrob. Agents Chemother.*, 22:323.

Dowding, J.E., 1979, A novel aminoglycoside modifying-enzyme from a clinical isolate of *Acinetobacter*, *J. Gen. Microbiol.*, 110:239.

Finland, M., 1979, Emergence of antibiotic resistance in hospital 1935-1975, *Rev. Infect. Dis.*, 1:4.

French, G.L., Casewell, M.W., Roncoroni, A.J., Knight, S., and Phillips, I., 1980, A hospital outbreak of antibiotic resistant *Acinetobacter anitratus*: epidemiology and control, *J. Hosp. Infect.*, 1:125.

Freney, J., Bouvet, P.J.M., and Tixier, C., 1989, Identification et détérmination de la sensibilité aux antibiotiques de 31 souches cliniques d'*Acinetobacter* autres que *A. baumannii*, *Ann. Biol. Clin.*, 47:41.

Garcia, I., Fainstein, V., Leblanc, B., and Bodey, G.P., 1983, *In vitro* activities of new beta-lactam antibiotics against *Acinetobacter* spp., *Antimicrob. Agents Chemother.*, 24:297.

Glew, R.H., Moellering, R.C., and Kunz, L.J., 1977, Infections with *Acinetobacter calcoaceticus (Herellea vaginicola)*. Clinical and laboratory studies, *Medicine*, 56:79.

Godineau-Gauthey, N., Lesage, D., Tessier, F., Kollia, D., and Daguet, G.L., 1988, *Acinetobacter calcoaceticus* variété *anitratus* ou *Acinetobacter baumannii*. Etude de la sensibilité aux antibiotiques de 65 souches hospitalières, *Med. Mal. Infect.*, 2bis:124.

Goldstein, G.W., Labigne-Roussel, A., Gerbaud, G., Carlier, C., Collatz, E., and Courvalin, P., 1983, Transferable plasmid mediated antibiotic resistance in Acinetobacter, *Plasmid*, 10:138.

Gomez-Lus, R., Larrad, L., Rubio-Calvo, M.C., Navarro, M., and Asierra, M.P., 1980, AAC(3) and AAC(6') enzymes produced by R plasmids isolated in general hospital, *in* "Antibiotic Resistance," p.295, S.Mitsushashi, L. Rosival and V. Kremery, eds, Springer-Verlag, Berlin.

Hancock, R.E.W., 1984, Alterations in outer membrane permeability, *Ann. Rev. Microbiol.*, 38:237.

Holton, J., 1982, A report of a further hospital outbreak caused by a multiresistant *Acinetobacter anitratus*, *J. Hosp. Infect.*, 3:305.

Hood, J., and Amyes, S.G.B., 1989, A novel method for the identification and distinction of the beta-lactamases of the genus *Acinetobacter*, *J. Appl. Bacteriol.*, 67:157.

Jacobs, M.R., Aronoff, S.C., Johenning, S., Shlaes, D.M., and Yamabe, S., 1986, Comparative activities of the beta-lactamase inhibitors YTR 830, clavulanate and sulbactam combined with ampicillin and broad spectrum penicillins against defined beta-lactamase-producing aerobic gram-negative bacilli, *Antimicrob. Agents Chemother.*, 29:980.

Joly-Guillou, M.L., and Bergogne-Bérézin, E., 1985a, Etude comparative *in vitro* de l'activité de cinq quinolones vis à vis d'*Acinetobacter calcoaceticus*, *Path. Biol.*, 33:416.

Joly-Guillou, M.L., and Bergogne-Bérézin, E., 1985b, Evolution d'*Acinetobacter calcoaceticus*, en milieu hospitalier, de 1971 à 1984, *Press. Med.*, 14:2331.

Joly-Guillou, M.L., and Bergogne-Bérézin, E., 1990, Présence d'une beta-lactamase à spectre élargi chez *Acinetobacter baumanii*, *Press. Med.*, 19:672.

Joly-Guillou, M.L., Bergogne-Bérézin, E., and Moreau, N., 1987, Enzymatic resistance to beta-lactams and aminoglycosides in *Acinetobacter calcoaceticus*, *J. Antimicrob. Chemother.*, 20:773.

Joly-Guillou, M.L., Bergogne-Bérézin, E., and Philippon, A., 1988, Distribution of beta-lactamases and phenotype analysis in clinical strains of *Acinetobacter calcoaceticus*, *J. Antimicrob. Chemother.*, 22:597.

King, A., and Phillips, I., 1986, The comparative *in vitro* activity of eight newer quinolones and nalidixic acid, *J. Antimicrob. Chemother.*, 18(Suppl.D):1.

Kitzis, M.D., Goldstein, F.W., Labia, P., and Acar, J.F., 1983, Activité du sulbactam et de l'acide clavulanique, seuls et associés, sur *Acinetobacter calcoaceticus*, *Ann. Microbiol. (Inst.Past.)*, 134A:163.

Knothe, H., and Dette, G.A., 1983, The current state of cephalosporin antibiotics : microbiological aspects, *Infection*, 11(Suppl.1):S12.

Labia, R., and Barthelemy, M., 1979, L'enzymogramme des beta-lactamases: adaptation en gel de la méthode iodométrique, *Ann. Microbiol. (Inst.Past.)*, 130B:295.

Lambert, T., Gerbaud, G., and Courvalin, P., 1988, Transferable amikacin resistance in *Acinetobacter* spp. due to a new type of 3′-aminoglycoside phosphotransferase, *Antimicrob. Agents Chemother.*, 32:15.

Lambert, T., Gerbaud, G., Bouvet, P., Vieu, J.F., and Courvalin, P., 1990, Dissemination of amikacin resistance in *Acinetobacter* spp. due to synthesis of a 3′-aminoglycoside phosphotransferase type VI, *Antimicrob.Agents Chemother.*, in press.

Loeffelholz, M.J., and Modrzankowski, M.C., 1986, Plasmid RP1-mediated susceptibility of *Acinetobacter calcoaceticus* to rat polymorphonuclear leukocyte granule contents, *Infect. Immun.*, 54:705.

Leoffelholz, M.J., and Modrzakowski, M.C., 1987, Outer membrane permeability of *Acinetobacter calcoaceticus* mediates susceptibility to rat polymorphonuclear leukocyte granule contents, *Infect. Immun.*, 55:2296.

McGowan, J.E., 1983, Antimicrobial resistance in hospital organisms and its relation to antibiotic use, *Rev. Infect. Dis.*, 5:1033.

Martin, P., Jullien, E., and Courvalin, P., 1988, Nucleotide sequence of *Acinetobacter baumannii* aphA-6 gene : evolutionary and functional implications of sequence homologies with nucleotide-binding proteins, kinases and other aminoglycoside-modifying enzymes, *Mol. Microbiol.*, 2:615.

Matthew, M., and Harris, A.M., 1976, Identification of beta-lactamases by analytical isoelectrofocusing; correlation with bacterial taxonomy, *J. Gen. Microbiol.*, 94:55.

Mayer, K.H., and Zinner, S.H., 1985, Bacterial pathogens of increasing sig nificance in hospital-acquired infections, *Rev. Infect. Dis.*, 7(Suppl.3):S371.

Medeiros, A.A., 1984, Beta-lactamases, *Br. Med. Bull.*, 40:18.

Morohoshi, T., and Saito, T., 1977, Beta-lactamase and beta-lactam antibiotic resistance in *Acinetobacter anitratus* (syn. *A. calcoaceticus*), *J. Antibiot.*, 30:969.

Muller-Serieys, C., Lesquoy, J.B., Perez, E., Fichelle, A., Bourjeois, B., Joly-Guillou, M.L., and Bergogne-Bérézin, E., 1989, Infections nosocomiales à Acinetobacter : épidémiologie et difficultés thérapeutiques, *Press. Med.*, 18:107.

Murray, B.E., and Moellering, R.C., 1980, Evidence of plasmid-mediated production of aminoglycoside-modifying enzyme not previously described in *Acinetobacter, Antimicrob. Agents Chemother.*, 17:30.

Nayler, J .H.C., 1987, Resistance to beta-lactams in gram negative bacteria: relative contributions of beta-lactamase and permeability limitations, *J. Antimicrob. Chemother.*, 19:713.

Neu, H.C., 1988, Bacterial resistance to fluoroquinolones, *Rev. Infect. Dis.*, 10 (Suppl.1):S57.

Obana, Y., Nishino, T., and Tanino, T., 1985, *In vitro* and *in vivo* activities of antimicrobial agents against *Acinetobacter calcoaceticus*, *J. Antimicrob. Chemother.* 15:441.

Paul, G., Joly-Guillou, M.L., Bergogne-Bérézin, E., Nevot, P., and Philippon, A., 1989, Novel carbenicillin-hydrolyzing beta-lactamase (CARB-5) from *Acinetobacter calcoaceticus* var. *anitratus*, *FEMS Microbiol. Lett.*, 59:49.

Philippon, A.M., Paul, G.C., and Nevot, P.A., 1980, Synergy of clavulanic acid with penicillins against ampicillin and carbenicillin-resistant gram-negative organisms, related to their type of beta-lactamase, *Curr. Chemother. Immunother., Proc. 11th ICC & 19th ICAAC*, 327.

Phillips, I., King, A., Shannon, K., and Warren, C., 1981, SQ 26.776: *in vitro* antibacterial activity and susceptibility to beta-lactamases, *J. Antimicrob. Chemother.*, 8(Suppl.E):103.

Piddock, L.J.V., and Wise, R., 1985, Newer mechanisms of resistance to beta-lactam antibiotics in gram-negative bacteria, *J. Antimicrob. Chemother.*, 16:279.

Piddock, L.J.V., and Wise, R., 1989, Mechanisms of resistance to quinolones and clinical perspectives, *J. Antimicrob. Chemother.*, 23:475.

Pinter, M., and Kantor, M., 1971, Quantitative antibiotic sensitivity pattern of *Acinetobacter* strains, *Ant. v. Leeuw.*, 37:197.

Quinn, J.P., Studemeister, A.E., Divincenzon C.A., and Lerner, S.A., 1988, Resistance to imipenem in *Pseudomonas aeruginosa*: clinical experience and biochemical mechanisms, *Rev. Infect. Dis.*, 10:892.

Retailliau, H.F., Hightower, A.W., Dixon, R.E., and Allen, J.R., 1979, *Acinetobacter calcoaceticus*: a nosocomial pathogen with an unusual seasonal pattern, *J. Infect. Dis.*, 139:371.

Rolston, K.V.I., Ho, D.H., Leblanc, B., Gooch, G., and Bodey, G.P., 1988, Comparative *in vitro* activity of the new difluoroquinolone temafloxacin (A-62254) against bacterial isolates from cancer patients, *Eur. J. Clin. Microbiol. Infect. Dis.*, 7:684.

Rudin, M.L., Michael, J.R., and Huxley, E.J., 1979, Community-acquired Acinetobacter pneumonia, *Amer. J. Med.*, 67:39.

Sanders, C.C., 1983, Novel resistance selected by the new expanded-spectrum cephalosporins: a concern, *J. Infect. Dis.*, 147:585.

Sawai, T., Kanno, M. and Tsukamoto, K., 1982, Characterization of eight beta-lactamases of gram-negative bacteria, *J. Bacteriol.*, 152:567.

Shannon, K.P., King, A., and Phillips, I., 1986, Beta-lactamases with high activity against imipenem and SCH 34343 from *Aeromonas hydrophila*, *J. Antimicrob. Chemother.*, 17:45.

Sirot, J., Chanal, C., Petit, A., Sirot, D., Labia, R., and Gerbaud, G., 1988, *Klebsiella pneumoniae* and other Enterobacteriaceae producing novel plasmid-mediated beta-lactamases markedly active against third generation cephalosporins: epidemiologic studies, *Rev. Infect. Dis.*, 10:850.

Towner, K.J., 1983, Transposon-directed mutagenesis and chromosome mobilization in *Acinetobacter calcoaceticus* EBF65/65, *Genet. Res.*, 41:97.

Towner, K.J., and Vivian, A., 1976, RP4-mediated conjugation in *Acinetobacter calcoaceticus*, *J. Gen. Microbiol.*, 93:355.

Uwaydah, M., and Taql-Eddin, A.R., 1976, Susceptibility of nonfermentative gram-negative bacilli to tobramycin, *J. Infect. Dis.*, 134(Suppl.):S28.

Vila, J., Almela, M., and Jimenez de Anta, M.T., 1989, Laboratory investigation of hospital outbreak caused by two different multiresistant *Acinetobacter calcoaceticus* subsp. *anitratus* strains, *J. Clin. Microbiol.*, 27:1086.

Wiedeman, B., Kliebe, E., and Kresken, M., 1989, The epidemiology of beta-lactamases, *J. Antimicrob. Chemother.*, 24(Suppl.B):1.

Wise, R., 1982, Beta-lactamase inhibitors, *J. Antimicrob. Chemother.*, 9(Suppl. B):31.

THE CHROMOSOMAL ß-LACTAMASES OF THE GENUS

ACINETOBACTER: ENZYMES WHICH CHALLENGE OUR

IMAGINATION

J. Hood[a] and S.G.B. Amyes[b]

[a]University Department of Bacteriology
Royal Infirmary
Glasgow G4 OSF, UK

[b]Department of Bacteriology, The Medical School
University of Edinburgh
Edinburgh EH8 9AG, UK

INTRODUCTION

At first sight it would appear that the properties of the presumed chromosomal ß-lactamases of the genus *Acinetobacter* have been fully characterised. Isoelectric points (pI) in polyacrylamide gel systems were first described by Matthew and Harris (1976) and reiterated by Sykes and Matthew (1976). Other workers since then have described the pIs of *Acinetobacter* ß-lactamases as ranging from 7.5 to >10 (Medeiros et al., 1985); >8 (Joly-Guillou et al., 1987) and 9.9 (Hikida et al., 1989). Similarly, the molecular mass (M_r) of *Acinetobacter* ß-lactamase has been described as 30,000 by Morohoshi and Saito (1977) and as 38,000 by Hikida et al. (1989).

The kinetics of hydrolysis to various ß-lactam antibiotics have been most fully described by Morohoshi and Saito (1977) and, subsequently, to newer agents by Hikida et al. (1989). These authors also examined the effect of a variety of ß-lactamase inhibitors. There is little doubt, however, from these data that the *Acinetobacter* ß-lactamases are cephalosporinases (Sykes and Matthew, 1976; Morohoshi and Saito, 1977; Medeiros, 1984; Bauernfeind, 1986; Neu, 1986; Joly-Guillou et al., 1987; Hikida et al., 1989).

The Biology of Acinetobacter, Edited by K. J. Towner *et al.*
Plenum Press, New York, 1991

The inducibility of these cephalosporinases by specific ß-lactam inducers has been far from clear. They have been described as being inducible by Morohoshi and Saito (1977) and quoted as inducible by Sykes and Matthew (1976), Medeiros (1984), Neu (1986) and Hikida et al. (1989). Bauernfeind (1986) classified them as inducible or constitutive, whereas Joly-Guillou et al. (1988) stated that ß-lactamase inducibility in Acinetobacter had not yet been clearly demonstrated. On the basis of the work by Morohoshi and Saito (1977), Bush (1989) classified the *Acinetobacter* chromosomal ß-lactamase in Group One, i.e. a cephalosporinase not inhibited by clavulanic acid (CEP-N). Subsequent substrate profile and inhibition studies by Hikida et al. (1989) suggested that this classification was correct.

This article presents evidence that the true character of the chromosomal ß-lactamase of the genus *Acinetobacter* is not as simple as has been proposed. Indeed there appears to be considerable heterogeneity, with at least four distinct ß-lactamases produced by this genus.

ISOELECTRIC POINTS ON POLYACRYLAMIDE ELECTROPHORESIS

Matthew and Harris (1976) listed pIs of around 8.6 for two ß-lactamases produced by strains of *Acinetobacter: A. inotti* 1786E and *Acinetobacter* sp. 1787E. These same two strains appear in Sykes and Matthew (1976) as *A. inotti* (1786E) pI 8.6 and *A. mallei* (1787E) pI 8.7. The latter strain *(A. mallei)* 1787E was not an *Acinetobacter* at all, but a *Pseudomonas mallei*. This organism had somewhat controversially been placed in the genus *Acinetobacter* by Cowan and Steel (1965) - it became *P. mallei* in subsequent editions (Cowan and Steel, 1974). Therefore, this enzyme can no longer be considered as one produced by the genus *Acinetobacter*.

The former strain *(A. inotti)* 1786E is equally interesting. An extensive taxonomic literature search has failed to find any previous or subsequent mention of the specific epithet 'inotti'. This strain is part of the Glaxo Research Laboratories' collection, and one of the original authors (Miss A.M. Harris) kindly sent us strain 1786E for further study. The freeze-dried vial was still marked as *A. inotti*, but it was identified by the API 20NE (API System. S.A. France) as *A. lwoffii* (new specific epithet: *A. junii*). On reflection, we suggest that 'lwoffi', when written, could easily have been wrongly transcribed to 'inotti' and perhaps this error took place somewhere between its original entry in the strain book and the subsequent publication.

Medeiros et al. (1985) were the first to mention that not all of the chromosomal ß-lactamases of *Acinetobacter* focused on conventional polyacrylamide isoelectric focusing (IEF) systems. Of 14 strains of *A. calcoaceticus*, the pI was stated to be 8.8 for one enzyme, 9.0 for one, 9.7 for one, 10.0 for four, > 10 for three and a 'blur' for four. Joly-Guillou et al.

(1987) described 30 strains of *Acinetobacter* with a chromosomal cephalosporinase of pI >8. This again suggests poor focusing of these enzymes on conventional polyacrylamide IEF systems. More recently, Hikida et al. (1989) described a pI of 9.9 for a purified cephalosporinase from *A. calcoaceticus* (ML 4961). However, these authors employed broad range ampholines of pH 3.5-10.0 in their IEF gel. This suggests that the enzyme had migrated to, or almost to, the cathode. Unfortunately, there was no photograph of the IEF gel.

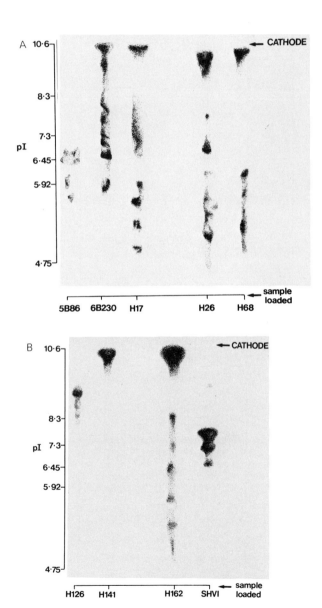

Fig. 1. Conventional polyacrylamide IEF gels of *Acinetobacter* chromosomal ß-lactamases and the plasmid ß-lactamase SHV-1. Enzyme activity identified with nitrocephin

Table 1. Properties of eight aztreonam-resistant strains of Acinetobacter

Acinetobacter strains				β-lactamase properties		
Strain No.	sub species	new species	Specimen type	Az^{a}MIC	M$_r^b$	ACE type^c
5B86	var. lwoffii	A. lwoffii	blood	16	>1,000,000	4
6B230	var. anitratus	A. junii	blood	32	>1,000,000	1
H17	var. anitratus	A. baumannii	urine	64	640,000	1
H26	var. anitratus	A. baumannii	sinus swab	32	>1,000,000	1
H68	var. anitratus	A. baumannii	urine	64	>1,000,000	1
H126	var. lwoffii	A. junii	wound swab	>256	32,500	3
H141	var. anitratus	A. baumannii	wound swab	64	>1,000,000	1
H162	var. anitratus	A. baumannii	wound swab	64	60,500	2

^{a}Az = aztreonam MIC in mg/l

^{b}M$_r$ = molecular mass

^{c}ACE = Acinetobacter chromosomal enzyme

We have studied eight partially purified ß-lactamases of *Acinetobacter,* obtained from aztreonam-resistant strains collected in the Royal Infirmary, Edinburgh (Table 1; Hood and Amyes, 1989). These ß-lactamases were applied to a conventional IEF system (Fig. 1), but seven of these enzymes did not focus in this system. Only the ß-lactamase from strain H126 *(A. lwoffii,* now *A. junii)* focused, with a pI of about 9.1. We postulated that the failure of these enzymes to focus resulted from two factors. Firstly, the basic nature of the enzyme (high pI), and secondly, their high molecular masses (see Table 1 and subsequent discussion on molecular masses). An agarose, urea and sorbitol (AUS) gel system was therefore devised for isoelectric focusing of these enzymes. This combination allowed the focusing of these basic enzymes (Fig. 2) and their classification into four groups, termed ACE-1 to ACE-4. ß-Lactamases from five of the eight strains studied were placed in the ACE-1 group. Four of these ACE-1 enzymes were derived from *A. baumannii* - the most prevalent *Acinetobacter* species isolated from clinical specimens (Bouvet and Grimont, 1987). A complete description of the AUS gel technique and further explanation of the IEF patterns has been published elsewhere (Hood and Amyes, 1989). Further confirmation of these groupings has been obtained from preliminary studies employing two additional techniques. Crude enzyme preparations from all eight strains were applied to either a MONO S or a MONO Q high performance ion-exchange column connected to a Fast Protein Liquid Chromatography (FPLC$^{(R)}$) system (Pharmacia, Uppsala, Sweden).

The FPLC method has been described fully elsewhere (Payne et al., 1990). Peaks of ß-lactamase activity obtained on the columns were detected by the nitrocephin spot test (Table 2) and demonstrated the clear differences between ACE-1, ACE-3 and ACE-4, while suggesting that ACE-2 is similar to ACE-1. In addition, the results suggest that there may be two subgroups within the ACE-1 group: one with a retention volume of 15-18 ml, the other with a retention volume of 12-14 ml. However, further work is required to verify these sub-groupings.

The purified enzymes obtained from the FPLC/ion exchange were also applied to a gel system on the PhastSystem$^{(R)}$ (Pharmacia, Uppsala, Sweden). This gel system was devised to resolve basic proteins on a polyacrylamide (PAGE) minigel. The exact sample preparation and running conditions were those described by Olsson and Tooke (1988). A Phastgel Homogeneous 20 (i.e. sodium dodecyl sulphate-free) was run with the reversed polarity electrodes. After electrophoresis, the enzymes were visualised by nitrocephin solution (500 mg/l). ACE-1 (e.g. H141) and ACE-2 (e.g. H162) preparations migrated similar distances from the cathode, whereas ACE-3 (H126) migrated slightly less far. Insufficient ACE-4 (5B86) was available to test on this system. Control preparations of the plasmid-mediated ß-lactamases SHV-1/TEM-1 and OXA-2/TEM-2 migrated to appropriate positions in this gel system, i.e. TEM-1 and TEM-2 stayed at the cathode, while SHV-1 and OXA-2 migrated to a point just behind the ACE enzymes. Further experimentation employing this technique is required. If it proves to be reproducible with crude ß-lactamase preparations, it would be an

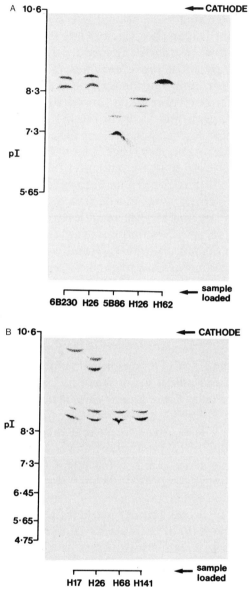

Fig. 2. Agarose, urea and sorbitol (AUS) IEF gels. Focused bands of *Acinetobacter* chromosomal ß-lactamase activity identified with nitrocephin

A. 6B230 = ACE-1
    H26 = ACE-1
   5B86 = ACE-4
  H126 = ACE-3
  H162 = ACE-2

B. H17 = ACE-1
  H26 = ACE-1
  H68 = ACE-1
H141 = ACE-1

Table 2. FPLC$^{(R)}$ of crude enzyme preparations on high performance ion exchange

Strain no.	Species	ACE type	Retention volume of major activity (ml)
Mono S (cation exchange) in 50 mM phosphate buffer pH7			
6B230	A. junii	1	15-16
H17	A. baumannii	1	17-18
H26	A. baumannii	1	18
H162	A. baumannii	2	15-16
H68	A. baumannii	1	12-13
H141	A. baumannii	1	13-14
Mono Q (anion exchange) in 25 mM Tris/HCl buffer pH8			
H126	A. junii	3	13
5B86	A. lwoffii	4	29

extremely useful rapid technique for the study of ß-lactamases obtained from clinical isolates of *Acinetobacter*.

MOLECULAR MASS ESTIMATIONS

Morohoshi and Saito (1977) gave the M_r of their cephalosporinase from *A. anitratum* NCTC 7844 as 30,000 when measured by gel filtration on a Sephadex G-75 column. Hikida et al. (1989) gave the M_r of their cephalosporinase from *A. calcoaceticus* ML4961 as 38,000. They estimated this M_r by SDS gel electrophoresis after purification of the enzyme with strong cation exchange and gel filtration.

We estimated the M_r of the ß-lactamases in our eight strains of *Acinetobacter* by gel filtration, initially on Sephadex G-75 and then, if required, on Sephacryl S-300 (Hood and Amyes, 1989). The results are listed in Table 1. One enzyme (ACE-3 from strain H126) had a similar M_r (35,000) to that described by Morohoshi and Saito (1977) and Hikida et al. (1989). The ACE-2 ß-lactamase from strain H162 had an M_r of 60,500, but the other six enzymes ranged from 640,000 to > 1,000,000. The results suggested that some of the ß-lactamases of *Acinetobacter* are either very large, or seem large because they are linked to other cell envelope components and the enzyme purification techniques used were incapable of removing these components. This raises the exciting hypothesis that these enzymes could be an intermediate step between penicillin binding proteins and the ß-lactamases. Alternatively, these enzymes may exist in multiple subunit form. Further experiments are being undertaken to examine these possibilities.

BIOCHEMISTRY

The specific activities of the eight ACE enzymes were measured for a selection of cephalosporins, penicillins and a monobactam (Table 3). The enzyme preparations were partially purified by gel filtration (Hood and Amyes, 1989). The spectrophotometric assay employed was that of O'Callaghan et al.(1969) and was carried out on a Pye-Unicam SP1800 UV/VIS spectrophotometer. Cephalosporins and the monobactam were used at concentrations of either 100 μM or 1000 μM (Table 3) and the penicillins at a concentration of 1000 μM. The specific activities of the eight ACE enzymes (Table 3) show clearly that these enzymes are cephalosporinases, although most of them also have some activity against the penicillins tested. The apparent detection of weak activity against cefuroxime in half the strains and weak activity against ampicillin and carbenicillin in one strain may be dismissed since the spectrophotometer was working at the limit of its sensitivity and, therefore, activity against these substrates may also be possible in the other strains. The only clear difference was the specific activity of ACE-4 obtained from strain 5B86, which was considerably lower than the others. Interestingly, ACE-1 from strain 6B230 and ACE-3 from H126 seemed similar - they are the only *A. junii* strains in the study.

Table 3. Specific activities[a] of ACE enzymes against a selection of cephalosporins and a monobactam (100 µM or 1000 µM) and penicillins (100 µM)

Strain no.	ACE type	AZ[b]	CAZ[b]	CTX[b]	CXM[b]	CER	CED	PENG	AMP	CARB
5B86	4	-	-	-	-	25	-	-	-	-
6B230	1	-	-	-	0.014	330	56	2.8	-	-
H17	1	-	-	-	-	93	69	4.7	-	-
H26	1	-	-	-	-	93	33	6.7	-	-
H68	1	-	-	-	0.13	31	19	6.0	0.7	0.46
H126	3	-	-	-	-	230	46	84	-	-
H141	1	-	-	-	0.16	49	17	4.7	-	-
H162	2	-	-	-	0.12	93	17	5.8	-	-

[a] µmoles of substrate hydrolysed/min/mg protein

[b] specific activities also measured with high substrate concentration (1000 µM)

- no detectable activity

AZ, aztreonam; CAZ, ceftazidime; CTX, cefotaxime; CXM, cefuroxime; CER, cephaloridine; CED, cephradine; PENG, penicillin; AMP, ampicillin; CARB, carbenicillin.

125

Table 4. MIC values (mg/1) for ACE-producing strains of a selection of cephalosporins, penicillins, and a monobactam

Strain no.	ACE type	AZ	CAZ	CTX	CXM	CER	CED	PENG	AMP	CARB
5B86	4	16	2	4	8	8	16	16	8	16
6B230	1	32	8	4	16	64	128	>32	128	64
H17	1	64	4	32	64	64	256	>32	32	32
H26	1	32	8	32	64	256	>256	>32	64	32
H68	1	64	8	16	32	64	256	>32	32	32
H126	3	>256	64	64	64	128	256	>32	128	128
H141	1	64	8	16	64	64	256	>32	64	32
H162	2	64	8	32	64	64	256	>32	32	32

Abbreviations as in Table 3.

Table 4 shows the MIC values of the substrates used in Table 3 for these strains. The results showed that all the *Acinetobacter* strains studied had a high degree of resistance to penicillins and first, second and third generation cephalosporins. There was some correlation between the resistance levels and the enzyme produced. All the strains encoding the ACE-1 enzyme had very similar MICs of all the drugs tested. H162, which encoded ACE-2, could not be distinguished from the ACE-1 strains on its resistance profile. However, strain H126, which encoded ACE-3, was generally more resistant to the drugs tested than the ACE-1 or ACE-2 producers. Specifically, H126 was more resistant to the third generation cephalosporins and the monobactam, aztreonam. On the other hand, strain 5B86, which encoded ACE-4, was generally less resistant than the other ACE producers. It was particularly less resistant to the penicillins and the first generation cephalosporins tested.

Table 5 shows the kinetics of hydrolysis for each of the enzymes, i.e. K_m values with nitrocephin and cephaloridine as substrates. This value was obtained by measuring the rate of hydrolysis at limiting substrate concentrations and Lineweaver-Burk plots. The K_m values obtained with both nitrocephin and cephaloridine were broadly similar to each other (Table 5). The K_m values to cephaloridine ranged from 150 μM to 710 μM and were similar to the values found by Morohoshi and Saito (1977) and Hikida et al. (1989) of 250 μM and 511 μM respectively.

These results show that all four ACE enzymes have moderate affinity for cephaloridine and nitrocephin. All the K_m values were within the same order of magnitude and could not convincingly be distinguished from one

Table 5.   Kinetic data for ACE enzymes

Strain no.	ACE type	$K_m{}^a(\mu m)$ Nitrocephin	$K_m{}^a(\mu M)$ Cephaloridine
5B86	4	220	710
6B230	1	280	150
H17	1	550	160
H26	1	250	150
H68	1	280	180
H126	3	150	380
H141	1	220	210
H162	2	250	260

[a]K_m, this value was obtained by measuring rates of hydrolysis at limiting substrate concentrations and determined by Lineweaver-Burk plots

Table 6. ID_{50}[a] of ACE enzymes with nitrocephin as substrate (μM)

Strain no.	ACE type	Aztreonam	Cloxacillin	Clavulanate
5B86	4	>100	0.18	>100
6B230	1	28	0.012	>100
H17	1	8	0.005	>100
H26	1	5	0.022	>100
H68	1	9	0.1	>100
H126	3	0.08	0.003	>100
H141	1	1.8	0.02	>100
H162	2	23	0.022	>100

[a]ID_{50} = amount of inhibitor required for 50% inhibition of nitrocephin hydrolysis

another. Thus none of the ACE enzymes demonstrated greater substrate affinity than the others and K_m values were a poor discriminator.

Table 6 shows the effect of the ß-lactamase inhibitors aztreonam, cloxacillin and clavulanic acid, expressed as the concentration required for 50% inhibition (ID_{50}) of enzyme activity. The enzymes from all the strains were readily inhibited by cloxacillin. Most were inhibited within the range of 0.003-0.022 μM. However, two enzymes required at least 0.1 μM cloxacillin for 50% inhibition. One of these enzymes was ACE-1, but this level of cloxacillin is still considered to be very low. Thus cloxacillin inhibition also seems to be a poor discriminator of the ACE enzymes, although a characteristic feature of them all is that they are cloxacillin sensitive. Similarly, the enzymes from all the strains were extremely resistant to clavulanic acid inhibition. The inability of any of these enzymes to be inhibited significantly by 0.1 mM clavulanic acid shows that clavulanic acid resistance is a characteristic feature of all these enzymes, but is incapable of being used as a discriminator.

Differences in inhibition were, however, identified when aztreonam was used as an inhibitor. There was little difference in the ID_{50} for the ACE-1 and ACE-2 enzymes, which were in the range of 1.8 to 28 μM (Table 6). The ACE-3 enzyme from strain H126 was much more sensitive to aztreonam inhibition than the others (ID_{50} = 0.08 μM), while the ACE-4 enzyme from strain 5B86 was much more resistant (ID_{50} = >100 μM). We believe that the inhibition profiles with aztreonam provide good discrimination of the ACE-3 and ACE-4 enzymes.

Table 7 shows the effect of EDTA, $HgCl_2$ and pCMB. All the enzymes were inhibited to the same extent by a fixed concentration of $HgCl_2$, but were

Table 7. Effect of inhibitors and metal ions on ACE enzymes with nitrocephin as substrate

| Strain no. | ACE type | Percentage inhibition | | |
		EDTA[a]	HgCl$_2$[b]	pCMB[c]
5B86	4	<20	89	0
6B230	1	<20	98	0
H17	1	<20	96	0
H26	1	<20	97	0
H68	1	<20	95	0
H126	3	<20	95	0
H141	1	<20	98	0
H162	2	<20	98	0

[a] 1 mM EDTA (inhibition variable: see text)
[b] 1 mM HgCl$_2$
[c] 0.1 mM p-chloromercuribenzoate

virtually unaffected by moderately high levels of pCMB. Thus no discrimination of these enzymes could be obtained from these profiles.

The effect of 1 mM EDTA on the enzymes with either nitrocephin or cephaloridine as the substrate was variable, although generally below 20%. Hikida et al. (1989) found that EDTA had no effect on their enzyme's activity. Since many of the enzymes described here are of large M$_r$ and, therefore, may exist in subunit form, it is possible that metal ions play a role in their function. This may explain the variable effects of EDTA on their activities. However, further experiments with more purified enzyme will be required to elucidate these findings.

The above biochemical data (EDTA apart) place all of these ß-lactamases firmly in Bush Group One (Bush, 1989).

INDUCTION EXPERIMENTS

Morohoshi and Saito (1977) described how the enzyme activities of five strains of A. anitratum "increased by about five to ten-fold" following treatment with 100 μg of benzylpenicillin, cephaloridine or 6-aminopenicillanic acid for 1 h at 37°C. Hikida et al. (1989) quote Morohoshi and Saito (1977) as having shown that Acinetobacter cephalosporinases are inducible, but they did not carry out induction experiments themselves.

We have carried out induction experiments with all eight strains, employing cefoxitin as the inducer at one quarter the MIC value for the culture. The method employed was that described by Minami et al. (1980). No discernible ß-lactamase induction was found with any of the strains studied.   In discussions with D. Livermore and C.C. Sanders (personal communications) it was stated that evidence of induction had not yet been found in any strain of *Acinetobacter* tested. This includes the use of cefoxitin, imipenem and penicillin as inducers. Much more convincing evidence would therefore be required before it can be claimed that these enzymes are inducible, since current evidence suggests they are not.

ROLE OF PRESUMED CHROMOSOMAL ß-LACTAMASES OF ACINETOBACTER IN ß-LACTAM RESISTANCE

Although it seems likely that these enzymes are largely responsible for the observed resistance to ß-lactam drugs, it is also probable that other factors, e.g. altered permeability or alterations in the penicillin binding proteins,may be equally important in some strains. The ultimate test must involve the cloning of the ß-lactamase gene from each of these strains into a suitable recipient, thereby allowing further study of the resistance mechanism. This would also enable their place on the bacterial chromosome, rather than on a plasmid, to be confirmed.

CONCLUSIONS

As one would expect from a genus that is very heterogeneous, with at least 17 genospecies (Grimont and Bouvet, this volume), the presumed chromosomal ß-lactamases are also heterogenous. At least four different ACE enzymes have been described in *Acinetobacter* on the basis of IEF in AUS gels, molecular mass and biochemical properties. As has been outlined above, these are very basic proteins with pIs probably greater than 10. So basic, in fact, that it is difficult to explain how the cell might secrete them into the periplasmic space; thus they may be bound to the cell membrane. As a group, they generally do not focus well, or at all, in conventional polyacrylamide IEF systems. It is possible that further different chromosomal enzymes exist, since we describe ACE-1 to ACE-4 in only three of the possible 17 genospecies, i.e. in *A. baumannii, A. junii* and *A. lwoffii.*

Further work is needed to explain the large M_r of the enzymes found in six of the eight strains. This may then shed light on the possible evolutionary place of these interesting enzymes. Future work on the eight enzymes should include further purification, perhaps by FPLC. This would allow closer comparison with the biochemical kinetic data obtained by Hikida et al. (1989) for their enzyme. It would be interesting to know the exact genospecies of both the Hikida and Morohoshi strains. It is clear, however, that the enzymes described in this article, together with those described by

Morohoshi and Saito (1977) and Hikida et al. (1989), are all cephalosporinases, not inhibited by clavulanic acid, and can be placed in Bush Group One (CEP-N).

ACKNOWLEDGEMENTS

We would like to thank E.R. Squibb and Sons Ltd. for financial support for some of this work; Mr R. Paton for technical assistance; Miss A.M. Harris of Glaxo Research Ltd. for kindly providing us with the strain of *A. inotti;* Miss Michaela Torrance of Pharmacia for the use of the FPLC system and the editor of the Journal of Applied Bacteriology for allowing us to reproduce some of our previous data.

REFERENCES

Bauernfeind, A., 1986, Classification of ß-lactamases, *Rev. Infect. Dis.*, 8 (suppl.5):S470.

Bouvet, P.J.M., and Grimont, P.A.D., 1987, Identification and biotyping of clinical isolates of *Acinetobacter, Ann. Microbiol.*, 138:569.

Bush, K., 1989, Classification of ß-lactamases: Groups 1, 2a, 2b and 2b¹, *Antimicrob. Agents Chemother.*, 33:264.

Cowan, S.T., and Steel, K.J., 1965, "Identification of Medical Bacteria," Cambridge University Press, Cambridge.

Cowan, S.T., and Steel, K.J., 1974, "Manual for the Identification of Medical Bacteria, 2nd edn.," Cambridge University Press, Cambridge.

Hikida, M., Yoshida, M., Mitsuhashi, S., and Inoue, M., 1989, Purification and properties of a cephalosporinase from *Acinetobacter calcoaceticus, J. Antibiot.*, 42:123.

Hood, J., and Amyes, S.G.B., 1989, A novel method for the identification and distinction of the ß-lactamases of the genus *Acinetobacter, J. Appl. Bacteriol.*, 67:157.

Joly-Guillou, M.L., Bergogne-Bérézin, E., and Moreau, N., 1987, Enzymatic resistance to ß-lactams and aminoglycosides in *Acinetobacter calcoaceticus, J. Antimicrob. Chemother.*, 20:773.

Joly-Guillou, M.L., Vallée, E., Bergogne-Bérézin, E., and Phillipon, A., 1988, Distribution of ß-lactamases and phenotype analysis in clinical strains of *Acinetobacter calcoaceticus, J. Antimicrob. Chemother.*, 22:597.

Matthew, M., and Harris, A.M., 1976, Identification of ß-lactamases by analytical isoelectric focusing: correlation with bacterial taxonomy, *J. Gen. Microbiol.*, 94:55.

Medeiros, A.A., 1984, ß-lactamases, *Br. Med. Bull.*, 40:18.

Medeiros, A.A., Hare, R., Papa, E., Adam, C., and Miller, G.H., 1985, Gram negative bacilli resistant to third-generation cephalosporins: ß-lactamase characterisation and susceptibility to Sch 34343, *J. Antimicrob. Chemother.*, 15(suppl.C):119.

Minami, S., Yotsuji, A., Inoue, M., and Mitsuhashi, S., 1980, Induction of ß-lactamase by various ß-lactam antibiotics in *Enterobacter cloacae, Antimicrob. Agents Chemother.*, 18:382.

Morohoshi, T., and Saito, T., 1977, ß-Lactamase and ß-lactam antibiotics resistance in *Acinetobacter anitratum, J. Antibiot.*, 30:969.

Neu, H.C., 1986, Antibiotic inactivating enzymes and bacterial resistance, *in* "Antibiotics in Laboratory Medicine," p.757, V. Lorian, ed., Williams and Wilkins, Baltimore.

O'Callaghan, C.H., Muggleton, P.W., and Ross, G.W., 1969, Effects of ß-lactamase from Gram negative organisms on cephalosporins and penicillins, *Antimicrob. Agents Chemother.*, 1968:57.

Olsson, I., and Tooke, N.E., 1988, PAGE of basic proteins, *in* "PhastSystem^R Application File No. 300," Pharmacia, Uppsala.

Payne, D.J., Hood, J., Marriott, M.S., and Amyes, S.G.B., 1990, Separation of plasmid mediated broad spectrum ß-lactamases by Fast Protein Liquid Chromatography (FPLC^R), *FEMS Microbiol. Lett.*, 69:195.

Sykes, R.B., and Matthew, M., 1976, The ß-lactamases of Gram negative bacteria and their role in resistance to ß-lactam antibiotics, *J. Antimicrob. Chemother.*, 2:115.

THE USE OF MOLECULAR TECHNIQUES FOR THE LOCATION

AND CHARACTERISATION OF ANTIBIOTIC RESISTANCE GENES

IN CLINICAL ISOLATES OF *ACINETOBACTER*

B.G. Elisha and L.M. Steyn

Department of Medical Microbiology
University of Cape Town & Groote Schuur Hospital
Observatory, 7925
South Africa

INTRODUCTION

Nosocomial isolates of *Acinetobacter*, particularly *A.baumannii*, are often highly resistant to a variety of antibiotics (Bergogne-Bérézin et al., 1987). Although much is known about the biochemical mechanisms of antibiotic resistance found in *Acinetobacter* (Bergogne-Bérézin and Joly-Guillou, this volume), little is known about the structure and regulation of the expression of these genes in *Acinetobacter* at the molecular level.

One of the reasons why there is a paucity of data on the primary structure of resistance genes in *Acinetobacter* is the difficulty encountered in locating these genes by conventional gene transfer experiments (Vivian et al., 1981; Chopade et al., 1985). For the same reason, data on the precise genetic location of these genes (i.e. plasmid, chromosome, or both) is often lacking (Murray and Moellering, 1980; Devaud et al., 1982; Goldstein et al., 1983; Divers et al., 1985). This article describes how molecular techniques can be used to locate and characterise antibiotic resistance genes in clinical isolates of *Acinetobacter*. In order to illustrate the use of these techniques, their application in characterising and determining the genetic location of chloramphenicol and aminoglycoside (gentamicin and tobramycin) resistance genes is described in detail. To facilitate the detection of these genes we first constructed two DNA probes: a 2"-O-adenylyltransferase (AAD(2")) gene probe and a chloramphenicol acetyltransferase (CATI) gene probe. The AAD(2") gene was chosen because it is the only enzyme with activity against gentamicin and tobramycin that has been demonstrated in local *Acinetobacter* isolates (unpublished observations). The construction and use of these probes to locate the corresponding resistance genes in a multi-resistant clinical isolate of *Acinetobacter* (strain SAK) is described. Subsequent cloning experiments not only confirmed the presence of these genes, but also revealed another aminoglycoside resistance gene with activity against gentamicin and tobramycin in strain SAK; this was further analysed using DNA sequencing techniques. Thus the studies described in this article not only illustrate the application of molecular techniques to the study of antibiotic resistance in *Acinetobacter*, but also serve to demonstrate the complex nature of the problems which may be encountered in multi-resistant clinical isolates of this genus.

CONSTRUCTION OF DNA PROBES

CAT Probe

The plasmid pA10cat2 (Rosenthal et al., 1983) contains the gene encoding CATI. The structural CATI gene is 650 bp long, of which the first 520 bp was used as a probe. To generate this fragment, pA10cat2 was digested with *Hin*dIII and *Nco*I, the fragments were separated by agarose gel electrophoresis and the 520 bp fragment was eluted from the gel (Seth, 1984). The eluted DNA was radio-labelled with ^{32}P-dCTP by nick translation.

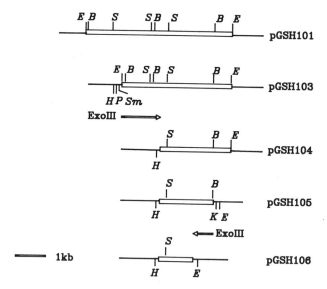

Fig.1. Construction of the AAD(2") gene probe. ⊏⊐ , insert DNA; ——— , vector DNA. B, *Bam*HI; E, *Eco*RI; K, *Kpn*I; P, *Pst*I; S, *Sac*I; Sm, *Sma*I; *Exo*III, exonuclease III. The vector restriction sites are shown below the bar. The *Bam*HI sites were mapped in pGSH103 and are therefore inferred in pGSH101.

The gentamicin/tobramycin resistance gene, encoding AAD(2"), was isolated from the multi-resistance plasmid, JR66 (Benveniste and Davies, 1971). JR66 DNA was cut with *Eco*RI and the restriction fragments cloned in the *Eco*RI site of pBR322. Gentamicin-resistant transformants were selected; all were resistant also to tobramycin and harboured an identical recombinant plasmid, designated pGSH101, which contained a 5.3 kb insert (Fig. 1). A *Sac*I fragment of 1.8 kb was deleted from pGSH101 by a partial *Sac*I digestion followed by re-ligation (forming pGSH102). The 3.5 kb *Eco*RI insert from pGSH102 was subcloned in pUC18 to form pGSH103. This subcloning was a prerequisite for controlled progressive deletion with exonuclease III (Henikoff, 1984) to reduce the size of the insert further while still retaining a functional AAD(2")gene. For the exonuclease III digestion, pGSH103 was cut with *Pst*I to generate a 3'-cohesive end adjacent to the vector DNA, while *Sma*I was used to generate an exonuclease III susceptible end adjacent to the insert. No gentamicin/tobramycin-resistant transformants were obtained when more than 1.2 kb of the insert DNA was deleted. One of the smallest gentamicin-resistant plasmids (pGSH104) contained an insert of 2.3 kb, from which a *Bam*HI/*Hin*dIII fragment (1.8 kb) was subcloned into pUC18 to form pGSH105. A second digestion with exonuclease III was performed to delete insert DNA at the opposite end of the AAD(2") gene. *Bam*HI was used to generate a susceptible site, while *Kpn*I was used to protect the vector; 0.7 kb of insert DNA was deleted before loss of AAD(2") activity occurred. The size of the reduced DNA insert containing the functional AAD(2") gene was 1.1 kb (pGSH106). This insert was radio-labelled by nick translation and used as a probe.

GENE LOCATION

Preparation of DNA for Hybridisation Experiments

DNA was isolated as described by Clewell and Helinski (1969). Plasmid DNA was purified by ultra-centrifugation in a CsCl/EtBr density gradient. The less-dense DNA band in the gradient was also isolated and designated chromosomal DNA. Plasmid and chromosomal DNA was filtered on to Hybond-N membranes in a dot-blot apparatus. When required, DNA was transferred from agarose gels to Hybond-N by the method of Southern (1975). Hybridisations were done as described by Johnson et al. (1984).

Hybridisation with the CAT Probe

The CAT probe was used to detect the CAT structural gene, and to determine its genetic location in the multi-resistant strain SAK. The probe gave no hybridisation signal with a sensitive strain of *Acinetobacter* (YON), but generated a signal with pA10cat2, the source of the probe fragment. The probe hybridised to both chromosomal and plasmid DNA preparations from SAK (Fig. 2).

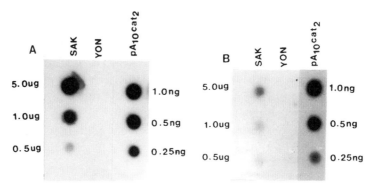

Fig.2. Chromosomal and plasmid DNA dot blots hybridised with the CAT gene probe. (A) Chromosomal DNA, (B) Plasmid DNA. Autoradiography was for 72 h.

Hybridisation with the AAD(2") Probe

The probe for the AAD(2") structural gene hybridised to JR66, the source of the AAD(2") gene. There was no hybridisation with DNA from a sensitive strain; however, a signal was obtained from the chromosomal DNA preparations (Fig. 3), but not from the plasmid preparations of strain SAK.

CLONING AND CHARACTERISATION OF THE ANTIBIOTIC RESISTANCE GENES

On the basis of the above results it was concluded that strain SAK contained the resistance genes encoding CAT and AAD(2"). Further support for this hypothesis was sought from cloning experiments.

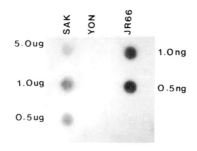

Fig.3. Chromosomal DNA dot-blot hybridised with the AAD(2") gene probe. Autoradiography was for 18 h.

Cloning of the CAT Gene

Chromosomal DNA was digested with either *Bam*HI, *Eco*RI, *Hind*III or *Pst*I and cloned into the respective sites in pUC18. *E. coli* transformants resistant to chloramphenicol were only obtained from the *Hind*III fragments. Of six recombinant plasmids, each of which contained an insert of 11.0 kb and hybridised to the CAT probe, one was selected for further manipulations. The insert was reduced in size by subcloning a 1.8 kb *Pst*I fragment which hybridised to the CAT probe into pUC18. This recombinant plasmid (pGSH201) carried a functional CAT gene.

A similar procedure was used to isolate the CAT gene from plasmid DNA. All of the recombinant plasmids contained a 5.0 kb *Hind*III insert and encoded resistance to chloramphenicol. Hybridisation experiments demonstrated that this 5.0 kb fragment was present in a 50-60 kb plasmid harboured by strain SAK (Fig. 4). None of the chloramphenicol-resistant recombinant plasmids derived from strain SAK encoded aminoglycoside resistance.

Identification of CAT Modifying Enzymes

Assays for CAT activity were performed as described by Gorman et al. (1982). Crude lysates prepared from strain SAK and *E. coli* (pGSH201) acetylated ^{14}C-chloramphenicol to its *O*-acetoxy derivatives (Fig. 5), thereby confirming the presence of a CAT gene in the *Acinetobacter* strain.

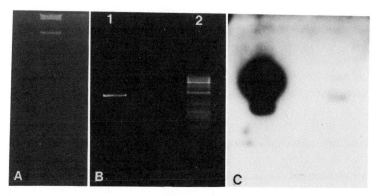

Fig.4. Plasmid DNA hybridised to the CAT gene probe. (A) Plasmid profile of strain SAK following electrophoresis on a 0.8% agarose gel for 16 h at 20 V (1 V/cm). (B) Lane 1: gel-purified 5.0 kb *Hind*III fragment which contains the CAT gene cloned from plasmid DNA; lane 2: *Hind*III digest of plasmid DNA. A 0.8% agarose gel was electrophoresed for 2 h at 70 V (7 V/cm). (C) Autoradiograph of DNA shown in (B) hybridised with CAT gene probe.

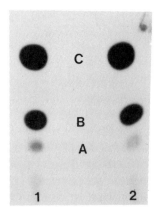

Fig.5. Assay of CAT activity. Lane 1: extract from strain SAK; lane 2:extract from *E.coli* (pGSH201). (A) 1-acetate chloramphenicol; (B) 3-acetate chloramphenicol; (C) 1,3-diacetate chloramphenicol.

Cloning of Aminoglycoside Resistance Genes

The partial library prepared above was also screened for the presence of gentamicin resistance genes. Recombinant plasmids encoding gentamicin resistance were obtained from all the digests. Five recombinants were screened with the AAD(2") probe: four gave strong hybridisation signals, but one, which contained a 5.8 kb *Hin*dIII insert, did not hybridise with the probe.

The recombinants which hybridised to the probe contained a 2.3 kb *Hin*dIII insert. This was reduced by subcloning a 1.9 kb *Bam*HI/*Hin*dIII fragment into pUC18. This recombinant plasmid was designated pGSH108 and still conferred gentamicin resistance. The lone recombinant which did not hybridise with the AAD(2") probe contained an insert of 5.8 kb. By digestion with exonuclease III, the size of the insert was reduced to 0.8 kb without altering the expression of gentamicin resistance. This recombinant plasmid was designated pGSH110. When the insert from pGSH110 was used to probe plasmid DNA prepared from strain SAK, a strong hybridisation signal was obtained (Fig. 6). It was possible to clone this plasmid gene on a 5.8 kb *Hin*dIII DNA fragment, but it is important to note that the restriction map of this 5.8 kb *Hin*dIII fragment was not similar to the 5.8 kb *Hin*dIII fragment cloned from the chromosomal DNA. *Pst*I digested the former recombinant plasmid into three fragments of 6.5 kb, 1.1 kb and 0.8 kb respectively, while the recombinant plasmid containing the plasmid-derived insert was digested into fragments of 3.8 kb, 3.5 kb and 1.1 kb. This insert was shown to be present in a 50-60 kb plasmid harboured by strain SAK (Fig. 6). Thus the 50-60 kb plasmid carries chloramphenicol and gentamicin resistance genes which are not closely linked, since the recombinant plasmids carry either chloramphenicol or gentamicin resistance, but not both.

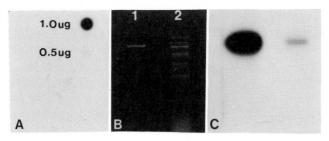

Fig.6. Plasmid DNA from strain SAK hybridised to the AAC(3) gene probe. (A) Plasmid DNA dot blot. (B) Lane 1: gel-purified 5.8 kb HindIII fragment containing the AAC(3) gene cloned from plasmid DNA; lane 2: HindIII digest of plasmid DNA. A 0.8% agarose gel was electrophoresed for 2 h at 7 V/cm; (C) Autoradiograph of DNA shown in (B) hybridised with the AAC(3) gene probe.

Identification of Aminoglycoside Modifying Enzymes

To identify the aminoglycoside-modifying enzyme encoded by pGSH110, and to confirm that pGSH108 encodes an adenylyltransferase, aminoglycoside modifying enzymes were assayed by the phosphocellulose paper binding assay (Shannon and Phillips, 1983). Crude lysates were prepared from strain SAK and from E.coli cells harbouring the recombinant plasmids. Sisomicin, a gentamicin derivitive, was used as a substrate in the screening of cell sonicates for AAC and AAD activity, while ^{14}C-acetyl coenzyme-A and ^{14}C-ATP were used as co-substrates in the respective enzymatic reactions. If the amount of radioactivity bound to the paper in the presence of sisomicin was double the background bound in the absence of sisomicin, it was assumed that the sisomicin had been modified. No modifying activity could be detected in lysates from strain SAK (Table 1); adenylyltransferase activity was detected in lysates prepared from E.coli (pGSH108) since there was an 83-fold increase in the ^{14}C-ATP counts bound in the presence of sisomicin. In contrast, the enzyme encoded by pGSH110 was shown to be an acetyltransferase as, in this case, there was a 107-fold increase in the radioactivity bound from ^{14}C-acetyl coenzyme-A. This enzyme was subsequently identified as an AAC(3) by Bristol Laboratories (personal communication). Thus strain SAK contains two resistance genes, AAD(2") and AAC(3), which have activity against gentamicin and tobramycin. The readily detectable AAD(2") and AAC(3) activity in the lysates prepared from E.coli (pGSH108) and E.coli (pGSH110) respectively is probably due to the high copy number of the plasmids and the concomitant high AAD(2") and AAC(3) gene dose.

Nucleotide Sequence of the AAC(3)

There are seven sub-classes of AAC(3): I, Ia, II, III, IV, V and VII (Vliegenthart et al., 1989). The sequences of AAC(3)I (Tenover et al.,

Table 1. Assay of Aminoglycoside-Modifying Enzymes.
Results are Expressed in Terms of cpm Bound to Phosphocellulose Paper

i) Assay of O-Adenylyltransferase activity with $[^{14}C]$-ATP as co-substrate

Bacterial strain	Substrate		Ratio
	None	Sisomicin	
SAK[a]	3291	3505	1.1
E.coli DK1 (pGSH106)	339	5432	16.00
E.coli DK1 (pGSH108)	133	11096	83.00
E.coli DK1 (pGSH110)	101	128	1.3

ii) Assay of N-Acetyltransferase activity with $[^{14}C]$-Acetyl-CoA as co-substrate

Bacterial strain	Substrate		Ratio
	None	Sisomicin	
SAK	612	713	1.2
E.coli DK1 (pGSH106)	2298	2129	0.9
E.coli DK1(pGSH108)	1823	1866	1.0
E.coli HB101 (pGSH110)	1024	9643	107

[a] This assay was performed with a radioactive substrate of higher specific activity than the other assays.

1989), AAC(3)II (Vliegenthart et al., 1989), AAC(3)III (Allmansberger et al., 1985) and AAC(3)IV (Brau et al., 1984) have been published. It is important to note that the published sequences for AAC(3)II and AAC(3)III are identical. AAC(3)I has been previously described in *Acinetobacter*, but to the best of our knowledge, none of the other sub-classes have been described in this organism.

To facilitate the classification of the AAC(3) derived from strain SAK, its corresponding gene was sequenced. Overlapping restriction fragments of the 0.8 kb DNA insert were subcloned into M13mp18 and M13mp19. The sequence of the insert was then determined by the dideoxy-chain termination

```
                                  -52  CAP                    -10           SD              -1
-80                                                                                                    SphI
GTCCTCCACCTCGGTCAAGCAGCGAGGAGGTAGTGATGATATCTCGATAGGATAGTGTTTCCAGTTTAGCAGATATCCGCA

                                         39
ATG CAT ACG CGG AAG GCA ATA ACG GAG GCA ATT CGA AAA CTC GGA GTC CAA ACC GGT GAC CTG TTG ATG GTC CAT GCC   78
MET His Thr Arg Lys Ala Ile Thr Glu Ala Ile Arg Lys Leu Gly Val Gln Thr Gly Asp Leu Leu Met Val His Ala

                                        117
TCA CTT AAA GCG ATT GGT CCG GTC GAA GGA GCG ACG GAG GTC GTT GCC TCC GCG TTA CGC TCC GCG GTT GGG CCG ACT   156
Ser Leu Lys Ala Ile Gly Pro Val Glu Gly Ala Thr Glu Val Val Ala Ser Ala Leu Arg Ser Ala Val Gly Pro Thr

                                        195
GGC ACT GTG GGA TAC GCG ATG TGG GAC TCA CGA TCA CCC TAC GAG GAG ACT CTG AAT GCC CCT CGG TTG GAT GAC AAA   234
Gly Thr Val Gly Tyr Ala Met Trp Asp Ser Arg Ser Pro Tyr Glu Glu Thr Leu Asn Gly Ala Arg Leu Asp Asp Lys

                                        273
GCC CGC CGT ACC TGG CCG CCG TTC GAT CCC GCA ACG GGG ACT TAC CGT GGC TTC GGG CTG CTG AAT CAA TTT CTG        312
Ala Arg Arg Thr Trp Pro Pro Phe Asp Pro Ala Thr Gly Thr Tyr Arg Gly Phe Gly Leu Leu Asn Gln Phe Leu

                                        351
GTT CAA GCC GCC CCG GCG AGC GGG CAC CCC GAT GCA TCG GTC ATG GGT CCG CTA GCT GAA ACG CTG                    390
Val Gln Ala Ala Pro Ala Ser Gly His Pro Asp Ala Ser Val Met Ala Val Gly Pro Leu Ala Glu Thr Leu

                               G        429
ACG GAG CCT CAC GAA CTC GGT CCG CCG TTG GGG AAA TCG GGG TCG CTC CCC GTC GAG GGG AAG GCC                    468
Thr Glu Pro His Glu Leu Gly Arg Pro Leu Gly Lys Ser Gly Ser Leu Pro Val Glu Gly Lys Ala

                                        507
CTG CTG TTG GGT GCG CCG CTA AAC TCC GTT ACC CAC TAC TAC GAG GCG GTT GCC GAT ATC CCC AAC AAA CGA            546
Leu Leu Leu Gly Ala Pro Leu Asn Ser Val Thr His Tyr Tyr Glu Ala Val Ala Asp Ile Pro Asn Lys Arg

                                        585
TGG GTG ACG TAT GAG ATG CCG ATG CTT GGA GAA GTC GCC TGG GCC TCA GAA TAC GAT TCA AAC                        624
Trp Val Thr Tyr Glu Met Pro Met Leu Gly Glu Val Ala Trp Ala Ser Glu Tyr Asp Ser Asn

                                        663
GGC ATT CTC GAT TGC TTT GCT GAA GGA AAG CCG GTC GAA GCT ATA GCA AAT GCC TAC CTG AAG CTC GGT                702
Gly Ile Leu Asp Cys Phe Ala Glu Gly Lys Pro Val Glu Ala Ile Ala Asn Ala Tyr Leu Lys Gly

                                        741
CGA CAT CGA GAA GGT GTC GGC GTT TTT GCT CAG TGC TAC TTC GAC CAG GCA CAC GCC CGC GTC ACC                    780
Arg His Arg Glu Gly Val Gly Val Phe Ala Gln Cys Tyr Leu Phe Asp Gln Ala His Ala Arg Val Thr

                                        819
TAT CTT GAG AAG CAC TTC GGA GCC ACT CCG ATC GTG CCA GCA CAC CAC GCC CAG CGC GGT TGC CCT TCC GGT            858
Tyr Leu Glu Lys His Phe Gly Ala Thr Pro Ile Val Pro Ala His His Ala Gln Arg Gly Cys Pro Ser Gly

TAG
***

SalI
AGGCCGTCGAC
```

Fig. 7. Nucleotide sequence and derived amino acid sequence of the AAC(3) gene. The -10 region, ribosome binding site (SD) and putative CAP binding site are underlined. The stop codon is indicated by ***. The G to A transition at position 425 is shown. The SphI site and the SalI site confining the gene fragment used as the probe are indicated.

method (Sanger et al., 1977), using ^{35}S-ATP and Sequenase (US Biochemicals). Internal oligonucleotide primers were used when necessary. The DNA sequence shown in Fig. 7 was confirmed by sequencing both strands of DNA. This sequence was compared with the published AAC(3) sequences; it showed 99% similarity with the DNA sequence of AAC(3)II, also designated AAC(3)III. The only difference in the structural gene sequence is a transition at position 425 of a G to an A, which results in the incorporation of lysine rather than glutamic acid (Fig. 7). The sequence similarity extends into the regulatory region, including the -10 region. However, unlike the published sequences, the strain SAK AAC(3) sequence has no sequence corresponding to the *E. coli* -35 consensus sequence, TTGACAT (Fig. 8). McClure (1985) suggested that promoters with a poor homology to the consensus sequence in the -35 region are frequently controlled by activator proteins. One such protein is CAP, which binds to cAMP to form an active complex. De Crombrugghe et al. (1984) have shown that 13 out of 17 regulatory sequences which bind the cAMP/CAP complex contain a conserved sequence, TGTGA, at variable distances from the -10 region. A similar sequence, AGTGA, is present in the AAC(3) sequence at -32 (Fig. 8).

EXPRESSION OF AMINOGLYCOSIDE RESISTANCE GENES IN STRAIN SAK

Since aminoglycoside modifying activity was not readily detectable in strain SAK, it was unclear whether the resistance could be attributed to one or both of the resistance genes. To clarify this, RNA was extracted from strain SAK and probed by Northern hybridisation with the AAD(2") probe and the cloned AAC(3) structural gene.

Detection of mRNA Transcripts

Total cellular RNA was extracted from strain SAK with hot phenol as described by Aiba et al. (1981). RNA from cells grown in the presence and absence of gentamicin (5 mg/l) was analysed. In addition, RNA from cells treated with rifampicin (200 mg/l) to block further initiation of transcription was included. RNA (10 μg) was electrophoresed in gels comprising 1% agarose and 35% formaldehyde with MOPS buffer (40 mM MOPS, 10 mM sodium acetate, 1.0 mM EDTA, pH 8.0) and transferred to Hybond-N in 20 x SSPE (1 x SSPE: 1.8 M NaCl, 10 mM Na$_2$HPO$_4$, 1.0 mM EDTA, pH 7.0). Pre-hybridisations and hybridisations were carried out as described by Johnson et al. (1984).

The AAC(3) probe hybridised to a mRNA transcript which remained detectable for more than 60 mins after treatment of the exponentially growing cells with rifampicin (Fig. 9). In contrast, the AAD(2") probe did not hybridise to mRNA from cells grown in the presence or absence of gentamicin, although it did hybridise to RNA prepared from *E. coli* harbouring pGSH108 (Fig. 9). Based on this, it appears that the AAC(3)

```
                                    -175                                                          -140
AAC(3)-III      TGATGCCGTATTTGCAGTACCAGCGTACGGCCCACAGAATGATGTCACGCTGAAAATGCCGGCCTT        AAC(3)-III


                                                                                    -70
                                                                         GTCCTCCACCTCG    AAC(3)(SAK)
                                                                          -35
AAC(3)-II                                                     TCACCACCGACTATTTGCAAC    AAC(3)-II
                                                              :::::::::::::::::::::
AAC(3)-III      TGAATGGGTTCATGTGCAGCTCCATCAGCAAAAGGGGATGATAAGTTATCACCACCGACTATTTGCAAC    AAC(3)-III


                          -36                                                         -1
AAC(3)(SAK)     GTCAAGCAGCGAGAGCTAGAGTGATATCTCGATAGGTATAGTGTTTTGCAGTTTAGAGGAGATATCGCG    AAC(3)(SAK)
                          ::::::::::::::::::::::::::::::::::::::::::::::::::::::::
AAC(3)-II       AGTGCCGGCCGAGAGGTAGAGTGATATCTCGATAGGTATAGTGTTTGCAGTTTAGAGGAGATATCGCG    AAC(3)-II
                ::::::                          ::::::::::::::::::::::::::::::::::::
AAC(3)-III      AGTGCC-------------TCTCGATAGGTATAGTGTTTTGCAGTTTAGAGGAGATATCGCG    AAC(3)-III
```

Fig. 8. Regulatory region of AAC(3) from SAK, AAC(3)II and AAC(3)III. The proposed -35 region of AAC(3)II and AAC(3)III are underlined, as are the almost perfect direct repeats spanning the -10 region in AAC(3) from SAK and AAC(3)II.

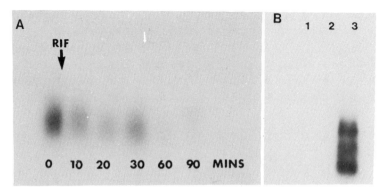

Fig.9. Northern blots hybridised with the AAC(3) and AAD(2") gene probes.
(A) RNA hybridised with the AAC(3) probe; RIF indicates the addition of
rifampicin to the cells; (B) RNA hybridised with the AAD(2") probe; lane
1:RNA from strain SAK; lane 2: RNA from strain SAK grown in the presence
of gentamicin (5 mg/l); lane 3: RNA from *E.coli* (pGSH108) which contains
the cloned AAD(2") gene from strain SAK.

gene alone is responsible for the gentamicin and tobramycin resistance
phenotype. In an attempt to try and "switch on" the AAD(2") gene, antibiotic
selective pressure was applied in the form of kanamycin. Kanamycin was
chosen because this antibiotic is included in the AAD(2"), but not the
AAC(3), substrate profile. Strain SAK (kanamycin MIC of 4 mg/l) was
cultured in media containing kanamycin, the concentration of which was
increased incrementally until strain SAK was resistant to 120 mg
kanamycin/l. Total cellular RNA was extracted from cells grown in the
presence of 120 mg kanamycin/l and probed with the AAD(2") and AAC(3)
DNA probes. Transcripts which hybridised only to the AAC(3) probe were
detected (data not shown).

DISCUSSION

        The aim of this article is to illustrate the use of techniques designed to
elucidate the molecular nature of antibiotic resistance genes in multi-
resistant clinical isolates of *Acinetobacter*. DNA probes for the AAD(2")
and CAT genes were prepared and used to probe chromosomal and plasmid
DNA preparations from a resistant strain of *Acinetobacter*. Both probes
hybridised to chromosomal DNA from the multi-resistant strain SAK, while
the CAT probe hybridised also to plasmid DNA in this strain. Effective
resistance to gentamicin or chloramphenicol could be transformed into
susceptible strains of *E.coli* after cloning strain SAK chromosomal
fragments into pUC18. Antibiotic modifying enzyme assays confirmed the
presence of AAD(2") and CAT activity in extracts prepared from cells
containing the respective recombinant plasmids.

The CAT gene was also shown to be carried on a 50-60 kb plasmid present in strain SAK. A chromosomal and plasmid locus for the CAT gene suggests the possibilty of a transposon. A transposon encoding chloramphenicol, gentamicin, streptomycin and sulphonomide resistance has been identified in *Acinetobacter* by Devaud et al. (1982), and the presence of inverted repeat sequences flanking a kanamycin resistance gene in the *Acinetobacter* plasmid pAV5 (Divers et al., 1985) indicates that this gene may also be part of a transposon. The probe specifies a Type I CAT enzyme that is also encoded by Tn9 and preliminary sequencing data show that the CAT gene derived from strain SAK is on a transposon which has sequence and structural similarity to Tn2670 (Womble and Rownd, 1988).

An additional aminoglycoside resistance gene was cloned from plasmid and chromosomal DNA prepared from strain SAK. This gene was identified as encoding an AAC(3) which, like AAD(2"), has activity against gentamicin and tobramycin. Like the CAT gene, the AAC(3) gene is also contained on a 50-60 kb plasmid harboured by strain SAK. Sequencing data demonstrated almost perfect sequence homology with the published sequences for AAC(3)II (Vliegenthart et al., 1989) and AAC(3)III (Allmansberger et al., 1985). However, the control region was different from both of the published sequences. Although the sequence derived from SAK has a consensus -10 sequence, there is poor similarity with the -35 sequence. A sequence located 12 bp from the -10 region, is, however, similar to the consensus sequence described for cAMP/CAP controlled promoters. Little is known about transcriptional control signals in *Acinetobacter*, so further studies are required to prove conclusively that the AAC(3) has a catabolite sensitive promoter in strain SAK. Catabolite repression of the AAC(3) was not anticipated since, to our knowledge, only two aminoglycoside modifying enzymes, AAD(3") (Harwood and Smith, 1971) and APH(3')II (Davies and Smith, 1978), are known to be subject to catabolite expression. It is well known that CAT is subject to catabolite repression in Gram-negative bacilli. Why the AAC(3) in strain SAK, or indeed any antibiotic resistance gene, should be subject to a form of regulation associated with the availability of energy sources is puzzling. Perhaps, as Shaw (1983) reasons, special metabolic demands are placed on microbes which inhabit ecological niches where antibiotics are found. *Acinetobacter* is a common soil organism and it may be that this control aspect in strain SAK is related to the origins of the organism.

It has been suggested by Prince and Jacoby (1982) that cloning a resistance gene into a high copy number plasmid is useful in elucidating the mechanism of aminoglycoside resistance in bacteria with little or no demonstrable enzyme activity. Although no AAD(2") or AAC(3) activity was detected in the extracts prepared from strain SAK, it was possible to identify these enzymes when the corresponding gene was cloned into a high copy number plasmid in *E.coli*. This study suggests, however, that cloning a functional resistance gene is not sufficient in itself to elucidate the mechanism of resistance. Judged by the cloning experiments alone, strain SAK contains two mechanisms of resistance to gentamicin/tobramycin, but the RNA transcript experiments indicated that only the AAC(3) gene is transcribed, and is therefore responsible for the resistance phenotype. It follows that it is insufficient to simply identify a resistance gene with a DNA probe; the gene must be shown to be expressed in the host strain before a role in the resistance phenotype can be assigned.

Why the AAD(2") is not expressed in strain SAK is not known, but it is possible that transcription of the AAD(2") is repressed by a protein which is absent in *E. coli*. The AAD(2") gene is currently being sequenced to see if there are any DNA sequences that could be involved in the control of transcription of this gene. One other possibility is that the AAC(3) gene is favoured because it has a stronger promoter.

The presence of an antibiotic provides a strong selective pressure which favours the survival of organisms that contain a resistance gene product which inactivates the antibiotic. Even when strain SAK was cultured in the presence of kanamycin, it was not possible to "switch on" the AAD(2") gene; only AAC(3) gene transcripts were detected. The mechanism of resistance to kanamycin in the derivative strains is being investigated. It may be that strain SAK contains yet another aminoglycoside resistance gene, or that the AAC(3) is responsible for the kanamycin resistance, in which case it may have undergone a mutation which broadens its substrate profile to include kanamycin.

Using a combination of DNA probing and cloning experiments it was possible to identify genes encoding CAT, AAD(2") and AAC(3) in a clinical isolate of *A. baumannii*. The gene probes will be used to probe current nosocomial isolates of *Acinetobacter*. It will be interesting to discover whether these later isolates also contain an unexpressed AAD(2") gene. The results described also serve to illustrate the molecular complexity of the resistance mechanisms which may be encountered in these increasingly common multi-resistant nosocomial isolates.

ACKNOWLEDGEMENTS

This work was supported by grants from the University of Cape Town and the Medical Research Council to L.M. Steyn and A.A. Forder. The computer programmes used in the analysis of the DNA sequences were obtained from the Molecular Biology Computer Research Resource, Harvard School of Public Health, Dana-Faber-Cancer Institute - D1154, Boston, MA, 02115.

REFERENCES

Aiba, H., Adhya, S., and De Crombrugge, B., 1981, Evidence for two functional gal promotors in intact *Escherichia coli, J. Biol. Chem.*, 256:11905.

Allmansberger, R., Brau, B., and Piepersberg, W., 1985, Genes for gentamicin-(3)-*N*-acetyltransferase III & IV, *Mol. Gen. Genet.*, 198:514.

Benveniste, R., and Davies, J., 1971, R-factor mediated gentamicin resistance: a new enzyme which modifies aminoglycoside antibiotics, *FEBS Lett.*, 14:293.

Bergogne-Bérézin, E., Joly-Guillou, M.L., and Vieu, J., 1987, Epidemiology of nosocomial infections due to *Acinetobacter calcoaceticus*, *J. Hosp. Infect.*, 10:105.

Brau, B., Pilz, U., and Piepersberg, W., 1984, Genes for gentamicin-(3)-*N*-acetyltransferases III & IV, *Mol. Gen. Genet.*, 193:179.

Chopade B.A., Wise, P.J., and Towner, K.J., 1985, Plasmid transfer and behaviour in *Acinetobacter calcoaceticus* EBF65/65, *J. Gen. Microbiol.*, 131:2805.

Clewell, D.B., and Helinski, D.R., 1969, Supercoiled circular DNA-protein complex in *Escherichia coli*: purification and induced conversion to an open circular DNA form, *Proc. Natl. Acad. Sci. USA*, 62:1159.

Davies, J., and Smith, D., 1978, Plasmid determined resistance to antimicrobial agents, *Ann. Rev. Microbiol.*, 32:469.

De Crombrugghe, B., Busby, S., and Buc, H., 1984, Cyclic AMP receptor protein: Role in transcription activation, *Science*, 224:831.

Devaud, M., Kayser, F.H., and Bachi, B., 1982, Transposon mediated multiple antibiotic resistance in *Acinetobacter* strains, *Antimicrob. Agents Chemother.*, 22:323.

Divers, M., Craven, P., and Vivian, A., 1985, Molecular analysis of an antibiotic resistance plasmid, pAV5, and its derivative plasmids in *Acinetobacter calcoaceticus*, *J. Gen. Microbiol.*, 131:3367.

Goldstein, F.W., Labigne-Roussel, A., Gerbaud, G., Carlier, C., Collatz E., and Courvalin P., 1983, Transferable plasmid-mediated antibiotic resistance in *Acinetobacter calcoaceticus*, *Plasmid*, 10:138.

Gorman, C.M., Moffat, L.F., and Harwood, B.H., 1982, Recombinant genomes which express chloramphenicol acetyltransferases in mammalian cells, *Mol. Cell. Biol.*, 2:1044.

Harwood, J., and Smith, D., 1971, Catabolite repression of chloramphenicol acetyltransferase synthesis in *E. coli* K12, *Biochem. Biophys. Res. Comm.*, 42:57.

Henikoff, S., 1984, Unidirectional digestion with exonuclease III creates targeted breakpoints for DNA sequencing, *Gene*, 28:351.

Johnson, D.A., Gautsch, J.W., Sportsman, J.R., and Elder, J.H., 1984, Improved technique utilizing non-fat dry milk powder for analysis of proteins and nucleic acids transferred to nitrocellulose, *Gene Analyt. Tech.*, 1:3.

McClure, W.R., 1985, Mechanism and control of transcription in prokaryotes. *Ann. Rev. Biochem.*, 54:171.

Murray, B.E., and Moellering, R.C., 1980, Evidence of plasmid-mediated production of aminoglycoside modifying enzymes not previously described in *Acinetobacter calcoaceticus.*, *Antimicrob. Agents Chemother.*, 17:30.

Prince, A.S., and Jacoby, G.A., 1982, Cloning the gentamicin resistance gene from a *Pseudomonas aeruginosa* plasmid in *Escherichia coli* enhances the detection of aminoglycoside modification, *Antimicrob. Agents Chemother.*, 22:525.

Rosenthal, N., Kress, M., Gruss, P., and Khoury, G., 1983, B.K. viral enhancer element and a human cellular homolog, *Science*, 222:749.

Sanger, F., Nicklen, S., and Coulsen, A.R., 1977, DNA sequencing with chain terminating inhibitors, *Proc. Natl. Acad. Sci. USA*, 74:5463.

Seth, A., 1984, A new method for linker ligation, *Gene Analyt. Tech.*, 1:99.

Shannon, K.P., and Phillips, I., 1983, Detection of aminoglycoside-modifying strains of bacteria, *in*: "Antibiotics: Assessment of Antimicrobial Activity and Resistance," p.183, A.D.Russell and L.B.Quesnel, eds., Academic Press, London.

Shaw, W., 1983, Chloramphenicol acetyltransferase: enzymology and molecular biology, *CRC Critical Rev. Biochem.*, 14:1.

Southern, E.M., 1975, Detection of specific sequences among DNA fragments separated by gel electrophoresis, *J. Mol. Biol.*, 98:503.

Tenover, F.C., Phillips, K.L.,Gilbert, T., Lockhart, P., O'Hara, P.J., and Plorde, J., 1989, Development of a DNA probe for the deoxyribonucleotide sequence of a 3-*N*-acetyltransferase (AAC(3)I) resistance gene, *Antimicrob. Agents Chemother.*, 33:551.

Vivian, A., Hinchliffe E., and Fewson, C.A., 1981, *Acinetobacter calcoaceticus:* some approaches to a problem, *J. Hosp. Infect.*, 2:199.

Vliegenthart, J.S., Ketelaar-van Gaalen, P.A.G. and Van De Klundert, J.A.M., 1989, Nucleotide sequence of the *aacC2* gene, a gentamicin resistance determinant involved in a hospital epidemic of multiply resistant members of the family Enterobacteriaceae, *Antimicrob. Agents Chemother.*, 33:1153.

Womble, D.D., and Rownd, R.H., 1988, Genetic and physical map of plasmid NRI: comparison with other IncFII antibiotic resistance plasmids, *Microbiol. Rev.*, 52:433.

PLASMID AND TRANSPOSON BEHAVIOUR IN *ACINETOBACTER*

K.J. Towner

Department of Microbiology & PHLS Laboratory
University Hospital
Nottingham NG7 2UH
UK

INTRODUCTION

Plasmids and transposons seem to be ubiquitous among bacteria and it has been increasingly recognised that these genetic elements play an important role in the overall biology of a wide range of prokaryotic organisms. Central to this role is the location on plasmids and transposons of genes encoding a variety of phenotypic properties, ranging from antibiotic resistance to unusual metabolic activities. In many instances the provision of these properties enables an organism to survive and prosper in an otherwise hostile environment. Transfer mechanisms associated with plasmids, combined with illegitimate recombination events specified by transposable elements, may also allow spread of such advantageous properties between cells of the same or different species.

In addition to these natural processes, the properties of plasmids and transposons can also be exploited in the laboratory to enable a detailed molecular analysis of gene structure and function. Transposons have been widely used as insertion mutagens and can be employed in this manner to identify replicons that have no readily scorable phenotype (e.g. cryptic plasmids). Apart from their involvement in illegitimate recombination events, they can also be used to provide mobile sites of homology upon which homologous recombination can act. On a wider basis, conjugation is still one of the most useful methods for the initial genetic analysis of an organism and seems to be invariably plasmid-mediated. Plasmids also have an essential role to play in the utilisation of in-vitro gene manipulation techniques. Although many such procedures involve the initial cloning of specific genes in an *Escherichia coli* K12 host, studies on gene regulation and expression often require the use of plasmid cloning vectors which are

The Biology of Acinetobacter, Edited by K. J. Towner *et al.*
Plenum Press, New York, 1991

capable of stable replication in the original parent strain. The purpose of this review is to describe the range of plasmids and transposons which can exist in *Acinetobacter*, either naturally or in the laboratory, and the behaviour and properties associated with them.

BEHAVIOUR OF ENTEROBACTERIAL R PLASMIDS AND TRANSPOSONS IN *ACINETOBACTER*

R Plasmids Capable of Transfer to Acinetobacter

Although the presence of extrachromosomal DNA in naturally occurring strains of *Acinetobacter* was reported by Christiansen et al. (1973), most of the early studies on plasmid behaviour in *Acinetobacter* were carried out using antibiotic resistance plasmids (R plasmids) which had originally been isolated from members of the Enterobacteriaceae and previously characterised in *E.coli* K12. Olsen and Shipley (1973) reported that the broad host range P1 incompatibility group plasmid R1822 (later shown to be indistinguishable from RP1, RP4 and RK2) could be transferred to *Acinetobacter* strain ACJ1. This observation was followed by the studies of Towner and Vivian (1976a, 1977) demonstrating that other related plasmids belonging to the P1 incompatibility group could be transferred to strain EBF65/65.

It was initially thought that members of the P1 incompatibility group were the only enterobacterial plasmids capable of transfer to *Acinetobacter*, but work by Hinchliffe et al. (1980) demonstrated that EBF65/65 contained a cryptic plasmid, pAV2, which was responsible for variations in the fertility of plasmids and probably determined a restriction-modification system in its host strain. A subsequent re-evaluation by Chopade et al. (1985), using a pAV2-minus derivative of EBF65/65 as recipient, showed that at least one plasmid from each of the incompatibility groups B, C, F_{IV}, H_2/S, Iα, Iδ, P1, W and X could be directly transferred by conjugation from *E.coli* K12 to *Acinetobacter*. Representative frequencies of transfer obtained using the filter mating method of Towner and Vivian (1976b) are listed in Table 1, but it is important to note that not all plasmids of a particular incompatibility group behave identically (Chopade et al., 1985). This may be due, at least in part, to the possible presence of a second restriction system in EBF65/65 and the availability of specific restriction sites on the plasmids being tested (Towner, 1980; Chopade et al., 1985).

In addition to the groups listed in Table 1, Barth et al. (1981) and Singer et al. (1986) have demonstrated that non-conjugative plasmids belonging to the Q incompatibility group can be introduced into *Acinetobacter* by either transformation or conjugal mobilisation. Singer et al. (1986) also showed that strain BD413 was a considerably more efficient recipient than strain HO1-N in mobilisation experiments, thereby

Table 1. Transfer frequencies of representative enterobacterial plasmids from *E.coli* K12 to *Acinetobacter* EBF65/65 (pAV2⁻) (data modified from Chopade et al., 1985)

Plasmid	Inc. group	Selection[a]	Frequency of transfer[b]
pUN65	B	Tp	5×10^{-7}
R57b	C	Cm	1×10^{-3}
R124	F_{IV}	Tc	1×10^{-5}
R478	H_2/S	Tc	4×10^{-4}
R64	$I\alpha$	Tc	9×10^{-6}
R821a	$I\delta$	Ap	3×10^{-5}
RP4	PI	Tc	6×10^{-4}
R388	W	Tp	3×10^{-7}
R6K	X	Ap	1×10^{-4}

[a]Ap, ampicillin; Cm, chloramphenicol;
Tc, tetracycline; Tp, trimethoprim
[b]transfer frequencies are expressed in terms of
tranconjugants / recipient cell following overnight
filter matings at $28^{\circ}C$

emphasising, as with EBF65/65, the importance of choice of recipient strain in plasmid transfer experiments.

Stability and Subsequent Behaviour of Enterobacterial Plasmids

All of the enterobacterial plasmids transferred to *Acinetobacter* by Chopade et al. (1985) were at least partially unstable in the EBF65/65 system. In each case it was found that stable maintenance required continual sub-culture on media containing appropriate selective antibiotics. Subsequent transfer experiments revealed varied and complex patterns of plasmid behaviour in EBF65/65. Some of the enterobacterial plasmids were capable of further rounds of transfer to other strains of *Acinetobacter* or of re-transfer to *E.coli* K12. In other cases it was observed that further transfer only occurred, if at all, following the introduction of an additional mobilising plasmid. Transposition of resistance genes to the mobilising plasmid was sometimes observed. Plasmids belonging to the same incompatibility group were found to behave differently, thereby emphasising the fact that tests with a single "representative" plasmid cannot always be regarded as conclusive. A later section of this review will consider the wider significance of these findings in relation to the possible role of *Acinetobacter* as a reservoir of plasmid-mediated antibiotic resistance genes in hospitals.

In addition to conferring antibiotic resistance, the acquisition of R plasmids by strains of *Acinetobacter* may also result in important outer membrane changes. Transfer of RP1 to the plasmid-free strain HO1-N has been shown to be associated with changes in the fatty acid composition of the lipopolysaccharide fraction and increased outer membrane permeability compared to cells without the plasmid. These changes seem to be linked with an increase in sensitivity to the antimicrobial activity of extracted contents from rat polymorphonuclear leukocyte granules (Loeffelholz and Modrzakowski, 1986; Loeffelholz et al., 1987). Plasmid acquisition and associated outer membrane changes may therefore play an important role in determining the susceptibility of *Acinetobacter* to host immune cell responses and phagocytosis.

Ability to Mobilise Chromosomal Genes

Plasmid-mediated mobilisation of *Acinetobacter* chromosomal genes was first reported by Towner and Vivian (1976a) using strain EBF65/65 and the P1 incompatibility group plasmid RP4. Transfer of chromosomal genes only occurred at a detectable frequency (10^{-5}-10^{-8} / recipient cell) on solid surfaces, not in liquid matings. Gene mapping and chromosome organisation in *Acinetobacter* has been reviewed by Vivian (this volume), but in summary, two RP4-carrying donor strains which donated the EBF65/65 chromosome in opposite directions from different origins were isolated and, as a result, it was possible to demonstrate that the linkage map of EBF65/65 is circular (Towner and Vivian, 1976b; Towner, 1978). Subsequently a number of other plasmids have also been shown to mobilise chromosomal genes of *Acinetobacter* (Towner and Vivian, 1977; Hinchliffe and Vivian, 1980b), but to date this property seems to be restricted to members of the P1 incompatibility group.

The precise mechanism by which chromosome mobilisation occurs in *Acinetobacter* has not yet been fully elucidated. RP4 and its relatives are known to mobilise the chromosomes of a variety of Gram-negative bacteria at a low frequency (Holloway, 1979). In *E.coli* K12 it has been proposed that this mobilisation may occur as a result of an integration event between two Tn1 copies, following the transposition of a Tn1 copy from RP4 to the host cell chromosome (Harayama et al., 1980). Although Tn1 can transpose readily between plasmids, chromosomal insertions are normally only obtained at a low frequency at a restricted number of sites (Sherratt, 1981), thereby possibly accounting for the relatively low frequencies of chromosome mobilisation observed. Other transposons insert into the *E.coli* K12 chromosome on a more random basis at a higher frequency (Sherratt, 1981) and therefore could possibly mediate higher frequencies of chromosome mobilisation in *Acinetobacter*. The behaviour and potential uses of transposons introduced into *Acinetobacter* from *E. coli* are considered in the next section.

Behaviour of Transposons Introduced into Acinetobacter

Transposons conferring antibiotic resistance have been shown to have considerable utility as mutagenic agents and genetic tools in a variety of microbial systems; however there have been few studies involving the direct introduction of these elements into strains of *Acinetobacter*. Towner (1980) demonstrated that P1 incompatibility group plasmids carrying the lysogenic bacteriophage Mu have great difficulty in becoming established in strain EBF65/65. This observation formed the basis of a "suicide" vector system which enabled Towner (1983) to use a RP4::Mu::Tn3171 vector (pUN308) to generate Tn3171 insertions in the chromosome of EBF65/65 at a frequency of 10^{-7}/ recipient cell. Tn3171 is probably identical to the well-characterised trimethoprim/streptomycin resistance transposon Tn7 (Towner 1981, 1983) and seemed to have an insertional preference for one or more arginine biosynthetic loci in EBF65/65. It is worth noting that Tn7 also has a single insertion site linked to an arginine biosynthetic gene in *Caulobacter crescentus* (Ely, 1982). Subsequent chromosome mobilisation experiments in EBF65/65, using either RP4 or RP4::Tn3171 as the mobilising vector, showed that the presence of Tn3171 in the chromosome enhanced the frequency of mobilisation of chromosomal genes adjacent to the original Tn3171 insertion, but that this process was not influenced by the presence of Tn3171 on the mobilising plasmid.

A similar suicide plasmid, pJB4JI, containing transposon Tn5 and phage Mu was used by Singer and Finnerty (1984) to introduce Tn5 into *Acinetobacter* strains BD413 and HO1-N. Tn5 was a particularly attractive candidate for use as an insertion mutagen in *Acinetobacter* since it has been shown in *E. coli* K12 to have a low insertional specificity and an ability to generate non-leaky polar mutations (Kleckner et al., 1977; Shaw and Berg, 1979). Although Tn5 insertions were obtained at frequencies of 10^{-6}-10^{-7}/ recipient for both *Acinetobacter* strains, screening of over 10,000 insertions failed to reveal any phenotypic mutations. Further analysis suggested the existence of a single Tn5-specific insertion site in HO1-N and two such sites in BD413. It seems, therefore, that *Acinetobacter* may provide an exception to the normal pattern of general insertion of Tn5 into numerous sites within prokaryotic genomes. However, in a subsequent study, Doten et al. (1987) were readily able to generate Tn5 insertions in five or six *pca* structural genes derived from *Acinetobacter* (encoding six enzymes that convert protocatechuate to citric acid cycle intermediates via ß-ketoadipate) which had been cloned in *E. coli*. One possibility is therefore that, in the absence of the hotspots, there may not be any barriers to generalised mutagenesis of *Acinetobacter* DNA with Tn5. The ease with which certain *Acinetobacter* strains can undergo transformation (Juni, 1972; Ahlquist et al., 1980) means that it should not be difficult to reintroduce such cloned genes containing Tn5 insertions into the *Acinetobacter* chromosome.

An alternative transposon delivery system, based on pRK2013, has been described by Ely (1985). Since pRK2013 has a broad host range transfer system linked to a CoIE1 replicon, it can be transferred to, but not replicated in, a variety of non-enteric Gram-negative bacteria. Delivery vectors carrying either Tn7 (pBEE7) or Tn10 (pBEE10 or pBEE104) generated transposon insertions at a frequency of 10^{-6}-10^{-8} in *Acinetobacter* strain BD413. The specificity of these insertions was not investigated, but in view of the results outlined above it would not be surprising if the insertional specificity of these elements differed in *Acinetobacter* from that previously observed with *E. coli* K12. Published data on the transposition frequencies of transposons introduced into *Acinetobacter* in the laboratory are summarised in Table 2.

INDIGENOUS PLASMIDS AND TRANSPOSONS

Plasmid Profiles in Acinetobacter

Following the first demonstration of extrachromosomal DNA in *Acinetobacter* by Christiansen et al. (1973), indigenous plasmids have been found in the majority of *Acinetobacter* isolates that have been examined. Although many of these are cryptic plasmids with as yet unidentified functions, others (some of which are discussed in more detail below) have been associated with such properties as antibiotic and heavy metal resistance, aromatic hydrocarbon degradation, conjugal fertility and restriction/modification systems. Many strains of *Acinetobacter* seem to carry multiple plasmids of variable molecular size and it has also been observed that unrelated isolates seem to have diverse plasmid profiles (Hartstein et al., 1988). Gerner-Smidt (1989) found indigenous plasmids in 75 of 93 strains of *Acinetobacter* examined, with a range of 0-20 plasmids / strain. The majority of plasmids (70%) were less than 23 kb in size, with two thirds of the strains containing two or more plasmids. As discussed by Noble (this volume), there is a need for suitable typing methods which can be applied to *Acinetobacter*. The determination of plasmid profiles may prove to be a valuable addition to the range of typing methods which can be used with clinical isolates of *Acinetobacter*.

Indigenous Antibiotic Resistance Plasmids

Although clinical isolates of *Acinetobacter* have shown widespread and increasing resistance to antibiotics, only slight evidence was obtained in early studies of resistant hospital strains to suggest the involvement of indigenous R plasmids (e.g. Murray and Moellering, 1979, 1980; French et al., 1980). Failure to detect plasmid-mediated transfer of antibiotic resistance genes in such studies may, however, simply reflect the absence of a suitable test system for plasmid transfer. For historical reasons, most attempts at plasmid transfer from clinical isolates of any Gram-negative species have initially used *E. coli* K12 as a recipient strain. The complex patterns and varied frequencies of plasmid transfer observed between

Table 2. Insertion frequencies of transposons introduced into *Acinetobacter* from *E.coli* K12 (data from Towner, 1983; Singer and Finnerty, 1984; Ely, 1985)

Transposon	Vector	Recipient strain	Frequency of transposition[a]
Tn5	pJB4JI	BD413	$10^{-6} - 10^{-7}$
Tn5	pJB4JI	HOI-N	$10^{-6} - 10^{-7}$
Tn7	pBEE7	BD413	10^{-6}
Tn10	pBEE10	BD413	10^{-8}
Tn10(HH104)[b]	pBEE104	BD413	10^{-7}
Tn3171	pUN308	EBF65/65	10^{-7}

[a] expressed in terms of insertions per recipient cell

[b] Tn10(HH104) is a mutant derivative of Tn10 which over-produces the Tn10 transposase in *E.coli* K12

Acinetobacter and *E.coli* K12 have already been outlined in this review, but it is pertinent to emphasise at this point that 17 of 31 self-transmissible trimethoprim R plasmids transferred from *E.coli* K12 to *Acinetobacter* EBF65/65 subsequently required the presence of an additional broad host range transmissible plasmid for re-transfer to occur (Chopade et al., 1985). Other studies have confirmed that although *Acinetobacter* can fairly readily acquire plasmids from *E.coli* K12, the opposite transfer seems to be a relatively rare event (Gomez-Lus et al., 1980; Goldstein et al., 1983). In addition, the observation that 70% of indigenous *Acinetobacter* plasmids are less than 23 kb in size (Gerner-Smidt, 1989) means that antibiotic resistance genes carried on such indigenous plasmids would, in any event, probably not be associated with the genetic information necessary for transfer. It may therefore be necessary to rigorously assess the potential for mobilisation of resistance genes from clinical isolates of *Acinetobacter* in order to clarify the role of plasmids in the increasing spread of such resistance.

In view of the above findings it is perhaps not surprising that in cases where indigenous transmissible antibiotic resistance has been identified in *Acinetobacter*, it has invariably been subsequently demonstrated to be associated with a plasmid belonging to a broad host-range incompatibility group. One of the most well-studied examples is that of the multi-resistant strain JC17, originally isolated from a hospital in Pretoria, South Africa. This strain carried a 128 kb conjugative plasmid which specified resistance to sulphonamides, designated pAV1, and a 23 kb non-conjugative plasmid which specified resistance to kanamycin/neomycin and tetracycline, designated pAV5 (Hinchliffe et al., 1980; Divers et al., 1985). pAV1 is an unusual example of a P1 incompatibility group plasmid with a restricted host range (Hinchliffe and Vivian, 1980a); however, although it is unable to transfer to *E.coli* K12, it is freely transmissible between several different strains of *Acinetobacter* and also serves to mobilise both pAV5 and the *Acinetobacter* chromosome (Hinchliffe and Vivian, 1980b; Divers et al., 1984).

The non-conjugative plasmid, pAV5, harboured by strain JC17 has also been the subject of more detailed investigation. Genetic transfer experiments revealed that the resistance determinants carried by pAV5 occasionally segregate on transfer, resulting in the apparent formation of independent, compatible replicons: pAV51 (11.8 kb), which specifies resistance to kanamycin/neomycin by means of an APH(3')-I inactivating enzyme, and pAV52 (17.6 kb) which specifies resistance to tetracycline (Divers et al., 1984). Detailed molecular analysis by Divers et al. (1985) showed that this segregation was associated with deletion of apparently non-overlapping segments of pAV5, with each of the two deleted segments having one terminus close to the kanamycin resistance gene (Fig.1). Electron microscopy of homoduplex structures revealed the presence of 0.81 kb inverted repeat sequences flanking this gene, while some limited genetic experiments suggested that the resistance marker was transposable. It therefore seems probable that the inverted repeats may be at least partially responsible for generating the observed deletions. Although resistance to either kanamycin/neomycin or tetracycline could be expressed in *E.coli* K12 following cloning of appropriate restriction fragments into pBR328, it isinteresting to note, in the wider context of antibiotic resistance in *Acinetobacter*, that pAV5 itself could not be maintained in *E.coli* following either transformation or mobilisation experiments.

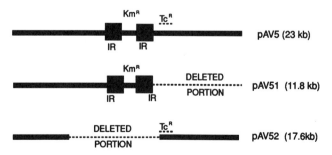

Fig. 1. Diagrammatic representation of the probable structures of pAV5 and its deletion derivatives (Divers et al., 1985)

A second well-characterised example of transferable plasmid-mediated antibiotic resistance in a multi-resistant isolate of *Acinetobacter* was provided by strain BM2500 isolated from a hospital in Paris, France. Goldstein et al. (1983) showed that resistance to ampicillin, chloramphenicol, kanamycin, streptomycin, sulphonamides and trimethoprim was encoded in this strain by a single 167 kb plasmid, designated pIP1031, belonging to incompatibility group C. Plasmids belonging to this group are known to possess a wide host range and have been shown to be transferable from *E.coli* to *Acinetobacter* (Chopade et al., 1985). pIP1031 was found to be transferable at an extremely low frequency (5×10^{-10}) from the original *Acinetobacter* isolate BM2500 to *E.coli* K12; however this transfer could only be obtained with the initial isolate. Detailed molecular studies indicated that loss of transferability was not associated with any major rearrangements to the plasmid genome, nor to integration of the resistance determinants into the host cell chromosome. This situation was in sharp contrast to the situation in *E.coli* K12, where pIP1031 was stable and could be easily transferred in further rounds of conjugation. It therefore seemed that pIP1031 was not well-adapted to carriage by strain BM2500, leading Goldstein et al. (1983) to suggest that pIP1031 had only been acquired extremely recently from an exogenous source and that it was probably not a true indigenous *Acinetobacter* plasmid. An additional interesting finding from the associated molecular studies was that pIP1031 carried an unusual transposable module, designated IS15, which can exist in two, apparently not alternative, structural configurations. IS15 can promote a range of genetic rearrangements and is found on R plasmids isolated from a large variety of Gram-negative bacteria, perhaps thereby indicating its important biological role in plasmid evolution (Labigne-Roussel and Courvalin, 1983).

A small number of other reports describing transferable antibiotic resistance in *Acinetobacter* have also appeared in the literature. Gomez-Lus et al. (1980) reported the transfer to *E.coli* K12, again at a low frequency, of a P1 incompatibility group plasmid from a strain of *Acinetobacter* isolated in Spain, while Chirnside et al. (1985) described a single clinical isolate of *Acinetobacter*, obtained from a London hospital in 1977, which transferred a 112 kb plasmid mediating trimethoprim and sulphonamide resistance to *E.coli* K12. In both of these cases it seems likely that the strain of *Acinetobacter* had only recently acquired its plasmid from enteric bacteria occupying the same ecological niche. More interestingly, Peng (1980) described transfer of several combinations of resistance genes from 13 strains of *Acinetobacter* isolated in Taiwan. Although these were also broad host range plasmids, they appeared to be well-adapted to existence in *Acinetobacter*, since intrageneric matings between strains of *Acinetobacter* occurred at frequencies $10-10^3$ times higher than intergeneric matings between *Acinetobacter* and *E.coli* K12. Of even greater interest was the description by Krcmery et al. (1985) of two strains of *Acinetobacter*, isolated from different hospitals in Czechoslovakia, the first of which transferred amikacin and ß-lactam resistance to *Pseudomonas aeruginosa*, but not *E.coli* K12, while the second strain transferred gentamicin and ß-lactam resistance to *E. coli* K12, but not *P. aeruginosa*. Unfortunately no further details of any of these plasmids are currently available.

Most recently, Lambert et al. (1988) described a 63 kb plasmid, designated pIP1841, which freely transferred resistance to kanamycin and structurally related antibiotics, including amikacin, between different taxonomic sub-groups of *Acinetobacter*, but not to *E. coli* K12. Cloning experiments, in which the resistance gene was transformed into *E. coli* K12 following in-vitro ligation with the vector plasmid pUC18, demonstrated that full expression of the resistance gene could be obtained in *E. coli* K12. It therefore seems probably that the apparent transfer barrier frequently observed between *Acinetobacter* and other organisms does not result from lack of expression of *Acinetobacter* genes in heterologous systems, but rather from defects in conjugation donor or recipient ability, or plasmid replication in the new host, or both (Lambert et al., 1988).

Chromosomally-Located Transposons

As described by Bergogne-Bérézin and Joly-Guillou (this volume), much of the antibiotic resistance found in clinical isolates of *Acinetobacter* seems to result from the presence of enzymic resistance mechanisms that are essentially identical to those known to be encoded by genes carried on plasmids in the Enterobacteriaceae and other major genera of bacteria. This review has already outlined the wide range of enterobacterial plasmids that are capable of transfer to *Acinetobacter*, even if not all are capable of subsequent stable maintenance. It therefore seems likely that many of the genes conferring antibiotic resistance in *Acinetobacter* were originally introduced by plasmid-mediated conjugation from other bacteria sharing the same environment. The fact that several such genes have been shown to be transposable in *E. coli* K12 means that, potentially, they could become stably inserted by transposition into the *Acinetobacter* chromosome, following their introduction on a plasmid which was itself unable to undergo stable replication in the *Acinetobacter* host. If such genetic events have occurred, they would provide an additional explanation as to why it is often difficult in clinical isolates of *Acinetobacter* to demonstrate an association between extrachromosomal DNA and resistance mechanisms which are normally thought of as being plasmid-encoded. It is therefore important that appropriate genetic and molecular tests to examine this possibility are included in studies of antibiotic-resistant clinical isolates.

To date there have been at least two published reports of chromosomally-located antibiotic resistance transposons in *Acinetobacter*. In both cases antibiotic resistance phenotypes were apparently neither self-transmissible nor associated with plasmid DNA, but could be mobilised to other strains by use of the conjugative broad host range P1 incompatibility group plasmid RP4, which underwent a size increase concomitant with mobilisation. Shimizu et al. (1981) described the mobilisation of a 5 kb DNA sequence encoding streptomycin/spectinomycin resistance from a strain of *Acinetobacter* isolated in Japan. In a similar study, Devaud et al. (1982) mobilised a 24 kb DNA sequence encoding resistance to chloramphenicol, gentamicin, streptomycin, sulphonamides and possibly tetracycline from a strain of *Acinetobacter* found in an intensive care unit in Switzerland. The occurrence of multi-resistant nosocomial isolates of *Acinetobacter* seems to be becoming more common, perhaps in response to the increasing usage of antibiotics in hospitals, and we have recently shown that a similar transposable DNA sequence encoding resistance to ß-lactams, aminoglycosides, tetracycline, chloramphenical and sulphonamides can be found in endemic hospital strains of *Acinetobacter* isolated in the United Kingdom (M. Nikolic and K.J. Towner, unpublished results). Experience suggests that *Acinetobacter* is a difficult organism to eradicate from the hospital environment, raising the worrying possibility that strains of *Acinetobacter* could act as reservoirs of antibiotic resistance genes for more important pathogens if such chromosomally-located transposons become widely distributed in clinical isolates.

Plasmids Conferring Heavy Metal Resistance

In addition to the role of plasmids in conferring antibiotic resistance, there have been several reports linking heavy metal resistance in *Acinetobacter* to indigenous plasmids. This is particularly true of strains of *Acinetobacter* isolated from polluted environments. A typical example is the study by Olson et al. (1979) describing a strain of *Acinetobacter*, designated W45, isolated from an estuarine environment in the USA. Strain W45 produced mercuric reductase, which conferred resistance to mercury salts, and contained three plasmids of molecular size 3.3, 7.1 and 124 kb respectively. Cured derivatives of W45 were found to have concomitantly lost the ability to produce mercuric reductase and had consequently regained their sensitivity to mercuric salts.

A more direct example of the involvement of plasmids in heavy metal resistance was provided by studies of bacteria contained in polluted soil surrounding the Khaidarkan mercury mine in the Kirghiz region of the USSR. (Khesin and Karasyova, 1984; Lomovskaya et al., 1986). One large (60 kb) and two small (7.5 and 4.5 kb) plasmids were found in all five strains of *Acinetobacter* examined. A mercury resistance determinant was found to be located on the 7.5 kb plasmid, designated pKL1, which could be mobilised by the large plasmid present in the same cell. Using conjugative crosses and transformation experiments, it was established that, although pKL1 was a broad host range plasmid, it could not be transferred to *E.coli*. Indeed, a characteristic feature of pKL1 was its instability in all foreign host cells after a few generations in the absence of selective pressure. Nevertheless, when the mercury resistance determinant from pKL1 was cloned into either pUC19 or pBR322 and transformed into *E.coli* K12, full expression of the mercury resistance determinant was obtained (Lomovskaya et al., 1986). Preliminary data also suggested that the large 60 kb conjugative plasmid encoded a mercury resistance determinant but was capable of replication in *Acinetobacter* only. As with antibiotic resistance, it therefore seems that the failure to mobilise heavy metal resistance to *E. coli* probably results from a transfer or replication barrier rather than an expression barrier.

An additional interesting finding from the above studies was that a plasmid apparently identical to pKL1 could be isolated from a strain of *Acinetobacter* obtained from a second mercury deposit at Bolshoi Shayan in the Transcarpathia region of the USSR (Lomovskaya et al., 1986). Transcarpathia is thousands of kilometres from Kirghiz, suggesting that pKL1 is widely distributed amongst strains of *Acinetobacter* in the USSR. A plasmid of similar size to pKL1, although not directly linked to mercury resistance, was also carried by strain W45 isolated in the USA (Olson et al., 1979).

Further evidence that small broad host range plasmids of the pKL1 type may be distributed in strains of *Acinetobacter* on a world-wide basis has also been provided by the study of Rochelle et al. (1988), which demonstrated that a 7.8 kb broad host range plasmid, designated pQM17, was responsible for encoding mercury resistance in epilithic strains of *Acinetobacter* isolated from the organically polluted River Taff in Wales. Other mercury-resistant *Acinetobacter* isolates from this river contained plasmids varying in size from 2.1 to 220 kb (Rochelle et al., 1988), but it seems that small non-conjugative plasmids of the pKL1/pQM17 type may play an important role in the maintenance and subsequent spread of mercury resistance genes in natural bacterial populations.

Metabolic Plasmids

One of the most important characteristics of bacteria belonging to the genus *Acinetobacter* is the ability of different strains to use a wide range of aromatic hydrocarbon and aliphatic compounds as a source of carbon and energy. As a consequence of their ability to degrade relatively recalcitrant compounds of this type, it has been suggested that strains of *Acinetobacter* may have an important role to play in the control of environmental pollution. Although most reports of metabolic plasmids have involved members of the *Pseudomonas* genus, there have also been several studies describing strains of *Acinetobacter* in which genes encoding significant degradative steps are plasmid-located.

An important group of environmental pollutants are the polychlorinated biphenyls (PCBs), normally considered to be non-degradable due to their unreactive structure and low water solubility. Furukawa and Chakrabarty (1982) reported the involvement of a 80 kb plasmid, designated pKF1, in the degradation of PCBs by *Acinetobacter* strain P6. This strain was able to dissimilate 33 different pure isomers of PCBs, with the formation of corresponding chlorobenzoic acids, but spontaneously lost this ability at a high frequency (4 to 8%) after overnight growth in nutrient broth. Although rigorous proof that pKF1 encoded the degradative pathway was lacking, loss of PCB-utilising activity was directly associated with the generation of small 2.4 kb pKF1 deletions. Attempts to transfer or transform pKF1 to other strains of *Acinetobacter* which could not utilise PCBs were unsuccessful, but an apparently identical plasmid was also observed in a strain of *Arthrobacter* present in the same original culture.

Plasmid-mediated degradation of 4-chlorobiphenyl (4CB) to 4-chlorobenzoate in *Acinetobacter* has also been described by Shields et al.(1985). Strains of *Acinetobacter* and *Alcaligenes* isolated from a mixed culture were both shown to carry a 53 kb plasmid, designated pSS50, which probably encoded a complete pathway for 4CB oxidation. Plasmid curing resulted in loss of the 4CB-utilising phenotype, and loss of even early 4CB metabolism from the *Acinetobacter* strain, but the phenotype was restored following filter-surface matings with the *Alcaligenes* strain. This indicated a direct role for pSS50 in 4CB oxidation, but the possibility also exists that pSS50 may activate other chromosomal genes partly involved in mineralisation.

A third important degradative plasmid was described by Winstanley et al. (1987), who investigated a benzene-utilising strain of *Acinetobacter* isolated from a selective soil enrichment and found that it carried a large (approximately 200 kb) transmissible plasmid, designated pWW174, encoding the enzymes necessary for catabolism of benzene via the ß-ketoadipate pathway. It was possible to demonstrate conjugal transfer of pWW174 (and the benzene-utilising phenotype) between different strains of *Acinetobacter* at frequencies varying from 10^{-3} to 10^{-7}, but no transfer occurred to *P.putida* (the species in which most reports of catabolic plasmids have been concentrated). An additional finding was that pWW174 was unstable in strains of *Acinetobacter* and was lost at high frequency in the absence of selective pressure. This ease of loss suggested that either pWW174 had a second alternative host in which it could replicate stably and from which it had only recently been transferred to *Acinetobacter*, or that it was a recently evolved recombinant plasmid in the soil from which it was isolated and had not been subjected to maintenance selection in the absence of benzene.

As a final example, there has also been a report of a strain of *Acinetobacter* with the ability to degrade crude oil (Rusansky et al., 1987). Strain RA57 contains four plasmids: pSR1 (5.1 kb), pSR2 (5.4 kb), pSR3 (10.5 kb) and pSR4 (20 kb). A proportion of colonies which were isolated at random after growth in the presence of acridine orange were found to have lost the ability to grow on and disperse crude oil in liquid cultures and to have concomitantly been cured of pSR4. Cells cured of pSR4 were still, however, capable of growth on hydrocarbon vapours, thereby indicating that pSR4 may encode genes that play a role in the physical interaction of the cell with the hydrocarbon substrate rather than actual catabolism. Initial results indicated that this role might be concerned with the process of uptake/internalisation of the hydrocarbon substrate in liquid culture prior to degradation. It would be of interest to examine the effect of pSR4 on the cell surface properties of other *Acinetobacter* strains which do not normally utilise hydrocarbons as the sole course of carbon and energy.

CLONING VECTORS FOR *ACINETOBACTER*

As outlined at the beginning of this review, most attempts to clone specific genes derived from *Acinetobacter* (or any other organism) invariably use *E.coli* K12 as an initial host strain in which to recognise recombinant plasmids. Nevertheless, in order to study gene expression and regulation, it is often necessary to subsequently reintroduce cloned genes into their original host strain. As an illustrative example of this sort of approach, Doten et al. (1987) cloned a 11 kb *Eco*RI DNA fragment derived from *Acinetobacter* into pUC18 and transformed the recombinant plasmid into *E.coli* K12. The 11 kb insert carried the *pca* structural genes encoding six enzymes responsible for converting protocatechuate to citric acid cycle intermediates via ß-ketoadipate. The recombinant plasmid was mutagenised with Tn5 in *E.coli* K12 and then used to transform appropriate *Acinetobacter* mutants in order to study the genetic organisation of the *pca* cluster. Similar cloning experiments with a variety of genes derived from *Acinetobacter* can be carried out using any one of the standard *E.coli* K12 host/vector systems that are now available. A major drawback, however, is that in many cases the recombinant plasmids cannot be stably maintained following their reintroduction into *Acinetobacter*. Thus, for example, vector plasmids carrying an origin of replication (*ori*) gene derived from pBR322 are not always stably maintained as plasmids in *Acinetobacter*, with the consequence that stable reintroduction of cloned genes can only be obtained following recombination with the chromosome (Goosen et al., 1987).

In an attempt to overcome this difficulty, the behaviour in *Acinetobacter* strains BD413 and HO1-N of several "broad host range" cloning vectors was examined by Singer et al. (1986). Plasmids derived from RSF1010 (Bagdasarian et al., 1979, 1981; Bagdasarian and Timmis, 1982), such as pKT230 (11.9 kb), pKT231 (13 kb) and pKT248 (12.4 kb),were found to be stably inherited after 20 generations of growth under non-selective conditions in both strains. These plasmids could not be reliably transformed into either BD413 or H01-N, but could be successfully introduced by means of a mobilisation system from *E.coli* K12 (Singer et al., 1986). Plasmid pTB107, a 10 kb derivative of R300B carrying a 1.5 kb kanamycin resistance determinant (Barth et al., 1981) was transformable into BD413 (not HO1-N), but was slightly unstable (15% plasmid loss) unless stabilised by low level antibiotic selection. Plasmid pRK290 is a 20 kb plasmid, derived from RK2, that can be mobilised into a variety of Gram-negative organisms including *Acinetobacter* (Ditta et al., 1980). pRK290 was, however, found to be highly unstable in both BD413 (67% plasmid loss) and HO1-N (>99% plasmid loss). This rather surprising instability of pRK290 limits its potential use as a cloning vector in *Acinetobacter* and may result from the deletion of a segment of RK2 DNA that may be required for stable plasmid maintenance (Ditta et al., 1980; Lanka and Barth, 1981).

As a consequence of these results, Singer et al. (1986) suggested that a vector system based on selected R300B or RSF1010 derivatives would be useful for routine cloning and complementation studies in *Acinetobacter*. Examples of the adoption of this approach include the cloning into pKT230 of genes encoding the ability of *Acinetobacter* to grow on benzene (Winstanley et al., 1987) and the construction of an *Acinetobacter* gene bank by cloning *Acinetobacter* DNA partially digested with *Sau*3A into a hybrid vector plasmid, pLV21, derived from RSF1010 and pUB110 (Goosen et al., 1987). A number of similar cloning projects have been successfully completed and it seems that these vector systems probably offer the best currently available approach to the use of in-vitro techniques for studying genetic organisation and regulation in the *Acinetobacter* genus. The construction and use of shuttle vectors in *Acinetobacter* for particular applications is described in more detail by Gutnick et al. (this volume).

CONCLUSIONS

Plasmids and transposons undoubtedly play an important role in the overall biology of *Acinetobacter*, but there is still a relative paucity of information about their detailed properties and behaviour in this genus. Much of the work described in this review has, either directly or indirectly, focused on the possibilities for exchange of genetic information between *Acinetobacter* and *E. coli*, largely as a consequence of the way in which the science of bacterial genetics has developed. It seems clear that, while certain types of plasmids are confined to the gene pool of *E. coli*, an "intermediate group" can cross the generic boundary more freely, although there may then be further barriers preventing establishment in the new host. Transposons probably play an important role in ensuring that particular genes (e.g. those conferring antibiotic resistance) are stably maintained even in the event of plasmid instability.

It also seems that there is a pool of plasmid-mediated genetic information which is largely confined to *Acinetobacter* and is not readily mobilisable to *E. coli* or other genera. Most "indigenous" plasmids so far characterised in *Acinetobacter* are those which have been capable of transfer/mobilisation to *E. coli* and which therefore belong to the "intermediate group". Many strains of *Acinetobacter* also contain several cryptic plasmids which presumably encode information which is useful to the cell. There seem to be few barriers to the expression of *Acinetobacter* genes in *E. coli*, but there are undoubtedly barriers to plasmid transfer and maintenance. In order to fully understand the role that these accessory elements play in the lifestyle of the cell, it will therefore be necessary to utilise in-vitro methods of gene cloning and analysis. It is hoped that the next few years will see major advances in this direction.

REFERENCES

Ahlquist, E.F., Fewson, C.A., Ritchie, D.A., Podmore, J., and Rowell, V., 1980, Competence for genetic transformation in *Acinetobacter calcoaceticus* NCIB 8250, *FEMS Microbiol. Lett.*, 7:107.

Bagdasarian, M., Bagdasarian, M.M., Coleman, S., and Timmis, K.N., 1979, New vector plasmids for gene cloning in *Pseudomonas, in:* "Plasmids of Medical, Environmental and Commercial Importance," p.411, K.N. Timmis and A. Puhler, eds., Elsevier/North Holland Biomedical Press, Amsterdam.

Bagdasarian, M., Lurz, R., Ruckert, B., Franklin, F.C.H., Bagdasarian, M.M., Frey, J., and Timmis, K.N., 1981, Specific purpose plasmid cloning vectors. Broad host range, high copy number, RSF1010-derived vectors, and a host-vector system for gene cloning in *Pseudomonas, Gene,* 16:237.

Bagdasarian, M., and Timmis, K.N., 1982, Host:vector systems for gene cloning in *Pseudomonas, Curr. Top. Micobiol. Immunol.*, 96:47.

Barth, P.T., Tobin, L., and Sharp, G.S., 1981, Development of broad host-range plasmid vectors, *in*: "Molecular Biology, Pathogenicity and Ecology of Bacterial Plasmids," p.439, S.B. Levy, R.C. Clowes and E.L. Koenig, eds., Plenum Publishing Corporation, New York.

Chopade, B.A., Wise, P.J., and Towner, K.J., 1985, Plasmid transfer and behaviour in *Acinetobacter calcoaceticus* EBF65/65, *J. Gen. Microbiol.*, 131:2805.

Christiansen, C., Christiansen, G., Leth Bak, A., and Stenderup, A., 1973, Extrachromosomal deoxyribonucleic acid in different enterobacteria, *J. Bacteriol.*, 114:367.

Devaud, M., Kayser, F.H., and Bachi, B., 1982, Transposon-mediated multiple antibiotic resistance in *Acinetobacter* strains, *Antimicrob. Agents Chemother.*, 22:323.

Ditta, G., Stanfield, S., Corbin, D., and Helinski, D.R., 1980, Broad host range DNA cloning system for gram-negative bacteria: construction of a gene bank of *Rhizobium meliloti, Proc. Natl. Acad. Sci. USA,* 77:7347.

Divers, M., Craven, P.L., and Vivian, A., 1985, Molecular analysis of an antibiotic resistance plasmid, pAV5, and its derivative plasmids in *Acinetobacter calcoaceticus, J. Gen. Microbiol.*, 131:3367.

Divers, M., Hinchliffe, E., and Vivian, A., 1984, Characterisation of an antibiotic resistance plasmid pAV5 and its constituent replicons in *Acinetobacter calcoaceticus, Genet. Res.*, 44:1.

Doten, R.C., Ngai, K.L., Mitchell, D.J., and Ornston, L.N., 1987, Cloning and genetic organisation of the *pca* gene cluster from *Acinetobacter calcoaceticus, J. Bacteriol.*, 169:3168.

Ely, B., 1982, Transposition of Tn7 occurs at a single site on the *Caulobacter crescentus* chromosome, *J. Bacteriol.*, 151:1056.

Ely, B., 1985, Vectors for transposon mutagenesis of non-enteric bacteria, *Mol. Gen. Genet.*, 200:302.

French, G.L., Casewell, M.W., Roncoroni, A.J., Knight, S., and Phillips, I., 1980, A hospital outbreak of antibiotic-resistant *Acinetobacter anitratus*: epidemiology and control, *J. Hosp. Infect.*, 1:125.

Furukawa, K., and Chakrabarty, A.M., 1982, Involvement of plasmids in total degradation of chlorinated biphenyls, *Appl. Environ. Microbiol.*, 44:619.

Gerner-Smidt, P., 1989, Frequency of plasmids in strains of *Acinetobacter calcoaceticus*, *J. Hosp. Infect.*, 14:23.

Goldstein, F.W., Labigne-Roussel, A., Gerbaud, G., Carlier, C., Collatz, E.,and Courvalin, P., 1983, Transferable plasmid-mediated antibiotic resistance in *Acinetobacter, Plasmid,* 10:138.

Gomez-Lus, R., Larrad, L., Rubio-Calvo, M.C., Navarro, M. and Assierra, M.P., 1980, AAC(3) and AAC(6') enzymes produced by R plasmids isolated in general hospital, *in*: "Antibiotic Resistance," p.295,S.Mitsushashi, L. Rosival and V. Krcmery, eds., Springer-Verlag, Berlin.

Goosen, N., Vermaas, D.A., and van de Putte P., 1987, Cloning of the genes involved in synthesis of coenzyme pyrroloquinolone-quinone from *Acinetobacter calcoaceticus, J. Bacteriol.*, 169:303.

Harayama, S., Tsuda, M., and Tino, T., 1980, High frequency mobilisation of the chromosome of *Escherichia coli* by a mutant of plasmid RP4 temperature-sensitive for maintenance, *Mol. Gen. Genet.*, 180:47.

Hartstein, A.I., Rashad, A.L., Liebler, J.M., Actis, L.A., Freeman, J., Rourke, J.W., Stibolt, T.B., Tomalsky, M.E., Ellis, G.R., and Crosa, J.H., 1988, Multiple intensive care unit outbreak of *Acinetobacter calcoaceticus* subspecies *anitratus* respiratory infection and colonisation associated with contaminated, reusable ventilator circuits and resuscitation bags, *Amer. J. Med., 85:624.

Hinchliffe, E., and Vivian, A., 1980a, Naturally occurring plasmids in *Acinetobacter calcoaceticus*: a P class R factor of restricted host range, *J. Gen. Microbiol.*, 116:75.

Hinchliffe, E., and Vivian, A., 1980b, Gene transfer in *Acinetobacter calcoaceticus*: fertility variants of the sex factor pAV1, *J. Gen. Microbiol.*, 119:117.

Hinchliffe, E., Nugent, M.E., and Vivan, A., 1980, Naturally occurring plasmids in *Acinetobacter calcoaceticus*: pAV2, a plasmid which influences the fertility of the sex factor pAV1, *J. Gen. Microbiol.*, 121:411.

Holloway, B.W., 1979, Plasmids that mobilise bacterial chromosome, *Plasmid,* 2:1.

Juni, E., 1972, Interspecies transformation of *Acinetobacter*: genetic evidence for a ubiquitous genus, *J. Bacteriol.*, 112:917.

Khesin, R.B., and Karasyova, E.V., 1984, Mercury-resistant plasmids in bacteria from a mercury and antimony deposit area, *Mol. Gen. Genet.*, 197:280.

Kleckner, N., Roth, J., and Botstein, D., 1977, Genetic engineering *in vivo* using translocatable drug-resistance elements, *J. Mol. Biol.*, 116:125.

Krcmery, V., Langsadl, L., Antal, M., and Seckarova, A., 1985, Transferable amikacin and cefamandole resistance: *Pseudomonas maltophilia* and *Acinetobacter* strains as possible reservoirs of R plasmids, *J. Hyg. Epidemiol. Microbiol. Immunol.*, 28:141.

Labigne-Roussel, A., and Courvalin, P., 1983, IS15, a new insertion sequence widely spread in R plasmids of Gram-negative bacteria, *Mol. Gen. Genet.*, 189:102.

Lambert, T., Gerbaud, G., and Courvalin, P., 1988, Transferable amikacin resistance in *Acinetobacter* spp. due to a new type of 3'-aminoglycoside phosphotransferase, *Antimicrob. Agents Chemother.*, 32:15.

Lanka, E., and Barth, P.T., 1981, Plasmid RP4 specifies a deoxyribonucleic acid primase involved in its conjugal transfer and maintenance, *J. Bacteriol.*, 148:769.

Loeffelholz, M.J., and Modrzakowski, M.C., 1986, Plasmid RP1-mediated susceptibility of *Acinetobacter calcoaceticus* to rat polymorphonuclear leukocyte granule contents, *Infect. Immun.*, 54:705.

Loeffelholz, M.J., Rana, F., Modrzakowski, M.C., and Blazyk, J., 1987, Effect of plasmid RP1 on phase changes in inner and outer membranes and lipopolysaccharide from *Acinetobacter calcoaceticus*: a Fourier transform infrared study, *Biochemistry*, 26:6644.

Lomovskaya, O.L., Mindlin, S.Z., Gorlenko, Zh.M., and Khesin, R.B., 1986, A nonconjugative mobilizable broad host range plasmid of Acinetobacter sp. that determines $HgCl_2$ resistance, *Mol. Gen. Genet.*, 202:286.

Murray, B.E., and Moellering, R.C., 1979, Aminoglycoside-modifying enzymes among clinical isolates of *Acinetobacter calcoaceticus* subsp. *anitratus (Herellea vaginicola)*: explanation for high-level aminoglycoside resistance, *Antimicrob. Agents Chemother.*, 15:190.

Murray, B.E., and Moellering, R.C., 1980, Evidence of plasmid-mediated production of aminoglycoside-modifying enzymes not previously described in *Acinetobacter, Antimicrob. Agents Chemother.*, 17:30.

Olsen, R.H., and Shipley, P., 1973, Host range and properties of the *Pseudomonas aeruginosa* R factor R1822, *J. Bacteriol.*, 113:772.

Olson, T., Barkay, T., Nies, D., Bellama, J.M., and Colwell, R.R., 1979, Plasmid-mediated mercury volatilization and methylation, *Develop. Indust. Microbiol.*, 20:275.

Peng, C-F., 1980, Studies on glucose nonfermentative Gram-negative bacilli. II. Transferable drug resistance in *Acinetobacter calcoaceticus, J. Formos. Med. Assoc.*, 79:583.

Rochelle, P.A., Day, M.J., and Fry, J.C., 1988, Occurrence, transfer and mobilization in epilithic strains of *Acinetobacter* of mercury-resistance plasmids capable of transformation, *J. Gen. Microbiol.*, 134:2933.

Rusansky, S., Avigad, R., Michaeli, S., and Gutnick, D.L., 1987, Involvement of a plasmid in growth on and dispersion of crude oil by *Acinetobacter calcoaceticus* RA57, *Appl. Environ. Microbiol.*, 53:1918.

Shaw, K.J., and Berg, C.M., 1979, *Escherichia coli* K12 auxotrophs induced by the insertion of the transposable element Tn5, *Genetics*, 92:741.

Sherratt, D., 1981, *In vivo* genetic manipulation in bacteria, *in*: "Genetics as a Tool in Microbiology," p.35, S.W. Glover and D.A. Hopwood, eds., Cambridge University Press, Cambridge.

Shields, M.S., Hooper, S.W., and Sayler, G.S., 1985, Plasmid-mediated mineralization of 4-chlorobiphenyl, *J. Bacteriol.*, 163:882.

Shimizu, S., Inoue, M., Mitsuhashi, S., Naganawa, H., and Kondo, S., 1981, Enzymatic adenylylation of spectinomycin by *Acinetobacter calcoaceticus* subsp. *anitratus*, *J. Antibiotics*, 34:869.

Singer, J.T., and Finnerty, W.R., 1984, Insertional specificity of transposon Tn5 in *Acinetobacter* sp., *J. Bacteriol.*, 157:607.

Singer, J.T., van Tuijl, J.J., and Finnerty, W.R., 1986, Transformation and mobilization of cloning vectors in *Acinetobacter* spp., *J. Bacteriol.*, 165:301.

Towner, K.J., 1978, Chromosome mapping in *Acinetobacter calcoaceticus*, *J. Gen. Microbiol.*, 104:175.

Towner, K.J., 1980, Behaviour of bacteriophage Mu in *Acinetobacter calcoaceticus* EBF65/65, *J. Gen. Microbiol.*, 121:425.

Towner, K.J., 1983, Transposon-directed mutagenesis and chromosome mobilisation in *Acinetobacter calcoaceticus* EBF65/65, *Genet. Res.*, 41:97.

Towner, K.J., and Vivian, A., 1976a, RP4-mediated conjugation in *Acinetobacter calcoaceticus*, *J. Gen. Microbiol.*, 93:355.

Towner, K.J., and Vivian, A., 1976b, RP4 fertility variants in *Acinetobacter calcoaceticus*, *Genet. Res.*, 28:301.

Towner, K.J., and Vivian, A., 1977, Plasmids capable of transfer and chromosome mobilization in *Acinetobacter calcoaceticus*, *J. Gen. Microbiol.*, 101:167.

Winstanley, C., Taylor, S.C., and Williams, P.A., 1987, pWW174: a large plasmid from *Acinetobacter calcoaceticus* encoding benzene catabolism by the ß-ketoadipate pathway, *Mol. Microbiol.*, 1:219.

PLASMID STABILITY AND THE EXPRESSION AND REGULATION

OF A MARKER GENE IN *ACINETOBACTER* AND

OTHER GRAM-NEGATIVE HOSTS

C. Winstanley[a], J.A.W. Morgan[b]
J.R. Saunders[a] and R.W. Pickup[b]

[a]Department of Genetics & Microbiology
University of Liverpool
Liverpool L69 3BX
UK

[b]Institute of Freshwater Ecology
Windermere Laboratories, Ambleside
Cumbria LA22 0LP
UK

INTRODUCTION

Much controversy has been raised over the possible intentional release of genetically engineered microorganisms (GEMs) into the natural environment. The potential for microbial detoxification of pollutants, crop protection and improved plant productivity make the proposition an attractive one. However, before any such release can be permitted, there is a need to understand the mechanisms of survival, expression, transfer and rearrangement of recombinant DNA in microbial communities.

The construction of safe GEMs has been a priority in order to lessen the potential risk of adverse consequences due to accidental release from laboratory or industry into the environment. Safeguards can include the use of either debilitated host bacteria or cloning vectors incapable of replication except under defined laboratory conditions. When the deliberate release of a GEM is considered, an important factor must be that the released strain should survive long enough to fulfill its intended function. Thus the same safeguards cannot be applied to the consideration of accidental and deliberate release. Although reports of actual releases are few in number (see for example, Drahos et al., 1986; Gaertner and Kim, 1988; Lindow and

The Biology of Acinetobacter, Edited by K. J. Towner *et al.*
Plenum Press, New York, 1991

Panopoulos, 1988; Stewart-Tull, 1988), it is possible to carry out some initial risk assessments using contained model environments in the laboratory. Among the questions which need to be asked are: (i) whether the organism or its recombinant DNA will survive; (ii) whether the organism can multiply and displace members of the natural population; (iii) whether the organism will become dispersed away from the intended release site. The potential consequences of transfer of recombinant DNA into the natural population must also be considered. Much work has concentrated on the possibility of gene transfer in the environment, but an equally important consideration is whether the recombinant DNA would be stable, expressed or regulated in potential recipients. The possible hazards posed by changes in the activity or control of genes introduced into natural systems need to be assessed.

To date, work on the study of gene survival and transfer in the environment has generally relied on antibiotic resistance markers carried on recombinant plasmids (Hughes and Datta, 1983; Devanas and Stotzky, 1986; Devanas et al., 1986). Attempts to monitor the transfer of such plasmids, particularly in aquatic systems, are likely to be hampered by the high numbers of naturally occuring antibiotic-resistant bacteria. Any study in freshwater systems is further complicated by the fact that only about 1% of cells are cultivable by available methods (Jones, 1977).

Table 1. Marker Plasmids used in this Study

Plasmid	Inc Group	Antibiotic Resistance[a]	Marker System
pLV1010	Q	Sm Ap	p_L-$xylE$
pLV1011	Q	Sm	p_L-$xylE$-cI_{857}
pLV1013	Q	Sm Km	p_R-$xylE$-cI_{857}
pLV1016	P	Ap Km Tc	p_R-$xylE$-cI_{857}
pLV1017	P	Ap Km Tc	p_L-$xylE$

[a]Sm: streptomycin; Ap: ampicillin; Km: kanamycin; Tc: tetracycline.

In an attempt to overcome these problems we have developed a versatile marker system designed to operate in a range of common freshwater and soil microorganisms (Winstanley et al., 1989). The system involves detection of the $xylE$ gene which encodes catechol 2,3 dioxygenase (C230). Colonies expressing $xylE$ can be detected on plates by spraying them with 1% catechol solution. A yellow coloration occurs due to the formation of 2-hydroxymuconic semialdehyde. The marker gene is expressed from the strong bacteriophage lambda promoters p_L or p_R which can be controlled by the presence of the temperature-sensitive lambda repressor cI_{857}. This enables the potentially deleterious metabolic burden imposed on the cell by high expression of $xylE$ to be countered. In parallel with the marker system we have also developed direct detection methods which preclude the necessity to culture the target organism (Morgan et al., 1989). The marker DNA has been introduced into broad host range plasmids to enable an assessment of $xylE$ expression and regulation in a range of common freshwater or soil bacteria. This has revealed significant differences in stability, expression and regulation of the plasmid marker systems in *Acinetobacter* compared with *Pseudomonas, Klebsiella, Aeromonas, Serratia* and *Escherichia coli* strains. These differences, and to a lesser extent differences observed in *Pseudomonas*, question whether it is possible to predict the behaviour of recombinant DNA in the event of gene transfer from one species to another in the natural environment.

STABILITY OF MARKER PLASMIDS

The marker plasmids which have been constructed are outlined in Table 1. The IncQ regulated marker plasmids (pLV1011 and pLV1013) and unregulated plasmid (pLV1010) are based on the plasmid pKT230 (Bagdasarian et al., 1981; Fig.1). The IncP regulated (pLV1016) and unregulated (pLV1017) marker plasmids are cointegrates based on R68.45 (Haas and Holloway, 1976; unpublished data). Both pKT230 and R68.45 are able to replicate in a wide range of Gram-negative microorganisms. R68.45 is naturally conjugative, whereas pKT230 may be mobilised by plasmid pNJ5000 which is unstable in the absence of tetracycline (Grinter, 1983). The marker plasmids were introduced into different host bacteria (Table 2) to assess the stability of the C230$^+$ phenotype. Cells were cultured on nutrient agar following growth under non-selective conditions (nutrient broth) for approximately 20 generations. Resulting colonies were then screened for C230$^+$ by spraying with catechol.

The IncP cointegrate plasmids pLV1016 and pLV1017 were stable (>99%) in all hosts except *K.pneumoniae* (85% stable). The IncQ marker plasmids showed marked differences in their stability depending on the plasmid and the host (Table 3). The unregulated system of pLV1010 was markedly less stable than the regulated pLV1013 system. However, there was a surprising difference in stability in *Pseudomonas* between the two regulated systems of pLV1013 and pLV1011. $xylE$ expression from p_L rather

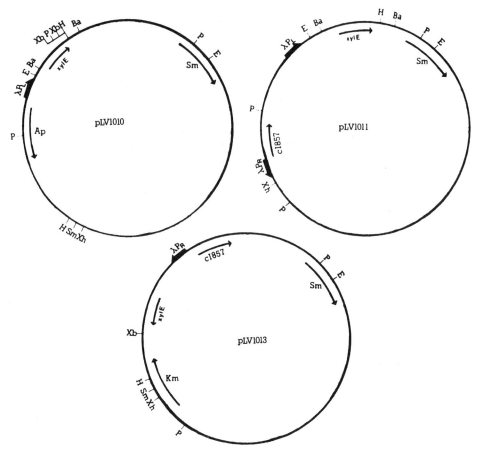

Fig. 1. IncQ marker plasmids. Abbreviations indicating cleavage sites for restriction enzymes are: Ba, *Bam*HI, E, *Eco*RI, H, *Hind*III, P, *Pst*I, Sm, *Sma*I, Xb, *Xba*I, Xh, *Xho*I. Ap, Km and Sm indicate resistance to ampicillin, kanamycin and streptomycin respectively.

than p_R apparently caused high instability despite the similar regulation of the two promoters by cI in bacteriophage lambda (Johnson et al., 1981). The largest difference of all was observed in *A.calcoaceticus* ADP1 where all three plasmids exhibited high degrees of instability. In particular the C230$^+$ phenotype of pLV1010 was retained, at best, by only 4% of the population. Similar high levels of instability were observed in two recent freshwater isolates, *P.putida* FBA11 and *P.fluorescens* FH1. Further observations revealed that the C230$^+$ phenotype was unstable in strains ADP1, FBA11 and FH1, even when antibiotics selecting for retention of the plasmid were present in the growth medium. In the presence of streptomycin (100 mg/l) only 4-5% of FBA11 colonies and 1-3% of FH1 colonies screened retained the C230$^+$ phenotype after overnight culture of pLV1010-containing strains. The addition of ampicillin to the growth medium made no difference. Screening of the DNA from four C230$^-$ Sm^{R}Ap^R colonies from each of FBA11 and FH1 revealed the presence of

Table 2. Host Bacteria used in this Study

Strain[a]	Plasmids Assessed
Acinetobacter calcoaceticus ADP1	All plasmids
Escherichia coli	
ED8654	pLV1010, pLV1011, pLV1013.
AB1157	pLV1016, pLV1017.
AB2463	pLV1016, pLV1017.
Pseudomonas putida	
PRS2000	All plasmids
PaW140	pLV1010, pLV1011, pLV1013.
PaW8	pLV1016, pLV1017.
FBA11	All plasmids.
Pseudomonas aeruginosa	
PAO1	All plasmids
NCIB8295	pLV1010, pLV1011, pLV1013.
Pseudomonas fluorescens	
FH1	All plasmids
Aeromonas hydrophila	
NCTC8049	All plasmids.
Klebsiella pneumoniae	
NC18418	All plasmids.
K2819	All plasmids.
Serratia rubidiae	All plasmids.

[a]most strains are described by Winstanley et al. (1989)

deleted pLV1010 derivatives. In each case the deletion included the loss of one *Bam*HI, one *Xba*I, one *Pst*I and one *Eco*RI restriction site. Since one of the two *Eco*RI sites on pLV1010 is located in the still functional Sm^R gene, it would appear that a deletion from an area including the *Eco*RI site upstream of *xyl*E to the *Xba*I site downstream of *xyl*E has occurred.

Table 3. Stability of IncQ Plasmid Constructs[a]

Strain	Plasmid		
	pLV1013	pLV1011	pLV1010
A.calcoaceticus ADP1	11%	30%	4%
E.coli ED8654	100%	99%	95%
P.putida PRS2000	98%	22%	90%
PaW8	100%	26%	98%
P.aeruginosa PAO1	90%	18%	51%
Aeromonas NCTC8049	100%	98%	94%
Klebsiella NCIB418	100%	99%	55%
S.rubidiae	100%	100%	67%

[a]stability is expressed as % retention of C230$^+$ after overnight growth in nutrient broth at 30°C. The mean of at least two experiments is shown.

Closer analysis of the stability of all three plasmids in *A.calcoaceticus* ADP1 has been carried out. Table 4 shows the result of an experiment to assess the distribution of the three phenotypes (C230$^+$Sm^{R}Ap^R, C230$^-$Sm^{R}Ap^R, C230$^-$Sm^{S}Ap^S). Colonies growing on nutrient agar were screened by patching them on to nutrient agar supplemented with streptomycin (100 mg/l) as well as nutrient agar alone. C230$^+$ colonies were then identified by spraying with catechol. In some cases, particularly when growth was carried out with streptomycin present, a significant proportion of colonies assumed to be C230$^-$ when grown on nutrient agar appeared to be C230$^+$ when grown on agar supplemented with streptomycin. Colonies isolated from the broth cultures containing the antibiotic selection also appeared to be C230$^-$ only on nutrient agar plates. This indicates that there is a certain amount of plasmid or *xyl*E gene

Table 4. Stability of C230[+] and Sm[R] Phenotypes in ADP1

Plasmid	Phenotype[a]	% Retention of Phenotype[b]	
		Growth Medium[c]	
		N broth	N broth+Sm
pLV1010	C230[+]Sm[R]	0.3%	6.2%
	C230[-]Sm[R]	7.3%	70.4%
	C230[-]Sm[S]	92.4%	23.4%
pLV1011	C230[+]Sm[R]	9.3%	47.4%
	C230[-]Sm[R]	0%	0.6%
	C230[-]Sm[S]	90.7%	52.0%
pLV1013	C230[+]Sm[R]	1.0%	74.0%
	C230[-]Sm[R]	0%	0%
	C230[-]Sm[S]	99.0%	26.0%

[a] Sm[R] and Sm[S] indicate streptomycin resistance and sensitivity respectively.
[b] results were based on an experiment in which a single colony, taken from a plate containing streptomycin, was inoculated into the growth medium. After overnight culture at 30 C, dilutions were plated on to nutrient agar and colonies were screened.
[c] N-broth indicates nutrient broth.

instability during growth on non-selective plates. C230[-] colonies were routinely re-screened after patching on to streptomycin-containing media to take account of this problem.

pLV1010 gave rise to high numbers of C230[-]Sm[R] cells, particularly under conditions selecting for antibiotic resistance. Fig.2 shows a restriction digest of DNA isolated from eight such colonies. The gel shows loss of a *Pst*I site and deletion of approximately 2.5-3.0 kb in each case. This was also associated with loss of *Bam*HI, *Xba*I and *Eco*RI sites. The deletions appear to be very similar to those observed in strains FH1 and FBA11. Three derivatives of *A. calcoaceticus* ADP1(pLV1011) with C230[-] Sm[R] phenotypes were also isolated. Analysis of their DNA revealed loss of one *Pst*I, one *Eco*RI and one *Bam*HI site from pLV1011. A deletion of approximately 4 kb including *xyl*E had occurred (data not shown). No

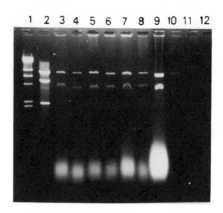

Fig. 2. Restriction digests of $Sm^RAp^RC230^-$ derivatives of pLV1010. Lane 1, HindIII digest of lambda DNA; lane 2, PstI digest of pLV1010; lanes 3-10, PstI digests of $Sm^RAp^RC230^-$ derivatives of pLV1010.

separation of the antibiotic resistance and $C230^+$ phenotypes was observed with strain ADP1(pLV1013).

EXPRESSION AND REGULATION OF *XYLE*

A simple assay for C230 activity (Sala-Trepat and Evans, 1971) has allowed *xyl*E expression and regulation to be assessed in a range of host strains containing the different marker plasmids. C230 specific activity is expressed in Units/mg protein.

Inc P Plasmids

Unregulated expression of *xyl*E from pLV1017 gave rise to high levels of C230 activity in all hosts tested (Table 2). The levels varied from 9.1 U/mg in *A. calcoaceticus* ADP1 to 25.4 U/mg in *E. coli* AB1157 (unpublished data). The regulation of *xyl*E expression from pLV1016 was assessed by incubation at 37°C after initial growth in nutrient broth at 28°C. Expression of *xyl*E was increased by incubation at the higher temperature in all the hosts tried. The Enterobacteriaceae showed a 1000-fold increase over 12 h, whereas lower induction was observed in *Pseudomonas* and *Acinetobacter* hosts. Uninduced C230 activity in these hosts showed a ten-fold increase over that observed in *E. coli*. The specific activity obtained after 12 h at 37°C varied from 0.5-2 U/mg in *Pseudomonas* to 4-5 U/mg in *E. coli* and *Klebsiella*. The level of C230 activity in *Acinetobacter* after the same period of incubation was 2.5 U/mg (unpublished data).

Inc Q Plasmids

Detailed analysis of the expression and regulation of *xyl*E from the IncQ marker plasmids has been carried out (Winstanley et al., 1989).

Unregulated C230 activity from pLV1010 was high in all hosts tested, with levels of 62 U/mg and 52 U/mg recorded in *P.putida* PaW140 and PRS2000 respectively. Levels in *P.aeruginosa* hosts varied from 22 U/mg (PA01) to 32 U/mg (NCIB8295), whilst good activity was also recorded in *E.coli*, *Klebsiella* and *Aeromonas* (14-24 U/mg). Levels of 9.5 U/mg were recorded in *S.rubidiae*, but the lowest activities were recorded in *P.putida* FBA11, *P.fluorescens* FH1 and *A.calcoaceticus* ADP1 (5.5-5.8 U/mg). It is no coincidence that these are the same strains in which pLV1010 is highly unstable, even in the presence of antibiotics. Where plasmids were known to be unstable in the host being investigated, antibiotics were included in the growth medium from which samples were taken for assay. However, the deletion of *xyl*E from pLV1010 in strains FBA11, FH1 and ADP1 inevitably leads to misleading levels of C230 activity being recorded in these strains.

Analysis of the regulated systems of pLV1011 and pLV1013 revealed complications (Winstanley et al., 1989). There was good induction of the p_R-*xyl*E-cI_{857} system of pLV1013 at 37°C over 25 h in *E.coli* (680-fold), *Klebsiella* (200 to 300-fold), *Serratia* (260-fold) and *Aeromonas* (210-fold). The regulation in *Pseudomonas* was also good (18 to 67-fold), but was hampered by a higher uninduced activity (86-290 mU/mg compared with 7.1 mU/mg for *E.coli*). Maximum activities observed varied from 16 U/mg in *P.aeruginosa* NCIB8295 to 2.3 U/mg for *S.rubidiae*. The worst induction was found in *A.calcoaceticus* where the uninduced activity of 560 mU/mg increased to a peak level of 3.4 U/mg over a 25 h period.

A number of differences were observed with the p_L-*xyl*E-cI_{857} system of pLV1011, despite the fact that the change in promoter from p_R to p_L is the only significant difference from the pLV1013 system. Much lower increases in C230 activity after incubation at 37°C were observed in all the hosts, but the most striking difference was found in *A.calcoaceticus* ADP1 where there appeared to be no regulation of *xyl*E expression. In some hosts (*E.coli*, *P.aeruginosa* and *Klebsiella*) an improvement in induction of C230 activity was achieved by raising the incubation temperature to 42°C. This effect was not observed with the pLV1013 system. The higher temperature appeared to have harmful effects on some host strains (*P.putida*, *P.fluorescens* and *Aeromonas*), but studies of the ADP1(pLV1013) system indicated that similar levels of induction occurred at both 37°C and 42°C, although the maximum was achieved earlier at the higher temperature (data not shown). This indicates that the pLV1011 system is genuinely unregulated in *A.calcoaceticus* ADP1, where the uninduced levels of C230 activity (2 U/mg) were the highest recorded.

The generally higher levels of C230 activity which can be achieved with IncQ plasmid systems can be accounted for by a higher copy number compared with the IncP plasmids. Higher uninduced C230 activity from pLV1013 in some species may be due to the presence of *Pseudomonas* promoters upstream of *xyl*E.

DISCUSSION

Considerable differences in the stability, expression and regulation of the *xyl*E marker systems were observed. Both the unregulated and regulated IncP plasmids pLV1017 and pLV1016 were apparently stable in the whole range of hosts tested. Expression and regulation of *xyl*E on these plasmids was also fairly consistent from species to species. Both IncP plasmid systems functioned adequately in *Acinetobacter*. The major differences were observed with the higher copy number IncQ marker plasmids. The unregulated plasmid pLV1010 was highly unstable in ADP1, FBA11 and FH1. *A. calcoaceticus* is commonly found in both freshwater and soil, whilst *P. putida* FBA11 and *P. fluorescens* FH1 were both recent freshwater isolates (Winstanley et al., 1989). Closer analysis revealed that *xyl*E was deleted from pLV1010 in these strains, even in the presence of antibiotic selection. Similar deleted derivatives of pLV1010 were obtained during growth of PRS2000(pLV1010) in sterile lake water model systems (Morgan et al., 1989), but the effect has not been observed when this strain is grown overnight in nutrient broth. Plasmid pLV1011, containing p_L-*xyl*E-cI_{857}, was also unstable in a number of hosts and the regulation of this system was inferior to the pLV1013 system. This was particularly striking in *A. calcoaceticus* ADP1 where no regulation was achieved. pLV1011 was also highly unstable in ADP1, with some evidence of loss of *xyl*E from the plasmid being apparent. More surprising was the high instability of pLV1013, although no evidence for deletion of *xyl*E from the plasmid was found.

There have been several reports of unstable naturally-occurring plasmids in *Acinetobacter* (Divers et al., 1984; Winstanley et al., 1987). However, pKT230, on which the IncQ plasmid systems are based, is known to replicate and be stably maintained in *Acinetobacter* (Singer et al., 1986); thus the reasons for the instability of the marker plasmids are unclear. Metabolic burden due to the high expression of *xyl*E could contribute to instability of pLV1010. However, C230 activity from pLV1011, despite being unregulated, would not appear to be high enough to account for its instability, particularly as uninduced levels of C230 activity from pLV1013 are higher in *A. calcoaceticus* ADP1 than other strains, but no greater than 1 U/mg. Indeed *xyl*E expression from pLV1017, which is perfectly stable in strain ADP1, gives rise to levels of 9.1 U/mg. Since both p_L and p_R are present on pLV1011 there is a possibility that titration of the cI_{857} repressor could cause reduced regulation because of interaction with both promoters. However, the introduction of p_L in *trans* on plasmid pPLc245 had no similar effect on regulation of the pLV1013 system (results not shown). It is possible that *cI* repressor is not fully effective in *Acinetobacter* since high levels of uninduced C230 activity were apparent with both the pLV1011 and pLV1013 systems. This would not account for the fact that some control by *cI* was apparent with pLV1013, but not pLV1011. There is also the evidence provided by the pLV1016 system, which is as well regulated in *Acinetobacter* as it is in *Pseudomonas*.

There are a number of examples demonstrating the potential of soil and aquatic bacteria to receive recombinant DNA in genetic exchanges (Stotzky and Babich, 1986; Bale et al., 1987; Krasovsky and Stotzky, 1987; Saye et al., 1987; Genthner et al., 1988; Zeph et al., 1988). There have also been cases of non-conjugative plasmids being mobilised by indigenous waste-water bacteria (McPherson and Gealt, 1986; Mancini et al., 1987). In the event of gene transfer, effects such as differences in gene regulation and expression or changes in gene stability, depending on the host, suggest that the behaviour of recombinant DNA following release into the natural environment would be difficult to predict. These kind of effects would also make detection of the released DNA more difficult. It may therefore be considered important to include an assessment of gene expression, regulation and stability in potential natural recipients in the consideration of the risk posed by the release of a GEM into the environment.

REFERENCES

Bagdasarian, M., Lurz, R., Ruckert, B., Franklin, F.C.H., Bagdasarian, M.M., Frey, J., and Timmis, K.N., 1981, Broad host range, high copy number, RSF1010-derived vectors and a host-vector system for gene cloning in *Pseudomonas, Gene*, 16:237.

Bale, M.J., Fry, J.C., and Day, M.J., 1987, Plasmid transfer between strains of *Pseudomonas aeruginosa* on membrane filters attached to river stones, *J. Gen. Microbiol.*, 133:3099.

Devanas, M.A., Rafaeli-Eshkol, D., and Stotzky, G., 1986, Survival of plasmid-containing strains of *Escherichia coli* in soil: effect of plasmid size and nutrients on survival of hosts and maintenance of plasmids, *Curr. Microbiol.*, 13:269.

Devanas, M.A., and Stotzky, G., 1986, Fate in soil of a recombinant plasmid carrying a *Drosophila* gene, *Curr. Microbiol.*, 13:279.

Divers, M., Hinchliffe, E., and Vivian, A., 1984, Characterisation of an antibiotic resistance plasmid pAV5 and its constituent replicons in *Acinetobacter calcoaceticus, Genet. Res.*, 44:1.

Drahos, D.J., Hemming, B.C., and McPherson, S., 1986, Tracking recombinant organisms in the environment: ß-galactosidase as a selectable non-antibiotic marker for fluorescent pseudomonads, *Biotechnology*, 4:439.

Gaertner, F., and Kim, L., 1988, Current applied recombinant DNA projects, *Trends Biotech.*, 6:S4.

Genthner, F.J., Chatterjee, P., Barkay, T., and Bourquin, A.W., 1988, Capacity of aquatic bacteria to act as recipients of plasmid DNA, *Appl. Environ. Microbiol.*, 54:115.

Grinter, N.J., 1983, A broad host-range cloning vector transposable to various replicons, *Gene*, 21:133.

Haas, D., and Holloway, B.W., 1976, R factor variants with enhanced sex factor activity in *Pseudomonas aeruginosa, Mol. Gen. Genet.*, 144:243.

Hughes, V.M., and Datta, N., 1983, Conjugative plasmids in bacteria of the pre-antibiotic era, *Nature*, 302:725.

Johnson, A.D., Poteete, A.R., Lauer, G., Sauer, R.T., Ackers, G.K., and Ptashne, M., 1981, Lambda repressor and cro-components of an efficient molecular switch, *Nature*, 294:217.

Jones, J.G., 1977, The effect of environmental factors on estimated viable and total populations of planktonic bacteria in lakes and experimental enclosures, *Freshwater Biol.*, 7:67.

Krasovsky, V.N., and Stotzky, G., 1987, Conjugation and genetic recombination in *Escherichia coli* in sterile and non-sterile soil, *Soil Biol. Biochem.*, 19:631.

Lindow, S.E., and Panopoulos, N.J., 1988, Field tests of recombinant Ice⁻ *Pseudomonas syringae* for biological frost control in potato, *in*: "The Release of Genetically Engineered Microorganisms," p.121, M.Sussman, C.H. Collins, F.A. Skinner and D.E. Stewart-Tull, eds., Academic Press, London.

McPherson, P., and Gealt, M.A., 1986, Isolation of indigenous wastewater bacterial strains capable of mobilizing plasmid pBR325, *Appl. Environ. Microbiol.*, 51:904.

Mancini, P., Fertels, S., Nave, D., and Gealt, M.A., 1987, Mobilization of plasmid pHSV106 from *Escherichia coli* HB101 in a laboratory-scale waste treatment facility, *Appl. Environ. Microbiol.*, 53:665.

Morgan, J.A.W., Winstanley, C., Pickup, R.W., Jones, J.G., and Saunders, J.R., 1989, Direct phenotypic and genotypic detection of a recombinant pseudomonad population released into lake water, *Appl. Environ. Microbiol.*, 55:2537.

Sala-Trepat, J.M., and Evans, W.C., 1971, The *meta* cleavage of catechol by *Azotobacter* species: 4-oxalocrotonate pathway, *Eur. J. Microbiol.*, 20:400.

Saye, D.J., Ogunseitan, O., Sayler, G.S., and Miller, R.V., 1987, Potential for the transduction of plasmids in a natural freshwater environment: effect of plasmid donor concentration and a natural microbial community on transduction in *Pseudomonas aeruginosa*, *Appl. Environ. Microbiol.*, 53:987.

Singer, J.T., Van Tuijl, J.J., and Finnerty, W.R., 1986, Transformation and mobilisation of cloning vectors in *Acinetobacter* spp., *J. Bacteriol.*, 165:301.

Stewart-Tull, D.E., 1988, Round table 5: case histories of deliberate release, *in*: "The Release of Genetically Engineered Microorganisms," p.245, M.Sussman, C.H. Collins, F.A. Skinner and D.E. Stewart-Tull, eds., Academic press, London.

Stotzky, G., and Babich, H., 1986, Survival of, and genetic transfer by, genetically engineered bacteria in natural environments, *Adv. Appl. Microbiol.*, 31:93.

Winstanley, C., Taylor, S.C., and Williams, P.A., 1987, pWW174: a large plasmid from *Acinetobacter calcoaceticus* encoding benzene catabolism by the ß-ketoadipate pathway, *Mol. Microbiol.*, 1:219.

Winstanley, C., Morgan, J.A.W., Pickup, R.W., Jones, J.G., and Saunders, J.R., 1989, Differential regulation of lambda p_L and p_R promoters by a cI repressor in a broad-host-range thermoregulated plasmid marker system. *Appl. Environ. Microbiol.*, 55:771.

Zeph, L.R., Onaga, M.A., and Stotzky, G., 1988, Transduction of *Escherichia coli* by bacteriophage P1 in soil, *Appl. Environ. Microbiol.*, 54:1731.

RANDOM INSERTIONAL MUTAGENESIS IN *ACINETOBACTER*

B. Vosman, R. Kok and K.J. Hellingwerf

Laboratorium voor Microbiologie
Universiteit van Amsterdam
Nieuwe Achtergracht 127
1018 WS Amsterdam
The Netherlands

INTRODUCTION

Mutant analysis is a powerful tool in the study of biological processes, the more so if it is combined with the relevant physiological experiments. Two basically different approaches can be used for the construction of mutants: (i) mutants can be obtained with chemical mutagens or by irradiation with UV or X-rays; (ii) mutants can be generated by insertional mutagenesis. The basic idea behind the latter method is the insertion of an easily selectable genetic element into the gene of interest. This has the major advantage that the gene is marked in such a way that it can be isolated easily.

MODES OF INSERTIONAL MUTAGENESIS

The most frequently used method of insertional mutagenesis involves the use of transposons (Kleckner et al., 1977; Simon et al., 1983). For *Escherichia coli*, bacteriophage lambda is the most commonly used delivery vehicle, whereas for most other Gram-negative bacteria a transposon is introduced by conjugation on a plasmid from *E.coli*. These plasmid vectors generally contain a ColEl-like origin of replication, which is unable to initiate replication in strains outside the enteric bacterial group (Simon et al., 1983).

In *Bacillus subtilis*, insertional mutagenesis can be accomplished by use of the transposon Tn917 (Youngman et al., 1983). Since *B.subtilis* is an organism that is competent for natural transformation, mutagenesis can also be accomplished by the insertion of a non-replicative plasmid into the gene of interest (Niaudet et al., 1982). For the latter method, pBR322 derivatives are normally used since they are not replicated in *B. subtilis*. Randomly generated chromosomal DNA fragments are cloned into these vectors and the recombinant molecules transformed into competent cells of *B. subtilis*. Plasmids containing inserts homologous to regions of the chromosome integrate efficiently into the chromosome of a wild-type recipient cell (Niaudet et al., 1982).

The insertion can occur via two different mechanisms, depending on the nature of the DNA insert. The first possibility is that the vector integrates by a Campbell-like mechanism (Campbell, 1962). This type of integration occurs when *one* DNA fragment is cloned into the carrier plasmid. After integration into the chromosome, the plasmid DNA is flanked by a direct repeat of the cloned fragment. This type of integration only creates a mutant phenotype if the DNA fragment carried by the vector does not contain either of the borders of a particular transcriptional unit. The second possibility is an integration via a double cross-over recombination event, also referred to as replacement recombination. The latter mechanism predominates when *two* non-contiguous DNA fragments are cloned into a plasmid. It results in deletion of the chromosomal DNA fragment located between the two cloned fragments.

REPLICATION OF pBR322 DERIVATIVES IN *A. CALCOACETICUS*

Singer et al. (1986) showed that plasmid pRK2013, which contains a ColEl-like origin of replication, can be transferred by conjugation from *E. coli* to *A. calcoaceticus*. Transfer frequencies with this plasmid (1.1 x 10^{-3}) were, however, approximately 500-fold lower than transfer frequencies using RSF1010 (8.2 x 10^{-1}), or derivatives thereof. After growing cells containing pRK2013 under non-selective conditions, the plasmid was rapidly lost. This indicates that plasmids containing a ColEl-like origin of replication do replicate, although poorly, in *A. calcoaceticus*.

We have transferred the transposon-containing plasmids pJFF350 (Fellay et al., 1989), pSUP2021 (Simon et al., 1983) and pSUP::Tn5-B20 (Simon et al., 1989), all of which contain a ColEl-like origin of replication, by conjugation from *E. coli* S17-1 to *A. calcoaceticus* AAC1. The latter strain is a rifampicin-resistant derivative of BD413. Rifampicin and kanamycin resistant transconjugants were selected. For plasmid pSUP::Tn5-B20 the transfer frequency was 1.2 x 10^{-3}, approximately the same frequency as found for pRK2013 by Singer et al. (1986).

Tn5-B20 carries a promoterless *lac*Z gene adjacent to $IS50_L$. With this construct, transposition should give rise to equal proportions of transconjugants that do or do not produce ß-galactosidase. This prediction is based on the assumption of random insertion of Tn5-B20 into the chromosome. However, all transconjugants obtained after transfer of pSUP::Tn5-B20 to strain AAC1 stained equally blue on Luria-Bertani (LB) plates containing 5-bromo-4-chloro-3-indolyl-ß-D-galactoside (X-gal) and kanamycin (15 mg/l). In addition to this, all colonies were resistant to gentamicin (10 mg/l). This resistance is encoded by the vector part of plasmid pSUP::Tn5-B20. These results indicated that all the transconjugants still contained the entire plasmid and that very few, if any, transposition events had taken place. Sub-culturing of several of these colonies in LB broth (without kanamycin), followed by plating on to agar containing X-gal and kanamycin, resulted in strains that *did* show differences in colour. This suggests that transposition takes place at a low frequency upon further sub-culturing. However, the latter phenotype was not accompanied by loss of gentamicin resistance. This shows that these cells still contained at least part of the vector moiety of pSUP::Tn5-B20. These observations indicated that straightforward transposon mutagenesis in *A. calcoaceticus* was not possible with the plasmids described above.

MUTAGENESIS OF *A. CALCOACETICUS* THROUGH RANDOM INSERTION OF A KANAMYCIN RESISTANCE MARKER INTO THE CHROMOSOME

The possibility of producing insertion mutations in *A. calcoaceticus* BD413 in a manner similar to that described above for *B. subtilis* was investigated. Plasmids such as pBR322 do replicate, although poorly, in *A. calcoaceticus*. Therefore the entire plasmid was not used, as for *B. subtilis*, but only an antibiotic resistance marker, in our case the *npt*II gene from pUC4K (Vieira and Messing, 1982) conferring resistance to kanamycin. This gene was isolated as a 1.3 kb *Bam*HI fragment from an agarose gel. Chromosomal DNA of *A. calcoaceticus* was cleaved simultaneously with *Bam*HI, *Bgl*II and *Bcl*I. This digestion yielded fragments ranging in size from approx. 5 kb to less than 0.5 kb, with an average of 1.3 kb. Chromosomal DNA was mixed with the DNA fragment carrying the kanamycin resistance gene in a one to one molar ratio and ligated (overnight at room temperature) at a final DNA concentration of 50 μg/ml.

The ligation mixture was transformed into competent cells of *A. calcoaceticus* strain BD413-*ivl*10 (Juni, 1972). For competence development, an overnight culture, grown at 30°C in minimal medium (Juni, 1974), was diluted 50-fold in the same medium and allowed to grow for 1 h. The culture was then concentrated ten-fold by centrifugation (10 min, 5,000 g) and 1 ml of the cell suspension incubated for 45 min at 30°C with 40 μl of the ligation mixture. To allow segregation and expression of the kanamycin resistance marker, 4 ml LB broth was added and the suspension was incubated for an additional 4 h. Portions (0.1 ml) of the cells were plated on to LB plates containing 15 mg kanamycin/l. Approximately 100 kanamycin-resistant colonies (transformation frequency: 4×10^{-6}) were obtained per plate. The collection of mutants was stored at -80°C after addition of glycerol (final conc. 15% v/v) until further use.

SELECTION OF SPECIFIC MUTANTS

   A.calcoaceticus is naturally transformable (Juni, 1972) and produces extracellular lipase(s) (Breuil and Kushner, 1975). Both characteristics are major research subjects in our department and mutants were therefore screened for defects in either of these two properties. Lipase production was tested on nutrient agar plates containing 1% egg yolk, 1% NaCl and 15 mg kanamycin/l. Among 9,000 kanamycin-resistant colonies tested, two (designated AAC300 and AAC302) were unable to form a precipitate on these plates. This indicated that these strains were mutated in a gene(s) involved in (phospho)lipase production. To select for transformation-deficient and auxotrophic mutants, 5,000 colonies were replica-plated from the nutrient agar plates on to minimal plates containing 15 mg kanamycin/l. In addition, 25 μg of wild-type chromosomal DNA was spread on these plates. Mutants that are unable to take up or integrate this DNA cannot acquire a wild-type allele of the *ivl* gene and are therefore unable to grow on these plates in the absence of isoleucine, leucine and valine. In addition, auxotrophic mutants which arise through insertion of the kanamycin resistance gene in a biosynthetic pathway gene are also unable to grow since the selection for kanamycin resistance was continued. One mutant (designated AAC211) was found to be completely transformation deficient. Two others (designated AAC213 and AAC214) were found to be severely impaired (with a transformation frequency more than 1,000-fold lower than the wild-type). In addition, 14 mutants were found to be auxotrophic (a frequency of 3×10^{-3}). Thus, the results showed that it was possible to generate a range of random mutants through insertion of a kanamycin resistance marker into the chromosome of *A.calcoaceticus*.

Further Characterisation of the Mutants

   The nature of the mutations in the two lipase-deficient strains (AAC300 and AAC302) and the transformation-deficient strain (AAC211) were further examined. To ensure that the observed mutant phenotype was caused by the insertion of the kanamycin resistance marker rather than an unlinked secondary mutation, *A. calcoaceticus* strain BD413-*ivl*10 was transformed with chromosomal DNA isolated from the mutant strains. The resulting kanamycin-resistant transformants all showed the mutant phenotype. This indicated that the mutations were closely linked to the kanamycin resistance marker and were probably caused by insertion of the kanamycin resistance marker into the genes of interest.

   This insertion was further verified by Southern hybridisation (Maniatis et al., 1982). The 1300 bp fragment, encoding kanamycin resistance, was used as a probe and labelled with digoxigenine (Boehringer-Mannheim, Germany). Since the fragment to be inserted was obtained by *Bam*HI digestion, it was still flanked by *Sal*I sites which could be used to excise the marker from the chromosomal DNA of the mutants. Fig. 1 shows that this fragment was present in mutant strains AAC211, AAC300 and AAC302 (compare lane 1 with lanes 3-5) and had undergone no major rearrangements. Lane 2, which contains wild-type DNA, did not show any hybridising fragment. The restriction enzyme *Bcl*I does not have a recognition site in the fragment encoding kanamycin resistance. *Bcl*I digestion of the mutant DNA should therefore provide information about the chromosomal sequences flanking the marker. Fig. 1 shows that the insertion in the two lipase mutants was not identical (compare lanes 7 and 8). This

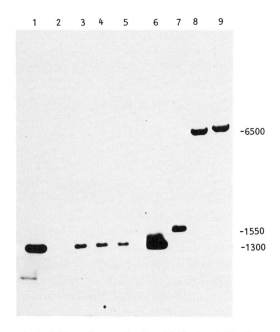

Fig. 1. Southern hybridisation of the 1300 bp DNA fragment from plasmid pUC4K, encoding kanamycin resistance, against chromosomal DNA of the mutant strains AAC300 (lanes 3 and 7), AAC302 (lanes 4 and 8), AAC211 (lanes 5 and 9) and wild-type *A.calcoaceticus* BD413 DNA (lane 2). Lanes 1 and 6 contain the fragment encoding the kanamycin resistance gene, isolated as a *Bam*HI fragment. Chromosomal DNA in lanes 2-5 was digested with *Sal*I; DNA in lanes 7-9 was digested with *Bcl*I. The molecular size of the fragments is indicated in basepairs (bp) in the right margin.

indicated that the marker in both mutants was situated at different positions and that the mutants resulted from independent insertions. Whether the same gene is involved remains to be elucidated. Similar results were obtained for strain AAC211 (lane 9). Although Fig. 1 shows that this mutant contained a *Bcl*I fragment of approximately the same size as mutant AAC302, the two strains exhibited different phenotypes.

CONCLUDING REMARKS

The results presented above show that it is possible to apply insertional mutagenesis to the naturally transformable *A. calcoaceticus* strain BD413 by use of an isolated antibiotic resistance marker. It is conceivable that all naturally transformable strains of *A. calcoaceticus* can be genetically manipulated in this way. We believe that the availability of such a reliable system for mutagenesis will simplify and strongly enhance genetic research in *A. calcoaceticus* and related organisms.

ACKNOWLEDGEMENTS

We thank Drs. E. Juni, R. Simon and J. Frey for providing us with the strains and plasmids used in this study.

REFERENCES

Breuil, C., and Kushner, D.J., 1975, Lipase and esterase formation by psychrophilic and mesophilic *Acinetobacter* species, *Can. J. Microbiol.*, 21:423.

Campbell, A., 1962, Episomes, *Adv. Genet.*, 11:101.

Fellay, R., Krissch, H.M., Prentki, P., and Frey, J., 1989, Omegon-Km: a transposable element designed for in vivo insertional mutagenesis and cloning of genes in Gram-negative bacteria, *Gene*, 76:215.

Juni, E., 1972, Interspecies transformation of *Acinetobacter*: genetic evidence for a ubiquitous genus, *J. Bacteriol.*, 112:917.

Juni, E., 1974, Simple genetic transformation assay for rapid diagnosis of *Moraxella osloensis*, *Appl. Microbiol.*, 27:16.

Kleckner, N., Roth, J., and Botstein, D., 1977, Genetic Engineering *in vitro* using translocatable drug-resistance elements, *J. Mol. Biol.*, 116:125.

Maniatis, T., Fritsch, E.F. and Sambrook, J., 1982, "Molecular Cloning. A Laboratory Manual," Cold Spring Harbor Laboratory, Cold Spring Harbor, NY.

Niaudet, B., Gose, A., and Ehrlich, S.D., 1982, Insertional mutagenesis in *Bacillus subtilis*: mechanism and use in gene cloning, *Gene*, 19:277.

Simon, R., Priefer, U., and Puhler, A., 1983, A broad host range mobilization system for *in vivo* genetic engineering: transposon mutagenesis in Gram-negative bacteria, *Biotechnology*, 1:784.

Simon R., Quandt, J., and Klipp, W., 1989, New derivatives of transposon Tn5 suitable for mobilization of replicons, generation of operon fusions and induction of genes in Gram-negative bacteria, *Gene*, 80:161.

Singer, J.T., van Tuijl, J.J., and Finnerty, W.R., 1986, Transformation and mobilization of cloning vectors in *Acinetobacter* spp., *J. Bacteriol.*, 165:301.

Vieira, J., and Messing, J., 1982, The pUC plasmids, an M13mp7-derived system for insertion mutagenesis and sequencing with synthetic universal primers, *Gene*, 19:259.

Youngman, P.J., Perkins, J.B., and Losick, R., 1983, Genetic transposition and insertional mutagenesis in *Bacillus subtilis* with the *Streptococcus faecalis* transposon Tn917, *Proc. Natl. Acad. Sci. USA*, 80:2305.

GENETIC ORGANISATION OF *ACINETOBACTER*

A. Vivian

Molecular Genetics Group
Science Department
Bristol Polytechnic
Bristol BS16 1QY
UK

INTRODUCTION

Perhaps in the rush to embrace the recent technological advances in gene manipulation, we have become mesmerised by organisation of the genome at the molecular level of the gene. Undoubtedly, the power afforded by the construction of gene libraries, coupled with transposon mutagenesis and marker exchange, has resulted in a tremendous increase in knowledge about the nature of genes and their products. Genetic studies of bacteria in the fifties and sixties concentrated on the identification of genes in terms of their phenotypes, together with attempts to map the order of genes on the bacterial chromosome. This phase of activity seemed to reach its zenith with the publication of the revised map of the *Escherichia coli* K12 chromosome (Taylor and Trotter, 1967). This map confirmed that many functionally-related genes were clustered in operons, although the disposition of operonic clusters on the map was less clear. Bacteria other than *E. coli* were studied on a lesser scale, but included Gram-negative bacteria such as *Salmonella typhimurium* (Sanderson, 1967) and *Pseudomonas aeruginosa* (Holloway, 1969), as well as the Gram-positive actinomycete *Streptomyces coelicolor* (Hopwood, 1967b).

While bacteria such as *E. coli* seemed to have little apparent order to the disposition of their genes on the circular map, *S. coelicolor* exhibited an intriguing degree of symmetry in the arrangement of biosynthetic genes (Hopwood, 1967a). Thus, loci for *cys* mutations were found to map at (approximate locations on the conventional map) 1 o'clock (*cysE*), 4 o'clock (*cysC,D*), 7 o'clock (*cysB*) and 10 o'clock (*cysA*). Another aspect of the map involved the 'blank' regions at 3 and 9 o'clock, which were apparently

The Biology of Acinetobacter, Edited by K. J. Towner *et al.*
Plenum Press, New York, 1991

devoid of genes. Hopwood (1967a) postulated that the curious features of this bacterium's genetic map might reflect repeated genome duplication during evolution, while the so-called blank regions (devoid of both auxotrophic and temperature-sensitive markers) could be artifacts of the conjugation process or recombinational hotspots (Hopwood, 1967b).

In *E. coli*, the reasonably precise understanding of the mechanism of gene exchange, via the F plasmid and derived Hfr strains, could be utilised to map genes by an interrupted mating technique that produced time-related entry of genetic markers into recipient bacteria (Hayes, 1968). In contrast, gene transfer was less well-defined for many of the other bacteria studied. In *P. aeruginosa* mapping was originally based on conjugation mediated by the FP2 sex factor present in some strains (Stanisich and Holloway, 1969) and transduction (Holloway, 1969), later superseded by IS-mediated conjugation involving broad-host range plasmids such as R68.45 (Haas and Holloway, 1976; Jacob et al., 1977). In the Gram-positive actinomycete *S. coelicolor*, a reliable chromosome mapping procedure, based on mixed growth of parental strains, was developed well in advance of any clear idea of the nature of the mechanisms underlying conjugation in this organism (Hopwood et al., 1973).

It was against this background that the first efforts were made (Towner and Vivian, 1976a) to establish a system of conjugal gene transfer which, it was hoped, would lay the foundations for a complete genetic analysis of *Acinetobacter calcoaceticus* and its many-faceted lifestyle. These efforts have subsequently been largely overtaken by the, possibly mistaken, belief that gene-mapping in bacteria is obsolete. Cloning and sequencing have revealed much about the fine detail of particular genes. The debt that most 'cloners' owe to the pioneers of *E. coli* genetics, who provided the basic map on which the detail could be assembled, is often forgotten. In many other bacteria, including *Acinetobacter*, there will be an increasing need to study the relationships that exist between groups of genes and their overall genetic organisation; perhaps, as with so many other aspects of scientific research, chromosome mapping will again become fashionable.

MECHANISMS OF GENE TRANSFER IN *ACINETOBACTER*

Transformation

The first indication that gene transfer could be effected in *Acinetobacter* was a report by Juni and Janik (1969) of transformation in strain BD4 and its micro-encapsulated derivative, BD413. The efficiency of transformation was exceptionally high and it was necessary merely to spread cells freshly produced by growth in broth, together with suitable DNA, on an agar plate and incubate to obtain transformants. Subsequently, Sawula and Crawford (1972) used this technique to map tryptophan genes identified in mutants of BD413. The seven *trp* genes mapped in three groups in an arrangement not previously found in other organisms. The nucleotide sequences of these groups have now been determined and potential regulatory elements identified (Haspel et al., this volume).

Cruze et al. (1979) improved and defined more precisely the parameters for efficient transformation of BD413. Modifications to the procedure for transforming BD413, including a plate-based method for detecting cloned fragments of *A.calcoaceticus* DNA in *E.coli*, were described by Neidle and Ornston (1986). It is worth noting that BD413 does not contain any discernible indigenous plasmids. Coupled with its potential for transformation, BD413 is therefore an extremely useful host for the study of plasmids and cloned genes in *A.calcoaceticus* (M. Divers, P.L. Craven and A. Vivian, unpublished results).

Organisms previously identified as *A. calcoaceticus* comprise a wide range of distinct types of bacteria. Perhaps nowhere is this more apparent than in relation to transformation. In spite of several attempts in our laboratory, strain EBF65/65 (see below) could not be transformed by either chromosomal or plasmid DNA. Conditions for the preparation of competent cells in the closely related strain NCIB 8250 were, however, described by Ahlquist et al. (1980). Maximum transformation frequencies were always lower (10^{-4} to 10^{-5}) than those found (0.7%) by Juni and Janik (1969). In NCIB 8250, competence occurred as a sharp peak at an early stage of batch culture.

Conjugation

The broad host range conjugative plasmid RP4 (Datta et al., 1971) was originally isolated from *P. aeruginosa*. This plasmid, in contrast to its behaviour in both *E. coli* and *P. aeruginosa*, was found to mediate chromosome transfer between auxotrophically-marked strains of EBF65/65 (Towner and Vivian, 1976a). The origin of transfer, as determined by a gradient of frequencies of marker inheritance in filter matings, was fixed for a given strain inheriting RP4 from *E. coli*. Subsequent transfer of RP4 from the initial donor strain (designated DO) to other derivatives of EBF65/65 conferred on the recipients the same donor characteristics. By screening initial RP4 transconjugants (from a cross with *E. coli*) for those with enhanced ability to mediate the transfer of an *ile-1* marker (normally transferred as a distal marker by the DO donor), a novel donor (designated D5) was also obtained. This enabled the circularity of the genetic map of EBF65/65 to be confirmed (Towner and Vivian 1976b; Fig.1).

In a subsequent paper, it was demonstrated that the ability to mobilise the chromosome of EBF65/65 was a feature of many incP group plasmids, with the notable exception of R68.45 (Towner and Vivian, 1977). With the benefit of hindsight and the intervening advances in molecular biology, it would be interesting to re-examine the phenomenon of chromosome mobilisation in EBF65/65 by RP4, and its relatives, with a view to determining whether some special feature of the *A. calcoaceticus*

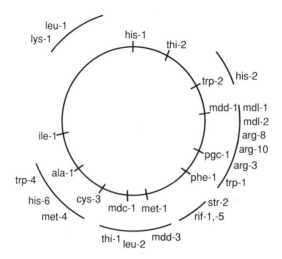

Fig. 1 Map of the chromosome of *Acinetobacter calcoaceticus* strain EBF65/65 (Vivian, 1987) generated by co-inheritance of markers (Table 1) from conjugative matings. Markers located on the circle are in the correct order; those located outside of and overlapping these markers are linked to them, but their precise order has not been confirmed. (Reproduced with permission from "Genetic Maps 1987, vol. 4," p.240, S.J. O'Brien, ed., Cold Spring Harbor Laboratory, New York).

chromosome, such as an insertion sequence, was involved, or whether mobilisation was solely dependent on the properties of RP4.

Hinchliffe and Vivian (1980a) also used the plasmid pAV1, originally isolated from a hospital in Pretoria, to mobilise the chromosome. This plasmid transferred at frequencies up to 10% (about 1000-fold higher than RP4) between strains of EBF65/65. Transfer of chromosomal genes was also up to 100 times more efficient, although it was not possible to discern any clear orientation of transfer (Hinchliffe and Vivian, 1980b). Nevertheless, the co-inheritance of markers transferred by pAV1 has been successfully used to map a number of catabolic genes on the chromosome of EBF65/65 (Vakeria et al., 1984; Fig. 1).

Since this time, regrettably, no further markers have been assigned to the genetic map of EBF65/65. However, many unanswered questions remain about the nature of both RP4- and pAV1-mediated conjugation in *A. calcoaceticus*. It is to be hoped that future changes in scientific priorities may permit these questions to be addressed

Table 1.   Marker designations (see Fig. 1)

Auxotrophic

ala-1	alanine
arg-3,-8,-10	arginine
cys-3	cysteine
his-1,-2,-6	histidine
ile-1	isoleucine
lys-1	lysine
met-1,-4	methionine
phe-1	phenylalanine
thi-1,-2	thiamine
trp-1,-2,-4	tryptophan

Catabolic

mdc-1	constitutive for mandelate R1 regulon
mdd-1,-3	unable to grow on D(-)-mandelate as sole carbon source
mdl-1,-2	unable to grow on L(+)-mandelate as sole carbon source
pgc-1	unable to grow on phenylglyoxylate as sole carbon source

Resistance

rif-1,-5	rifampicin
str-2	streptomycin

Transduction

Bacteriophages, as mediators of gene transfer, have been extremely useful for fine structure mapping of bacterial genomes (e.g. Hartman et al., 1960).   Their occurrence in *A. calcoaceticus* has been documented by several authors (Twarog and Blouse, 1968; Ackermann et al., 1973; Vieu et al., 1979; Pines and Gutnick, 1981) and sought by others.   While phages attacking strains such as JC17 (Hinchliffe and Vivian, 1980a) can be readily obtained from fresh activated sludge effluent, some strains seem to be reluctant hosts, including pAV2⁻ strains (lacking a restriction/modification system) of EBF65/65; indeed, no phages active against EBF65/65, have yet been obtained from a wide range of potential sources (M. Divers, E. Hinchliffe and A. Vivian, unpublished results).   Hermann and Juni (1974) noted that, while they had little difficulty in obtaining phages, all ten phages isolated were restricted in their host range to the strain on which they were originally isolated.

Twarog and Blouse (1968) described a small short-tailed phage, morphologically similar to the coliphages T3 and T7, which was thought to transduce arginine and histidine genes in strain ATCC 15150;   however, a personal communication from R. Twarog, cited in Hermann and Juni (1974), discounts this observation. The latter authors described a generalised transducing phage (P78) for *A. calcoaceticus* strain 78. Frequencies of transduction were low and no evidence for cotransduction of markers was observed.

Strain NCIB 8250

An apparently novel gene transfer system was found in *A. calcoaceticus* NCIB 8250 by Vakeria et al. (1985). This system did not appear to involve a conjugative plasmid, although the strain is known to harbour two cryptic plasmids, pAV10 (approx. 100 kb) and pAV11 (approx. 15 kb). Mixed culture on membrane filters generated recombinants for chromosomal genes at frequencies up to 10^{-5}. Any strain derived from the wild type was capable of acting as a donor of markers, but crosses were frequently observed to be strongly polarised in favour of one or other parent. This system of gene transfer was quite distinct from the transformation system in NCIB 8250 and was resistant to treatment with deoxyribonuclease (Vakeria et al., 1985).

Transposon-Mediated Conjugation

Towner (1980) demonstrated that plasmids carrying the lysogenic bacteriophage Mu have great difficulty establishing in *A. calcoaceticus* EBF65/65. Making use of an RP4::Mu vector (pUN308), it was possible to insert the trimethoprim/streptomycin transposon, Tn3171, into the chromosome of EBF65/65 close to the previously mapped *ile-1* marker (Fig. 1). The insertion resulted in the appearance of arginine auxotrophy in the resulting strain (C4508). An approximate ten-fold increase in the frequency of transfer of the *ile-1* marker was observed in crosses using C4508 carrying RP4, with or without the presence of a second Tn3171 on the mobilising plasmid. This technique clearly has a useful potential for enhancing the frequencies of gene transfer over restricted regions of the genome (Towner, 1983).

MUTATION AND MAPPING STUDIES

A variety of treatments has been described for the isolation of mutants in different strains of *A. calcoaceticus*. Towner & Vivian (1976a) used UV and *N*-methyl-*N'*-nitro-*N*-nitrosoguanidine (NTG) to obtain auxotrophs of strain EBF65/65. In the same strain, ethyl methane sulphonate (EMS) was used to isolate mutants defective in mandelate catabolism (Vakeria et al., 1984). A further procedure for EMS mutagenesis was described by Goosen et al. (1987) for use with strain LMD79.41. In Ac78, Herman and Juni (1974) used NTG mutagenesis, together with an enrichment procedure using cycloserine and ampicillin, to obtain auxotrophs. In a comprehensive comparison of treatments with UV, NTG, EMS and 8-methoxypsoralen plus near-uv irradiation, Ahlquist et al. (1975) concluded that EMS and NTG were the most effective methods for the generation of auxotrophs. These authors used an enrichment procedure involving vancomycin and penicillin V with strain NCIB 8250.

There appear to have been very few attempts to use transposons as mutagenic agents in *A. calcoaceticus* (reviewed by Towner, this volume). Tn5, which has proven to be so useful in a wide range of Gram-negative bacteria, was found to show insertional specificity in strains HO-1N and BD413 (Singer and Finnerty, 1984). Towner (1983) also noted a preferred site of insertion in an *arg* gene in EBF65/65 for the Tn7-like transposon, Tn3171. However, Doten et al. (1987) obtained Tn5 insertions in a number of *pca* genes cloned from strain ADP161 (a derivative of BD413), using mutagenesis mediated by bacteriophage lambda in *E. coli*. This would appear to indicate that the specificity of insertion observed in the genome of BD413 does not operate in *E. coli*.

Towner (1978) produced the first definitive map of the circular genome of strain EBF65/65. Since then, further genes have been mapped, particularly those concerned with the catabolism of mandelate (Vivian, 1987); these are listed in Table 1 and included in the current genetic map (Fig. 1).

CONCLUSIONS

Two opposing types of model to account for the disposition of genes on the chromosomes of bacteria have been suggested. One type invokes chromosome duplication, followed by progressive loss of duplicated functions, while the other involves the acquisition of gene functions, perhaps mediated by plasmids and transposons (Riley and Anilionis, 1978). As already mentioned, duplication has been proposed for *S. coelicolor* (Hopwood, 1967a), while Zipkas and Riley (1975) detected a similar symmetry of functionally related genes in *E. coli*. Thus, four clusters located at 90° contained virtually all the genes concerned with the catabolism of glucose in *E. coli* (Riley and Anilionis, 1978). Knowledge of the genome organisation of *Acinetobacter* remains incomplete and, unfortunately, it is not possible, as yet, to discern any overall features in the arrangement of the genome. Nevertheless, there are sufficient indications to conclude that further work aimed at defining the gross topological structure would be of considerable comparative interest in terms of the evolution and functional significance of bacterial chromosomes. The tools for such investigations are available, particularly the efficient conjugation and transformation systems available for several strains. The prerequisites for genetic manipulation in *A. calcoaceticus* are also in place to enable the fine structure of genes to be determined. Many scientists are perhaps still inclined to feel that all bacteria are essentially like *E. coli*; this should not prevent further work to discover just how different *A. calcoaceticus* might be!

REFERENCES

Ackermann, H-W., Brochu, G., and Cherchel, G., 1973, Structure de trois nouveaux phages de *Bacterium anitratum* (groupe B5W), *J. Microscopie*, 16:215.

Ahlquist, E.F., Fewson, C.A., and Ritchie, D.A., 1975, The induction of mutants of *Acinetobacter calcoaceticus* NCIB 8250 and their selection by vancomycin, *J. Gen. Microbiol.*, 91:338.

Ahlquist, E.F., Fewson, C.A., Ritchie, D.A., Podmore, J., and Rowell, V., 1980, Competence for genetic transformation in *Acinetobacter calcoaceticus* NCIB 8250, *FEMS Microbiol. Lett.*, 7:107.

Cruze, J.A., Singer, J.T., and Finnerty, W.R., 1979, Conditions for quantitative transformation in *Acinetobacter calcoaceticus*, *Curr. Microbiol.*, 3:129.

Datta, N., Hedges, R.W., Shaw, E.J., Sykes, R.B., and Richmond, M.H., 1971, Properties of an R factor from *Pseudomonas aeruginosa*, *J. Bacteriol.*, 108:1244.

Doten, R.C., Ngai, K-L., Mitchell, D.J., and Ornston, L.N., 1987, Cloning and genetic organization of the *pca* gene cluster from *Acinetobacter calcoaceticus*, *J. Bacteriol.*, 169:3168.

Goosen, N., Vermaas, D.A.M., and Van de Putte, P., 1987, Cloning of the genes involved in synthesis of coenzyme pyrrolo-quinoline-quinolone from *Acinetobacter calcoaceticus*, *J.Bacteriol.*, 169:303.

Haas, D., and Holloway, B.W., 1976, R factor variants with enhanced sex factor activity in *Pseudomonas aeruginosa*, *Mol.Gen.Genet.*, 144: 243.

Hartman, P.E., Loper, J.C., and Serman, D., 1960, Fine structure mapping by complete transduction between histidine-requiring *Salmonella* mutants, *J. Gen. Microbiol.*, 22:323.

Hayes, W., 1968, "The Genetics of Bacteria and their Viruses," Blackwell Scientific Publications, Oxford.

Herman, N.J., and Juni, E., 1974, Isolation and characterization of a generalized transducing bacteriophage for *Acinetobacter*, *J. Virol.*, 13:46.

Hinchliffe, E., and Vivian, A., 1980a, Naturally occurring plasmids in *Acinetobacter calcoaceticus*: a P class R factor of restricted host range, *J. Gen. Microbiol.*, 116:75.

Hinchliffe, E., and Vivian, A., 1980b, Gene transfer in *Acinetobacter calcoaceticus:* fertility variants of the sex factor pAV1, *J. Gen. Microbiol.*, 119:117.

Holloway, B.W., 1969 Genetics of *Pseudomonas*, *Bacteriol. Rev.*, 33:419.

Hopwood, D.A., 1967a, In discussion to F.W. Stahl's paper: Circular genetic maps, *J. Cell. Physiol.*, 70(suppl.1):1.

Hopwood, D.A., 1967b, Genetic analysis and genome structure in *Streptomyces coelicolor*, *Bacteriol. Rev.*, 31:373.

Hopwood, D.A., Chater, K.F., Dowding, J.E., and Vivian, A., 1973, Advances in *Streptomyces coelicolor* genetics, *Bacteriol. Rev.*, 37:371.

Jacob, A.E., Cresswell, J.M., and Hedges, R.W., 1977, Molecular characterization of the P group plasmid R68 and variants with enhanced chromosome mobilizing ability, *FEMS Microbiol. Lett.*, 1:71.

Juni, E., and Janik, A., 1969, Transformation of *Acinetobacter calco-aceticus (Bacterium anitratum)*, *J.Bacteriol.*, 98:281.

Neidle, E.L., and Ornston, L.N., 1986, Cloning and expression of *Acinetobacter calcoaceticus* catechol 1,2-dioxygenase structural gene *catA* in *Escherichia coli*, *J.Bacteriol.*, 168:815.

Pines, O., and Gutnick, D.L., 1981, Relationship between phage resistance and emulsan production, interaction of phages with the cell-surface of *Acinetobacter calcoaceticus* RAG-1, *Arch. Microbiol.*, 130:129.

Riley, M., and Anilionis, A., 1978, Evolution of the bacterial genome, *Ann. Rev. Microbiol.*, 32:519.

Sanderson, K.E., 1967, Linkage map of *Salmonella typhimurium*, *Bacteriol. Rev.*, 31:341.

Sawula, R.V., and Crawford, I.P., 1972, Mapping of the tryptophan genes of *Acinetobacter calcoaceticus* by transformation, *J. Bacteriol.*, 112:797.

Singer J.T., and Finnerty, W.R., 1984, Insertional specificity of transposon Tn5 in *Acinetobacter* sp., *J.Bacteriol.*, 157:607.

Stanisich, V.A., and Holloway, B.W., 1969, Conjugation in *Pseudomonas aeruginosa*, *Genetics*, 61:327.

Taylor, A.L., and Trotter, C.D., 1967, Revised linkage map of *Escherichia coli*, *Bacteriol. Rev.*, 31:332.

Towner, K.J., 1978, Chromosome mapping in *Acinetobacter calcoaceticus*, *J. Gen. Microbiol.*, 104:175.

Towner, K.J., 1980, Behaviour of bacteriophage Mu in *Acinetobacter calcoaceticus* EBF65/65, *J. Gen. Microbiol.*, 121:425.

Towner, K.J., 1983, Transposon-directed mutagenesis and chromosome mobilization in *Acinetobacter calcoaceticus* EBF65/65, *Genet. Res.*, 41:97.

Towner, K.J., and Vivian, A., 1976a, RP4-mediated conjugation in *Acinetobacter calcoaceticus*, *J. Gen. Microbiol.*, 93:355.

Towner, K.J., and Vivian, A., 1976b, RP4 fertility variants in *Acinetobacter calcoaceticus*, *Genet. Res.*, 28:301.

Towner, K.J., and Vivian, A., 1977, Plasmids capable of transfer and chromosome mobilization in *Acinetobacter calcoaceticus*, *J. Gen. Microbiol.*, 101:167.

Twarog, R., and Blouse, L.E., 1968, Isolation and characterization of transducing bacteriophage BP1 for *Bacterium anitratum (Achromobacter* sp.), *J. Virol.*, 2:716.

Vakeria, D., Vivian, A., and Fewson, C.A., 1984, Isolation, characterization and mapping of mandelate pathway mutants of *Acinetobacter calcoaceticus*, *J. Gen. Microbiol.*, 130:2893.

Vakeria, D., Fewson, C.A., and Vivian, A., 1985, Gene transfer in *Acinetobacter calcoaceticus* NCIB 8250, *FEMS Microbiol. Lett.*, 26:141.

Vieu, J.F., Minck, R., and Bergogne-Berezin, E., 1979, Bacteriophages et lysotypie de *Acinetobacter, Ann. Microbiol.*, 130A:405.

Vivian, A., 1987, *Acinetobacter calcoaceticus* strain EBF65/65, *in*: "Genetic Maps 1987, vol. 4," p.240, S.J. O'Brien, ed., Cold Spring Harbor Laboratory, New York.

Zipkas, D., and Riley, M., 1975, Proposal concerning mechanism of evolution of the genome of *Escherichia coli, Proc. Natl. Acad. Sci. USA.*, 72:1354.

EVOLUTION OF GENES FOR THE ß-KETOADIPATE PATHWAY IN *ACINETOBACTER CALCOACETICUS*

L.N. Ornston[a][*] and E.L. Neidle[b]

[a]Department of Biology, Yale University
New Haven, CT 06511, USA

[b]Department of Microbiology, Medical School
University of Texas at Houston
Houston, TX 77225, USA

INTRODUCTION

The ß-ketoadipate pathway (Fig. 1) is widely distributed in the microbial world (Stanier et al., 1966; Ornston and Ornston, 1972; Stanier and Ornston, 1973; Cain, 1980; Parke and Ornston, 1984) and was one of the first subjects of physiological investigation of enzyme induction in bacteria (Stanier, 1951). Enzymes associated with the pathway proved to be inducible in fluorescent *Pseudomonas* species, and later studies revealed that mechanisms of transcriptional control were conserved in this biological group (Ornston, 1966; Kemp and Hegeman, 1968). Representatives of *Acinetobacter* (formerly *Moraxella*) share induction patterns unlike those found in *Pseudomonas* (Ornston, 1966; Canovas and Stanier, 1967; Canovas et al., 1967; Stanier and Ornston, 1973). Investigation of the genetic basis for the different forms of transcriptional regulation exercised in the two genera has given insight into processes underlying their evolutionary divergence.

This review summarises evidence pointing to five general conclusions. First, structural genes for isofunctional enzymes from *Acinetobacter* and *Pseudomonas* share common ancestors. Where examined, enzymes from the two biological groups share amino acid sequence identity that usually exceeds 40%. The level of amino acid sequence similarity in the enzymes is remarkable because, with few exceptions, the GC content of the structural genes for the enzymes differs by about 20%. The difference in GC content is reflected in highly distinctive patterns of codon usage. Second, extensive

[*]corresponding author

The Biology of Acinetobacter, Edited by K. J. Towner *et al.*
Plenum Press, New York, 1991

Fig. 1. Designations for structural genes associated with the ß-ketoadipate pathway in *Acinetobacter*. The protocatechuate 3,4-dioxygenase gene, formerly *pcaA*, now is indicated by *pcaG* and *pcaH* which respectively encode its nonidentical α and ß subunits. Designations *catI* and *catJ* or *pcaI* and *pcaJ* represent the α and ß subunits respectively of ß-ketoadipate succinyl CoA transferase. Structural genes for this enzyme were formerly designated *catE* and *pcaE*. Gene designations correspond to enzymes as follows: *pobA*, p-hydroxybenzoate hydroxylase (EC 1:14:13:2); *pcaG* and *pcaH*, protocatechuate 3,4-dioxygenase (EC 1.99.2.3); *pcaB*, ß-carboxy-cis,cis-muconate cycloisomerase (EC 5.5.1.2); *pcaC*, γ-carboxymuconolactone decarboxylase (EC 4.1.1.44); *pcaD* or *catD*, ß-ketoadipate enol-lactone hydrolase (EC 3.1.1.24); *pcaI* and *pcaJ* or *catI* and *catJ*, ß-ketoadipate:succinyl-CoA transferase (EC 2.8.3.6); *pcaF* or *catF*, ß-ketoadipyl CoA thiolase (EC 2.3.1.16); *benABC*, reductive benzoate dioxygenase (EC 1.13.99.2); *benD*, 1,2-dihydro-1,2-dihydroxybenzoate dehydrogenase (EC 1.3.1.25); *catA*, catechol 1,2-dioxygenase (EC 1.13.11.1); *catB*, cis,cis-muconate cycloisomerase (EC 5.5.1.1); *catC*, muconolactone isomerase (EC 5.3.3.4).

METABOLITES

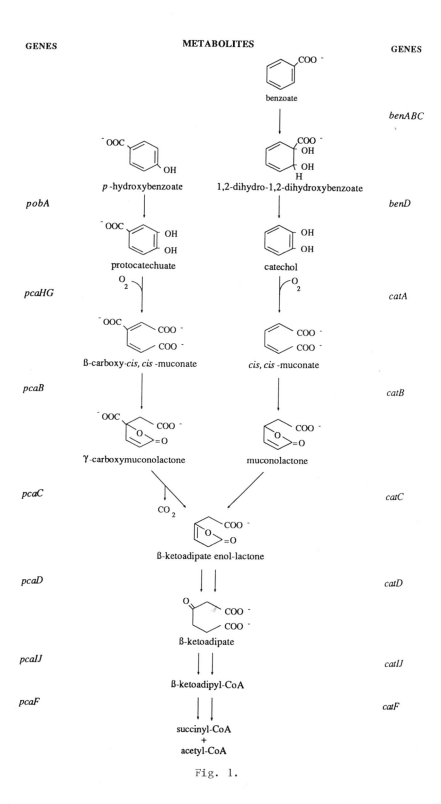

Fig. 1.

gene rearrangements took place as the ß-ketoadipate pathway diverged in *Acinetobacter* and *Pseudomonas*. In some cases, gene rearrangements resulted in divergent transcriptional controls. In other instances, the amount of gene rearrangement exceeds the level required for divergent transcriptional regulation. Third, genes for physiologically interdependent steps in the ß-ketoadipate pathway tend to be linked in supraoperonic clusters within the chromosomes of *Acinetobacter* and *Pseudomonas*. Selective pressures favouring this clustering have not been defined, but the genetic properties of *Acinetobacter* open this problem to analysis. Fourth, genes carried in the *Acinetobacter* chromosome possess the potential ability to interact. Some of the interactions are physiological and involve the intermediacy of either regulatory proteins or inducer-metabolites. Other interactions are genetic, as exemplified by sequence exchange between nearly identical regions of DNA encoding isofunctional enzymes associated with the ß-ketoadipate pathway. Fifth, entirely different patterns of DNA sequence repetition were acquired as homologous structural genes diverged in *Acinetobacter* and *Pseudomonas*. The patterns suggest that sequence exchange between slipped DNA strands caused mutations leading to genetic divergence; repair events involving sequence exchange between slipped DNA strands may contribute to conservation of the evolved sequences.

The extraordinary competence of one *Acinetobacter* strain for natural transformation (Juni, 1972) opens opportunities for analysis of genes and gene clusters. This review concludes with a summary of some procedures that may facilitate future examination of the genetic properties of *Acinetobacter*.

HOMOLOGOUS GENES WITH ANALOGOUS FUNCTIONS IN *ACINETOBACTER* AND *PSEUDOMONAS*

In so far as has been determined, genes with analogous functions in *Acinetobacter* and *Pseudomonas* share common ancestry (McCorkle et al., 1980; Yeh et al., 1980b; Yeh and Ornston, 1981; Kaplan et al., 1984; Neidle et al., 1988; Hartnett et al., 1990; Ornston et al., 1990b). Homologous proteins from the two genera possess a level of amino acid sequence identity that is particularly striking in the light of divergence at the level of DNA that has produced a difference of about 20% in GC content (Neidle et al., 1988; Hartnett et al., 1990). These observations are illustrated by comparison of the respective *catMBC* and *catRBC* genes from *Acinetobacter* and *Pseudomonas* (Table 1). The *catB* and *catC* genes encode enzymes catalysing consecutive metabolic transformations of muconate (Fig. 1). Common ancestry of the *catB* and *catC* genes from the two genera is demonstrated by conservation of amino acid sequence identities of about 50% in the respective gene products (Aldrich et al., 1987; Aldrich and Chakrabarty, 1988; J. Houghton and J. Hughes, unpublished observations). The crystal structures of the protein products of the *P. putida catB* and *catC* gene products have been determined (Goldman et al., 1985; Katz et al., 1985; Katti et al., 1988; Allewell, 1989); many of the

Table 1. Properties of homologous genes from *Acinetobacter* and *Pseudomonas*

	Acinetobacter			*Pseudomonas*		
	catM[a] *catB*		*catC*	*catR*[a] *catB*		*catC*
GC in DNA	44.6	44.7	41.6	66.2	63.6	61.8
ATT codons for Ile[b]	72.0	63.6	57.1	22.2	26.7	33.3
TTA codons for Leu[b]	6.0	29.6	37.5	2.4	0.0	0.0
AGA codons for Arg[b]	40.0	0.0	0.0	0.0	0.0	0.0

[a]the *catM* and *catR* regulatory gene products recognise muconate as an inducer and share an amino acid sequence identity of 40%. The genes differ in that *catM* encodes a repressor whereas *catR* encodes an activator of transcription.
[b]Expressed as a percentage of all the codons

amino acid residues conserved with those of the corresponding *Acinetobacter* gene products make essential contributions to protein structure (Katti et al., 1988).

The crystal structures of *Acinetobacter* CatM and *Pseudomonas* CatR are unknown, but the 40% amino acid sequence identity shared by these proteins indicates that they have experienced selective pressure for extensive structural conservation (Neidle et al., 1989; Rothmel et al., 1990; J. Houghton, unpublished observations). Some regions of conservation, notably those associated with a potentially DNA-binding helix-turn-helix, are conserved with other members of the wide ranging LysR family (Henikoff et al., 1988) of regulatory proteins (Fig.2). Other regions of sequence conservation are unique to CatM and CatR; some of these regions may have been selected on the basis of their common function, binding of the inducer muconate so as to govern expression of the *cat* structural genes.

In general, CatM and CatR resemble each other more closely than they resemble a consensus sequence for other members of the LysR family (Henikoff et al., 1988). As illustrated in Fig. 2, the sequence similarities are widely distributed through the primary structures of CatM and CatR, and it thus seems probable that the muconate-binding regulatory proteins shared their evolutionary history after divergence from other transcriptional regulators. The inferred common evolutionary history for the muconate-binding proteins is remarkable because they acquired opposite mechanisms

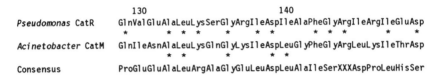

A

	20	30	40
Pseudomonas CatR	ArgAlaAlaGluLeuLeuHisIleAlaGlnProProLeuSerArgGlnIleSerGlnLeuGlu		
	* * *· * * * * * * * * * * *		
Acinetobacter CatM	LysAlaAlaGluLysLeuCysIleAlaGlnProProLeuSerArgGlnIleGlnLysLeuGlu		
	* * * * * * * * * * *		
Consensus	AlaAlaAlaArgAlaLeuHisLeuSerGlnProAlaIleSerArgGlnIleAlaArgLeuGlu		

B

	130	140	
Pseudomonas CatR	GlnValGluAlaLeuLysSerGlyArgIleAlaAspIleAlaPheGlyArgIleAlaArgIleGluAsp		
	* * * * * * * * * * * *		
Acinetobacter CatM	GlnIleAsnAlaAlaLeuLysGlnGlyLysIleAspLeuGlyPheGlyArgLeuLysIleThrAsp		
	* * * * * *		
Consensus	ProGluGluAlaLeuArgAlaGlyGluLeuAspLeuAlaIleSerXXXAspProLeuHisSer		

C

	230	240	
Pseudomonas CatR	GluLeuGlnThrAlaIleGlyLeuValAlaAlaValGlyValGlyValThrLeuValProAlaSer		
	* * * * * * * * * * * * *		
Acinetobacter CatM	GluIleGlnLeuAlaLeuGlyLeuValAlaAlaGlyGlyGlyValCysIleValAlaProAlaSer		
	* * * * *		
Consensus	GlySerValXXXMetValValMetLeuAlaAlaGlyValGlyIleAlaAlaLeuProLeuVal		

Fig. 2. Conserved regions in the deduced amino acid sequences of *Acinetobacter* CatM and *Pseudomonas* CatR. Numerals indicate positions of amino acid residues in the CatR sequence; asterisks indicate amino acid sequence identity of CatM with either CatR or a consensus sequence (Henikoff et al., 1988). The consensus sequence, a rough representation of an ancestral sequence, contains the most frequently shared amino acid residues in the LysR family of transcriptional regulators. (A) All three sequences share substantial similarity in the helix-turn-helix region that is likely to be associated with DNA binding. In this region, and in internal regions (exemplified by (B) and (C)) the CatR and CatM proteins resemble each other more closely than they resemble the putative ancestral sequence.

of transcriptional control: CatM represses gene expression in *Acinetobacter* (Neidle et al., 1989) whereas CatR activates transcription in *Pseudomonas* (Wheelis and Ornston, 1972; J. Hughes et al., unpublished observations). A possibility raised by these observations is that activation is a preferred mode of regulation for the GC-rich genes of *Pseudomonas*, whereas repression may be a favoured mechanism for control of the AT-rich genes in *Acinetobacter*.

Selection for different mechanisms of transcriptional control was one force favouring divergence of *catM* and *catR*. Another selective force, acting generally upon genes in bacteria, appears to have been a demand for codon usage consonant with the distribution of transfer RNA molecules in the host cell (Grantham et al., 1981; Ikemura, 1981a,b). An illustration of differences in codon usage, shown in Table 1, is the predominance of ATT codons for isoleucine in the relatively AT-rich genes from *Acinetobacter*.

Consistent with the notion that the translational process exerts a powerful selective force on gene divergence, patterns of codon usage for different genes within a species tend to be remarkably constant (Kaplan et al., 1984; West and Iglewski, 1988). Constancy of GC content also tends to be observed (Chargaff, 1950), and it is difficult to escape the conclusion that selection for codon usage exerts a profound effect on the GC content of genes (Grantham et al., 1981; Ikemura, 1981a,b). However, it should be noted that selective demand for codon usage may not be the sole factor influencing the GC content of genes. As described below, gene conversion mutations may shift DNA sequences within the chromosome and therefore may serve as a genetic source providing to a gene a GC content consistent with that of surrounding genes.

There are significant exceptions to general patterns of codon usage. Known exceptions include regulatory genes (Konigsberg and Godson, 1983; Sharp and Li, 1986), and an additional example is the *Acinetobacter catM* gene (Neidle et al., 1989). As shown in Table 1, this gene has resisted the tendency to relatively high usage of the TTA codon illustrated by the *Acinetobacter catB* and *catC* genes. Even more striking is the presence of AGA as four of the ten arginine codons in the *Acinetobacter catM* gene. The AGA codon is found in none of the seven *cat* structural genes from *Acinetobacter,* yet occupies two of the first seven codons translated during synthesis of the *catM* encoded repressor molecule (Fig. 3). One interpretation of the unusual codon usage patterns in regulatory genes is that their low levels of expression have not challenged pools of activated transfer RNA molecules to the extent demanded by highly expressed structural genes (Konigsberg and Godson, 1983; Sharp and Li, 1986). The unusual distribution of AGA codons in *catM* prompts the additional speculation that the exceptional codon usage pattern may have been selected, perhaps in part as a mechanism for controlling expression of the repressor gene. It is

```
Pseudomonas catR        ATGGAGCTGCGCCACCTGCGTTACTTCAAG
                        MetGluLeuArgHisLeuArgTyrPheThr
                        *  *  *  *  *  *  *  *  *
                        MetGluLeuArgHisLeuArgTyrPheVal
Acinetobacter catM      ATGGAACTAAGACACCTCAGATATTTTGTG
```

Fig. 3. The translational origin of the *Pseudomonas catR* and *Acinetobacter catM* genes. The amino acid sequences of the translated gene products are nearly identical in this region. Highlighted in bold and underlined are two of the four AGA arginyl codons in *catM*. The presence of these codons, absent in the *Pseudomonas catRBC* region (Table 1) and unrepresented in the seven *Acinetobacter cat* structural genes, suggests that the normally infrequent AGA codons may have been selected during divergence of the *catM* gene.

noteworthy that a mutation in the cognate tRNA for AGA selectively lowers expression of AGA-rich genes in *Escherichia coli* (Chen et al., 1990).

GENE REARRANGEMENTS DURING DIVERGENCE OF THE ß-KETOADIPATE PATHWAY

Without known exception, genes for isofunctional enzymes from *Acinetobacter* and *Pseudomonas* can be traced to a common ancestor. This knowledge opens opportunities to explore how genes were rearranged during evolution and, as described later, to analyse how DNA sequences within genes changed during their divergence.

Extensive gene rearrangements took place during evolution of the ß-ketoadipate pathway. As shown in Fig. 4A, the *catIJFD* region, unrepresented in fluorescent *Pseudomonas* species (Ornston, 1966; Kemp and Hegeman, 1968; Hughes et al., 1988), is fused to the *catBC* genes in *Acinetobacter* (Shanley et al., 1986; A. Harrison et al., unpublished observations). The *cat* genes share an element of organisation common to many operons governed by members of the LysR family of transcriptional regulators (Henikoff et al., 1988): the regulatory genes lie upstream and are divergently transcribed from the structural genes that they govern (Fig. 4A).

Organisation of the *Pseudomonas pca* genes is somewhat fragmentary as compared with the tight linkage of these genes in *Acinetobacter*. Depicted in Fig. 4B, the *pcaBDC* operon represents the most extensive linkage of *pca* genes in *P. putida* (Hughes et al., 1988). In contrast, all of the *Acinetobacter pca* structural genes are linked in a single operon (Doten et al., 1987a). Comparison of the two operons reveals that the order of the neighbouring *pcaD* and *pcaB* genes was reversed during their divergence in *Acinetobacter* and *Pseudomonas* (Fig. 4B).

Rearrangement of genes during evolution of the ß-ketoadipate pathway within *Acinetobacter* appears to have occurred because the *catIJFD* region, lying at the end of the *cat* operon (Shanley et al., 1986), is represented by its homologue *pcaIJFD* placed at the beginning of the *pca* operon (Doten et al., 1987a; Fig. 4C). DNA sequences of the *catIJF* and *pcaIJF* regions are nearly identical and, as described below, the similarity appears to afford opportunity for DNA sequence exchange between the two regions.

SUPRAOPERONIC CLUSTERING

Transductional analysis of fluorescent *Pseudomonas* strains (Rosenberg and Hegeman, 1969; Wheelis and Stanier, 1970) gave the first indications that independently regulated genes for aromatic catabolism were clustered in the bacterial chromosome. The clustered genes frequently participate in the same overall physiological function, such as the utilisation

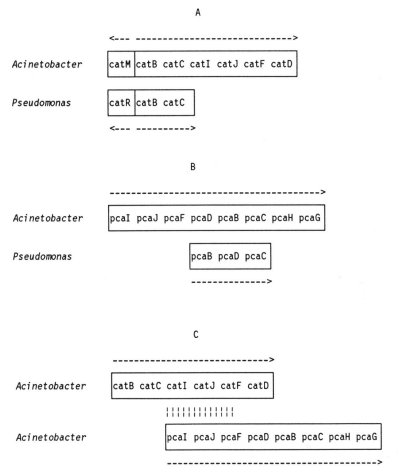

Fig. 4. Organisation of *cat* and *pca* genes. Genes within operons are
enclosed in boxes, and arrows indicate the direction of transcription.
(A) In *Acinetobacter*, the *catBC* sequence forms an operon with the
downstream *catIJFD* region. A direct counterpart of the catIJFD region is
not present in fluorescent *Pseudomonas* strains such as *P. putida*. In
these organisms, the physiological function of the region is fulfilled by
the *pcaI*, *pcaJ*, *pcaF* and *pcaD* genes. (B) All of the structural genes
specifically required for metabolism of protocatechuate (Fig. 1) are
linked within the *pca* operon of *Acinetobacter*. Organisation of *pca* genes
in *P. putida* is relatively fragmentary. In the longest *P. putida pca*
operon, the transcriptional order of the linked *pcaB* and *pcaD* genes is
the reverse of that found in *Acinetobacter*. (C) In *Acinetobacter*, the
catIJFD region appears at the end of the *cat* operon whereas the
homologous region lies at the beginning of the *pca* operon. Vertical
lines indicate near identity of the *catIJF* and the *pcaIJF* genes in DNA
sequence. The *catD* and *pcaD* genes have diverged substantially in
Acinetobacter (Yeh et al., 1980a).

of benzoate. The most fully characterised supraoperonic cluster, associated with benzoate metabolism in *Acinetobacter* (Neidle et al., 1987) is depicted in Fig. 5A. The complete DNA sequence of this 16 kb cluster has been determined. It contains three open reading frames (ORF1, ORF2 and ORF3; Fig. 5A). The functions of these open reading frames are yet to be clearly established.

Positioned to be expressed in a transcript downstream from *catA*, the structural gene for catechol oxygenase, are ORF2 and ORF3 (Fig. 5A). Genetic disruption of these open reading frames does not prevent growth with either anthranilate or benzoate, compounds that are metabolised via catechol (Neidle et al., 1989). As described below, some evidence suggests that these open reading frames may contribute to transcriptional regulation.

The *catM* gene, the only regulatory gene to be characterised in the 16 kb segment, represses expression of both *catA* and the *catBCIJFD* operon (Neidle et al., 1989). As is often the case with members of the *lysR* family of regulatory genes (Henikoff et al., 1988), the *catM* gene is transcribed divergently from the genes it controls. The orientation of the genes may contribute to regulation of their expression, and this possibility can be explored by examining the physiological consequences of reversing the orientation of the *catA* gene or the *catBCIJFD* operon.

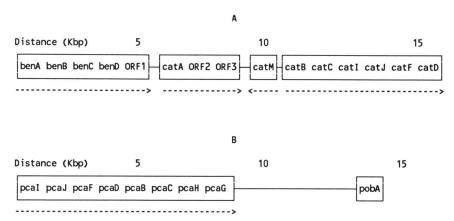

Fig. 5. Supraoperonic clustering. The distances are in kb from the translational origin of the first operon in the cluster. Arrows indicate the direction of transcription. (A) The *ben* and *cat* genes are clustered within 16 kb of DNA and are transcribed as four separate transcriptional units. The *catM* gene encodes a repressor of both *catA* and the *catBCIJFD* operon. Functions are yet to be assigned to the open reading frames designated ORF1, ORF2 and ORF3. (B) The direction of transcription of *pobA* is yet to be determined; this gene lies about 5 kb downstream from the end of the *pcaIJFDBCHG* operon.

As described below, transcriptional control of the *ben-cat* genes is complex, and it is evident that additional regulatory genes lie outside the 16 kb segment. The most persuasive evidence in support of this conclusion is the observation that the cloned *catA* gene, carried on a broad host-range plasmid in an *Acinetobacter* strain from which the 16 kb *ben-cat* segment has been deleted, remains subject to induction in response to either benzoate or muconate (Neidle et al., 1989).

All of the eight structural genes specifically associated with protocatechuate metabolism are contained within the *pcaIJFDBCHG* operon (Doten et al., 1987a; Hartnett et al., 1990). As shown in Fig. 5B, the independently transcribed *pobA* gene lies about 5 kb downstream from the *pcaG* gene (B. Averhoff and G. Hartnett, unpublished observations). Thus an additional supraoperonic cluster in the *Acinetobacter* chromosome has been identified. Other genes that may prove to be associated with this cluster are those concerned with conversion of hydroaromatic growth substrates, quinate and shikimate, to protocatechuate (Tresguerres et al., 1970). These genes, like the *pcaIJFDBCHG* operon, are expressed in response to the inducer protocatechuate, and all of the genes governed by protocatechuate appear to share a common regulatory gene (Canovas et al., 1968b).

Thus far we have discussed chromosomal genes, and genes for the ß-ketoadipate pathway generally are associated with the chromosome. Exceptions exist (Hewetson et al., 1978). Of particular interest is the full set of *cat* genes carried on an *Acinetobacter* plasmid associated with benzene metabolism (Winstanley et al., 1987). The distance between *catA* and *catB* is greater in the plasmid than in the chromosome, and it is possible that there are other differences in gene organisation. It is not known if the plasmid carries a counterpart of the regulatory gene *catM*.

Clustering of plasmid-borne genes can be readily interpreted as the consequence of a demand for their cotransfer during conjugation (Wheelis, 1975). This argument is less effective when applied to chromosomal genes that, as described above, have demonstrated marked potential for gene rearrangement during their divergence. In the absence of selection, rearrangement mutations would be expected to lead to randomisation of the chromosomal location of independently transcribed genes. Absence of such randomisation, exemplified by closer clustering of *catA* and *catB* on the chromosome than on a plasmid, indicates that selective demands can maintain close linkage of genes that contribute to an overall nutritional function such as metabolism of benzoate. One set of demands may be physiological: examples might be either global transcriptional controls exercised over linked operons or the formation of DNA loops that may influence gene expression (Schleif, 1988). Other selective demands may be genetic. DNA sequence exchange may buffer genes against change by repairing mutations (Edelman and Gally, 1970; Nagylaki and Petes, 1982; Xiong et al., 1988; Meagher et al., 1989). The possibility exists that such genetic interactions contribute to the stabilisation of genes within linkage groups (Ornston et al., 1990a,b). Physiological and genetic interactions between chromosomal *Acinetobacter* genes for the ß-ketoadipate pathway are discussed below.

PHYSIOLOGICAL AND GENETIC INTERACTIONS BETWEEN DIFFERENT CHROMOSOMAL DNA SEGMENTS

Many of the physiological interactions between different chromosomal DNA segments are indirect because they are mediated in part by inducer-metabolites and regulatory proteins. Such interactions can be interpreted as the consequence of the movement of regulatory molecules from one chromosomal region to another. Other controls exhibit complexity, raising the possibility that one region of the chromosome may influence regulation exercised in another. Direct physical interaction between different regions of the chromosome is indicated by observed repair of a mutation in the *pcaJ* gene, a process that depends upon the presence of the unlinked *catJ* sequence (Doten et al., 1987b).

Physiological Interactions

Mutant strains that constitutively express ß-ketoadipate-:succinyl CoA transferase. Among the first regulatory mutant strains isolated from *Acinetobacter* were organisms that had lost the ability to form protocatechuate 3,4-dioxygenase (Canovas and Stanier, 1967). These strains were selected with ß-ketoadipate, a growth substrate that demands constitutive synthesis of ß-ketoadipate:succinyl CoA transferase and ß-ketoadipyl CoA thiolase. Further analysis revealed that loss of the oxygenase was accompanied by constitutive expression of the other *pca* genes, including those that encode the transferase and thiolase activities (Canovas et al., 1968a). The apparent cause of constitutivity was metabolic formation of the inducer-metabolite protocatechuate from endogenously synthesised shikimate (Fig. 6).

3-Hydroxy-4-methylbenzoate, a chemical analogue of protocatechuate, acts as an anti-inducer and prevents constitutive expression of the remaining functional *pca* genes in mutant organisms lacking protocatechuate-oxygenase (Canovas et al., 1968b). When these strains are exposed to ß-ketoadipate in the presence of the protocatechuate analogue, secondary mutant strains with altered regulation can be selected. Mutations in these strains change the specificity of induction so that the analogue, rather than the natural substrate protocatechuate, acts as inducer of the *pca* enzymes. The gene associated with altered inducer-specificity is yet to be mapped.

Selection for growth of *Acinetobacter* with ß-ketoadipate yields additional strains in which *cat* genes are expressed constitutively. Some of these organisms carry mutations in the repressor gene *catM* (Neidle et al., 1989). Others among the mutants constitutively express the *catCIJFD*

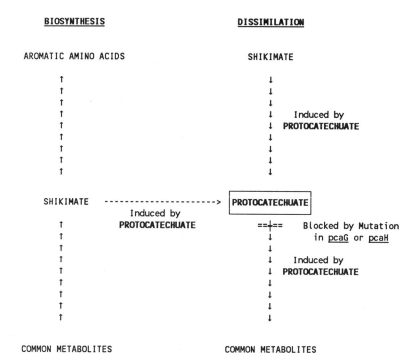

Fig. 6. The dual role of shikimate as a biosynthetic intermediate and as a growth substrate. Shikimate, formed as an intermediate in the biosynthesis of aromatic amino acids, can serve as a growth substrate for *Acinetobacter*. Protocatechuate is an intermediate in the dissimilation of shikimate and induces synthesis of all of the enzymes associated with this metabolic pathway. Mutations in *pcaG* or *pcaH* block synthesis of protocatechuate oxygenase and result in the metabolic accumulation of protocatechuate. Such mutants constitutively form other enzymes associated with shikimate catabolism. Apparently, catabolism of shikimate formed biosynthetically in the mutant strains causes accumulation of endogenous inducer in quantities sufficient to trigger synthesis of the full set of catabolic enzymes.

region, but retain inducible control over *catA* and *catB* (Canovas and Johnson, 1968; Patel et al., 1975). An interpretation of this finding is that these mutant strains have acquired a promoter within or near *catB*. This proposal has not been explored genetically.

Demand for growth of *Acinetobacter* strains with ß-ketoadipate yields one set of mutant organisms that are constitutive for neither *pca* nor *cat* genes. Biochemical investigation revealed that such strains constitutively form a ß-ketoadipate:succinyl CoA transferase that exhibits relatively low affinity for ß-ketoadipate and high affinity for adipate as a competitive inhibitor (Canovas and Johnson, 1968). Evidently this transferase is associated with adipate metabolism, yet contains substrate specificity sufficiently broad to allow growth with ß-ketoadipate. The properties of these strains suggest that they constitutively form a ß-ketoadipyl CoA thiolase that is not encoded by *pcaF* or by *catF*. The existence of such a thiolase has been confirmed by demonstration that growth with adipate is unimpeded in *Acinetobacter* strains from which the *pcaF* and *catF* genes have been deleted (L. Gregg, unpublished observations).

Regulation of the expression of the ben genes. The *benABC* region contains genes for the three proteins that catalyse the reductive oxygenation of benzoate to the corresponding 1,2 dihydrodiol (Fig. 1). This metabolite is converted to catechol by a dehydrogenase encoded by *benD* (Fig. 1). DNA sequence evidence (E.Neidle and C.Hartnett, unpublished observations) suggests that the three *ben* genes are expressed as a single transcript that is likely to include ORF1 (Fig. 5). The terminus of *benC* is separated by only 18 bp from the translational origin of *benD*, and there is no discernible sequence evidence for separation of transcriptional control of *benD* from *benABC*. The activity of the NAD-dependent dehydrogenase encoded by *benD* is relatively easy to determine, and measurement of this activity reveals complexity of transcriptional control.

A mutation within *benABC* prevents metabolism of benzoate, but does not prevent this metabolite from inducing full levels of *benD* expression in *Acinetobacter* (Neidle et al., 1987). Thus benzoate is identified as an inducer of the putative *benABCD* operon. Benzoate dihydrodiol, the product of the *benABC*-encoded enzyme, does not support growth of the wild-type cells. This may be because the diol does not induce *benD* expression, but the possibility that the compound does not permeate the cell membrane at a rate sufficient to support growth has not been excluded.

Physiological complexity arises from the fact that growth with muconate, a compound three metabolic steps removed from benzoate (Fig. 1), elicits expression of *benD* but not of *benC* (Neidle et al., 1987). Furthermore, muconate-grown cells do not oxidise benzoate (Canovas and Stanier, 1967). Thus induction with muconate separates the physiological expression of *benABC* from expression of *benD*. The possibility that benzoate and muconate induce separate enzymes with the *benD*-encoded activity is eliminated by the fact that mutations within *benD* prevent its expression in response to either benzoate or muconate (Neidle et al., 1987).

Physiological separation of the *benABC* and *benD* encoded activities does not require that the genes be transcribed separately. It is possible that control is exercised by inactivation of either the *benABC* transcript or the protein gene products. This possibility is heightened by the fact that mutations within *benD* lead to minimal expression of the oxygenative function associated with the upstream *benABC* genes (Neidle et al., 1987). These genes express a reductive benzoate oxygenase at high levels when placed under a *lac* promoter in *E. coli*. Thus elements within the genetic background of *Acinetobacter* shut down *benABC* expression in the absence of metabolism of the benzoate diol, the end-product of their encoded enzymes and the substrate of the enzyme encoded by the possibly cotranscribed *benD* gene. Similar observations have been reported with genes associated with benzoate oxidation in *Pseudomonas* (Harayama et al., 1984).

The fact that *benD* is expressed in response to muconate raises the possibility that CatM, a repressor protein that responds to muconate, contributes to control of this structural gene. This possibility is rendered unlikely by the fact that *benD* remains under inducible control in mutants carrying a dysfunctional *catM* gene. Such mutants constitutively express enzymes encoded by *catA* and by the *catBCIJFD* operon (Neidle et al., 1989).

Regulation of the expression of catA. Several genes contribute to regulation of the expression of *catA*. This structural gene has a regulatory role in its own right because the oxygenase it encodes gives rise to muconate, the inducer of all of the *cat* genes in *Acinetobacter* (Canovas and Stanier, 1967). These genes are expressed when interaction of muconate with CatM in wild-type cells relieves repression. The wild-type *catM* gene appears to be dominant over a mutated variant because repression can be exerted from a plasmid-borne *catM* gene in a strain in which the chromosomal *catM* gene is dysfunctional. Selection for constitutive expression of the *cat* structural genes in such a strain selects for loss of the plasmid (Neidle et al., 1989).

Inferences concerning the regulation of *catA* can be drawn from examination of the influence of flanking regions of *Acinetobacter* DNA upon expression of the gene in *E. coli* (Neidle and Ornston, 1986). As shown in Table 2, *catA* can be expressed inducibly in *E. coli* from the *lac* promoter in pUC19. In the absence of other *Acinetobacter* genes, levels of *catA* expression are relatively low (Table 2 - pIB1345). Presence of the *Acinetobacter* open reading frame ORF2 downstream from *catA* increases its uninduced expression ten-fold and its IPTG-induced expression 20-fold (Table 2 - pIB1343). Expression of *catA* was lowered somewhat by inclusion of ORF3 within the insert (Table 2 - pIB1341) and almost eliminated by the presence of upstream DNA including ORF1 (Table 2 - pIB1362).

One interpretation is that ORF2, coexpressed with *catA* when the neighbouring genes undergo in-vitro transcription-translation (Neidle et al., 1989), contributes to the expression of *catA*. This contribution cannot be essential for gene expression because a deletion introduced into ORF2 does not discernibly impede the growth of *Acinetobacter* with benzoate, anthranilate or muconate (Neidle et al., 1989). Similar observations were obtained with *Acinetobacter* strains carrying insertions within ORF3. The possibility that ORF2 and ORF3 may assist in expression of *catA* was strengthened by the observation that a mutation in ORF2 resulted in catechol accumulation during growth with benzoate, and similar accumulation was observed when a strain carrying a mutation in ORF3 was grown with anthranilate (Neidle et al., 1989). It is possible that unexplored physiological conditions, such as nitrogen limitation, might reveal more dramatically regulation exerted by ORF2 and ORF3.

Table 2. Expression in *E.coli* of *catA* carried in *Acinetobacter* DNA
fragments inserted downstream from a pUC19 *lac* promoter

Recombinant Plasmid	IPTG Absent[a]	IPTG Present[a]	Presence (+) or Absence (-) of			
			ORF1	*catA*	ORF2	ORF3
pIB1345	0.05	0.11	-	+	-	-
pIB1343	0.48	2.0	-	+	+	-
pIB1341	0.16	0.8	-	+	+	+
pIB1362	0.004	0.02	+	+	+	+

[a]the specific activity of the *catA* gene product was measured as units of
catechol 1,2-dioxygenase activity / mg protein and is presented as the
ratio of observed activity to the activity found in extracts of benzoate-
grown *Acinetobacter* cultures. All of the recombinant plasmids carried
DNA inserts containing *catA* in the multiple cloning site of pUC19 and
orientated so that its direction of transcription lies downstream from
the *lac* promoter. The *Acinetobacter* DNA inserts in pIB1345, pIB1343 and
pIB1341 begin at the same site upstream from *catA*. The inserts in
pIB1341 and pIB1362 end at the same site downstream from ORF3 (Neidle and
Ornston, 1986; Neidle et al., 1989).

The lowering of *catA* expression in the presence of upstream DNA
extending as far as ORF1 (Table 2 - pIB1362) could be attributed to the
presence of a possible transcriptional terminator (ATTCGCTTTATTCTAA-
GGCGAATTTTTCTTTTTT) 40 bp downstream from this open reading
frame. Transcriptional termination upstream from *catA* would demand the
presence of promoters upstream from the gene, and the presence of such
promoters can be inferred from the response of DNA fragments, including
catA, to inducers when carried on a broad host range plasmid in
Acinetobacter strains from which the entire *ben-cat* region has been
deleted (Neidle et al., 1989). When placed under control of the *lac*
promoter of pRK415, an Acinetobacter DNA segment containing *catA* and
50 bp upstream from the gene expressed the gene at levels corresponding to
about 10% of the specific activity observed in fully induced wild-type cells;
expression of the gene from this plasmid, probably in response to the *lac*
promoter of pRK415, was not influenced by the presence of benzoate or
muconate in the growth medium. Derivatives of pRK415 containing *catA*

and 1 kb or more of *Acinetobacter* DNA upstream from this gene expressed its product at relatively high levels corresponding to about 50% of the specific activity in fully induced cells. This level of gene expression, independent of the orientation of the insert with respect to the *lac* promoter, was elevated six-fold when either benzoate or muconate was included in the growth medium. Thus the 1 kb of DNA upstream from *catA*, a region that does not contain a complete open reading frame, appears to contain a promoter that recognises one or more regulatory molecules that respond to either benzoate or muconate as an inducer (Neidle and Ornston, 1987). The regulatory molecules must be encoded by DNA lying outside the 16 kb of DNA which, containing known *ben* and *cat* genes, is lacking in the *Acinetobacter* deletion mutant that acted as host for the recombinant plasmids (Neidle et al., 1989).

Induction of plasmid-borne *catA* in the deletion mutant can take place in the absence of *catM*, the known repressor of the structural gene, and in the absence of ORF2, a gene that was implicated in control of the *catA* gene when it was cloned in *E. coli* (Neidle et al., 1989). The general pattern that emerges from these observations is that *catA* is subject to overlays of transcriptional control to which a number of regulatory genes contribute. It is possible that genetic interactions associated with these control mechanisms contribute to conservation of the supraoperonic clustering of *ben-cat* genes. This possibility can be explored by examining the physiological consequences of transposition of *catA* and neighbouring DNA fragments from their present locus to other loci in the *Acinetobacter* chromosome.

Expression of Acinetobacter genes in heterologous hosts. Differences in codon usage pattern (Table 1) present no barrier to expression of *Acinetobacter* genes in *Pseudomonas*. The *Acinetobacter catBCIJFD* operon, contained as a 5.0 kb *Eco*R1 insert in the broad host range plasmid pKT230, was selected initially on the basis of its ability to complement a *pcaIJ (pcaE)* mutation in *P. putida* (Shanley et al., 1986). The recombinant plasmid also allowed growth of complemented *P. putida* strains carrying mutations in *catB, catC* or *pcaD*. In every case, the complementing *Acinetobacter* gene was expressed at levels comparable to those found in fully induced wild-type cells.

When placed under control of a *lac* promoter in pUC13 and induced by growth of host *E. coli* cells in the presence of IPTG, the products of the *catBCIJFD* genes represent more than 10% of the cell protein (Shanley et al., 1986). Even higher levels of expression, with *Acinetobacter* enzymes accounting for about half of the soluble protein in *E. coli,* are produced when the *Acinetobacter pcaIJFDBCHG* operon under the control of a *lac* promoter is expressed in response to IPTG (Doten et al., 1987a). In these cells, levels of most of the protocatechuate-metabolising enzymes are ten to 30-fold higher than those found in fully-induced *Acinetobacter* cultures. The sole exception to this pattern is protocatechuate oxygenase which, at a three-fold higher level than in induced *Acinetobacter* cultures, represents about 10% of the *E. coli* soluble protein. The oxygenase purified from *E. coli* contains only 20% of the specific activity of the same gene product purified from *Acinetobacter*. It is possible that *E. coli* lacks factors that are required for effective assembly of the oxygenase when it is expressed at extremely high levels. Perhaps overproduction of the oxygenase demands that iron be supplied at a rate that exceeds the limit which can be supplied physiologically.

The *pca* genes encode all of the enzymes required for metabolism of protocatechuate to citric acid cycle intermediates, and *E. coli* cultures in which the enzymes are expressed might be expected to grow at the expense of protocatechuate (Doten et al., 1987a). Growth of such cultures is not observed. Rather, the cultures rapidly and quantitatively convert protocatechuate to ß-ketoadipate. Evidently ß-ketoadipate succinyl CoA transferase, expressed at high levels in the *E. coli* cultures, is physiologically ineffective in these cells. A possible reason for inactivity of the expressed enzyme is the relatively low level of succinyl CoA, the required thioester donor for ß-ketoadipate metabolism, in *E. coli* (Jackowski and Rock, 1986).

As described above, the *benABC* genes can be expressed effectively in *E. coli* cells in which the encoded enzymes efficiently convert benzoate to the corresponding dihydrodiol. Such conversion is not observed when the *benABC* genes are expressed in *Acinetobacter* cells carrying a dysfunctional *benD* gene (Neidle et al., 1987). This evidence suggests the presence in *Acinetobacter* of regulatory mechanisms that govern the rate of conversion of benzoate to the dihydrodiol. Thus examination of *Acinetobacter* gene regulation in *E. coli* can give an insight that might not be gained by examination of *Acinetobacter* alone. A further example is the demonstration that ORF2 contributes to the expression of *catA* (Table 2), an observation that might have been obscured by knowledge that mutations within ORF2 do not prevent growth of *Acinetobacter* with substrates that demand *catA* expression for their metabolism.

Genetic Interaction

Participation of the *catIJ* region in repair of a mutation in the *pcaIJ* region. In *Acinetobacter* strains containing the wild type *catIJ* genes, the *pcaJ3125* mutation is repaired at a frequency of 10^{-4}. Deletion of a region containing the *catIJ* region from the chromosome causes this frequency to drop 300-fold to 3×10^{-7}. Thus it appears that the *catIJ* region contributes DNA sequences that repair the *pcaJ3125* lesion (Doten et al., 1987b). The repair process depends upon a functional *recA* gene and is not inhibited by deoxyribonuclease at concentrations that prevent natural transformation of the mutation to wild type by exogenous DNA (L. Gregg and D. Elsemore, unpublished observations). Therefore it seems likely that the repair process is mediated by direct intrachromosomal DNA sequence transfer from the *catIJ* region into the *pcaIJ* region. The linear chromosomal distance between these two regions is not known, but it appears to exceed 15 kb. Therefore folding of DNA must bring the regions into close physical proximity so that the sequence exchange may take place.

DNA sequence exchange as a mechanism for conservation of exceptional DNA sequences in catIJF and pcaIJF. Repair of mutations in the *pcaIJ* region by *catIJ* sequences may serve as a mechanism for gene stabilisation: genetic variation can be counterbalanced by substitution of the altered DNA with a sequence copy carried elsewhere on the chromosome. The effectiveness of such a genetic process may be assessed by the degree to which it has overcome selective pressures for genes to take on the GC content and codon usage patterns that are characteristic of the host chromosome (Grantham et al., 1981; Ikemura, 1982; Muto et al., 1984; Ohama et al., 1987). The existence of such pressures is evident from the closeness with which these acquired traits are shared by most genes associated with catabolism of aromatic compounds in *Acinetobacter*. As summarised in Table 3, most genes for aromatic catabolism, derived from separate ancestors, vary only slightly in GC content and in codon usage pattern. These genes, the reference genes in Table 3, have no close homologues in the *Acinetobacter* chromosome, but share similar GC contents of 44±2% and, with rare exception, exhibit codon usage patterns that are typical for *Acinetobacter* genes.

Genes in the *catIJF* and *pcaIJF* regions, unlike other *Acinetobacter* genes, contain DNA sequences that are nearly identical: comparison of 1.2 kb encoding *catIJ* and *pcaIJ* revealed only six base pair differences in DNA sequence (A. Benson and J. Houghton, unpublished observations). It seems likely that this close sequence similarity underlies the capacity of the genes to exchange sequence information, as revealed by *catJ*-mediated repair of the *pcaJ3125* mutation. Most remarkable is the set of genetic properties shared by *catI*, *catJ* and *catF*. As shown in Table 3, these genes share closely similar GC contents ranging from 54% to 57%, values more than 10% higher than the average of those shared by all other known *Acinetobacter* genes. Furthermore, the *catI*, *catJ* and *catF* genes share similar patterns of codon usage that distinguish them from the other *Acinetobacter* genes (Table 3). For example, the codon CTG accounts for roughly one sixth of the leucyl codons in the reference genes and represents half or more of the leucyl codons in *catI*, *catJ* and *catF* (Table 3).

The presence in *catI*, *catJ* and *catF* of highly distinctive properties, traits shared with the nearly identical *pcaI*, *pcaJ* and *pcaF* genes, is open to three possible interpretations. One proposal is that both sets of genes are recent evolutionary acquisitions that have not undergone the selective pressures that conferred characteristic GC contents and codon usage patterns upon the other *Acinetobacter* genes. This interpretation seems somewhat unlikely because it is difficult to account for evolution of genes catalysing most reactions in the ß-ketoadipate pathway (the reference genes in Table 3) in the absence of genes for the enzymes that most directly yield the selective benefit of the pathway. The suggestion that the nearly identical *catIJF* and *pcaIJF* regions replaced *Acinetobacter* genes relatively recently in evolutionary history requires the assumption that highly

Table 3.　GC content and codon usage frequency in *A. calcoaceticus* genes

	reference	*catI*	*catJ*	*catF*
GC content	44+2	57	55	54
TTA (Leu)	24+8	5	10	7
TTG (Leu)	23+7	0	0	7
CTG (Leu)	17+10	55	65	50
ATT (Ile)	68+10	22	29	34
ATC (Ile)	29+12	67	64	62
TCC (Ser)	03+04	33	29	6
CCT (Pro)	31+17	8	8	6
CCG (Pro)	20+16	38	50	61
ACC (Thr)	35+14	69	69	52
GCT (Ala)	20+7	4	5	5
GCC (Ala)	21+10	56	59	24
TAC (Tyr)	18+14	71	33	56
CAG (Gln)	39+15	50	57	80
AAC (Asn)	31+20	73	60	53
GAC (Asp)	28+8	82	53	45
CGT (Arg)	69+14	43	20	38
CGC (Arg)	18+9	43	60	48
AGC (Ser)	15+10	33	29	31
GGC (Gly)	31+17	48	46	49

All values are given as percentages.　The reference values are the average observed with the structural genes *pcaC*, *pcaG*, *pcaH*, *benA*, *benB*, *benC*, *benD*, *catA*, *catB*, *catC* and *catD*.　These values do not differ significantly from those found with biosynthetic structural genes from *A. calcoaceticus*　(Kaplan et al., 1984).

improbable events happened twice. A second interpretation is that some intrinsic property of CoA transferase and thiolase genes demands conservation of properties at the level of DNA sequence. This interpretation is rendered unlikely by the fact that widely divergent thiolase genes from other organisms have been sequenced (Peoples and Sinskey, 1989). Furthermore, the CoA transferase and thiolase genes that are associated with adipate metabolism in *Acinetobacter* have diverged significantly, as evidenced by their failure to hybridise with DNA containing the *catIJF* or the *pcaIJF* genes (Shanley et al., 1986; Doten et al., 1987a). The third interpretation, and in our view the most likely, is that the documented capability for sequence exchange between the *catIJF* and the *pcaIJF* regions has allowed repair to buffer them against events that conferred characteristics typical of the *Acinetobacter* chromosome upon other genes.

The proposal that sequence exchange between *catIJF* and *pcaIJF* contributes to their conservation can be explored by examination of the corresponding regions from divergent *Acinetobacter* strains. A major contribution of sequence exchange to conservation would be reflected in concerted divergence so that the two genetic regions retained their sequence similarity as they diverged in different organisms. The natural transformation system of *Acinetobacter* should allow replacement of either *catIJF* or *pcaIJF* with a divergent region from another strain; genetic interplay between the regions could be monitored in the recombinant organism.

A major question to be addressed is the basis for the genetic stability of the nearly identical *catIJF* and *pcaIJF* regions in the *Acinetobacter* chromosome. Recombinational events lead to rapid excision of most repeated DNA sequences from the prokaryotic chromosome (Horuichi et al., 1962; Jackson and Yanofsky, 1973; Rigby et al., 1974; Anderson and Roth, 1977). Notable exceptions are ribosomal genes (Nomura et al., 1984; Widom et al., 1988) and now, *catIJF* and *pcaIJF*. It is possible that the chromosomal location of these two regions allows their conservation in the *Acinetobacter* chromosome. This possibility can be explored by selection for recombinants in which the genes have been rearranged. For example, the *catBCIJFD* operon, completely deleted from the chromosome of strain ADP161, is carried on a 5.0 kb *Eco*R1 fragment (Shanley et al., 1986). Insertion of this fragment in the correct orientation into *pcaHG*, the genes encoding protocatechuate oxygenase, would inactivate the oxygenase and result in constitutive expression of the inserted genes (Hartnett et al., 1990). A recombinant that had acquired this construct could be selected by demanding growth of strain ADP161 with muconate after transformation with engineered DNA containing *catBCIJFD* within *pcaHG*. The recombinant would carry the nearly identical *catIJF* and *pcaIJF* DNA segments within 3 kb of each other, and the genetic stability of this arrangement could be examined by monitoring the ability of the recombinant's progeny to grow with muconate.

DNA SEQUENCE EXCHANGE AS A GENETIC MECHANISM FOR EVOLUTIONARY DIVERGENCE

In contrast to the remarkable conservation of *catIJF* and *pcaIJF* sequences, substantial sequence variation accompanied divergence of the isofunctional *catD* and *pcaD* genes in *Acinetobacter* (Yeh et al., 1980a). The DNA sequence of the *pcaD* gene has not yet been determined, but divergence of the NH_2-terminal amino acid sequences of the enol-lactone hydrolases encoded by *Acinetobacter catD* and *pcaD* is so extreme that inclusion of the amino acid sequence encoded by the *Pseudomonas pcaD* gene was required to provide evidence for common ancestry of all of the hydrolases (McCorkle et al., 1980). Thus, somewhere before the terminal gene in *catIJFD* and *pcaIJFD*, the evolutionary pattern shifts from extreme conservation to extensive divergence.

The NH$_2$-terminal amino acid sequences of the enol-lactone hydrolases show patterns of internal repetition (McCorkle et al., 1980) and also give evidence of sequence exchange between their structural genes and *catC* during their evolutionary divergence (Yeh et al., 1978; Yeh and Ornston, 1980). This evidence fostered the hypothesis that gene conversion, the substitution of a DNA sequence from one chromosomal region into another, may have produced extensive genetic variation during evolutionary divergence (Ornston and Yeh, 1979, 1982). DNA sequences for some of the relevant genes are now available, and they support the conclusion that such sequence exchange played a significant role in the acquisition of DNA sequences that distinguish homologous genes for the ß-ketoadipate pathway (Neidle et al., 1988; Hartnett et al., 1990). The evolved genes seem to possess the potential ability to form sets of slippage structures created by hybridisation between misaligned DNA strands. An inference from this finding is that the slippage structures, formed during evolutionary divergence, may make a contribution to maintenance of the evolved DNA sequence through mismatch repair between slipped DNA strands. In such events, complementary components of repeated DNA sequences are envisaged as hybridising in slippage structures and exchanging sequence information so that the wild type repetitions are maintained (Ornston et al.,

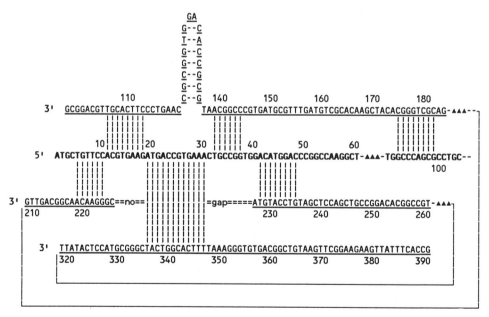

Fig. 7. An apparent slippage structure formed by the *P.putida catC* sequence. The beginning of the 5' strand of the structural gene shows the potential ability to enter into hybridisation in three different alignments with the slipped 3' strand. The 5' strand is shown in bold, and the 3' strand is underlined. Vertical lines indicate where six or more contiguous residues in each strand exhibit perfect complementarity and may hybridise.

1990a,b). According to this view, mutations causing nucleotide substitutions within one component of a repetition could be eliminated by mismatch repair. Thus, as suggested for *catIJF* and *pcaIJF*, events taking place at the level of DNA metabolism could account for the maintenance of sequence information.

The most clearly elucidated slippage structures are those associated with the *P. putida catC* gene (Aldrich et al., 1987; Aldrich and Chakrabarty, 1988; J. Houghton, unpublished observations). As illustrated in Fig. 7, the beginning of the 5' strand of this gene appears to have the capacity to hybridise to three different regions of the 3' strand. Thus the slippage structure can be envisaged as a three-dimensional structure in which the 3' strand coils around so that different portions of its sequence can achieve hybridisation with the 5' strand. The central role of the 5' strand in the structure allows its other components to be discerned relatively directly. Other slippage structures are less apparent, and are reflected largely in the presence of different patterns of sequence repetition that were acquired as genes diverged.

Some sets of sequence repetitions lie close to each other in positions that might facilitate ratcheting of DNA out of its conventional duplex structure so that single strands may hybridise in slippage structures elsewhere. Examples, shown in Fig. 8, are segments of *catC* encoding the middle of the primary sequence of muconolactone isomerase from *Acinetobacter* and *Pseudomonas*. As shown in Fig. 8A, the amino acid sequences of the isomerases are quite similar, especially in light of substantial divergence in the GC content of the structural genes (Table 1). Clusters of similar DNA sequence remain in the divergent genes (Fig. 8B), but they have acquired quite different patterns of sequence repetition (Fig. 8C).

Additional examples of genetic events accompanying evolutionary divergence emerge from comparison of genes encoding intradiol dioxygenases. The crystal structure of *Pseudomonas* protocatechuate 3,4-dioxygenase has been solved (Ohlendorf et al., 1988). The protein is formed by association of equal quantities of α and ß subunits, the respective products of the *pcaG* and *pcaH* genes. The ß subunit binds ferric ion and catalyses the oxygenative reaction. The α subunit is homologous with the ß subunit and the two subunits bind to each other to form a complementary array in the enzyme. An arginyl side chain in the α subunit contributes to binding of the negatively-charged carboxylate of protocatechuate in the active site of the enzyme. The *Acinetobacter pcaHG* region has been sequenced (Hartnett et al., 1990), and the encoded proteins share amino acid sequence identity of about 50% with their homologues from *Pseudomonas*. Residues indicated by crystallography to be essential have been conserved, and comparison of the *pcaH* gene with *Acinetobacter catA* (the catechol 1,2-oxygenase structural gene) and with the *Pseudomonas clcA* (the chlorocatechol 1,2-dioxygenase structural gene) gives some insight into events accompanying their evolutionary divergence. For example, the

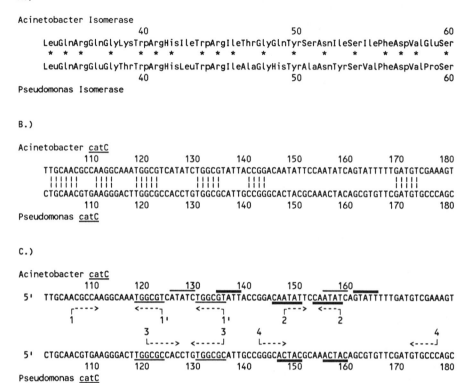

Fig. 8. Sequence conservation and divergence within *catC*, the structural gene for muconolactone isomerase from *Acinetobacter* and *Pseudomonas*. (A) Identical amino acids in the aligned sequences are marked with an asterisk. (B) Conserved regions of nucleotide sequence are marked with vertical lines where four or more contiguous bases in the 5′ strand are identical. (C) Different patterns of repetition were acquired as the *catC* gene diverged in *Acinetobacter* and *Pseudomonas*. Direct repetitions of five or more contiguous base pairs are marked by single or double underlining or overlining. Direct slippage of 5′ and 3′ strands followed by hybridisation between the complementary components of sequence repetitions could help to melt DNA into single strandedness so that slippage structures such as that depicted in Fig. 7 could be formed. Inverted repetitions, marked with numbered arrows, could facilitate the localised melting of DNA by forming loops within strands.

catechol oxygenases are formed by self-association of a single subunit, the evolutionary counterpart of the ß subunit of protocatechuate oxygenase (Frantz and Chakrabarty, 1987; Neidle et al., 1988; Ngai and Ornston, 1988). Thus divergence of protocatechuate and catechol oxygenase was

225

	Tyr	His His
Substrate		

Protocatechuate ProGlyPro**Tyr**ProTrpArgAsnArgIleAsnGlu----------TrpArgProAla**His**Ile**His**PheSerLeuIleAlaAspGly

Catechol ProAlaGly**Tyr**GlyCysProProGluGlyProThrGlnGlnLeuLeuAsnGlnLeuGlyArgHisGlyAsnArgProAla**His**Ile**His**TyrPheValSerAlaAspGly

Chlorocatechol ProValPro**Tyr**GlnIleProTyrGlyGlyProThrHrGlyArgLeuLeuGlyHisLeuGlyHisLeuSerHisThrTrpArgProAla**His**IleVal**His**PheLysValArgLysAspGly

Fig.9. Conserved amino acid residues in the iron-binding region of intradiol dioxygenases. Highlighted are two histidyl residues that bind ferric ion within the enzymes. The amino acid sequence of the β subunit of *Acineto-bacter* protocatechuate oxygenase is aligned with the corresponding sequences of *Acinetobacter* catechol oxygenase and *Pseudomonas* chlorocatechol oxygenase. Asterisks mark identical residues shared by the catechol oxygenase sequence and either of the other sequences.

A

Catechol Oxygenase
```
                    210             220        His224 His226  230
          GlyProThrGlnGlnLeuLeuAsnGlnLeuGlyArgHisGlyAsnArgProAlaHisIleHisTyrPheValSerAlaAspGly
           *  *  *        *           *  *  *  *        *  *        *  *  *  *  *  *  *
Chlorocatechol Oxygenase
          GlyProThrGlyArgLeuLeuGlyHisLeuGlySerHisThrTrpArgProAlaHisValHisPheLysValArgLysAspGly
                    170             180        His188 His190
```

B
```
Acinetobacter catA
5' GGTCCAAGCGACAACAGTTGCTGAATCAGTTGGGCGTCATGGTAACCCCCTGCCACATTCACTATTTGTTTCGCCGATGGA 3'
   620       630       640       650       660       670       680       690
     |||||  |||| ||||||        |||||    ||||  |||||||||||||      |||||  |||
Pseudomonas clcA
5' GGGCCGACTGGGCGTCTGCTGGGCCACCTGGGCAGCCATACCTGGCGTCCGGCCACTTCAAGGTGCCAAGGACGGT 3'
   510       520       530       540       550       560       570       580       590
```

C
```
Acinetobacter catA                                                              680       690
5' GGTCCAAGCGACAACAGTTGCTGAATCAGTTGGGCGTCATGGT------AACCCCCTGCCACATT  CACTATTTGTTTCTGCCGATG
   620       630       640       650              660       670                             G
     ||||||||||        |||||||       |||||||                  ||||||                        A
                                                                                             C
Acinetobacter catA:same strand                                                   TTAAACGCATCAATCAAACGCCA
3' CCCAACGTATTCGTTTGCAGTATCCACATGCCTAGCGG--TCGGTGTAA                   720       710       700
   760       750       740       730
```

D
```
Pseudomonas clcA                                            <-----380
5' GGGCCGACTGGGCGTCTGCTGGGCCAC--CTGGGCAGCCATACCTGGCGTCCGGGCCACGTGCAC 3'
   510       520       530       540       550       560       570
     |||||  |||| ||||||       ||||||      |||||
Pseudomonas clcA:opposite strand
3' CCCGGCT--GACCCGCAGACGACCCG------GTGGACCCGTCGGTATGGACCGGCCAGGCCGCGTGCACGTG 5'
   510       520       530       540       550       560       570
                    <------124
```

Fig. 10. Different DNA slippage structures in the divergent *Acinetobacter catA* and *Pseudomonas clcA* genes. Numerals above or below residues indicate their position in primary sequence. (A) A portion of the iron-binding region in the two proteins. Asterisks mark identical residues occupying the same position in the aligned sequences. (B) DNA sequences corresponding to the aligned amino acid sequences. Vertical bars indicate positions in which four or more contiguous DNA bases are identical in the aligned sequences. (C) A potential loop formed by the 5' *Acinetobacter catA* DNA strand corresponding to the compared region. Direct sequence repetitions that could serve as a basis for alignment with the slipped 3' strand are underlined. (D) A potential slippage structure formed between misaligned 5' and 3' *Pseudomonas clcA* DNA strands corresponding to the compared region. Arrows indicate DNA that is pinched into single strandedness in this slippage structure and which has the potential to hybridise with DNA in the same strand at positions indicated by the numerals.

227

accompanied by gain or loss of the α subunit that specifically binds the protocatechuate carboxylate, a functional group that is absent in the catechols.

Amino acid sequences associated with three of the four iron-binding ligands in the oxygenases are aligned in Fig. 9. It is evident that the two catechol oxygenases resemble each other more closely than they resemble protocatechuate oxygenase. Particularly noticeable is the gap required to align the latter sequence with the former sequences. The gap, lying near a loop in the protocatechuate oxygenase structure (Ohlendorf et al., 1988), is filled with an 11 amino residue amino acid sequence that, judging by its conservation, appears to serve a significant function in the catechol oxygenases. Conservation of the amino acid sequences masks considerable divergence at the level of DNA. As shown in Fig. 10, *Acinetobacter catA* and *Pseudomonas clcA* acquired entirely different slippage structures during their evolution (Neidle et al., 1988; Ornston et al., 1990b). A portion of the aligned amino acid sequences of the two genes is depicted in Fig. 10A. Fig 10B indicates the remaining DNA sequence homology between the 5′ strands encoding the amino acid sequences. As shown in Fig. 10C, the *Acinetobacter catA* sequence has diverged to form a slippage structure in which an intrastrand loop is the predominant component, whereas hybridisation between the 5′ and 3′ strands forms the major portion of the slippage structure in the *Pseudomonas clcA* sequence.

It is difficult to account for how the distinctive slippage structures of divergent genes could have been formed without hybridisation between their complementary strands. It is likely that mismatch repair (Kunkel and Soni, 1988; Meselson, 1988) helped to create the repeated sequences, and a similar process may contribute to maintaining the repetitions in the evolved sequences. One avenue to exploring the process would be to disrupt slippage structures by site-directed mutagenesis (Kunkel, 1985) and to examine the effects of sequence repetitions on repair. A more general approach, probably more warranted at our present stage of enlightenment about slippage structures, would be to examine DNA sequences that undergo spontaneous mutation and repair. This approach would take into account the fact that, as exemplified by repair events mediated between *catJ* and *pcaJ*, genetic interactions may involve DNA sequences that are separated by a substantial linear distance on the chromosome.

It is possible that sequence interactions between separate genes may be a selective force accounting for supraoperonic clustering of genes for the ß-ketoadipate pathway. If so, gene rearrangements, readily created in the naturally transformable strain of *Acinetobacter*, may influence patterns of spontaneous mutagenesis and repair. An example, described above, would be design of a rearrangement that brings the nearly identical *catIJF* and *pcaIJF* regions into linear proximity. In light of the general tendency of duplicated regions to be unstable in prokaryotic chromosomes, it seems possible that such a rearrangement would elicit genetic instability.

Analysis of the ß-ketoadipate pathway has raised questions about the genetic and physiological consequences of gene arrangement. Natural transformation (Juni, 1972) should facilitate investigation of these and other questions that are raised by the fascinating biology of *Acinetobacter*. This article concludes with a general summary of insights that may be gained by application of relatively simple genetic procedures to the naturally transformable strain of *Acinetobacter*.

Selection of Spontaneous Mutant Strains

As noted in a preceding section, ß-ketoadipate selects for strains carrying dysfunctions in *pcaG* and *pcaH*, structural genes for protocatechuate oxygenase(Canovas and Stanier, 1967). Such strains accumulate the inducer protocatechuate from endogenously-formed shikimate and consequently express the *pca* genes constitutively. An alternative and more specific procedure for mutant selection emerged from the observation that a mutation in *pcaD* prevents growth with succinate in the presence of compounds that are metabolised via protocatechuate (G. Hartnett and B. Averhoff, unpublished observations). The physiological basis for the growth inhibition is unknown; accumulation of the chemically reactive ß-ketoadipate enol-lactone (Ornston and Stanier, 1966), the substrate of the enzyme encoded by *pcaD*, may be toxic to cells. Secondary mutants that resist the toxic effect can be isolated readily, and most of these are blocked in early steps of the protocatechuate pathway (G. Hartnett, unpublished observations). Thus exposure of *pcaD* mutants to *p*-hydroxybenzoate yields mutants blocked in *pobA*, the structural gene for *p*-hydroxybenzoate hydroxylase (Fig. 1). Such mutants remain unable to grow in the presence of quinate or shikimate, compounds that are metabolised to protocatechuate independently of *p*-hydroxybenzoate. The initial selection also yields some strains that grow in the presence of quinate or shikimate; these strains usually carry mutations in the structural genes for protocatechuate oxygenase.

Selection for secondary mutants derived from a *pcaD* deficient strain opens the genetics of the early steps of the protocatechuate pathway to analysis. It also allows selection of recombinant strains carrying inserts within the structural genes for protocatechuate oxygenase. Because such strains endogenously accumulate the inducer protocatechuate, they constitutively express the inserted DNA.

Analysis of the genetic properties of mutants lacking protocatechuate oxygenase is facilitated if they carry a wild-type *pcaD* gene, and this gene can be introduced from a cloned DNA fragment by transformation. For successful selection of the wild-type recombinant, the initial recipient carries mutations in both *catD* and *pcaD* (G. Hartnett, unpublished observations). The wild-type *pcaD* gene, constitutively expressed in the recombinant because of the dysfunction in protocatechuate oxygenase, is selected by demanding its complementation of the *catD* mutation during growth with benzoate. Thus it is possible to select and maintain mutants lacking protocatechuate oxygenase in a genetic background containing a mutant *pcaD* gene. This gene can be replaced with the wild-type gene when genetic analysis necessitates separation of the protocatechuate oxygenase mutations from other *pca* mutations. This procedure should facilitate investigation of unstable protocatechuate oxygenase mutations and may give an insight into the processes underlying repair of such mutations.

Location of Mutations and Identification of Cloned DNA

As noted above for *pcaD,* wild-type *Acinetobacter* DNA carried in recombinant *E. coli* plasmids can be used to replace mutations in the *Acinetobacter* chromosome (Neidle and Ornston, 1986; Neidle et al., 1987). This procedure permits mutations to be located within known DNA fragments carried in plasmids. Restriction inactivates donor DNA if the site of cleavage lies within 100 bp of the mutant allele (Hartnett et al., 1990). Thus sets of donor DNA fragments can be used to determine the approximate position of a mutant allele, and further restriction of donor DNA can locate mutations within distances that can be resolved on a single DNA sequencing gel.

Strains carrying known mutations in the *Acinetobacter* chromosome can be used as recipients in natural transformation to identify the corresponding wild-type DNA either in gels or in *E. coli* colonies carrying recombinant plasmids. DNA from gel slices containing an *Acinetobacter* gene can be placed directly upon a lawn of recipient cells in which the gene is dysfunctional and prevents growth on the selective growth medium (Neidle and Ornston, 1986; Neidle et al., 1987). Recombinant colonies appear where DNA containing the wild-type gene comes into contact with the plate. *E. coli* colonies carrying the *Acinetobacter pobA* gene in pUC18 were identified by replica-plating of colonies containing a library of cloned *Acinetobacter* DNA within the plasmid directly upon a lawn of recipient *Acinetobacter* strains, deficient in *pobA,* and spread upon *p*-hydroxybenzoate plates (B.Averhoff, unpublished observations). Growth of transformants, produced from donor DNA released from lysed cells within the *E. coli* colonies, revealed the colonies that contained the wild type *pobA* gene.

Design of Strains Carrying Deletions and Gene Rearrangements

DNA that has undergone genetic engineering can be introduced into *Acinetobacter* by natural transformation (Doten et al., 1987b). This procedure is so efficient that recombinants which have acquired the engineered DNA arise with a frequency of several percent of the population after transformation has taken place in the absence of selection. Thus a successful screening procedure is all that is required to isolate recombinants containing designed mutations. Such transformations were used to examine the effect of *catIJ* on repair of *pcaIJ* mutations (Doten et al., 1987b) and to create *Acinetobacter* strains carrying deletions in *pcaD* (G. Hartnett, unpublished observations). Successful selection procedures can simplify genetics, and it should be noted that the *pcaD* mutants provide a genetic background in which constitutively expressed DNA inserts within *pcaG* or *pcaH* can be selected.

Recovery of DNA from the Acinetobacter Chromosome

Genetic analysis of *Acinetobacter* is facilitated by procedures that allow the direct recovery of specified DNA fragments from the chromosome, and such a technique is presented by gap repair (Orr-Weaver et al., 1983). This technique rests upon the necessity for circularisation of a plasmid in order to achieve its replication. A donor plasmid contains a copy of *Acinetobacter* DNA extending beyond the chromosomal target for recovery. The plasmid is linearised by introduction into its copy of *Acinetobacter* DNA of a gap encompassing the chromosomal target. The linearised DNA is introduced into a recipient *Acinetobacter* strain by transformation, and strains containing a circularised plasmid are selected by demanding a plasmid-encoded function for growth. Recombination of target DNA from the chromosome into the plasmid allows its circularisation. After their selection in *Acinetobacter,* plasmids carrying the target DNA can be introduced into *E. coli,* and thus specified chromosomal DNA segments can be isolated for analysis (L. Gregg, unpublished observations).

Gap repair has been applied successfully to recover mutations from the *Acinetobacter* chromosome (L. Gregg, unpublished observations) and also to obtain DNA lying between *pobA* and the *pca* region in the supraoperonic cluster containing these genes (B. Averhoff, unpublished observations). There is every reason to believe that gap repair and other techniques made possible by the natural transformation of *Acinetobacter* will open much of the biology of this extraordinary genus to genetic investigation.

ACKNOWLEDGEMENTS

Research from our laboratory has been supported by grants from the Army Research Office, the National Science Foundation and the National Institutes of Health. Support from the Celgene Corporation has been particularly valuable because it has allowed development of *Acinetobacter* genetics. Kind permission to report unpublished research from our laboratory was given by B. Averhoff, A.Benson, D. Elsemore, L. Gregg, G. Hartnett, A. Harrison, J. Houghton, J. Hughes, D. Mitchell, E. Neidle, R. Parales and M. Shanley. DNA sequences from the *benABCD* region, determined by E. Neidle and C. Hartnett, were analysed in collaboration with A. Bairoch, M. Rekik and S. Harayama from the University of Geneva, Switzerland. This manuscript was organised and prepared under the expert guidance of Susan Voigt.

REFERENCES

Aldrich, T. L., and Chakrabarty, A. M., 1988, Transcriptional regulation, nucleotide sequence, and localization of the promoter of the *catBC* operon in *Pseudomonas putida, J. Bacteriol.,* 170:1297.

Aldrich, T. L., Frantz B., Gill B., Kilbane, J. F., and Chakrabarty, A. M., 1987, Cloning and complete nucleotide sequence determination of the *catB* gene encoding *cis, cis*-muconate lactonizing enzyme, *Gene* 52:185.

Allewell, N., 1989, Evolving to dissimilate hydrocarbons, *Trends in Biochem. Sci.,* 168:473.

Anderson, R. P., and Roth, J. R., 1977, Tandem genetic duplications in phage and bacteria, *Ann. Rev. Microbiol.,* 31:473.

Cain, R. B., 1980, The uptake and catabolism of lignin-related aromatic compounds and their regulation in microorganisms, *in* "Lignin Biodegradation: Microbiology, Chemistry and Potential Applications, vol. 1," p.21, T. Kent Kirk, T. Higuchi, and H. Chang, eds., CRC Press, Boca Raton FL.

Canovas, J. L., and Johnson, B. F., 1968, Regulation of the enzymes of the ß-ketoadipate pathway in *Moraxella calcoacetica* 4. Constitutive synthesis of ß-ketoadipate succinyl-CoA transferases II and III, *Eur. J. Biochem.,* 3:312.

Canovas, J. L., and Stanier, R. Y., 1967, Regulation of the enzymes of the ß-ketoadipate pathway in *Moraxella calcoaceticus, Eur. J. Biochem.,* 1:289.

Canovas, J. L., Ornston, L. N., and Stanier, R. Y., 1967, Evolutionary significance of metabolic control systems, *Science* 156:1695.

Canovas, J. L., Wheelis, M. L., and Stanier, R. Y., 1968a, Regulation of the enzymes of the ß-ketoadipate pathway in *Moraxella calcoacetica* 2. The role of protocatechuate as inducer, *Eur. J. Biochem.* 3:293.

Canovas, J. L., Johnson, B. F., and Wheelis, M. L. 1968b, Regulation of the enzymes of the ß-ketoadipate pathway in *Moraxella calcoacetica* 3. Effects of 3-hydroxy-4-methylbenzoate on the synthesis of enzymes of the protocatechuate branch, *Eur. J. Biochem.,* 3:305.

Chargaff, E., 1950, Chemical specificity of nucleic acids and mechanism of their enzymatic degradation, *Experimentia,* 6:201.

Chen, K.-S., Peters, T. C., and Walker, J. R., 1990, A minor arginine tRNA mutant limits translation preferentially of a protein dependent on the cognate codon, *J. Bacteriol.,* 172:2504.

Doten, R. C., Ngai, K.-L., Mitchell, D. J., and Ornston, L. N., 1987a, Cloning and genetic organization of the *pca* gene cluster from *Acinetobacter calcoaceticus, J. Bacteriol.,* 169:3168.

Doten, R. C., Gregg, L. A., and Ornston, L. N., 1987b, Influence of the *catBCE* sequence on the phenotypic reversion of a *pcaE* mutation in *Acinetobacter calcoaceticus, J. Bacteriol.,* 169:3175.

Edelman, G. M., and Gally, J. A., 1970, Arrangement and evolution of eukaryotic genes, *in* "The Neurosciences: Second Study Program," p.962, F. O. Schmitt, ed., Rockefeller University Press, New York.

Frantz, B., and Chakrabarty, A. M., 1987, Organization and nucleotide sequence determination of a gene cluster involved in 3-chlorocatechol degradation, *Proc. Natl. Acad. Sci. USA,* 84:4460.

Goldman, A., Ollis, D. L., Ngai, K.-L., and Steitz, T. A., 1985, Crystal structure of muconate lactonizing enzyme at 6.5 A resolution, *J. Mol. Biol.,* 182:353.

Grantham, R., Gautier, C., Gouy, M., Jacobzone, M., and Mercier, R., 1981, Codon catalog usage is a genome strategy modulated for gene expressivity, *Nucl. Acids Res.,* 9:r43.

Harayama, S., Lehrbach, P. R., and Timmis, K. N., 1984, Transposon mutagenesis analysis of meta-cleavage pathway operon genes of the TOL plasmid of *Pseudomonas putida* mt-2, *J.Bacteriol.,* 160:251.

Hartnett, C.S., Neidle, E.L., Ngai, K.-L., and Ornston, L.N., 1990, DNA sequences of genes encoding *Acinetobacter calcoaceticus* protocatechuate 3,4-dioxygenase: Evidence indicating shuffling of genes and of DNA sequences within genes during their evolutionary divergence, *J. Bacteriol.,* 172:956.

Henikoff, S., Haughn, G. W., Calvo, J. M., and Wallace, J. C., 1988, A large family of activator proteins, *Proc. Natl. Acad. Sci. USA,* 85:6602.

Hewetson, L., Dunn, H. M., and Dunn, N. W., 1978, Evidence for a transmissible catabolic plasmid in *Pseudomonas putida* encoding the degradation of *p*-cresol via the protocatechuate *ortho* cleavage pathway, *Genet. Res.,* 32:249.

Horuichi, T., Tomizawa, J., and Novick, A., 1962, Isolation and properties of bacteria capable of high rates of ß-galactosidase, *Biochim. Biophys.Acta,* 55:152.

Hughes, J., Shapiro, M., Houghton, J., and Ornston, L.N., 1988, Cloning and expression of *pca* genes from *Pseudomonas putida* in *Escherichia coli, J. Gen. Microbiol.,* 134:2877.

Ikemura, T., 1981a, Correlation between the abundance of *E.coli* transfer RNAs and the occurrence of the repetitive codon in protein genes, *J. Mol. Biol.,* 146:1.

Ikemura, T., 1981b, Correlation between the abundance of *E. coli* transfer RNAs and the occurrence of the repetitive codon in protein genes: a proposal for a synonymous codon choice that is optimal for *E. coli* translational system, *J. Mol. Biol.,* 151:389.

Ikemura, T., 1982, Correlation between the abundance of yeast transfer RNAs and the occurrence of the repetitive codon in protein genes: differences in synonymous codon choice patterns of yeast and *E. coli* with references to the abundance of isoacceptor transfer RNAs, *J. Mol. Biol.,* 158:573.

Jackowski, S., and Rock, C.O., 1986, Consequences of reduced intracellular coenzyme A content in *Escherichia coli.*, *J. Bacteriol.*, 166:866.

Jackson, E. N., and Yanofsky, C., 1973, Duplication-translocations of tryptophan operon gene in *Escherichia coli*, *J. Bacteriol.*, 116:33.

Juni, E., 1972, Interspecies transformation of *Acinetobacter:* Genetic evidence for a ubiquitous genus, *J. Bacteriol.*, 112:917.

Kaplan, J. B., Goncharoff, P., Seibold, A. M., and Nichols, B. P., 1984, Nucleotide sequence of the *Acinetobacter calcoaceticus trpGDC* gene cluster, *Mol. Biol. Evol.*, 1:456.

Katti, S.K., Katz, B., and Wyckoff, H.W., 1988, Crystal structure of muconolactone isomerase at 3.3 A resolution, *J. Mol. Biol.*, 205:557.

Katz, B., Ollis, D. L., and Wyckoff, H. W., 1985, Low resolution crystal structure of muconolactone isomerase, *J. Mol. Biol.*, 184:311.

Kemp, M. B., and Hegeman, G. D., 1968, Genetic control of the ß-ketoadipate pathway in *Pseudomonas aeruginosa*, *J. Bacteriol.*, 96: 1488.

Konigsberg, W., and Godson, G. N., 1983, Evidence for use of rare codons in the dnaG gene and other regulatory genes of *Escherichia coli*, *Proc. Natl. Acad. Sci. USA*, 80:687.

Kunkel, T., 1985, Rapid and efficient site specific mutagenesis without phenotypic selection, *Proc. Natl. Acad. Sci. USA*, 82:488.

Kunkel, T. A., and Soni, A., 1988, Mutagenesis by transient misalignment, *J. Biol. Chem.*, 263:14784.

McCorkle, G. M., Yeh, W. K., Fletcher, P., and Ornston, L. N., 1980, Repetitions in the NH_2-terminal amino acid sequence of ß-ketoadipate enol-lactone hydrolase from *Pseudomonas putida*, *J. Biol. Chem.*, 255:6335.

Meagher, R., Berry-Lowe, S., and Rice, K., 1989, Molecular evolution of the small subunit of ribulose bisphosphate carboxylase: Nucleotide substitution and gene conversion, *Genetics* 123:845.

Meselson, M., 1988, Methyl directed repair of DNA mismatches, *in* "Recombination of Genetic Material," p.91, K.B. Low, ed., Academic Press, New York.

Muto, A., Kawauchi, Y., Yamao, F., and Osawa, S., 1984, Preferential use of A- and U-rich codons for *Mycoplasma capricolom* ribosomal proteins S8 and L6, *Nucl. Acids Res.*, 12:8209.

Nagylaki, T., and Petes, T.D., 1982, Intrachromosomal gene conversion and the maintenance of sequence homogeneity among repeated genes, *Genetics*, 100:315.

Neidle, E.L., and Ornston, L.N., 1986, Cloning and expression of the *Acinetobacter calcoaceticus* catechol 1,2-dioxygenase structural gene, *catA*, in *Escherichia coli*, *J. Bacteriol.*, 168:815.

Neidle, E.L., and Ornston, L.N., 1987, Benzoate and muconate, structurally dissimilar metabolites, induce expression of *catA* in *Acinetobacter calcoaceticus*, *J. Bacteriol.*, 169:414.

Neidle, E.L., Shapiro, M., and Ornston, L. N., 1987, Cloning and expression in *Escherichia coli* of *Acinetobacter calcoaceticus* genes for benzoate degradation, *J. Bacteriol.*, 169:5496.

Neidle, E. L., Hartnett, C., Bonitz, S., and Ornston, L. N., 1988, DNA sequence of the *Acinetobacter calcoaceticus* catechol 1,2-dioxygenase I structural gene *catA:* evidence for evolutionary divergence of intradiol dioxygenases by acquisition of DNA repetitions, *J. Bacteriol.*, 170:4874.

Neidle, E. L., Hartnett, C. S., and Ornston, L. N., 1989, Characterization of *Acinetobacter calcoaceticus catM*, a repressor gene homologous in sequence to transcriptional activator genes, *J.Bacteriol.*, 171:5410.

Ngai, K.-L., and Ornston, L. N., 1988, Abundant expression of *Pseudomonas* genes for chlorocatechol metabolism, *J. Bacteriol.*, 170:2412.

Nomura, M., Gourse, R., and Baughman, G., 1984, Regulation of the synthesis of ribosomes and ribosomal components, *Ann. Rev. Biochem.*, 53:75.

Ohama, T., Yamao, F., Muto, A., and Osawa, S., 1987, Organization and codon usage of the streptomycin operon in *Micrococcus luteus*, a bacterium with a high genomic G+C content, *J. Bacteriol.*, 169:4770.

Ohlendorf, D.H., Lipscomb, J. D., and Weber, P. C., 1988, Structure and assembly of protocatechuate 3,4-dioxygenase, *Nature*, 336:403.

Ornston, L.N., 1966, The conversion of catechol and protocatechuate to ß-ketoadipate by *Pseudomonas putida*, IV. Regulation, *J. Biol. Chem.*, 241:3800.

Ornston, M. K., and Ornston, L. N., 1972, The regulation of the ß-ketoadipate pathway in *Pseudomonas acidovorans* and *Pseudomonas testosteroni, J. Gen. Microbiol.*, 73:455.

Ornston, L.N., and Stanier, R.Y., 1966, The conversion of catechol and protocatechuate to ß-ketoadipate by *Pseudomonas putida*. I. Biochemistry, *J. Biol. Chem.*, 241:3776.

Ornston, L. N., and Yeh, W.K., 1979, Origins of metabolic diversity: Evolutionary divergence by sequence repetition, *Proc. Natl. Acad. Sci. USA*, 76:3996.

Ornston, L.N., and Yeh, W. K., 1982, Recurring themes and repeated sequences in metabolic evolution, *in* "Biodegradation and Detoxification of Environmental Pollutants," p.105, A.M. Chakrabarty, ed., CRC Press, Miami.

Ornston, L. N., Neidle, E. L., and Houghton, J.E., 1990a, Gene rearrangements, a force for evolutionary change; DNA sequence arrangements, a source of genetic constancy, *in* "The Bacterial Chromosome," p.325, M. Riley and K. Drlica, eds., American Society for Microbiology, Washington, D.C.

Ornston, L. N., Houghton, J. E., Neidle, E. L., and Gregg, L. A., 1990b, Subtle selection and novel mutation during evolutionary divergence of the ß-ketoadipate pathway, *in* "Pseudomonas: Biotransformation, Pathogenesis and Evolving Biotechnology," p.207, S.Silver et al., eds., American Society for Microbiology. Washington, D.C.

Orr-Weaver, T. L., Szostak, J. W., and Rothstein, R. J., 1983, Genetic applications of yeast transformation with linear and gapped plasmid, *Meth. Enzymol.*, 101:228.

Parke, D., and Ornston, L. N., 1984, Nutritional diversity of Rhizobiaceae revealed by auxanography, *J. Gen. Microbiol.*, 130:1743.

Patel, R. N., Mazumdar, S., and Ornston, L. N., 1975, ß-Ketoadipate enol-lactone hydrolases I and II from *Acinetobacter calcoaceticus*, *J. Biol. Chem.*, 250:6567.

Peoples, O. P., and Sinskey, A. J., 1989, Poly-ß-hydroxybutyrate biosynthesis in *Alcaligenes eutrophus* H16: Characterization of the genes encoding ß-ketothiolase and acetoactyl-CoA reductase, *J. Biol. Chem.*, 264:15293.

Rigby, P. W. J., Burleigh, B. D., and Hartley, B. S., 1974, Gene duplication in experimental enzyme evolution, *Nature*, 251:200.

Rosenberg, S. L., and Hegeman, G. D., 1969, Clustering of functionally related genes in *Pseudomonas aeruginosa*, *J. Bacteriol.*, 108:1270.

Rothmel, R. K., Aldrich, T. L., Houghton, J. E., Coco, W. M., Ornston, L. N., and Chakrabarty, A.M., 1990, Nucleotide sequencing and characterization of *Pseudomonas putida catR:* A positive regulator of the *catBC* operon is a member of the LysR family, *J. Bacteriol.*, 172:922.

Schleif, R., 1988, DNA looping, *Science*, 240:127.

Shanley, M. S., Neidle, E. L., Parales, R. E., and Ornston, L. N., 1986, Cloning and expression of *Acinetobacter calcoaceticus catBCDE* genes in *Pseudomonas putida* and *Escherichia coli*, *J.Bacteriol.*, 165:557.

Sharp, P.M., and Li, W.-H., 1986, Codon usage in regulatory genes in *Escherichia coli* does not reflect selection for "rare" codons, *Nucl. Acids Res.*, 14:7737.

Stanier, R. Y., 1951, Enzymatic adaptation in bacteria, *Ann. Rev. Microbiol.*, 5:35.

Stanier, R. Y., and Ornston, L. N., 1973, The ß-ketoadipate pathway, *in* "Advances in Microbial Physiology, vol.9," p.89, A. H. Rose and D. W. Tempest, eds., Academic Press, London.

Stanier, R. Y., Palleroni, N. J., and Doudoroff, M., 1966, The aerobic Pseudomonads: A taxonomic study, *J. Gen. Microbiol.*, 43:159.

Tresguerres, M. E. F., de Torrontequi, G., Ingledew, W. M., and Canovas, J. L.,1970, Regulation of enzymes of the ß-ketoadipate pathway in *Moraxella*: Control of quinate oxidation by protocatechuate, *Eur. J. Biochem.*, 14:445.

West, S.E.H., and Iglewski, B., 1988, Codon usage in *Pseudomonas aeruginosa*, *Nucl. Acids Res.*, 16:9323.

Wheelis, M.L., 1975, The genetics of dissimilarity pathways in *Pseudomonas aeruginosa*, *Ann.Rev.Microbiol.*, 29:505.

Wheelis, M. L., and Ornston, L. N., 1972, Genetic control of enzyme induction in the ß-ketoadipate pathway of *Pseudomonas putida:* Deletion mapping of *cat* mutations, *J. Bacteriol.,* 108:790.

Wheelis, M. L., and Stanier, R. Y., 1970, The genetic control of dissimilatory pathways in *Pseudomonas putida, Genetics,* 66:245.

Widom, R. L., Jarvis, E. D., LaFauci, G., and Rudner, R., 1988, Instability of rRNA operons in *Bacillus subtilis, J. Bacteriol.,* 170:605.

Winstanley, C., Taylor, S. C., and Williams, P. A., 1987, pWW174: A large plasmid from *Acinetobacter calcoaceticus* encoding benzene catabolism by the ß-ketoadipate pathway, *Mol. Microbiol.,* 1:219.

Xiong, Y., Sakaguchi, B., and Eickbush, T. H., 1988, Gene conversions can generate sequence variants in the late chorion multigene families of *Bombyx mori, Genetics,* 120:221.

Yeh, W. K., and Ornston, L. N., 1980, Origins of metabolic diversity: Substitution of homologous sequences into genes for enzymes with different catalytic activities, *Proc. Natl. Acad. Sci. USA,* 77:5365.

Yeh, W. K., and Ornston, L. N., 1981, Evolutionarily homologous $\alpha_2\beta_2$ oligomeric structures in ß-ketoadipate succinyl CoA transferases from *Acinetobacter calcoaceticus* and *Pseudomonas putida, J. Biol. Chem.,* 256:1565.

Yeh, W.K., Davis, G., Fletcher, P., and Ornston, L.N., 1978, Homologous amino acid sequences in enzymes mediating sequential metabolic reactions, *J. Biol. Chem.,* 253:4920.

Yeh, W. K., Fletcher, P., and Ornston, L. N., 1980a, Evolutionary divergence of coselected ß-ketoadipate enol-lactone hydrolases in *Acinetobacter calcoaceticus, J. Biol. Chem.,* 255:6342.

Yeh, W. K., Fletcher, P., and Ornston, L. N., 1980b, Homologies in the NH_2-terminal amino acid sequence of ɣ-carboxymuconolactone decarboxylases and muconolactone isomerases, *J. Biol. Chem.* 255:6347.

ORGANISATION, POTENTIAL REGULATORY ELEMENTS

AND EVOLUTION OF *TRP* GENES IN

ACINETOBACTER

G. Haspel, V. Kishan and W. Hillen[*]

Lehrstuhl für Mikrobiologie
Friedrich Alexander Universität Erlangen-Nürnberg
Staudtstrasse 5
D-8520 Erlangen
Germany

INTRODUCTION

This article discusses the evolutionary and regulatory implications derived from the nucleotide sequence of *trp* genes from *Acinetobacter calcoaceticus*. The genetics of tryptophan biosynthesis have been studied in a wide variety of organisms (Crawford, 1989). As depicted in Fig. 1, organisation of *trp* genes differs considerably among the prokaryotes studied so far. For instance, *Escherichia coli* has five genes in a single operon for tryptophan biosynthesis (Yanofsky et al., 1981), *Bacillus subtilis* has six genes in a single operon (Henner et al., 1984) and *A. calcoaceticus* has seven genes in three unlinked clusters (Sawula and Crawford, 1972, 1973). Unlinked clusters of *trp* genes have also been reported in *Pseudomonas aeruginosa, P. putida* (Holloway et al., 1979; Shinomiya et al., 1983) and *Rhizobium meliloti* (Johnston et al., 1978). Sequence information is available for *trp* genes from *E. coli* (Yanofsky et al., 1981), *B. subtilis* (Henner et al., 1984), *R. meliloti* (Bae et al., 1989), *P. putida* (Crawford and Eberly, 1989; Essar et al., 1990a) and *P. aeruginosa* (Crawford et al., 1986; Essar et al., 1990b). From *A. calcoaceticus* the *trpGDC* operon (Kaplan et al., 1984), the *trpE* gene (Haspel et al., 1990) and the *trpFB* operon (Ross et al., 1990; Kishan and Hillen, in preparation) have been sequenced. Using a newly developed shuttle vector (Hunger et al., 1990) for *A. calcoaceticus*, the promoter structures of the *trpE* and *trpFB* genes have been determined. This article compares the organisation of *trp* genes

---

[*]corresponding author

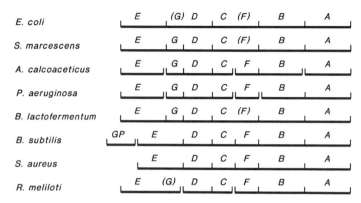

Fig. 1. Genetic organisation of *trp* genes in different prokaryotes. Universal designations for the *trp* genes are used. A solid unbroken line indicates a unit of transcription. Gene fusions are indicated by brackets. The organisms are specified on the left side of the figure. C = indoleglycerol phosphate synthase; D = phosphoribosyltransferase; E, G = subunits of anthranilate synthase; F = *N*-phosphoribosylanthranilate isomerase. The gene arrangements are taken from: *E. coli*: Yanofsky et al. (1981); *S. marcescens*: Yanofsky (1984); *A. calcoaceticus*: this work, Kaplan et al. (1984); *P. aeruginosa*: Essar et al. (1990b); *B.lactofermentum*: Matsui et al. (1986), Sano and Matsui (1987); *B.subtilis*: Henner et al. (1984); *S. aureus*: Crawford (1975); *R.meliloti*: Bae et al. (1989).

in *A. calcoaceticus* with that in other organisms and discusses potential common regulatory sequences for their expression. Striking similarities of *trp* genes from *P. putida* and *A. calcoaceticus* with respect to these features are presented and discussed.

THE ORGANISATION OF *TRP* GENES IN *ACINETOBACTER*

It has been reported previously that the *A. calcoaceticus trp* genes are dispersed in three unlinked clusters, namely *trpE, trpGDC* and *trpFBA* (Sawula and Crawford, 1972, 1973). We have recently isolated and sequenced the *trpFB* operon from *A. calcoaceticus* (Kishan and Hillen, in preparation) and found no indication for the presence of the *trpA* gene in that operon. Thus, we propose that *trpA* is distinct from the *trpFB* operon. The comparison with the other *trp* gene organisations in Fig. 1 reveals that *A. calcoaceticus* and *P. aeruginosa* both contain four unlinked clusters; however, the distribution of genes in the clusters is different. While *P. aeruginosa* contains a single *trpF* gene and a *trpBA* operon, *A. calcoaceticus* contains a *trpFB* operon and an unrelated *trpA* gene. The organisation of *trp* genes in *P. putida* seems to be the same as in *P. aeruginosa* (Hadero and Crawford, 1986). Three clusters are found in *R. meliloti*, two in *B. subtilis* and the other organisms show single

Table 1. Codon usage in *A.calcoaceticus trp* genes

Codon	AA	trpE	trpG	trpD	trpC	trpF	trpB
GGG	Gly	9	2	4	0	1	3
GGA	Gly	9	4	2	0	5	3
GGT	Gly	9	5	16	10	8	25
GGC	Gly	9	5	10	0	2	10
GAG	Glu	8	5	3	2	5	4
GAA	Glu	27	8	20	19	4	19
GAT	Asp	25	3	16	16	14	15
GAC	Asp	6	5	1	4	0	11
GTG	Val	14	5	9	6	6	10
GTA	Val	2	1	5	5	6	8
GTT	Val	11	5	3	6	8	6
GTC	Val	14	3	6	5	2	5
GCG	Ala	11	0	7	1	5	9
GCA	Ala	14	5	15	9	12	17
GCT	Ala	5	2	8	7	5	6
GCC	Ala	7	1	7	5	5	7
·AGG	Arg	0	0	0	0	0	0
AGA	Arg	1	1	0	0	0	0
AGT	Ser	1	1	6	2	3	6
AGC	Ser	2	2	5	2	1	1
AAG	Lys	6	3	3	6	3	6
AAA	Lys	17	4	17	11	9	11
AAT	Asn	10	7	11	7	2	10
AAC	Asn	7	2	7	2	0	5
ATG	Met	11	5	15	4	2	15
ATA	Ile	2	0	0	1	1	0
ATT	Ile	22	11	24	12	10	20
ATC	Ile	8	6	6	2	2	7
ACG	Thr	5	2	4	0	3	4
ACA	Thr	10	2	1	1	1	4
ACT	Thr	5	2	4	5	2	4
ACC	Thr	10	3	5	2	2	7
TGG	Trp	4	2	1	1	3	2
TGA	End	1	0	1	0	0	0
TGT	Cys	3	3	1	4	3	2
TGC	Cys	0	1	1	1	2	0
TAG	End	0	0	0	0	0	0
TAA	End	0	1	0	1	1	1
TAT	Tyr	17	6	5	1	3	14
TAC	Tyr	1	1	2	2	0	4
TTG	Leu	13	3	6	9	3	6
TTA	Leu	9	8	8	12	3	5
TTT	Phe	17	8	7	9	6	6
TTC	Phe	1	1	1	2	1	4
TCG	Ser	4	2	2	0	1	1
TCA	Ser	9	2	4	6	2	5
TCT	Ser	7	2	5	4	1	4
TCC	Ser	4	3	3	0	1	0
CGG	Arg	1	0	1	2	0	0
CGA	Arg	3	2	1	1	3	1
CGT	Arg	17	2	7	8	3	16
CGC	Arg	5	1	4	4	2	7
CAG	Gln	11	7	2	7	5	11
CAA	Gln	12	4	14	13	14	6
CAT	His	9	8	6	5	5	12
CAC	His	3	2	2	0	1	2
CTG	Leu	10	1	6	4	4	6
CTA	Leu	1	0	2	2	1	2
CTT	Leu	7	3	8	5	4	10
CTC	Leu	7	1	3	2	3	3
CCG	Pro	2	2	3	0	0	1
CCA	Pro	12	3	2	3	5	6
CCT	Pro	10	6	1	8	3	9
CCC	Pro	1	0	1	1	2	0

operons. In *E. coli* the GD and CF functions are encoded in single polypeptides, while in *Serratia marcescens* and *B. lactofermentum* only the CF functions are encoded in a single polypeptide. In *R. meliloti* the EG functions are located on a single polypeptide. The other examples shown in Fig. 1 encode a separate polypeptide for every function. Thus, the organisation of *trp* genes in *A. calcoaceticus* shows most similarity to those in *P. aeruginosa* and *P. putida*. This result is also supported by the amino acid homologies (see below).

NUCLEOTIDE SEQUENCE OF THE TRP GENES ENCODED BY *ACINETOBACTER*

The *trpGDC* (Kaplan et al., 1984), *trpE* (Haspel et al., 1990) and *trpFB* (Ross et al., 1990; Kishan and Hillen, in preparation) determinants have been isolated and sequenced. The results constitute a data base for the analysis of codon usage in *A. calcoaceticus* in general, and in the *trp* genes in particular. The codon usage frequencies are presented in Table 1. The codons with maximal AT content are generally preferred over the ones with GC nucleotides in their wobble positions. This preference is particularly pronounced for the His, Pro, Phe, Tyr, Cys, Ile, Asn, Lys, Asp and Glu codons in all *trp* genes presented in Table 1. Among the Ile codons ATA seems to be avoided and, as a result, the AT preference in this case is almost exclusively achieved by use of the ATT codon. No clear preference for AT rich codons is found for Gly, Val, Ser, Thr, Leu and Gln. Interestingly the codon preference of *trpE* differs for some amino acids from that of the other *trp* genes. In particular, in the cases of Leu and Val, a preference for GC-rich codons is found. Since *trpE* is unlinked to any other *trp* gene, this could reflect a different evolutionary source of *trpE* (see below). The amino acid Arg can be encoded by six codons. Among these, AGG is not used at all, and AGA only at two positions. CGG is used only four times while CGC is used 23 times. CGA and CGT are used 11 and 53 times, respectively. This shows a clear tendency to avoid certain codons. The reason for this is not clear; however, they are under-represented in many other prokaryotic genes as well (Grantham et al., 1980).

We have recently reported the *trpE* nucleotide sequence from *A. calcoaceticus* (Haspel et al., 1990). Here we extend this sequence by about 850 nucleotides upstream of *trpE*. The sequence is displayed in Fig. 2. It contains a reading frame with a maximal coding capacity for 234 amino acids constituting a potential protein of 30,817 Dal. We do not have any data regarding the expression of this reading frame. In addition, other start codons appear in the *N*-terminal region which could lead to smaller proteins. The 3′-end of the reading frame overlaps with the *trpE* promoter as determined by primer extension analysis (Haspel and Hillen, in preparation) and indicated in Fig. 2. This argues strongly against expression of this reading frame because no primer extension product of the corresponding length has been found. The nucleotide sequencing result from the *trpE* gene of *P. putida* revealed the same organisation with a reading frame

```
             10                  30                  50
GTTGTTTTTTAAATAATCTTGCCTGAGCTAGACAAACGTCTGTAAAGGCATGCGAAAAGC
                                                    M  R  K  A

             70                  90                 110
ATGCCTTTACTTTTTTATAAGGTGCAATATGTCTTTAGTACAATTAGAACGTCGTCAACT
 C  L  Y  F  F  I  R  C  N  M  S  L  V  Q  L  E  R  R  Q  L

            130                 150                 170
GGTCTTGTTTGATCTAGATGGAACATTGGTCGATACAGCATCAGATATGTATCGTGCGAT
 V  L  F  D  L  D  G  T  L  V  D  T  A  S  D  M  Y  R  A  M

            190                 210                 230
GAATCTGACTTTGGATCATCTGGGGTGGTCTCGTGTTACAGAAGCTCAAATTCGCCAATG
 N  L  T  L  D  H  L  G  W  S  R  V  T  E  A  Q  I  R  Q  W

            250                 270                 290
GGTTGGACAGGGTACTGGCAAGTTATGCGATGCAGTTTTAAAACATTTGTTTGAAGAAGT
 V  G  Q  G  T  G  K  L  C  D  A  V  L  K  H  L  F  E  E  V

            310                 330                 350
AGAACCTGCAAAGCATCAAATGCTGTTAACCACTTATCTCGAAATTTATGCACAAGAGTT
 E  P  A  K  H  Q  M  L  L  T  T  Y  L  E  I  Y  A  Q  E  L

            370                 390                 410
GTGTGTGACCAGTCGCTTATTTGAAGGTGTGCAAGCATTTCTTGATGAGTGTAAGGCGCG
 C  V  T  S  R  L  F  E  G  V  Q  A  F  L  D  E  C  K  A  R

            430                 450                 470
TAAAATCGAGATGGCTTGTGTAACCAACAAGCCTGAGCAATTGGCGAGAAATTTATTGGA
 K  I  E  M  A  C  V  T  N  K  P  E  Q  L  A  R  N  L  L  E

            490                 510                 530
AACACTTAAGATTGGCGATTACTTTGACTTGGTCGTGGGTGGGGATACTTTGCCTGTACG
 T  L  K  I  G  D  Y  F  D  L  V  V  G  G  D  T  L  P  V  R

            550                 570                 590
AAAACCAGACCCGCTTCCATTGCTACATAGTGTGCAGGTGATGAAGACCACTATTGAGAA
 K  P  D  P  L  P  L  L  H  S  V  Q  V  M  K  T  T  I  E  N

            610                 630                 650
TACCTTGATGATCGGTGATTCAAAAAATGATGTTGAAGCAGCCAGACGTGCGGGTATCGA
 T  L  M  I  G  D  S  K  N  D  V  E  A  A  R  R  A  G  I  D

 -35    670              -10                        710
TTGTATTGTAGTCAGCTATGGCTATAATCATGGAGAAAATATTTATGATAGTCATCCGCA
 C  I  V  V  S  Y  G  Y  N  H  G  E  N  I  Y  D  S  H  P  Q

            730                 750                 770
AGAAGTTGTGGATCGTCTGGATCAATTGATTTAATCATTGAGGAGTGGGGAATGTATCAC
 E  V  V  D  R  L  D  Q  L  I  *

            790                 810                 830
CTTATGGTTTTTCGATGGCGATAATCGCGTCTTAAATCACCTTTATCGTACTTAACTTAA

            850                 870                 890
AACCATCGAGAATACTCATGACATCACTAACTCAATTTGAACAGCTTAAAAACAGCAGGCT
                    M  T  S  L  T  Q  F  E  Q  L  K  T  A  G
```

Fig. 2. Nucleotide and deduced amino acid sequence of 850 bp upstream of the *A. calcoaceticus* trpE gene. The amino acid sequence is given in the one letter abbreviation. The -10 and -35 regions of the *trpE* promoter are indicated by the lines above the sequence. The *trpE* gene starts at position 858.

upstream of *trpE* (Essar et al., 1990a). A comparison of the encoded amino acid sequence is shown in Fig. 3. The *P. putida* N-terminus shows homology beginning at amino acid number 31 of the *A. calcoaceticus* sequence. At the C-terminal end the *P. putida* reading frame extends for a further 36 codons. Both encoded amino acid sequences have 39% identical amino acids, which are mainly clustered in the C-terminal half. This is far above background and indicates clearly that both *trpE* genes originate from the same source, which may be different from that of the other *trpE* genes. This conclusion is also supported by the homology in the primary amino acid structures of both *trpE* genes. They contain about 53% identical amino acids and only three gaps need to be introduced to reach this strong homology (Deveraux et al., 1984). The average homology among *trpE* genes

```
 31 GTLVDTASDMYRAMNLTLDHLGWSRVTEAQIRQWVGQGTGKLCDAVLKHL  80
    ::: :     :   :   :  ::         :     : :   :   :
  1 GTLIDSVPDLAAAVDRMLLELGRPPADLEAVRHWVGNGAQVLVRRALAGG  50

 81 FEE..VEPAKHQMLLTTYLEIYAQELCVTSRLFEGVQAFLDECKARKIEM 128
     :        :    : :::  :  :           ::  :       ::
 51 IEHDAVDDVLAEQGLALFMEAYAQSHELTV.VYPGVKDTLRWLQKQGVEM  99

129 ACVTNKPEQLARNLLETLKIGDYFDLVVGGDTLPVRKPDPLPLLHSVQVM 178
     :  :::::: ::   :::  ::  :::::: : ::: ::  :  : : :
100 ALITNKPERFVAPLLDQMKIGRYFRWMIGGDTLPQKKPDPAALLFVMQMA 149

179 KTTIENTLMIGDSKNDVEAARRAGIDCIVVSYGYNHGENIYDSHPQEVVD 228
     : :   : :::  ::   :: :   :: :  ::::::  :   :   :
150 GVTPQQSLFVGDSRSDVLAAKAAGVQCVGLTYGYNHGRPIHDETPSLVID 199

229 RLDQLI 234
     :: ::
200 DLRALL 205
```

Fig. 3. Homology of the open reading frames upstream of *trpE* in *A. calcoaceticus* and *P. putida*. The deduced amino acid sequences are compared according to Deveraux et al. (1984). Only identical amino acids are marked by dots between the sequences. The upper sequence originates from *A. calcoaceticus*.

Strain	Gene	Promoter Sequence			
E.c.	trp	TGAGCTG	**TTGACA**	ATTAATCATCGAACTAG	**TTAACT** AGTACGCA*
E.c.	trpR	GTCGTTA	**CTGATC**	CGCACGTTTATGATATGC	**TATCGT** ACTCTTTA*
S.t.	trp	ATAGGTG	**TTGACA**	TTATTCCATCGAACTAG	**TTAACT** AGTACGA*
A.c.	trpE	GTATCGA	**TTGTAT**	TGTAGTCAGCTATGGC	**TATAAT** CATGGA*
A.c.	trpFB	TTCAACT	**TTGACG**	CAAAGCACAAAAATTGCAT	**TACAAT** ACTTA*
S.m.	trp	AAGAGGG	**TTGACT**	TTGCCTTCGCGAACCAG	**TTAACT** AGTACACA*
B.l.	trp	CGGAAAC	**TACACA**	AGAACCCAAAAATGATT	**AATAAT** TGAGACAA
B.s.	trp	TTTATCA	**TTGACA**	AAAAATACTGAATTGTAA	**TACGAT** AAGAACA*
Consensus			**TTGACA**		**TATAAT**

Fig. 4. Promoter structures of different *trp* genes. The nucleotide sequences of different *trp* promoters are shown. The nucleotides for initiation of transcription are indicated by asterisks. The consensus boxes are indicated in bold print. Sequences are aligned with respect to their consensus boxes. The abbreviations used and the respective references are: A.c.: *A. calcoaceticus*; B.s.: *B. subtilis*, Henner et al. (1984); B.l.: *Brevibacterium lactofermentum*, Matsui et al. (1986); E.c.: *E. coli*, Yanofsky et al. (1981), Gunsalus and Yanofsky (1980); S.m.: *Serratia marcescens*, Miozzari et al. (1978); S.t.: *S.typhimurium*, Bennett et al. (1978).

ranges between 30% and 40% (Haspel et al., 1990). Only the *trpE* gene encoded by *Clostridium thermocellum* shows a similar homology (49% identical amino acids). However, it does not contain an upstream reading frame (Sato et al., 1989).

COMPARISON OF *TRP* PROMOTERS FROM *A. CALCOACETICUS* WITH THOSE OF OTHER ORGANISMS

In Fig. 4 *trp* promoters from different organisms are compared with those from *A. calcoaceticus* as determined by primer extension analyses (Haspel and Hillen, in preparation; Kishan and Hillen, in preparation). Generally, the *A. calcoaceticus* promoters seem to use the same consensus sequence as the other organisms (Hawley and McClure, 1983). Among the *trp* promoters shown in Fig. 4 they have the best conservation of the -10 regions, while in the -35 regions only the TTG sequence is strictly conserved and seems to be most important. It is noteworthy, that the *A. calcoaceticus* promoters show the most extreme spacings between the -35 and -10 regions, with only 16 bp for the *trpE* and as many as 19 bp for the *trpFB* promoter. Both spacings would not be optimal for *E. coli* (Hawley and McClure, 1983). Furthermore, the distances of the -10 regions to the start nucleotides of transcription are very short. It remains to be shown in the future whether these features are peculiarities of these two promoters or whether they are characteristic for *A. calcoaceticus* promoters in general.

POTENTIAL REGULATORY FEATURES OF THE *A. CALCOACETICUS TRP* GENES

Fig. 5 shows a comparison of the upstream sequences from the *A. calcoaceticus trpGDC, trpE* and *trpFB* determinants. A conserved sequence is found at different distances with respect to the start codons. A similar element is also found in the *trpE* and *trpGDC* sequences of *P. putida* (Essar et al., 1990a). It is not clear whether this sequence is involved in regulation; however, in *trpE* and *trpFB* of *A. calcoaceticus* it is located just downstream of the -10 promoter regions. Thus, it may well be a *cis* active regulatory site for these promoters. Promoter studies for the *trpGDC* operon of *A. calcoaceticus* and the *trpE* and *trpGDC* determinants of *P. putida* have not been published so far. However, it is most unlikely that the conservation of this sequence has occurred by chance in the upstream region of five functionally linked genes. Blumenberg and Yanofsky (1982) have shown consensus regions in the *trp* leader sequences of some Enterobacteriaceae, which are most probably involved in attenuation of transcription. This mechanism involves eight blocks of homologous sequences, six of which are involved in the formation of alternative secondary structures of the mRNA, depending on the extent of translation of a leader peptide. Most interestingly, the consensus sequence found here for the *trp* genes of *A. calcoaceticus* and *P. putida* is nearly identical to two of the homology blocks (designated 6 and 3) described for

```
trpFB   A.c. -  91  TTTGACGCAAAGCACAAAATTGCATTACAATACTTAGCCCAATGATGGATAGATCGGCTGTCTGTCAGGCA -  20
                                           ::::::  ::::::: : :::::::::
trpE    A.c. - 180  ATGGCTATAATCATGGAGAAATATTTATGATAGTCATCCGCAAGAAGTTGTGGATCGTCTGGATCAATTGA - 109
                    :::::: ::::::::  :::   :::::: :::::::::::::::   :::::::::::::::::   :::
trpE    P.p. - 142  ACCGTGGCGGTTACTGGCAAATTCTGGATGAAAGTCATCAAGGCTCTGGCCCGTTGGCCGTTGGCGCCTGA -  71
                                       CCTCATAAGTGGGCTATCATGATAGTTAGCCCGAACTCATAGTCTAATGTATAAAAAATAGTG - 122
                                       :::::: :  ::::::: :: ::::: :::::::
trpGDC  A.c. - 184  TCAGGGAACATCCTTTCTTTGTTTTATGAGGTACTGCCCGATCGACCTTAGCAAATCGCCGTGCATCCCT - 299
trpGDC  P.p. - 370
                                       tTtATGAtAgTcAgCCCgA
```

Fig. 5. Homology of upstream sequences of *trp* genes from *A. calcoaceticus* and *P. putida*. Upstream sequences, numbered from the respective start codon and specified on the left side using the same abbreviations as in Fig.4, are shown for various *trp* genes. Identical nucleotides of neighbouring sequences are indicated by dots. The conserved sequence is boxed and printed in bold letters. A concensus sequence is given at the bottom. A capital letter indicates conservation in four or more genes, while a small letter indicates conservation in three genes.

246

the Enterobacteriaceae. However, the other elements necessary to cause attenuation of transcription are not present. This indicates that the evolutionary source of the *trp* genes may have been regulated by attenuation. Why most of the necessary sequence elements except the two shown in Fig. 5 were removed by evolutionary drift is not clear. Their conservation would imply, however, an important function for these elements. If they exhibited a regulatory function one would have to assume that the *trp* genes are co-regulated; indeed Cohn and Crawford (1976) showed that the expression of *A. calcoaceticus*-encoded *trp* genes is subject to repression. However, according to these results, *trpGDC* and *trpE* are co-regulated, while *trpFB* and *trpA* show a different type of regulation. A previous report has focused on the feedback inhibition of the first two enzymes of the pathway and on repression of anthranilate synthase (Twarog and Liggins, 1970). Thus, the question of regulation of *trp* genes in *A. calcoaceticus* has not yet been fully resolved.

REFERENCES

Bae, Y.M., Holmgren, E., and Crawford I.P., 1989, *Rhizobium meliloti* anthranilate synthase gene: Cloning, sequence and expression in *Escherichia coli*, *J. Bacteriol.*, 171:3471.

Bennett, G.N., Brown, K.D., and Yanofsky, C., 1978, Nucleotide sequence of promoter-operator region of the tryptophan operon of *Salmonella typhimurium*, *J. Mol. Biol.*, 121:139.

Blumenberg, M., and Yanofsky, C., 1982, Regulatory region of the *Klebsiella aerogenes* tryptophan operon, *J. Bacteriol.*, 152:49.

Cohn, W., and Crawford, I.P., 1976, Regulation of enzyme synthesis in the tryptophan pathway of *Acinetobacter calcoaceticus*, *J. Bacteriol.*, 127:367.

Crawford, I.P., 1975, Gene rearrangements in the evolution of the tryptophan synthetic pathway, *Bacteriol. Rev.*, 39:87.

Crawford, I.P., 1989, Evolution of a biosynthetic pathway: the tryptophan paradigm, *Ann. Rev. Microbiol.*, 43:567.

Crawford, I.P., and Eberly, L., 1989, DNA sequence of the tryptophan synthase genes of *Pseudomonas putida*, *Biochimie*, 71:521.

Crawford, I.P., Wilde, A., Yelverton, E.M., Figurski, D., and Hedges, R.W., 1986, Structure and regulation of the anthranilate synthase genes in *Pseudomonas aeruginosa*. II. Cloning and expression in *Escherichia coli*, *Mol. Biol. Evol.*, 3:449.

Deveraux, J., Haeberli, P., and Smithies, O., 1984, A compressive set of sequence analysis programs for the VAX, *Nucl. Acids Res.*, 12:387.

Essar, D.W., Eberly, L., and Crawford, I.P., 1990a, Evolutionary differences in chromosomal locations of four early genes of the tryptophan pathway in fluorescent pseudomonads: DNA sequences and characterization of *Pseudomonas putida trpE* and *trpGDC*, *J. Bacteriol.*, 172:867.

Essar, D.W., Eberly, L., Han, C.J., and Crawford, I.P., 1990b, DNA sequences and characterization of four early genes of the tryptophan pathway in *Pseudomonas aeruginosa*, *J. Bacteriol.*, 172:853.

Grantham, R., Gautier, C., and Gouy, M., 1980, Codon frequencies in 119 individual genes confirm consistent choices of degenerate bases according to genome type, *Nucl. Acids Res.*, 8:1893.

Gunsalus, R.P., and Yanofsky, C., 1980, Nucleotide sequence and expression of *Escherichia coli trpR*, the structural gene for the *trp* aporepressor, *Proc. Natl. Acad. Sci. USA.*, 77:717.

Hadero, A., and Crawford, I.P., 1986, Nucleotide sequence of the genes for tryptophan synthase in *Pseudomonas aeruginosa*, *Mol. Biol. Evol.*, 3:191.

Haspel, G., Hunger, M., Schmucker, R., and Hillen, W., 1990, Identification and nucleotide sequence of the *Acinetobacter calcoaceticus* encoded *trpE* gene, *Mol. Gen. Genet.*, 220:475.

Hawley, D.K., and McClure, W.R., 1983, Compilation and analysis of *Escherichia coli* promoters, *Nucl. Acids Res.*, 11:2237.

Henner, D.J., Band, L., and Shimotsu, H., 1984, Nucleotide sequence of the *Bacillus subtilis* tryptophan operon, *Gene*, 34:169.

Holloway, B.W., Krishnapillai, V., and Morgan, A.F., 1979, Chromosomal genetics of *Pseudomonas*, *Microbiol. Rev.*, 43:73.

Hunger, M., Schmucker, R., Kishan, V., and Hillen, W., 1990, Analysis and nucleotide sequence of an origin of DNA replication for *Acinetobacter calcoaceticus* and its use for shuttle plasmids with *Escherichia coli*, *Gene*, 87:45.

Johnston, A.W.B., Bibb, M.J., and Beringer, J.E., 1978, Tryptophan genes in *Rhizobium* - their organization and their transfer to other bacterial genera, *Mol. Gen. Genet.*, 165:323.

Kaplan, J.B., Goncharoff, P., Seibold, A.M., and Nichols, B.P., 1984, Nucleotide sequence of the *Acinetobacter calcoaceticus trpGDC* gene cluster, *Mol. Biol. Evol.*, 1:456.

Matsui, K., Sano, K., and Ohtsubo, E., 1986, Complete nucleotide and deduced amino acid sequences of the *Brevibacterium lactofermentum* tryptophan operon, *Nucl. Acids Res.*, 14:10113.

Miozzari, G.F., and Yanofsky, C., 1978, The regulatory region of the *trp* operon of *Serratia marcescens*, *Nature*, 276:684.

Ross, C.M., Kaplan, J.B., Winkler, M.E., and Nichols, B.P., 1990, An evolutionary comparison of the *Acinetobacter calcoaceticus trpF* gene with *trpF* genes of several organisms, *Mol. Biol. Evol.*, 7:74.

Sano, K., and Matsui, K., 1987, Structure and function of the *trp* operon control regions of *Brevibacterium lactofermentum*, a glutamic acid producing bacterium, *Gene*, 53:191.

Sato, S., Nakada, Y., Hon-Nami, K., Yasui, K., and Shiratsuchi, A., 1989, Molecular cloning and nucleotide sequence of the *Clostridium thermocellum trpE* gene, *J. Biochem.*, 105:362.

Sawula, R.V., and Crawford, I.P., 1972, Mapping of the tryptophan genes of *Acinetobacter calcoaceticus* by transformation, *J. Bacteriol.*, 112:797.

Sawula, R.V., and Crawford, I.P., 1973, Anthranilate synthetase of *Acinetobacter calcoaceticus*: separation and partial characterization of subunits, *J. Biol. Chem.*, 248:3573.

Shinomiya, T., Shiga, S., and Kageyama, M., 1983, Genetic determinant of pyocin R2 in *Pseudomonas aeruginosa* PAOI. Localization of the pyocin R2 gene cluster between the *trpCD* and *trpE* genes, *Mol. Gen. Genet.*, 189:375.

Twarog, R., and Liggins, L., 1970, Enzymes of the tryptophan pathway in *Acinetobacter calcoaceticus*, *J. Bacteriol.*, 104:254.

Yanofsky, C., 1984, Comparison of regulatory and structural regions of genes of tryptophan metabolism, *Mol. Biol. Evol.*, 1:143.

Yanofsky, C., Platt, T., Crawford, I.P., Nichols, B., Christie, G., Horowitz, H., van Clemput, M., and Wu, A., 1981, The complete nucleotide sequence of the tryptophan operon of *Escherichia coli*, *Nucl. Acids Res.*, 9:6647.

CODON USAGE IN *ACINETOBACTER* STRUCTURAL GENES

P.J. White[a], I.S. Hunter[a] and C.A. Fewson[b] [*]

Departments of [a]Genetics and [b]Biochemistry
University of Glasgow
Glasgow G12 8QQ
UK

INTRODUCTION

The first gene sequence from an *Acinetobacter* was described by Kaplan et al. (1984). An increasing number of such sequences have been reported subsequently, reflecting the growing interest in understanding the molecular biology of this organism. It is now possible to analyse codon usage from a substantial number of *Acinetobacter* genes and this is important for several reasons (Andersson and Kurland, 1990). In particular: (a) it forms a valuable basis for evolutionary speculations; (b) it is an important consideration when attempting the expression of heterologous genes, either of *Acinetobacter* genes in *Escherichia coli* and other hosts, or of foreign genes in *Acinetobacter*; (c) DNA may be analysed and potential protein coding regions identified from a knowledge of codon usage; (d) knowledge of codon usage can aid in the design of effective synthetic oligonucleotide probes used to clone genes when partial amino acid sequences are known.

This article presents an analysis of the sequences of various structural genes from a number of acinetobacters, including the codon usage and overall base composition at each of the three codon positions.

CODON USAGE AND OVERALL BASE COMPOSITION

The codon usage patterns for all *Acinetobacter* structural gene sequences in the EMBL and NBRF (nucleic) data bases and in the published

---

[*]corresponding author

The Biology of Acinetobacter, Edited by K. J. Towner *et al.*
Plenum Press, New York, 1991

Table 1. *Acinetobacter* structural genes used in codon analysis

Gene	Phenotype or gene product	Bacterial strain	Ref.
aphA-6	kanamycin resistance	A. baumannii BM 2580	1
gdhA	glucose dehydrogenase	A. calcoaceticus LMD 79.41	2
gdhB	glucose dehydrogenase	A. calcoaceticus LMD 79.41	3
-	cyclohexanone monooxygenase	Acinetobacter sp. NCIB 9871	4
est	esterase	A. calcoaceticus RAG-1	5
catA	catechol 1,2-dioxygenase	A. calcoaceticus BD 413	6
mro	mutarotase	A. calcoaceticus DSM 30007	7
pqq (I,II,III)	PQQ biosynthesis	A. calcoaceticus LMD 79.41	8
trpG,D,C	tryptophan biosynthesis	A. calcoaceticus BD 413	9
trpF	tryptophan biosynthesis	A. calcoaceticus BD 413	10
gltA	citrate synthase	A. anitratum NCTC 7844	11
pcaC	carboxymuconolactone decarboxylase	A. calcoaceticus BD 413	12
pcaG,H	protocatechuate 3,4-dioxygenase	A. calcoaceticus BD 413	12
trpB	tryptophan biosynthesis	A. calcoaceticus BD 413	13
trpE	anthranilate synthase	A. calcoaceticus BD 413	14

References: (1) Martin et al., 1988; (2) Cleton-Jansen et al., 1988; (3) Cleton-Jansen et al., 1989; (4) Chen et al., 1988; (5) Reddy et al., 1989; (6) Neidle et al., 1988; (7) Gatz and Hillen, 1986; (8) Goosen et al., 1989; (9) Kaplan et al., 1984; (10) Ross et al., 1990; (11) Donald and Duckworth, 1987; (12) Hartnett et al., 1990; (13) Haspel et al., this volume; (14) Haspel et al., 1990.

literature were determined. Twenty gene sequences were available for analysis in May 1990 (Table 1). The University of Wisconsin Genetics Computer Group (UWGCG) program CODONFREQ (Devereux et al., 1984) was used to calculate codon frequencies for each of the genes and the results for all 20 genes were summed (Table 2). These data were then used

Table 2. Codon usage in *Acinetobacter* structural genes

Amino acid	Codon	Total	%	Amino acid	Codon	Total	%
Gly	GGG	39	8	Met	ATG	154	100
	GGA	70	14				
	GGT	286	55	Ile	ATA	14	3
	GGC	120	23		ATT	286	69
					ATC	116	28
Glu	GAG	78	22				
	GAA	284	78	Thr	ACG	64	17
					ACA	94	25
Asp	GAT	307	74		ACT	130	35
	GAC	107	26		ACC	88	23
Val	GTG	118	25	Trp	TGG	91	100
	GTA	101	22				
	GTT	164	35	Cys	TGT	43	72
	GTC	83	18		TGC	17	28
Ala	GCG	92	18	Tyr	TAT	178	74
	GCA	195	38		TAC	63	26
	GCT	142	27				
	GCC	90	17	Leu	TTG	105	18
					TTA	200	35
Arg	AGG	3	1		CTG	65	11
	AGA	12	4		CTA	46	8
	CGG	8	3		CTT	108	19
	CGA	33	11		CTC	54	9
	CGT	185	62				
	CGC	56	19	Phe	TTT	182	71
					TTC	75	29
Ser	AGT	65	18				
	AGC	52	15	Gln	CAG	117	35
	TCG	28	8		CAA	216	65
	TCA	99	28				
	TCT	96	27	His	CAT	109	70
	TCC	17	5		CAC	46	30
Lys	AAG	88	24	Pro	CCG	54	16
	AAA	274	76		CCA	156	47
					CCT	105	32
Asn	AAT	218	63		CCC	16	5
	AAC	129	37				
				Term	TGA	6	30
					TAG	2	10
					TAA	12	60

Table 3. Percentage base composition of *Acinetobacter* structural genes

Codon position

Base	First	Second	Third
A	26.9	33.4	27.0
G	34.3	16.2	16.6
C	20.7	22.1	17.0
T	18.0	28.2	39.3
A+T	44.9	61.6	66.3

to calculate the overall base composition at each of the three codon positions in the *Acinetobacter* genes analysed (Table 3).

The results of these analyses show that A and T predominate in the third codon position (over 66% of third position bases). Similarly, A and T are again favoured over G and C in the second codon position, comprising nearly 62% of the bases used. In contrast, there is no bias toward A and T in the first codon position. The overall A+T content of the 20 genes, as determined by the average composition of the three codon positions, is 57.6%. These conclusions are in agreement with those of authors who have observed a predominance of A and T bases in the third position of codons in individual genes (e.g. Chen et al., 1988; Haspel et al., this volume). The overall A+T content of these 20 genes is also in agreement with the predicted A+T content of the *Acinetobacter* genome (55-62%; Henriksen, 1976). There is a preference for G in the first codon position (34.3%) and a paucity of G in the second codon position (16.2%). This pattern of dinucleotide usage has been observed in other protein coding sequences and may be responsible for monitoring the correct reading frame during translation (Trifonov, 1987).

Several codons seem to be used very rarely by acinetobacters. The codons AGG, AGA and CGG for arginine and the ATA codon for isoleucine represent less than 5% of the codons used from specific synonymous codon groups.

Only one of the gene sequences listed in Table 1 showed any gross differences in codon usage when compared with the calculated mean codon usage. This was the *gltA* gene (citrate synthase) from *A. anitratum* NCTC 7844, which showed differences, especially in the preferences for codons for phenylalanine, leucine, histidine, asparagine, aspartate and tyrosine. This difference in codon usage may be a reflection of the level of expression of this gene, or of its evolution, or it may reflect some other feature of the genome of this particular strain.

The abundance of A and T bases in the degenerate position of amino acid codons in *Acinetobacter* genes should help in the construction of oligonucleotide probes based upon amino acid sequences. For example, the codon for glutamic acid is found to be GAA in 78% of all cases. Similarly the codon for aspartate is predominantly GAT.

OTHER GENES AND SEQUENCES

The *catM* gene, which encodes a transcriptional repressor, has been sequenced (Neidle et al., 1989). Its codon usage shows some differences from that of the structural genes listed in Table 1, for instance four of its ten arginine codons are AGA. The overall codon usages of the structural genes summarised in Table 2 seem to be very similar to those of 11 of the structural genes of the ß-ketoadipate pathway of *Acinetobacter* (Ornston and Neidle, this volume). However, all of them are different from the codon usages of the structural genes *catI*, *catJ* and *catF*. The special evolutionary factors and selective forces that may have been involved in the emergence of the ß-ketoadipate pathway of *A. calcoaceticus* are discussed by Ornston and Neidle (this volume).

The sequence of the mutarotase gene indicated the existence of a 20 amino acid leader sequence not found in the mature enzyme (Gatz and Hillen, 1986), but no sequence corresponding to a signal peptide for protein export was found within the sequence for the *est* gene, even though the enzyme may be located external to the cytoplasmic membrane (Reddy et al., 1989).

Several *Acinetobacter* genes have been expressed in *E. coli* and *Pseudomonas putida* (e.g. Neidle et al., 1987; Chen et al., 1988; Reddy et al., 1989; Ornston and Neidle, this volume). Although it is too early to generalise about promoter sequences, ribosome binding sites and other regulatory elements in *Acinetobacter*, the sequences involved in the regulation of expression of a few of the genes are now being clarified (e.g. Haspel et al., this volume; Ornston and Neidle, this volume).

CONCLUSIONS

This article provides an analysis of the codon usage and base composition of 20 *Acinetobacter* structural genes. The distribution of genes amongst different strains and metabolic pathways should be sufficiently random to ensure that the results are representative of *Acinetobacter* structural genes in general. The analysis therefore provides a reference for comparison of individual genes and a basis for the construction of oligonucleotide probes.

REFERENCES

Andersson, S.G.E., and Kurland, C.G., 1990, Codon preferences in free-living microorganisms, *Microbiol. Rev.*, 54:198.

Chen, Y.-C.J., Peoples, O.P., and Walsh, C.T., 1988, *Acinetobacter* cyclohexanone monooxygenase: gene cloning and sequence determination, *J. Bacteriol.*, 170:781.

Cleton-Jansen, A.-M., Goosen, N., and van de Putte, P., 1988, Nucleotide sequence of the gene coding for quinoprotein glucose dehydrogenase from *Acinetobacter calcoaceticus, Nucl. Acids Res.*, 16:6228.

Cleton-Jansen, A.-M., Goosen, N., Vink, K., and van de Putte, P., 1989, Cloning, characterization and DNA sequencing of the gene encoding the M$_r$50,000 quinoprotein glucose dehydrogenase from *Acinetobactercalcoaceticus, Mol. Gen. Genet.*, 217:430.

Devereux, J., Haeberli, P., and Smithies, O., 1984, A comprehensive set of sequence analysis programmes for the VAX, *Nucl. Acids Res.*, 12:387.

Donald, L.J., and Duckworth, H.W., 1987, Expression and base sequence of the citrate synthase gene of *Acinetobacter anitratum, Biochem. Cell. Biol.*, 65:930.

Gatz, C., and Hillen, W., 1986, *Acinetobacter calcoaceticus* encoded mutarotase: nucleotide sequence analysis of the gene and characterisation of its secretion in *Escherichia coli, Nucl. Acids Res.*, 14:4309.

Goosen, N., Horsman, H.P.A., Huinen, R.G.M., and van de Putte, P., 1989, *Acinetobacter calcoaceticus* genes involved in biosynthesis of the coenzyme pyrrolo-quinoline-quinone: nucleotide sequence and expression in *Escherichia coli* K-12, *J. Bacteriol.*, 171:447.

Hartnett, C., Neidle, E.L., Ngai, K., and Ornston, L.N., 1990, DNA sequences of genes encoding *Acinetobacter calcoaceticus* protocatechuate 3,4-dioxygenase: Evidence indicating shifting of genes and of DNA sequences within genes during their evolutionary divergence, *J. Bacteriol.*, 172:956.

Haspel, G., Hunger, M., Schmucher, R., and Hillen, W., 1990, Identification and nucleotide sequence of the *Acinetobacter calcoaceticus* encoded *trpE* gene, *Mol. Gen. Genet.*, 220:475.

Henriksen, S.D., 1976, *Moraxella, Neisseria, Branhamella* and *Acinetobacter, Ann. Rev. Microbiol.*, 30:63.

Kaplan, J.B., Goncharoff, P., Siebold, A.M., and Nichols, B.P., 1984, Nucleotide sequence of the *Acinetobacter calcoaceticus* *trpGDC* gene cluster, *Mol. Biol. Evol.*, 1:456.

Martin, P., Jullien, E., and Courvalin, P., 1988, Nucleotide sequence of *Acinetobacter baumannii aphA-6* gene: evolutionary and functional implications of sequence homologies with nucleotide-binding proteins, kinases and other aminoglycoside-modifying enzymes, *Mol. Microbiol.*, 2:615.

Neidle, E.L., Shapiro, M.K., and Ornston, L.N., 1987, Cloning and expression in *Escherichia coli* of *Acinetobacter calcoaceticus* genes for benzoate degradation, *J. Bacteriol.*, 169:5496.

Neidle, E.L., Hartnett, C., Bonitz, S., and Ornston, L.N., 1988, DNA sequence of the *Acinetobacter calcoaceticus* catechol 1,2-dioxygenase 1 structural gene *catA*: evidence for evolutionary divergence of intradiol dioxygenases by acquisition of DNA sequence repetitions, *J. Bacteriol.*, 170:4874.

Neidle, E.L., Hartnett, C., and Ornston, L.N., 1989, Characterization of *Acinetobacter calcoaceticus catM*, a repressor gene homologous in sequence to transcriptional activator genes, *J. Bacteriol.*, 171:5410.

Reddy, P.G., Allon, R., Mevarech, M., Mendelovitz, S., Sato, Y., and Gutnick, D.L., 1989, Cloning and expression in *Escherichia coli* of an esterase-coding gene from the oil-degrading bacterium *Acinetobacter calcoaceticus* RAG-1, *Gene*, 76:145.

Ross, C.M., Kaplan, J.B., Winkler, M.E., and Nichols, B.P., 1990, An evolutionary comparison of *Acinetobacter calcoaceticus trpF* with *trpF* genes of several organisms, *Mol. Biol. Evol.*, 7:74.

Trifonov, E.N., 1987, Translation framing code and frame-monitoring mechanism as suggested by the analysis of mRNA and 16 S rRNA nucleotide sequences, *J. Mol. Biol.*, 194:643.

THE OUTER MEMBRANE OF *ACINETOBACTER*:

STRUCTURE-FUNCTION RELATIONSHIPS

P. Borneleit and H.-P.Kleber

Sektion Biowissenschaften
Bereich Biochemie
Karl-Marx-Universität Leipzig
Talstrasse 33
7010 Leipzig
Germany

INTRODUCTION

The fine structure of the cell envelope of *Acinetobacter* has been determined from electron micrographs of thin sections, negatively-stained and freeze-etched preparations of whole cells, and of envelopes treated to remove the various layers (Thornley and Glauert, 1968; Thorne et al., 1973; Sleytr et al., 1974; Thornley and Sleytr, 1974; Aurich et al., 1977). These early studies revealed the presence of an outer membrane typical of Gram-negative bacteria. Furthermore, there are numerous reports of superficial material covering the outer surface of several *Acinetobacter* strains; such material includes crystalline surface layers of protein (S-layers) (Thornley et al., 1973, 1974), and polysaccharide capsules (Taylor and Juni, 1961; Pines et al., 1983). These additional surface layers are of tremendous importance in determining the surface properties of the cells, and the polysaccharide capsular material has proven to be a potent extracellular emulsifying agent (reviewed by Rosenberg and Kaplan, 1987).

This article concentrates on the outer membrane of *Acinetobacter* in the strict sense. Since the early seventies, much knowledge has accumulated on the structure and function of outer membranes, but interest has focused mainly on the Enterobacteriaceae (for reviews see Lugtenberg and van Alphen, 1983; Hancock, 1984; Nikaido and Vaara, 1985). Much less is known about non-enteric bacteria, and it has been speculated that their outer membrane organisation may be completely different. In this review we attempt to summarise the known facts about the outer membrane of *Acinetobacter* which, in comparison to the Enterobacteriaceae, is characterised by some special permeability properties. These are, as we will show, most probably related to unusual features of its lipopolysaccharide (LPS) constituent.

PERMEABILITY PROPERTIES

One of the main functions of the outer membrane is to serve as a selective barrier towards the external medium. Small solutes may pass through the outer membrane via water-filled proteinaceous diffusion channels termed porins (reviewed by Benz, 1985). On the other hand, in enteric bacteria the outer membrane has developed into a very effective barrier giving protection against the penetration of hydrophobic molecules and the lytic action of detergents (Nikaido, 1979).

There are several lines of evidence that suggest the existence of a functioning so-called hydrophobic pathway in *Acinetobacter* spp. First, their usual habitats, soil, water and the skin and mucous membranes of humans (Henriksen, 1973; Juni, 1978) do not require the protection mechanisms effective in enteric bacteria living in the intestinal tract of animals. Second, many *Acinetobacter* strains are capable of using hydrophobic growth substrates such as long-chain hydrocarbons (Baumann et al., 1968), which, prior to degradation, have to pass the outer membrane. Strains especially well-characterised in this respect are *Acinetobacter* sp. HO1-N, growing on *n*-alkanes (C_{10} to C_{20}) (Stewart et al., 1959), primary aliphatic alcohols (C_{12} to C_{16}) and aliphatic aldehydes (C_{12} to C_{16}) (Singer et al., 1985), *A. calcoaceticus* 69-V (Kleber et al., 1973) and *A. calcoaceticus* RAG-1 (Reisfeld et al., 1972). According to a screening test performed by Kleber et al. (1983), 12 of 13 *Acinetobacter* strains investigated were capable of degrading hydrocarbons with chain lengths longer than C_{10}. Finally, the inclusion of *Acinetobacter* in the family Neisseriaceae (Juni, 1984) also renders the existence of a hydrophobic pathway highly likely. For this bacterial family several cases of hypersensitivity towards hydrophobic antibacterial agents such as erythromycin, fusidic acid, Triton X-100, acridine orange (Sarubbi et al., 1975), crystal violet (Guymon and Sparling, 1975) and hydrophobic steroid hormones (Lysko and Morse, 1981) have been reported.

However, apart from these general lines of investigation, a systematic analysis of the permeability properties of the outer membrane of *Acinetobacter* has not been performed. We therefore now present some additional data concerning the functioning of the hydrophobic pathway (in part recently published in Borneleit et al., 1990). In contrast to *Escherichia coli* and *Salmonella typhimurium*, ten of 11 *Acinetobacter* strains tested were found to be sensitive to the hydrophobic antibiotic erythromycin, with minimal inhibitory concentrations of between 1.2 and 50 mg/l (Table 1). Furthermore, for two of the strains, sensitivity towards novobiocin and gentian violet was observed. Interestingly, in *A. calcoaceticus* strain CCM 5593, sensitivity to hydrophobic agents was

Table 1. Minimal inhibitory concentrations (mg/l) of antibacterial agents

Strain		Erythromycin	Novobiocin	Gentian violet
Escherichia coli K-12		>150.0	>150.0	>2.0
Salmonella typhimurium LT 2		>150.0	nd	>2.0
Acinetobacter calcoaceticus CCM 2355[b]		1.2	nd	nd
Acinetobacter sp. EB 10d[c]		5.0	nd	nd
EB 114[c]		5.0	nd	nd
EB 6[c]		5.0	nd	nd
Acinetobacter lwoffii CCM 5572[b]		7.5	nd	nd
Acinetobacter sp. EB 104[c]		7.5	nd	nd
EB 113[c]		10.0	nd	nd
Acinetobacter calcoaceticus 69-V[d]		10.0	50.0	0.25
Acinetobacter sp. EB 110[c]		15.0	nd	nd
Acinetobacter calcoaceticus CCM 5595[b]		50.0	nd	nd
CCM 5593[b]		150.0	100.0	1.5

[a]minimal inhibitory concentrations were determined in nutrient broth containing varying concentrations of the agent tested and were defined as the lowest concentration which inhibited visible growth
[b]Czechoslovak Collection of Microorganisms, Brno, Czechoslovakia
[c]Strain Collection of the Institute of Biotechnology, Leipzig, Germany
[d]isolated from soil specimens and identified as described by Kleber et al. (1973)
nd: not determined.

not as pronounced as in strain 69-V. According to Kleber et al. (1983), these strains differ greatly in their ability to grow on long-chain hydrocarbons. In agreement with these data, we were able to demonstrate uptake of the hydrophobic dye gentian violet by the hydrocarbon-degrading strain 69-V, but not by strain CCM 5593 or the enteric bacteria *E. coli* and *S. typhimurium* (Fig. 1). Finally, we found *A. calcoaceticus* strains to be

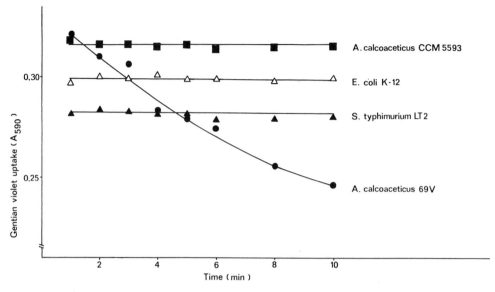

Fig. 1. Gentian violet uptake by intact cells of *A. calcoaceticus* 69-V and CCM 5593, *E. coli* K-12 and *S. typhimurium* LT-2. Cells were grown in nutrient broth to the early logarithmic phase and uptake rates determined at 30°C as described by Gustafsson et al. (1973).

much more susceptible to the lytic action of the strong ionic detergent sodium dodecyl sulphate (SDS) than Enterobacteriaceae. While *E. coli* has been reported to tolerate as much as 50 g/l in the growth medium (Lugtenberg and van Alphen, 1983), we could observe visible growth of *Acinetobacter* spp. only up to concentrations of about 0.05 g SDS/l.

PROTEIN EXPORT

Another important function of the outer membrane of Gram-negative bacteria is to confine the periplasmic enzymes and other proteins to the periplasmic space. The enzymes localised in the periplasm are mainly hydrolytic, with functions that include enabling the cell to digest macromolecular molecules or low molecular mass organic compounds as nutrients and conferring protection against certain antibacterial agents. Several hydrolytic enzymes have been described in *Acinetobacter* spp. (Table 2), while the outer membrane itself has been shown to contain a phospholipase A_1 (Torregrossa et al., 1978) and a palmitoyl-CoA esterase (Claus et al., 1985).

Most interestingly, however, some *Acinetobacter* spp. have been found to produce extracellular hydrolytic enzymes. An extracellular lipase has been described for *A. lwoffii* 016 (Breuil and Kushner, 1975a,b; Breuil et al., 1978) and *A. calcoaceticus* 69-V (Haferburg and Kleber, 1982, 1983; Fischer and Kleber, 1987). An extracellular phospholipase C (haemolysin)

Table 2. Hydrolytic enzymes in *Acinetobacter* spp.

Enzyme	Reference
Aminopeptidase I	Ludewig, 1983
Endopeptidase	Fricke et al., 1983,1986
Exopeptidase	Jahreis and Aurich, 1989
Collagenase	Monboisse et al., 1979a,b
α-amylases I and II	Onishi and Hidaka, 1978; Onishi et al., 1979
ß-Glucanase	Katohda et al., 1979
Lipases	Breuil and Kushner, 1975a,b;
	Haferburg and Kleber, 1982;
	Kalogridov-Vassiliadov, 1984
Phospholipase A_1	Torregrossa et al., 1977
Phospholipase A_2	Thorne et al., 1976
Phospholipase C	Lehmann, 1971a,b; Lehmann, 1972, 1973a;
	Zaikina and Robakidze, 1976
Esterase	Claus et al., 1985; Fischer, 1986
Penicillin amidase	Baker, 1984
ß-Lactamase	Joly-Guillou et al., 1988; Paul et al., 1989

was found in six *A. calcoaceticus* and two *A. lwoffii* strains (Lehmann 1973a,b), while an extracellular ß-lactamase is produced by *A. calcoaceticus* strains 69-V and CCM 5593 (Blechschmidt et al., 1989). The regularly arranged surface protein A of *Acinetobacter* sp. 199A has also been shown to have phospholipase A_2 activity, and about half of this protein is released into the growth medium (Thorne et al., 1976).

Protein export across the outer membrane of Gram-negative bacteria has, in the past, been considered a rare phenomenon, and has attracted attention only recently (Pugsley and Schwartz, 1985). In view of the numerous reports of extracellular enzymes (see above), the outer membrane of *Acinetobacter* seems to be especially well-equipped for this process. As to the underlying mechanism, very few investigations have been performed, but a genuine selective protein export system is obviously involved.

Non-specific periplasmic leakage could be excluded for the export of ß-lactamase by *A. calcoaceticus* 69-V because other periplasmic enzymes, such as glucose dehydrogenase or alkaline phosphatase, were shown to be retained within the cells (Blechschmidt et al., 1989). When the time course of lipase location was followed during the growth of *A. calcoaceticus* 69-V, an accumulation of the enzyme on the outer membrane was observed prior to export (Fischer et al., 1987). According to a model based on these results, the lipase is released from the "overloaded" outer membrane into the surrounding medium without cell lysis or membrane blebbing. The latter statement is in contrast to our observations concerning the export of ß-lactamase by *A. calcoaceticus* CCM 5593. In this strain the appearance of an extracellular ß-lactamase is clearly accompanied by a release of the outer membrane constituent LPS (Fig. 2). However, as postulated by Pugsley and Schwartz (1985), cells may easily have developed different strategies to overcome the outer membrane permeability barrier.

COMPOSITION AND STRUCTURE

The composition of the outer membrane of *Acinetobacter* is basically the same as that of the Enterobacteriaceae, with proteins, phospholipids and LPS as the main constitutents.

With regard to the protein component, a major protein species of approximate M_r 40,000 was demonstrated in *Acinetobacter* sp. HO1-N (Scott et al., 1976) and *A. calcoaceticus* 69-V (Borneleit et al., 1988), and was additionally shown to be heat-modifiable (*A. calcoaceticus* 69-V:

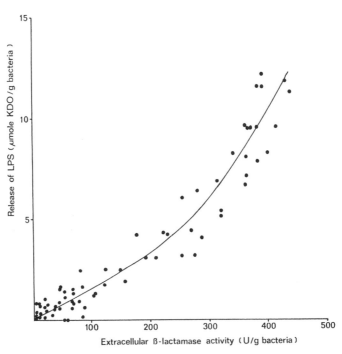

Fig. 2. LPS release as a function of extracellular ß-lactamase production by *A. calcoaceticus* CCM 5593. Cells were grown on complex growth media or minimal media (with acetate as carbon source) supplemented with yeast extract. ß-Lactamase production was varied by adding different concentrations (0-750 mg/l) of ß-lactam inducers (ampicillin or cefotaxime).

Fischer et al., 1984; *Acinetobacter* sp. ATCC 23055 and its radiation-resistant mutant FO-1: Nishimura et al., 1986), thus being comparable to the structural OmpA protein of *E. coli* (Benz, 1985). Furthermore, a pore-forming peptidoglycan-associated protein (porin) of M_r 53,000 was identified (Fischer et al., 1984), analogous to the OmpC/F porins in *E. coli* (Benz, 1985). A distinguishing feature in the protein composition of the outer membrane of the Neisseriaceae might be the absence of a small (M_r 10,000) lipoprotein which is present in the Enterobacteriaceae in large amounts (Hebeler et al., 1978). The presence of a similar protein in *A. calcoaceticus* 69-V was, however, claimed by Fischer et al. (1984) and Borneleit et al. (1988).

As to the phospholipid composition, in all *Acinetobacter* outer membranes isolated so far, the typical phospholipids of bacterial membranes, i.e. phosphatidylethanolamine (main component), phosphatidylglycerol and cardiolipin, were found. In addition, the lysophospholipids lysocardiolipin and lysophosphatidylethanolamine were detected (Scott et al., 1976; Borneleit et al., 1988), as would be expected from the presence of an outer membrane phospholipase A especially active against cardiolipin (Torregrossa et al., 1977).

The LPS component in early studies was a matter of debate because it was assumed to be practically devoid of the LPS-specific sugar constituent 3-deoxy-D-*manno*-2-octulosonic acid (KDO) (Adams et al., 1970; Scott et al., 1976). Later, the presence of KDO was demonstrated by the usual thiobarbituric acid assay in whole cells of several *A. calcoaceticus* strains (McDougall et al., 1983), in isolated outer membranes (Fischer et al., 1984; Borneleit et al., 1988), and in isolated LPS (Thorne et al., 1973; Brade and Galanos, 1982; Brade et al., 1987; Borneleit et al., 1989a).

Close examination of the chemical composition and structure of the *Acinetobacter* LPS has shown that it is unusual. First, in contrast to the enteric bacteria, the LPS isolated so far from *Acinetobacter* spp. has been of the R-type, i.e. it lacks an O-specific sugar chain (Rietschel et al., 1987). There is, however, some evidence that a small fraction of S-LPS containing an O-antigenic chain is also present in *A. calcoaceticus* 69-V cells (Borneleit et al., 1990). Second, in *A. calcoaceticus* NCTC 10305, the core oligosaccharide and lipid A component have been shown to be interlinked by a 2-octulosonic acid instead of the KDO found in Enterobacteriaceae (Kawahara et al., 1987). As a consequence, the respective ketosidic linkage is highly resistant to acid hydrolysis, and this might be advantageous in view of the ability of many *Acinetobacter* strains to form organic acids from several sugars (Juni, 1978). Furthermore, in *A. calcoaceticus* NCTC 10305 LPS, a previously undescribed 3-deoxyaldulosonic acid (2-keto-3-deoxy-1,7-dicarboxyheptonic acid) was detected (Brade and Rietschel, 1985). As another unusual feature, C_{12} fatty acids were found to be attached to the lipid A region (Brade and Galanos, 1982; Borneleit et al., 1989a) instead of the C_{14} fatty acids predominant in lipid A from many Enterobacteriaceae, and this is probably a characteristic of the Neisseriaceae (Horisberger and Dentan, 1980).

The special structure of the *Acinetobacter* LPS might be expected to exert some effect on the physico-chemical properties of the molecule and its behaviour as an integral constituent of the outer membrane. Indeed, the LPS of different *Acinetobacter* strains has aroused interest because, in contrast to the Enterobacteriaceae, it has been found to be liberated from the cells into the surrounding medium in considerable amounts (Brade and Galanos, 1982). This LPS release is especially pronounced when cells are grown on hydrophobic substrates such as long-chain hydrocarbons (Käppeli and Finnerty, 1979; Claus et al., 1984; Borneleit et al., 1988). In such circumstances, LPS has been shown to be released in the form of sedimentable membrane vesicles over the whole surface of the cell. This process is, however, more complex than the simple blebbing of the complete outer membrane structure that is sometimes observed in Enterobacteriaceae (Nikaido and Vaara, 1985), if only because the composition of the outer membranes and the membrane vesicles is completely different. While the outer membranes contain about 24% LPS (weight % of the three main constituents: protein, phospholipid, LPS), the vesicles are composed of about 90% LPS (Borneleit et al., 1988), and, on the basis of the KDO content /mg protein, are about 360-fold enriched in LPS over their respective outer membranes (Kappeli and Finnerty, 1979).

With regard to the mechanism of LPS release, a disturbance of the interaction between LPS and outer membrane proteins is most probably not the crucial factor. The use of an affinity electrophoresis system showed that there was a lower dissociation constant (indicative of a stronger interaction) for an LPS-protein complex from the LPS vesicle-forming *A. calcoaceticus* 69-V than for a similar complex from *A. calcoaceticus* CCM 5593 where no vesicle formation has been observed (Borneleit et al., 1989b). Instead, fluorescence investigations indicated a clustering of the LPS molecules, leading to the formation of large aggregates which are probably no longer stabilised within the membrane structure (Borneleit et al., 1990).

In Enterobacteriaceae, the structural basis for the barrier properties of the outer membrane against hydrophobic molecules, and the lytic action of detergents or digestive enzymes, resides in the tightly sealed outer LPS leaflet which covers the cells (Nikaido, 1979). Therefore, as a consequence of a disturbance of the intact lateral organisation of this leaflet by the shedding of LPS, severe changes in the permeability properties are easily conceivable, and susceptibility to phospholipases has been reported for vesicle-forming cells of *A. calcoaceticus* 69-V (Borneleit et al., 1988).

The LPS release, mentioned above, which is observed during ß-lactamase export across the outer membrane of *A. calcoaceticus* CCM 5593 is, in our opinion, in agreement with the general view that *Acinetobacter* LPS is easily removed from the membrane structure, leading to a disruption of the permeability barrier. Both disruptive processes may occur together. This may explain the increase in production of extracellular enzymes observed when cells are grown on hydrophobic carbon sources (Breuil et al., 1978; Haferburg and Kleber, 1983).

REFERENCES

Adams, G.A., Oadling, C., Yaguchi, M., and Tornabene, T.G., 1970, The chemical composition of cell wall lipopolysaccharides from *Moraxella duplex* and *Micrococcus calcoaceticus, Can. J. Microbiol.*, 16:1.

Aurich, H., Sorger, H., and Müller, H., 1977, Isolierung und Charakterisierung der Zellgrenzschichten von *Acinetobacter calcoaceticus, Zeits. Allg. Mikrobiol.*, 17:333.

Baker, W.L., 1984, A sensitive procedure for screening microorganisms for the presence of penicillin amidase, *Aust. J. Biol. Sci.*, 37:257.

Baumann, P., Doudoroff, M., and Stanier, R.Y., 1968, A study of the *Moraxella* group. II. Oxidase-negative species (genus *Acinetobacter*), *J. Bacteriol.*, 95:1520.

Benz, R., 1985, Porin from bacterial and mitochondrial outer membranes, *CRC Crit. Rev. Biochem.*, 19:145.

Blechschmidt, B., Borneleit, P., Emanouilidis, I., Lehmann, W., and Kleber, H.-P., 1989, Extracellular location of a ß-lactamase produced by *Acinetobacter calcoaceticus, Appl. Microbiol. Biotech.*, 32:85.

Borneleit, P., Hermsdorf, T., Claus, R., Walther, P., and Kleber, H.-P., 1988, Effect of hexadecane-induced vesiculation on the outer membrane of *Acinetobacter calcoaceticus, J. Gen. Microbiol.*, 134:1983.

Borneleit, P., Blechschmidt, B., Blasig, R., Franke, P., Günther, P., and Kleber, H.-P., 1989a, Preparation of R-type lipopolysaccharide of *Acinetobacter calcoaceticus* by EDTA-salt extraction, *Curr. Microbiol.*, 19:77.

Borneleit, P., Blechschmidt, B., and Kleber, H.-P., 1989b, Interaction between lipopolysaccharide and outer membrane proteins of *Acinetobacter calcoaceticus* studied by an affinity electrophoresis system, *Electrophoresis*, 10:234.

Borneleit, P., Binder, H., and Kleber, H.-P., 1990, Outer membrane vesiculation of *Acinetobacter calcoaceticus, Acta Biotech.*, 10:127.

Brade, H., and Galanos, C., 1982, Isolation, purification and chemical analysis of the lipopolysaccharide and lipid A of *Acinetobacter calcoaceticus* NCTC 10305, *Eur. J. Biochem.*, 122:233.

Brade, H., and Rietschel, E.T., 1985, Identification of 2-keto-3-deoxy-1,7-dicarboxyheptonic acid as a constituent of the lipopolysaccharide of *Acinetobacter calcoaceticus* NCTC 10305, *Eur. J. Biochem.*, 153:249.

Brade, H., Tacken, A., and Christian, R., 1987, Isolation and identification of a rhamnosyl-rhamnosyl-3-deoxy-D-*manno*-octulosonic acid trisaccharide from the lipopolysaccharide of *Acinetobacter calcoaceticus, Carbohyd. Res.*, 167:295.

Breuil, C., and Kushner, D.J., 1975a, Lipase and esterase formation by psychrophilic and mesophilic *Acinetobacter* species, *Can. J. Microbiol.*, 21:423.

Breuil, C., and Kushner, D.J., 1975b, Partial purification and characterization of the lipase of a facultatively psychrophilic bacterium (*Acinetobacter* 016), *Can. J. Microbiol.*, 21:434.

Breuil, C., Shindler, D.B., Sijher, J.S., and Kushner, D.J., 1978, Stimulation of lipase production during bacterial growth on alkanes, *J. Bacteriol.*, 133:601.

Claus, R., Käppeli, O., and Fiechter, A., 1984, Possible role of extracellular membrane particles in hydrocarbon utilization by *Acinetobacter calcoaceticus* 69-V, *J. Gen. Microbiol.*, 130:1035.

Claus, R., Fischer, B.E., and Kleber, H.-P., 1985, An esterase as marker enzyme of the outer membrane of *Acinetobacter calcoaceticus*, *J. Basic Microbiol.*, 25:299.

Fischer, B.E., 1986, Localization of hydrolytic enzymes in *Acinetobacter calcoaceticus*, *J. Basic Microbiol.*, 26:9.

Fischer, B.E., and Kleber, H.-P., 1987, Isolation and characterization of the extracellular lipase of *Acinetobacter calcoaceticus* 69-V, *J. Basic Microbiol.*, 27:427.

Fischer, B.E., Claus, R., and Kleber, H.-P., 1984, Isolation and characterization of the outer membrane of *Acinetobacter calcoaceticus*, *J. Biotech.*, 1:111.

Fischer, B.E., Koslowski, R., Eichler, W., and Kleber, H.-P., 1987, Biochemical and immunological characterization of lipase during its secretion through cytoplasmic and outer membrane of *Acinetobacter calcoaceticus* 69 V, *J. Biotech.*, 6:271.

Fricke, B., Jahreis, G., and Sorger, H., 1983, Characterization of membrane-bound endoprotease activities in *Acinetobacter*, *Wiss. Beitr. Martin-Luther-Univ. Halle-Wittenberg*, 52:50.

Fricke, B., Jahreis, G., Sorger, H., and Aurich, H., 1986, Cell envelope-bound proteinase activities in *Acinetobacter calcoaceticus*, *Biomed. Biochim. Acta*, 45:257.

Gustafsson, P., Nordström, K., and Normark, S., 1973, Outer penetration barrier of *Escherichia coli* K-12: kinetics of the uptake of gentian violet by wild type and envelope mutants, *J. Bacteriol.*, 116:893.

Guymon, L.F., and Sparling, P.F., 1975, Altered crystal violet permeability and lytic behaviour in antibiotic-resistant and -sensitive mutants of *Neisseria gonorrhoeae*, *J. Bacteriol.*, 124:757.

Haferburg, D., and Kleber, H.-P., 1982, Extracellular lipase from *Acinetobacter calcoaceticus*, *Acta Biotechnol.*, 2:337.

Haferburg, D., and Kleber, H.-P., 1983, Regulation of extracellular lipase of an alkane-utilizing *Acinetobacter* strain, *Acta Biotechnol.*, 3:185.

Hancock, R.E.W., 1984, Alterations in outer membrane permeability, *Ann. Rev. Microbiol.*, 38:234.

Hebeler, B.H., Morse, S.A., Wong, W., and Young, F.E., 1978, Evidence for peptidoglycan-associated protein(s) in *Neisseria gonorrhoeae*, *Biochem. Biophys. Res. Comm.*, 81:1011

Henriksen, S.D., 1973, *Moraxella*, *Acinetobacter* and *Mimeae*, *Bacteriol. Rev.*, 37:522.

Horisberger, M., and Dentan, E., 1980, Chemical composition and ultrastructure of cellular and extracellular *Moraxella glucidolytica* lipopolysaccharides, *Arch. Microbiol.*, 128:12.

Jahreis, G., and Aurich, H., 1989, Identification and localization of exopeptidase activities bound to membranes of *Acinetobacter calcoaceticus, Biomed. Biochim. Acta*, 48:517.

Joly-Guillou, M.L., Valleé, E., Bergogne-Bérézin, E., and Philippon, A., 1988, Distribution of ß-lactamase and phenotype analysis in clinical strains of *Acinetobacter calcoaceticus, J. Antimicrob. Chemother.*, 22:597.

Juni, E., 1978, Genetics and physiology of *Acinetobacter, Ann. Rev. Microbiol.*, 32:349.

Juni, E., 1984, Genus III. *Acinetobacter* Brisou et Prévot 1954, *in* "Bergey's Manual of Systematic Bacteriology, vol.1," p.303, N.R.Greig and J.G. Holt, eds., Williams and Wilkins, Baltimore.

Kalogridov-Vassiliadov, D., 1984, Lipolytic activity and heat-resistance of extracellular lipases of some gram-negative bacteria, *Milchwissenschaft*, 39:601.

Käppeli, O., and Finnerty, W.R., 1979, Partition of alkane by an extracellular vesicle derived from hexadecane-grown *Acinetobacter, J. Bacteriol.*, 140:707.

Katohda, S., Suzuki, F., Katsuki, S., and Sato, T., 1979, Studies on microbial enzyme active in hydrolyzing yeast polysaccharides. Part II. Purification and some properties of endo-beta-1,6-glucanase from *Acinetobacter* sp., *Agric. Biol. Chem.*, 43:2029.

Kawahara, K., Brade, H., Rietschel, E.T., and Zähringer, U., 1987, Studies on the chemical structure of the core-lipid A region of the lipopolysaccharide of *Acinetobacter calcoaceticus* NCTC 10305. Detection of a new 2-octulosonic acid interlinking the core oligosaccharide and lipid A component, *Eur. J. Biochem.*, 163:489.

Kleber, H.-P., Schöpp, W., and Aurich, H., 1973, Verwertung von n-Alkanen durch einen Stamm von *Acinetobacter calcoaceticus, Zeits. Allg. Mikrobiol.*, 13:445.

Kleber, H.-P., Claus, R., and Asperger, O., 1983, Enzymologie der n-Alkanoxidation bei *Acinetobacter, Acta Biotech.*, 3:251.

Lehmann, V., 1971a, Production of phospholipase C in *Acinetobacter calcoaceticus, Acta Path. Microbiol. Scand.*, B79:789.

Lehmann, V., 1971b, Phospholipase activity of *Acinetobacter calcoaceticus, Acta Path. Microbiol. Scand.*, B79:372.

Lehmann, V., 1972, Properties of purified phospholipase C from *Acinetobacter calcoaceticus, Acta Path. Microbiol. Scand.*, B80:827.

Lehmann, V., 1973a, Nature of phospholipase C from *Acinetobacter calcoaceticus*. Effects of whole red cells and red cell membranes, *Acta Path. Microbiol. Scand.*, B81:419.

Lehmann, V., 1973b, Hemolytic activity of various strains of *Acinetobacter, Acta Path. Microbiol. Scand.*, B81:427.

Ludewig, M., 1983, An aminopeptidase bound to cell envelopes of *Acinetobacter calcoaceticus, Wiss. Beitr. Martin-Luther-Univ. Halle-Wittenberg*, 52:47.

Lugtenberg, B., and van Alphen, L., 1983, Molecular architecture and functioning of the outer membrane of *Escherichia coli* and other gram-negative bacteria, *Biochim. Biophys. Acta*, 737:51.

Lysko, P.G., and Morse, S.A., 1981, *Neisseria gonorrhoeae* cell envelope: permeability to hydrophobic molecules, *J. Bacteriol.*, 145:946.

McDougall, G.J., Fixter, L.M., and Fewson, C.A., 1983, The occurrence of 2-keto-3-deoxy-octonic acid in *Acinetobacter calcoaceticus, FEMS Microbiol. Lett.*, 18:79.

Monboisse, J.-C., Labadie, J., and Gouet, P., 1979a, Attachment of a collagenolytic bacteria to its substrate, *Ann. Microbiol.*, 130A:435.

Monboisse, J.-C., Labadie, J., and Gouet, P., 1979b, Purification and physiochemical properties of collagenase synthesized by *Acinetobacter* sp. bacteria, *Biochimie*, 61:1169.

Nikaido, H., 1979, Permeability of the outer membrane of bacteria, *Ang. Chemie*, 18:337.

Nikaido, H., and Vaara, M., 1985, Molecular basis of bacterial outer membrane permeability, *Microbiol. Rev.*, 49:1.

Nishimura, Y., Ino, T., and Iizuka, H., 1986, Isolation and characterization of the outer membrane of radiation resistant *Acinetobacter* sp. FO-1, *J. Gen. Appl. Microbiol.*, 32:177.

Onishi, H., and Hidaka, O., 1978, Purification and properties of amylase produced by a moderately halophilic *Acinetobacter* species, *Can. J. Microbiol.*, 24:1017.

Onishi, H., Hidaka, O., and Nishimura, A., 1979, Halophilic alpha-amylase I and II, *Jap. Kokai Tokkyo Koho*, 9.

Paul, G., Joly-Guillou M.L., Bergogne-Bérézin, E., Nevot, P., and Philippon, A., 1989, Novel carbenicillin-hydrolizing ß-lactamase (CARB-5) from *Acinetobacter calcoaceticus* var. *anitratus, FEMS Microbiol. Lett.*, 59:45.

Pines, O., Bayer, E., and Gutnick, D., 1983, Localization of emulsan-like polymers associated with the cell surface of *Acinetobacter calcoaceticus, J. Bacteriol.*, 154:893.

Pugsley, A.P., and Schwartz, M., 1985, Export and secretion of proteins by bacteria, *FEMS Microbiol. Rev.*, 32:3.

Reisfeld, A., Rosenberg, E., and Gutnick, D., 1972, Microbial degradation of crude oil: factors affecting the dispersion in sea water by mixed and pure cultures, *J. Appl. Microbiol.*, 24:363,

Rietschel, E.T., Brade, L., Schade, U., Seydel, U., Zähringer, U., Kusumoto, S., and Brade, H., 1987, Bacterial endotoxins: Properties and structure of biologically active domains, *in* "Surface Structures of Microorganisms and their Interaction with the Mammalian Host," p.1, E. Schirmer, M.H. Richmond, G. Seibert and U. Schwarz, eds., Hoechst, Ringberg.

Rosenberg, E., and Kaplan, N., 1987, Surface-active properties of *Acinetobacter* exopolysaccharides, *in* "Bacterial Outer Membranes as Model Systems," p. 311, M. Inouye, ed., Wiley, New York.

Sarubbi, F., Sparling, P.F., Blackman, E., and Lewis, E., 1975, Loss of low-level antibiotic resistance in *Neisseria gonorrhoeae* due to *env* mutations, *J. Bacteriol.*, 124:750.

Scott, C., Makula, R.A., and Finnerty, W.R., 1976, Isolation and characterization of membranes from a hydrocarbon-oxidizing *Acinetobacter* sp., *J. Bacteriol.*, 127:469.

Singer, M.E., Tyler, S.M., and Finnerty, W.R., 1985, Growth of *Acinetobacter* sp. strain HO1-N on hexadecanol: physiological and ultrastructural characteristics, *J. Bacteriol.*, 162:162.

Sleytr, U.B., Thornley, M.J., and Glauert, A.M., 1974, Location of the fracture faces within the cell envelope of *Acinetobacter* species strain MJT/F5/5, *J. Bacteriol.*, 118:693.

Stewart, J.E., Kallio, R.E., Stevenson, D.P., Jones, A.C., and Schissler, D.O., 1959, Bacterial hydrocarbon oxidation. I. Oxidation of n-hexadecane by a gram-negative coccus, *J. Bacteriol.*, 78:441

Taylor, W.H., and Juni, E., 1961, Pathways for biosynthesis of bacterial capsular polysaccharide. I. Characterization of the organism and polysaccharide, *J. Bacteriol.*, 81:688.

Thorne, K.J.I., Thornley, M.I., and Glauert, A.M., 1973, Chemical analysis of the outer membrane and other layers of the cell envelope of *Acinetobacter* sp., *J. Bacteriol.*, 116:410.

Thorne, K.J.I., Oliver, R.C., and Heath, M.F., 1976, Phospholipase A2 activity of the regularly arranged surface protein of *Acinetobacter* sp. 199A, *Biochim. Biophys. Acta*, 450:335.

Thornley, M.J., and Glauert, A.M., 1968, Fine structure and radiation resistance in *Acinetobacter*: studies on a resistant strain, *J. Cell Sci.*, 3:273.

Thornley, M.J., and Sleytr, U.B., 1974, Freeze-etching of the outer membranes of *Pseudomonas* and *Acinetobacter*, *Arch. Microbiol.*, 100:409.

Thornley, M.J., Glauert, A.M., and Sleytr, U.B., 1973, Isolation of outer membranes with an ordered array of subunits from *Acinetobacter*, *J. Bacteriol.*, 115:1294.

Thornley, M.J., Thorne, K.J.I., and Glauert, A.M., 1974, Detachment and chemical characterization of the regularly arranged subunits from the surface of *Acinetobacter*, *J. Bacteriol.*, 118:654.

Torregrossa, R.E., Makula, R.A., and Finnerty, W.R., 1977, Outer membrane phospholipase A from *Acinetobacter* sp. HO1-N, *J. Bacteriol.*, 131:493.

Torregrossa, R.E., Makula, R. A., and Finnerty, W.R., 1978, Hydrolysis of phosphatidylethanolamine and phosphatidylglycerol by outer membrane phospholipase A from *Acinetobacter* sp. HO1-N, *J. Bacteriol.*, 136:803.

Zaikina, N.A., and Robakidze, T.N., 1976, Phospholipase C of fungi and staphylococci, *Mikrobiologiya*, 45:466.

ENERGY RESERVES IN *ACINETOBACTER*

L.M. Fixter and M.K. Sherwani

Department of Biochemistry
University of Glasgow
Glasgow, G12 8QQ
UK

INTRODUCTION

The first systematic investigation of the occurrence of energy reserves in *Acinetobacter* formed part of the delineation of the genus by Baumann et al. (1968). Their study indicated that poly-ß-hydroxybutyrate was not present in any of the 106 strains examined. They examined two strains in detail and concluded that they did not accumulate special non-nitrogenous reserve materials. However, since this work, polyphosphates, poly-ß-hydroxybutyrate and simple wax esters, a novel energy reserve, have been identified in a large number of acinetobacters. In several of these organisms, the compounds show many, but not all, of the characteristics of energy reserves. As the study of these compounds arose largely from investigations of other aspects of acinetobacter microbiology, this review will touch on these areas as well. The biochemistry and molecular biology of energy reserves in *Acinetobacter* has not been extensively studied and so this review will attempt to relate this area to what is known of the general biochemistry of bacterial energy reserves.

GENERALITIES: DEFINITION AND FUNCTIONS OF ENERGY RESERVES

Wilkinson (1959) proposed three criteria which compounds must fulfil to be considered as energy reserves. First, they should accumulate when growth is limited by a nutrient in the presence of excess carbon and energy source. Second, they should be utilised when there is insufficient exogenous carbon and energy source for growth or the maintenance of viability. Third, their catabolism must yield energy. Certain overflow metabolites meet at least some of these criteria and therefore only those compounds which

The Biology of Acinetobacter, Edited by K. J. Towner *et al.*
Plenum Press, New York, 1991

accumulate intracellularly and meet these criteria are generally regarded as true energy reserves. Such materials have an ancillary characteristic, insolubility in water. This is an advantage as they do not disturb the osmotic balance of the bacterium in which they occur and, because of this, they are found as inclusions within bacteria (reviewed by Shively, 1974). Glycogen, glycogen-like polymers, poly-ß-hydroxybutyrate and other poly-ß-hydroxyalkanoates meet these criteria in many bacteria (reviewed by Dawes and Senior, 1973; Dawes, 1986; Preiss, 1989). A general role for polyphosphate as an energy reserve is more problematical (reviewed by Harold, 1966; Kulaev and Vagabov, 1983) and it serves simply as a phosphate store in many bacterial species.

Energy reserves may have secondary metabolic functions depending on their chemical structure. In *Azotobacter vinelandii*, poly-ß-hydroxybutyrate acts as a sink for reducing equivalents when the oxygen supply is limited. In the presence of oxygen, its breakdown provides substrates for the high rate of respiration which helps to protect the oxygen-sensitive nitrogenase of this organism (Dawes and Senior, 1973). *Euglena gracilis* utilises the difference in oxidation state between its two energy reserves, paramylon (a glucose polymer) and wax esters, to allow it to carry out wax fermentation (Inui et al., 1982). Upon transition from aerobic to anaerobic conditions, this organism converts its paramylon to the more highly reduced wax esters, obtaining the use of a sink for reducing equivalents. In the reverse transition, waxes are reconverted to paramylon and the reducing equivalents released are used to generate energy. Polyphosphate is a polyanion at intracellular pH and is normally found complexed with polyamines or metal ions. Thus, it may function as a store of ions such as magnesium (Kulaev and Vagabov, 1983).

Recent work on poly-ß-hydroxybutyrate in *Escherichia coli* has shown that a compound which is generally regarded as an energy reserve may serve a quite different role which is critically dependent on its chemical structure. In this bacterium, only small amounts of poly-ß-hydroxybutyrate are found and it is associated with the cytoplasmic membrane (Reusch et al., 1986, 1987). Its content is increased in conditions which increase the potential of this organism for genetic transformation. Poly-ß-hydroxybutyrate can adopt a helical configuration in hydrophobic conditions (Dawes and Senior, 1973). Two mechanisms have been suggested to explain how the competent state is induced. Poly-ß-hydroxybutyrate helices in the membrane may cause changes in membrane structure (Reusch et al. 1987), allowing the entry of DNA (Reusch et al. 1987), or transmembrane helices, possibly in association with polyphosphate, may form channels for DNA import (Reusch and Sadoff, 1988).

The accumulation of energy reserves when bacterial growth is nutrient-limited in the presence of excess carbon and energy source has been studied in batch and continuous culture. In such batch cultures, the content of glycogen, poly-ß-hydroxybutyrate, or polyphosphate increases at about the time of cessation of exponential growth and during the stationary phase. In continuous cultures, the concentration of energy reserves increases at low specific growth rates when the culture is limited by a nutrient other than the carbon or energy source (Dawes and Senior, 1973). Similar results are usually obtained irrespective of whether the cultures are limited by the supply of nitrogen, sulphur or phosphorus. In such experiments it is also often found that at low growth rates the specific activities of the enzymes involved in energy reserve metabolism are increased. There are, however, exceptions to these generalisations. Glycogen accumulation can occur during exponential growth in some bacteria (reviewed by Dawes and Senior, 1973; Preiss, 1984). Polyphosphate accumulation probably shows the greatest number of exceptions because of the "phosphate overplus" phenomenon which is obtained when bacteria are transferred from phosphate-limited conditions to phosphate-excess conditions (Harold, 1966; Dawes and Senior, 1973). This phenomenon consists of a rapid uptake of inorganic phosphate by the phosphate-starved bacteria on transfer to a phosphate-rich medium containing a carbon and energy source. Some of the inorganic phosphate taken up is converted to polyphosphate which can be seen as volutin granules within the cell. After a short period of growth this polyphosphate is degraded and utilised to supply phosphate for synthesis of nucleic acids by the growing bacteria. In cases where two carbon-containing energy reserves are present in an organism, their accumulation usually occurs under the same type of conditions. Thus in *Bacillus megaterium*, which contains both glycogen and poly-ß-hydroxybutyrate, the choice of energy reserve accumulated depends on the carbon and energy source, with acetate favouring poly-ß-hydroxybutyrate synthesis and glucose favouring glycogen synthesis, but the reserve accumulates only when growth is limited in the presence of the carbon and energy source (Dawes and Senior, 1973). In *Alcaligenes eutrophus*, poly-ß-hydroxybutyrate and polyphosphate, which in this bacterium does not serve as an energy reserve, accumulate together under oxygen-limiting conditions (Doi et al. 1989).

There are examples of bacteria in which accumulated carbon-containing energy reserves maintain viability or growth during carbon and energy source starvation. In *E. coli*, glycogen increases the survival rate of the bacterium (Dawes and Senior, 1973). A high content of poly-ß-hydroxybutyrate in the nitrogen-fixing bacterium *Azospirillum brasilense* even allows the organism to grow when starved in a phosphate buffer (Tal and Okon, 1985). There are, however, observations showing that the presence of energy reserves in certain starvation conditions does not help to maintain viability. In *Sarcina lutea*, the presence of the polyglucose energy reserve reduces viability when the organism is starved in phosphate buffer (Dawes and Senior, 1973). Bacteria with high levels of energy reserves have a lower rate of degradation of proteins and RNA than bacteria with low levels of energy reserves (Dawes and Senior, 1973). This suggests that one of the mechanisms by which energy reserves maintain a cell's viability during carbon and energy source starvation is by providing an alternative energy source, thereby reducing the need to catabolise essential cellular components. The processes that yield energy by degrading proteins and RNA also result in the production of ammonia, which itself may have deleterious effects on the survival of the cell (Dawes and Senior, 1973).

ACINETOBACTER ENERGY RESERVES

Simple Wax Esters

The first investigations of wax ester metabolism in *Acinetobacter* were those of Stewart and Kallio (1959), who described the production of wax esters during alkane oxidation by "*Micrococcus cerificans*" HO1-N, which was later reclassified as an acinetobacter (Baumann et al., 1968). During alkane oxidation, cultures of this strain can produce up to 4.3 g wax esters/l and they are largely extracellular. The difficulty of separating biomass from extracellular wax esters prevented the identification of intracellular wax esters. Later work showed that the bacteria in such cultures contained about 9 mg wax esters/g dry wt (Scott et al., 1976). The distribution of wax esters between the intracellular and extracellular compartments in such cultures is difficult to derive from the results presented (Stewart and Kallio, 1959; Makula et al. 1975). Using a molar growth yield of 20.1 g dry wt/mol acetate (Fewson, 1985) to calculate the biomass present in cultures, taken together with the results of Makula et al. (1975) for diauxic growth of this strain with acetate and hexadecane as substrates, 97% of the wax esters produced can be estimated to be extracellular. Thus wax esters produced during alkane oxidation are overflow metabolites rather than energy reserves because they do not accumulate to a significant extent within the organism. This strain contained about 9 mg intracellular wax esters/g dry wt when grown in the absence of alkanes (Finnerty, 1977), supporting the idea that a large proportion of the wax esters present in alkane-grown cultures is extracellular. Attempts to increase wax ester yield from alkanes have produced mutants of this strain which accumulate larger amounts of wax esters than the parent strain after growth on both alkane and non-alkane feedstocks (Geigert et al., 1984; Neidelman, 1987).

Gallagher (1971) provided the first report of endogenous wax esters as part of a taxonomic study of acinetobacters which had been isolated from lamb carcasses. The composition of the intact waxes that were isolated was not reported, but the waxes contained saturated and unsaturated alkan-1-ols which had three to five carbon atoms or eight to 20 carbon atoms. The proportion of short and long chain alkanols, and their degree of unsaturation, was dependent on the strain and the growth temperature. Subsequently, *Acinetobacter* NCIB 8250 was shown to contain wax esters (about 1 mg/g dry wt) when grown with succinate or other carbon sources in carbon-limited batch cultures (Fixter and Fewson, 1974). The waxes of this strain have chain lengths of between 30 and 36 carbon atoms and are largely saturated. The fatty acid and alkan-1-ol composition of the waxes, in terms of chain length and degree of unsaturation, is very similar to the fatty acid composition of the phospholipids of this strain, with 16 and 18 carbon saturated species predominant. It was shown that the wax content increased to 40-60 mg/g dry wt in the stationary phase of nitrogen-limited batch cultures. There was no large change in the composition of the wax esters, and the predominant waxes contained 32 carbon atoms. The increase in the wax content of nitrogen-limited stationary phase cultures, compared with similar carbon-limited cultures, is typical of energy reserve compounds and, solely on this basis, it was proposed that wax esters in this strain may be energy reserves. In taxonomic investigations, Bryn and co-workers extended the number of acinetobacters known to contain wax esters (Jantzen et al., 1975; Bryn et al., 1977). The quantitative data presented in the latter paper showed that only four of nine strains analysed contained wax esters. In these strains the wax esters had chain lengths of between 32 and 36 carbon atoms and were largely mono-unsaturated species. The total wax esters of strains grown on complex solid media ranged from 1-5 mg/g dry wt. This value for wax content is very similar to that found in other studies for either exponentially growing or carbon-limited cultures (Fixter and McCormack, 1976; Fixter et al. 1986). These more extensive studies showed that waxes with similar general composition occurred in a further 17 different strains of *Acinetobacter* (Fixter et al., 1986). However, there was one significant difference in that this work showed that strain NCTC 10306, which is identical to strain ATCC 17907 included in the study of Bryn et al. (1977), contained significant amounts of wax in carbon-limited conditions (2.3 mg/g dry wt), while the previous authors found that this and a related strain contained no waxes. The taxonomic potential of waxes or their alkan-1-ols is still being investigated (Moss et al., 1988), but it appears that clinical isolates of *Acinetobacter* lack hexadecan-1-ol and octadecan-1-ol, and this apparently indicates an absence of wax ester. This observation should be treated with caution as it is based on the analysis of cultures grown on a complex solid medium. Bacterial colonies isolated from this medium would be expected to be a mixture of growing bacteria and bacteria in carbon-limited stationary phase. The very low levels of wax esters found in such bacteria may not yield detectable amounts of alkan-1-ols.

The composition of endogenous wax esters from *Acinetobacter* has been examined in detail for only a few strains (Bryn et al., 1977; Fixter et al., 1986; Neidelman, 1987). The composition of wax esters isolated from two strains grown into nitrogen-limited stationary phase is shown in Table 1; the wax esters are predominantly saturated 32 and 34 carbon atom species. Determination of the isomeric composition of each wax ester species by gas liquid chromatography / mass spectrometry for strain NCIB 8250 (Fixter et al., 1986) showed that, in terms of specific alkanol-fatty acid combinations, the composition found was that expected from the random combination of alkanols and acids present in the wax esters. The random composition combination of wax esters has been reported in another species of wax ester-containing bacteria (Russell and Volkman, 1980). A comparison of the composition of waxes from strains studied by Bryn et al. (1977) indicates that, in these strains, mono-unsaturated and di-unsaturated species constitute about 80% of the wax esters. This may simply reflect the different growth temperatures, 30°C and 22°C, used in the two studies. It has been shown that, in strain H01-N, the degree of unsaturation of the wax esters is dependent on growth temperature (Neidelman, 1987). In the case of cultures grown with ethanol as substrate, the percentage of mono-and di-unsaturated species increases from 24% when the growth temperature is 30°C to 90%

Table 1. The composition of wax esters isolated from *Acinetobacter* NCIB 8250 and *Acinetobacter* NCIB 10487 grown into stationary phase in nitrogen-limited batch culture. Wax ester analysis was done using capillary column gas liquid chromatography and the identity of the waxes was confirmed by mass spectrometry. Tc, Trace (<1% of the total). Data from Fixter et al., (1986); by permission of the Editors of the Journal of General Microbiology.

Wax ester	Composition (% by weight) of total wax esters	
	Strain NCIB 8250	Strain NCIB 10487
30:0	2.9	2.2
30:1	0.4	0.6
31:0	Tc	Tc
32:0	44.9	33.3
32:1	4.8	9.8
32:2	0.4	0.8
33:0	Tc	Tc
34:0	26.9	23.1
34:1	8.5	13.6
34:2	1.9	1.9
35:0	Tc	Tc
36:0	4.0	5.7
36:1	3.1	6.2
36:2	2.1	2.8

when the growth temperature is 17°C. In terms of chain length, as opposed to degree of unsaturation, there is little difference caused by changes in growth temperature. In the case of the wax esters produced during growth on alkanes, the major determinant of their composition is the chain length of the alkane substrate (Stewart and Kallio, 1959; Makula et al., 1975; DeWitt et al. 1982). The wax esters produced during alkane oxidation are composed largely of alkan-1-ols of the same chain length as the alkane, esterified to fatty acids of same chain length or two or four carbon atoms shorter (Neidelman, 1987). This again indicates that such wax esters are best regarded as overflow metabolites of alkane oxidation.

Fixter and McCormack (1976) and Fixter et al. (1986) showed that 16 of the 17 strains which they studied had increased (5-20 fold) wax ester contents in nitrogen-limited, compared with carbon-limited, stationary phase cultures. This indicated the possibility that simple wax esters may be widely used as energy reserves within this genus. Interestingly, the two strains examined in detail by Baumann et al. (1968) for poly-ß-hydroxybutyrate, strains ATCC 17902 and NCIB 8250, are among these wax-accumulating strains, and this is possibly the reason for the failure to detect poly-ß-hydroxybutyrate. The earliest report showing an energy reserve, possibly a carbohydrate, in an acinetobacter (Clifton, 1937) was on the neotype strain ATCC 23055, which is also a wax ester-accumulating strain (Fixter et al., 1986). In strain NCIB 8250, more detailed studies (Fixter et al., 1986) showed that bacterial wax content increased dramatically as growth ceased in nitrogen-limited batch culture, and then increased more slowly in stationary phase (Fig. 1). In continuous nitrogen-limited cultures, wax content increased as specific growth rate decreased, while in carbon-limited cultures, wax levels were lower and increased with specific growth rate. Limitation by oxygen in continuous culture did not greatly change the wax content of this strain (Dawes, 1986), which means that this strain is unable to use this highly reduced reserve as a sink for reducing equivalents. Wax esters in these 16 strains meet the first criterion of energy reserves. Further, as wax esters are degraded to carbon dioxide and water-soluble products during carbon and energy source starvation in strain NCIB 8250 (Fixter et al. 1986), there is evidence that wax esters meet the second criterion, since their catabolism yields energy by the same final pathway as alkane catabolism (Asperger and Kleber, this volume). Based on the measured rates of wax ester degradation and the assumption that the major water-soluble product is acetate, the rapid

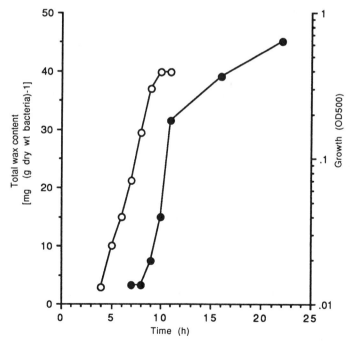

Fig. 1. The accumulation of simple wax esters in a nitrogen-limited culture of *Acinetobacter* strain NCIB 8250. •, Wax content; o, growth. Reproduced from Fixter et al., (1986) by permission of the Editors of the Journal of General Microbiology.

initial phase of wax breakdown would give about 40% of the maintenance energy of this strain, calculated on the basis of the maintenance oxygen requirement (Hardy and Dawes, 1985). The final slower phase of wax ester degradation, when the rate is only 10% of the initial rate, makes an insignificant contribution to maintenance energy needs.

Wax esters at concentrations greater than 20 mg/g dry wt in acinetobacters cause the appearance of characteristic inclusions (Fig. 2; Fixter, 1976; Singer et al., 1984). The inclusions are plate-like and have varying thicknesses. This means that they appear either as discs, when their major plane is parallel to the plane of the section, or as sharp rectilinear inclusions, when their major plane is perpendicular to the plane of the section (Fig. 2a). Freeze-fracture techniques show the inclusions to be multi-layered (Fig. 2b). The thicknesses of the inclusions are multiples of 4 nm, which corresponds to the thickness of a bilayer composed of wax esters with 32-36 carbon atoms (Aleby et al. 1971). Analysis of the inclusions has shown that wax esters are their major component in strain H01-N grown on hexadecanol (Singer et al., 1984). The presence of phospholipids in these inclusions may mean that they are bounded by membranes. Thus wax esters, like other energy reserves, occur as characteristic inclusions within the bacterium.

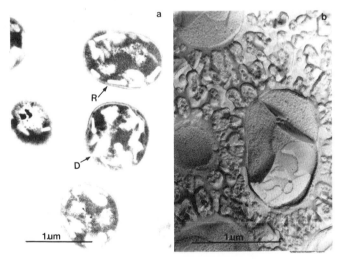

Fig.2. Inclusions in *Acinetobacter* strain NCIB 8250. The strain was grown into stationary phase in nitrogen-limited batch culture where it had a wax ester content of 56.8 mg /g dry wt. Samples were prepared for transmission electron micrography and freeze fracture electron micrography. Transmission electron micrography (Fig. 2a) showed numerous inclusions as discs (D) or rectilinear inclusions (R). Freeze fracture micrography (Fig. 2b) showed the plate-like nature of the inclusions and that they were layered structures. Bar mark = 1 μm.

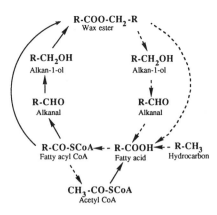

Fig.3. Proposed pathway for the metabolism of simple wax esters in *Acinetobacter*. Synthetic reactions are indicated by solid lines and degradative reactions by dashed lines.

The biochemistry of endogenous wax metabolism in *Acinetobacter* is largely unknown, but there is evidence for the pathway in Fig. 3 which is based on the metabolism of alkanes in acinetobacters (Aperger and Kleber, this volume), wax metabolism in other bacteria (Albro, 1976; Lloyd and Russell, 1983), and the pattern of labelling of the intermediates with radioactive fatty acids and alkan-1-ols in strain NCIB 8250 (Nagi, 1981). The source of fatty acids is synthesis *de novo* which, in most bacteria, yields fatty acyl-ACP derivatives rather than fatty acyl-CoAs, and there is evidence that, in *E. coli*, lipid biosynthesis uses fatty acyl-ACPs rather than CoA derivatives (Rock and Cronan, 1985). Therefore, it is possible that ACP derivatives are used in wax biosynthesis rather than the CoA esters. Although there are well-studied enzyme systems for alkane catabolism in acinetobacters, there are very few published observations which are directly relevant to endogenous wax synthesis. A constitutive NAD-dependent long chain alkanal (fatty aldehyde) dehydrogenase and an inducible NADP-dependent alkanal dehydrogenase have been shown to occur in the membranes of strain H01-N (Singer and Finnerty, 1985a). In some strains there is a constitutive NADP-linked alcohol dehydrogenase which may reduce long chain aldehydes to alkanols for wax synthesis (Fixter and Nagi, 1984), and similar activities have been found in other wax ester-producing bacteria (De Bruyn et al., 1981). The inducible NAD-dependent dehydrogenase specific for long chain alkan-1-ols in strain H01-N (Singer and Finnerty, 1985b) appears to be involved in exogenous alkane and alkan-1-ol degradation because, although there is a low activity of whole-cell hexadecanol oxidation in uninduced cultures, this enzyme could not be detected. The breakdown of wax esters is presumably initiated by hydrolysis of the ester link, and esterolytic activities have been identified in several acinetobacters (Breuil and Kushner, 1975; Shabtai and Gutnick, 1985; Sherwani and Fixter, 1989). In most of the organisms studied, the activities appear to be secreted forms involved in the breakdown of extracellular lipids (Breuil and Kushner, 1975; Shabtai and Gutnick, 1985), but there are intracellular forms in several strains (Sherwani and Fixter, 1989). There is evidence that the regulation of wax ester content is by changes in the rate of wax degradation (Nagi and Fixter, 1981) rather than by changes in the rate of synthesis, and this would be compatible with the proposal that mutants of strain H01-N which accumulate larger amounts of wax esters are unable to degrade wax esters (Geigert et al., 1984).

Polyphosphates

Biological removal of phosphate from waste-water by conventional sewage treatments removes somewhat more than half the phosphate input (Nesbitt, 1969). Modification of sewage treatment to include an anaerobic treatment stage increases the efficiency of phosphate removal (Yall et al., 1970). In these more efficient phosphate-removing plants there have been numerous investigations of the sewage sludge ecology to determine which microorganisms are responsible for phosphate removal (Wentzel et al., 1988). Acinetobacters have been isolated from this sludge which are capable of polyphosphate synthesis (reviewed by Stephenson, 1987) and, on this basis, models of the phosphate-removing process in sewage sludge have been proposed (Fuhs and Chen, 1975; Wentzel et al., 1986). The data upon which models and research programmes have been built pose a number of problems. Although there is a consensus opinion that acinetobacters are important in phosphate removal (Lotter et al., 1986; Lotter, 1987a; Streichan et al., 1990), and the general view of their application in treatment of waste-water has been summarised by Gutnick et al. (this volume), there are opposing views about the role of acinetobacters in this process (Brodisch and Joyner, 1983; Suresh et al., 1985; Roberts, 1987). Even though novel approaches, such as estimation of the type and quantity of quinones have been used (Hiraishi et al., 1989), enumeration of the bacterial types within the activated sludge remains a problem (Streichan et al., 1990). Even the identification of acinetobacters must be treated with caution when this is based on a limited number of phenotypic tests (Duncan et al., 1988; Beacham et al., 1990). Initially, the mechanism by which anaerobic pre-treatment increases phosphate removal in a subsequent aerobic phase centred upon the role of an anaerobic stress in inducing extra phosphate uptake (Nicolls and Osborn, 1979). However, more recent studies have emphasised the significance of the production of volatile short chain fatty acids during the anaerobic phase, and the subsequent utilisation of these fatty acids by aerobic organisms during the aerobic phosphate-removing phase (Fuhs and Chen, 1975; Brodisch, 1985; Ye et al., 1988).

The acinetobacters responsible for phosphate removal accumulate polyphosphate as typical volutin granules (Deinema et al., 1980). The general view is that the phosphate removed is converted to polyphosphate within the bacterium. Whether this is the major mechanism of phosphate removal is open to doubt as quantitative studies on activated sludge have shown that only a third of the phosphate removed could be accounted for by polyphosphate granules in acinetobacters (Cloete and Steyn, 1988). The removal of phosphate and accumulation of polyphosphate in such studies shows a typical feature of energy reserves in that the presence of a carbon and energy source, whether added as acetate or already present as high biological oxygen demand nutrients in the sewage, can improve the uptake of phosphate and production of polyphosphate granules (Fuhs and Chen, 1975; Brodisch, 1985; Ye et al., 1988).

Pure cultures of acinetobacters isolated from activated sludge have been used to investigate the culture conditions in which polyphosphate accumulates (Deinema et al., 1980, 1985; Murphy and Lotter, 1986). Experiments in which the phosphate content of the bacteria is derived by phosphate uptake from the medium, and where no evidence is presented that the accumulated phosphate is in the polyphosphate form (Hao and Chang, 1987), must be regarded critically as it is well known that changes in growth rates change the amount of RNA, largely ribosomal RNA, within bacteria and hence increase their phosphate content (Tempest, 1969). In nitrogen-limited batch cultures, the polyphosphate content of acinetobacters increases on cessation of growth (Halvorson et al., 1987). There is some evidence that acetate as carbon source, rather than other carbon and energy sources, gives the greatest accumulation of polyphosphate (Murphy and Lotter, 1986). Rather confusing results have been obtained in continuous culture experiments. The phosphate content of one strain (210A) does not increase at low dilution rates in nitrogen-limited cultures, but does increase in sulphur-limited cultures (Van Groenestijn et al., 1989b). The effects of limitation of potassium and magnesium ions showed that potassium ions played a role in promoting phosphate uptake, and similar limitation in continuous culture reduced the phosphate content of strain 210A. Although there is good evidence that magnesium ions are the predominant counterions for polyphosphate in acinetobacters (Van Groenestijn et al., 1988), magnesium limitation in continuous culture did not reduce phosphate content dramatically, and an explanation for this is that other ions, particularly calcium ions, may substitute for magnesium ions (Buchan, 1983; Beacham et al., 1990). Overall, accumulation of polyphosphate in acinetobacters occurs as expected for an energy reserve compound, but as explained below, it is probably simply a phosphate store in some strains.

The nature and distribution of acinetobacter polyphosphate has been studied only in strain JW11, a phosphate-removing strain isolated from activated sludge (Suresh et al., 1985). The highly polymeric polyphosphate isolated from this strain contains more than 200 phosphate residues / molecule (Halvorson et al. 1987), which is similar to the polyphosphate isolated from *E. coli* (Rao et al., 1987). Using the effect of EDTA on ^{31}P NMR signals and the metachromatic shift of a non-penetrating dye, toluidine blue (Suresh et al., 1985; Halvorson et al., 1987), it was shown that there were two forms of polyphosphate in this strain. The bulk of the polyphosphate was cytoplasmic, with a small fraction (about 2%) associated with the outer membranes or periplasmic space. This bi-phasic distribution is also found in other bacteria (Rao et al., 1987). It is interesting to note that some of the strains examined by Streichan et al. (1990) show electron-dense material associated with the bacterial wall, and this may be the periplasmic polyphosphate. Suresh et al. (1986) showed that the small pool of polyphosphate was utilised for cadmium ion transport into strain JW11, but that the large cytoplasmic pool was not used for ion transport.

Important activities in bacterial polyphosphate metabolism include polyphosphatases, ATP:polyphosphate phosphotransferases (polyphosphate kinases), polyphosphate glucokinases and polyphosphate:AMP phosphotransferases (reviewed by Kulaev and Vagabov, 1983; Wood and Clark, 1988). The role and the existence of 1,3-*bis*phosphoglycerate polyphosphate phosphotransferase is doubtful (Wood and Clark, 1988), while the polyphosphate:NAD kinase is not considered to be central to polyphosphate metabolism because of its restricted distribution among bacteria and the fact that it uses cyclic rather than linear polyphosphates (Wood and Clark, 1988). There is evidence for polyphosphatase activity, polyphosphate:AMP phosphotransferase, 1,3-*bis*phosphoglycerate polyphosphate phospho-transferase and polyphosphate kinase in a small number of acinetobacters (T'Seyen et al., 1985; Van Groenestijn et al., 1989a; Vasiliadis et al., 1990). Attempts to detect polyphosphate glucokinase and polyphosphate:NAD kinase in several of the strains in these studies proved fruitless. The activities of all the enzymes of polyphosphate metabolism detected are low, in the 0-100 nmol/min/mg protein range, and in many cases this has prevented studies of the properties of the enzymes in crude extracts. Strain 210A, an extensively studied strain (Van Groenestijn et al., 1987), contains polyphosphate:AMP phosphotransferase and polyphosphatase activity. However, strain 210A lacks the enzyme polyphosphate kinase, which is known to be responsible for polyphosphate biosynthesis in other genera of bacteria (Wood and Clark, 1988), and so polyphosphate:AMP phosphotransferase (Eqn.1) may play a synthetic role as well as a degradative role in this strain. The polyphosphatase activity is enhanced by ammonium ions and, in crude extracts, 300 mM ammonium chloride increases the activity 20-fold. The stimulated activity is about half that of the phosphotransferase reaction in the direction of polyphosphate degradation. Thus, in certain circumstances, polyphosphatase may be relatively important in polyphosphate breakdown. This strain, together with other acinetobacters, contains significant adenylate kinase activity (Eqn.2). The two enzymes allow this organism to generate ATP from polyphosphate:

$$(P)_{n+1} + AMP \longrightarrow (P)_n + ADP \quad \ldots\ldots(Eqn.1)$$

$$2ADP \longrightarrow ATP + AMP \quad \ldots\ldots\ldots\ldots(Eqn.2)$$

If the large pool of polyphosphate in the cytoplasmic compartment of the bacterium is to serve as an energy reserve in this strain, these two reactions are the only ones available. Calculations based on the activity of this system *in vitro* show that it could supply about 25% of the maintenance energy requirement (Van Groenestijn et al., 1987). Adenylate kinase activity would also be essential for the synthesis of polyphosphate in this organism as only ADP functions as a phosphate donor, and in fact there is increased phosphate removal by sludges with high adenylate kinase activities. Other strains of *Acinetobacter*, such as NCIB 8250, contain a polyphosphate kinase and a polyphosphatase which are responsible for their polyphosphate metabolism (Van Groenstijn et al. 1989a), and in these strains polyphosphate is simply a phosphate reserve. The most recent work (Vasiliadis et al., 1990) has shown a high activity of 1,3-*bis*phosphoglycerate polyphosphate phosphotransferase in certain acinetobacters, and indeed the activities reported (55-92 nmoles/min/mg protein are approximately 100-fold higher than those found in other species of bacteria (Wood and Clark, 1988). It would seem, therefore, that these types of *Acinetobacter* contain a very unusual polyphosphate-metabolising system. Polyphosphate kinase appears to be the major enzyme responsible for the synthesis of polyphosphate in these strains because its activity is much higher than the polyphosphate:AMP phosphotransferase activity present. Vasiliadis et al. (1990) propose that the 1,3-*bis*phosphoglycerate polyphosphate phosphotransferase, acting in the direction 3-phosphoglycerate to 1,3-*bis*phosphoglycerate, plays a role in the utilisation of polyphosphate. This proposal is based upon the fact that the strains were grown on acetate and would be carrying out gluconeogenesis which requires the conversion of 3-phosphoglycerate to 1,3-*bis*phosphoglycerate. In view of the difficulties which other workers have experienced in assaying 1,3-*bis*phosphoglycerate polyphosphate phosphotransferase (Wood and Clark, 1988), the purification and characterisation of this enzyme would be essential to confirm the ideas put forward by Vasiliadis et al. (1990). The factors which regulate polyphosphate metabolism in acinetobacters have not been identified.

Poly-ß-Hydroxybutyrate

Poly-ß-hydroxybutyrate accumulation is associated with the polyphosphate-storing acinetobacters isolated from sewage (Fuhs and Chen, 1975; Deinema et al., 1980; Lotter et al., 1986; Lotter, 1987b; Heyman et al. 1989), but not all these strains contain poly-ß-hydroxybutyrate. Although ß-hydroxybutyrate, ß-hydroxyvalerate and other ß-hydroxyalkanoates are found in the "poly-ß-hydroxybutyrate" of sewage sludge (Odham et al., 1986), only ß-hydroxybutyrate was detected in the one acinetobacter analysed using gas liquid chromatography (Deinema et al., 1980). Poly-ß-hydroxybutyrate has been postulated to play a special role in phosphate accumulation by acinetobacters (Fuhs and Chen, 1975; Lotter et al., 1986). The exact nature of this role is not clear as some authors feel that poly-ß-hydroxybutyrate is important in the anaerobic zone of activated sludge systems, while others feel that its importance is in the aerobic phase. In the first case, the synthesis of poly-ß-hydroxybutyrate during the anaerobic phase allows its use in the subsequent aerobic phase to provide oxidisable substrate for phosphate uptake and polyphosphate synthesis (Lotter, 1987b). This process depends upon the ability of the acinetobacter to use nitrate as an electron acceptor in anaerobic conditions. In the second case, the production of volatile fatty acids during the anaerobic phase provides the conditions of excess carbon and energy source required in the aerobic phase for the accumulation of energy reserves, with the result that both poly-ß-hydroxybutyrate and polyphosphate accumulate.

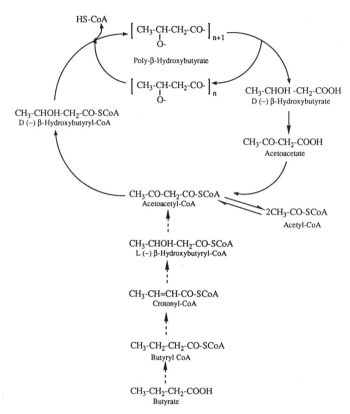

Fig. 4. Bacterial poly-ß-hydroxybutyrate metabolism. Dashed lines indicate the conversion of butyrate to poly-ß-hydroxybutyrate without its metabolism to acetate.

The pathway of poly-ß-hydroxybutyrate metabolism (Fig. 4; Dawes and Senior, 1973) has been modified by the finding that butyrate can be incorporated directly without being converted to acetyl CoA (Doi et al., 1989). Only in two species, *Azotobacter beijerinckii* and *Alcaligenes (Hydrogenomonas) eutrophus* have there been detailed studies of control of poly-ß-hydroxybutyrate metabolism at the enzyme level. The important regulated steps are probably the conversion of acetyl CoA to acetoacetyl CoA by 3-ketothiolase, which is inhibited by CoA (Oeding and Schlegel, 1973; Haywood et al., 1988), and the conversion of D(-)ß-hydroxybutyrate to acetoacetate by an NAD-dependent dehydrogenase which is inhibited by NADH, pyruvate and either oxaloacetate or 3-oxoglutarate (Senior and Dawes 1973). The first reaction plays a key role in regulating synthesis and the second in regulating breakdown. A ß-hydroxybutyrate dehydrogenase has been found in several acinetobacters (Lotter and Dubery, 1989), but this activity is not well-characterised as all assays were done with D,L-ß-hydroxybutyrate. The relationship of this activity to the stereospecific dehydrogenase of poly-ß-hydroxybutyrate breakdown should be determined. A partially purified preparation from the strain with the highest activity was used to investigate the effects of acetyl CoA. NADH and oxaloacetate on the oxidation of ß-hydroxybutyrate, and these compounds did indeed inhibit the activity, but at high concentrations (Lotter and Dubery, 1989). The fact that this enzyme activity has similar properties to the D(-)ß-hydroxybutyrate dehydrogenase from other poly-ß-hydroxybutyrate-storing bacteria may indicate that this enzyme is involved in poly-ß-hydroxybutyrate breakdown, but this is not certain.

SUMMARY

It would appear that there is good evidence that simple wax esters, poly-ß-hydroxybutyrate and polyphosphate occur in acinetobacters, or bacterial strains at least partially characterised as acinetobacters. Wax esters meet the three criteria which are used to define energy reserves: they accumulate in conditions of carbon and energy source excess when growth is limited by another nutrient; they are degraded during carbon and energy source starvation; and their metabolism by the pathways known to exist in acinetobacters would yield energy. On the basis of the known role of poly-ß-hydroxybutyrate as an energy reserve in other bacteria, this polymer can be presumed to fulfil a similar role in acinetobacters. There are, however, few studies on the physiology and biochemistry of this compound in well-characterised acinetobacter strains. Polyphosphate can act as an energy reserve in certain strains, as either the enzymes required for ATP synthesis from polyphosphate are present, or there is evidence that polyphosphate is utilised for ion transport in these acinetobacters. Other acinetobacters would appear to use it simply as a phosphate store because they lack any means of energy generation from polyphosphate. The fact that there are two polyphosphate pools in some organisms, one of which can serve as an energy reserve and the other as a simple phosphate store, means that generalisations about the functions of this compound are difficult without detailed investigations of the strains in question.

The occurrence of more than one carbon-containing energy reserve within a single bacterial species is not unusual. Glycogen and poly-ß-hydroxybutyrate are found in pseudomonads and bacilli, and in some cases in the same strain. This may, in the case of the bacilli, be a reflection of the metabolic capacity of these organisms since it is known that, with acetate as substrate, poly-ß-hydroxybutyrate synthesis is favoured, while with glucose as substrate, glycogen synthesis predominates. It is possible that, in other cases, the choice of energy reserve depends on environmental factors which may favour the selection of poly-ß-hydroxybutyrate because it can also serve as a sink for reducing equivalents. Are considerations such as these influencing the choice of energy reserves among acinetobacters ? It is certainly possible because, in very simple terms, the wax ester-containing strains were isolated from soil, water and clinical materials, while the poly-ß-hydroxybutyrate-containing strains were isolated from a single ecological niche, activated sewage sludge. There is no evidence in the case of wax esters that their synthesis is increased in oxygen-limiting conditions, and so they would provide no extra advantage to acinetobacters from sewage plants which cycle bacteria between aerobic and anaerobic compartments. Perhaps poly-ß-hydroxybutyrate synthesis in these strains can serve as a transient sink for reducing equivalents during the aerobic to anaerobic transition, and this has led to the selection of poly-ß-hydroxybutyrate-containing acinetobacters in this particular niche. It would certainly be of interest to learn more about the physiological factors regulating poly-ß-hydroxybutyrate synthesis in this group of acinetobacters. More subtle selection pressures may also be playing a role. The density of poly-ß-hydroxybutyrate is greater than that of water, and the density of bacteria increases significantly as their poly-ß-hydroxybutyrate content increases (Pedros-Alio et al., 1985). In contrast, wax esters have densities of about 0.8 g/cm^3, much less than that of water (Sargent, 1978), and thus an increase in the wax content of organisms would give them a greater buoyancy. The acinetobacters from sewage sludge exist in a bacterial community associated with particulate material. It is possible that the greater buoyancy of wax-containing strains could make it more difficult for them to become or remain attached to this material.

The genus *Acinetobacter* contains a wide range of strains, and it could be that the type of carbon-containing energy reserve is simply a reflection of a strain's position within the genus, and thus energy reserves could be determined by the taxonomic position of the strain within the genus. Again, there are not really enough data to reach any firm conclusions about this, and some of the existing results are contradictory. The survey of *Acinetobacter* for the use of wax esters as reserves (Fixter et al., 1986) was based on the classification scheme of Baumann et al. (1968), and wax-accumulating organisms were found in both of the original major phenotypic groups and their sub-divisions. More recent and extensive taxonomic work has changed the internal sub-divisions considerably (Bouvet and Grimont, 1986; Grimont and Bouvet, this volume). Only a small number of strains from the survey of wax ester-containing acinetobacters are included in the later classification scheme and this reduces the value of the survey. The wax ester-containing strains are found in genospecies 1, 2, 4, 5, 7, 9 and 11 of the scheme proposed by Bouvet and Grimont (1986). The taxonomic position of the polyphosphate and poly-ß-hydroxybutyrate-accumulating strains is more problematic. The acinetobacters from sewage sludge belong almost exclusively to genospecies 5, 7 and 8/9 (Duncan et al., 1988; Beacham et al., 1990). As polyphosphate and poly-ß-hydroxybutyrate-accumulating strains are isolated from this environment, they are presumably members of these genospecies, but there is no direct evidence for this. Thus it is possible that the taxonomic distribution of these two energy reserves is more restricted. The lack of a study focusing on the distribution of all three energy reserves in well-characterised strains represents a major gap in our knowledge of the genus and its physiology and biochemistry.

REFERENCES

Albro, P.W., 1976, Bacterial waxes, *in* "Chemistry and Biochemistry of

Natural Waxes," p.419, P.E. Kolattukudy, ed., Elsevier, Amsterdam.

Aleby, S., Fischmeister, I., and Iyengar, B.T.R., 1971, The infrared spectra and polymorphism of long chain esters, *Lipids*, 6:421.

Baumann, P., Doudoroff, M., and Stanier, R.Y., 1968, A study of the Moraxella group. II Oxidative-negative species (genus *Acinetobacter*), *J. Bacteriol.*, 95:1520.

Beacham, A.M., Seviour, R.J., Lindrea, K.C., and Livingston, I., 1990, Genospecies diversity of *Acinetobacter* isolates from a biological nutrient removal pilot plant of a modified UCT configuration, *Water Res.*, 24:23.

Bouvet, P.J.M., and Grimont, P.A.D., 1986, Taxonomy of the genus *Acinetobacter* with the recognition of *Acinetobacter baumannii* sp. nov., *Acinetobacter haemolyticus* sp. nov., *Acinetobacter johnsonii* sp. nov., and *Acinetobacter junii* sp. nov., and emended descriptions of *Acinetobacter calcoaceticus* and *Acinetobacter lwoffii*, *Int. J. Syst. Bacteriol.*, 36:228.

Breuil, C., and Kushner, D.J., 1975, Lipase and esterase formation by psychrophilic and mesophilic *Acinetobacter* species, *Can. J. Microbiol.*, 21:423.

Brodisch, K.E.U., 1985, Interaction of different groups of micro-organisms in biological phosphate removal, *Water Sci. Tech.*, 17:89.

Brodisch, K.E.U., and Joyner, S.J., 1983, The role of microorganisms other than *Acinetobacter* in biological phosphate removal in activated sludge processes, *Water Sci. Tech.*, 15:117.

Bryn, K., Jantzen, E., and Bovre, K., 1977, Occurrence and patterns of waxes in *Neisseriaceae*, *J. Gen. Microbiol.*, 102:33.

Buchan, L., 1983, Possible biological mechanism of phosphorus removal, *Water Sci. Tech.*, 15:87.

Clifton, C.E., 1937, On the possibility of preventing assimilation in respiring cells, *Enzymologia*, 4:246.

Cloete, T.E., and Steyn, P.L., 1988, The role of *Acinetobacter* as a phosphorus removing agent in activated sludge, *Water Res.*, 22:971

Dawes, E.A., 1985, The effect of environmental oxygen concentration on the carbon metabolism of some aerobic bacteria, *in* "Environmental Regulation of Microbial Metabolism," p.121, I.S. Kulaev, E.A. Dawes and D.W. Tempest, eds., Academic Press, London.

Dawes, E.A., 1986, "Microbial Energetics," p.145, Blackie, London.

Dawes, E.A., and Senior, P.J., 1973, The role and regulation of energy reserve polymers in microorganisms, *Adv. Microb. Physiol.*, 10:136.

De Bruyn, J., Johannes, A., Weckx, M., and Breumer-Jochmans, M-P., 1981, Partial purification and characterization of an alcohol dehydrogenase of *Mycobacterium tuberculosis* var. *bovis* (BCG), *J. Gen. Microbiol.*, 124:359.

Deinema, M.H., Habets, L.H.A., Schalten, J., Turkstra, E., and Webers, H.A.A.M., 1980, The accumulation of polyphosphate in *Acinetobacter* spp. *FEMS Microbiol. Lett.*, 9:275.

Deinema, M.H., Van Loosdrecht, M., and Schalten, A., 1985, Some physiological characteristics of *Acinetobacter* spp accumulating large amounts of phosphate, *Water Sci. Tech.*, 17:119.

DeWitt, S., Ervin, J.L., Howes-Orchison, D., Daletas, D., Neidelman, S.L., and Geigert, J., 1982, Saturated and unsaturated wax esters produced by *Acinetobacter* sp. HO1-N grown on C_{16}-C_{20} n-alkanes, *J. Amer. Oil Chem. Soc.*, 59:69.

Doi, Y., Kawaguchi, Y., Nakamura, Y., and Kunioka, M., 1989, Nuclear magnetic resonance studies of poly(3-hydroxybutyrate) and polyphosphate metabolism in *Alcaligenes eutrophus*, *Appl. Environ. Microbiol.*, 55:2932.

Duncan, A., Vasiliadis, G., Bayly, R.C., May, J.W., and Rapor, W.G.C., 1988, Genospecies of *Acinetobacter* isolated from activated sludge showing enhanced removal of phosphate during pilot scale treatment of sewage, *Biotech. Lett.*, 10:831.

Fewson, C.A., 1985, Growth yields and respiratory efficiency of *Acinetobacter calcoaceticus*, *J. Gen. Microbiol.*, 131:865.

Finnerty, W.R., 1977, The biochemistry of microbial alkane oxidation, new insights and perspectives, *Trends Biochem. Sci.*, 2:73.

Fixter, L.M., 1976, Ultrastructural studies of *Acinetobacter* strains grown in carbon and nitrogen limiting batch cultures, *Proc. Soc. Gen. Microbiol.*, 4:177.

Fixter, L.M., and Fewson, C.A., 1974, The accumulation of waxes by *Acinetobacter calcoaceticus* NCIB 8250, *Biochem. Soc. Trans.*, 2:944.

Fixter, L.M., and McCormack, J.G., 1976, The effect of growth conditions on the wax content of various strains of *Acinetobacter*, *Biochem. Soc. Trans.*, 4:504.

Fixter, L.M., and Nagi, M.N., 1984, The presence of an NADP-dependent alcohol dehydrogenase in *Acinetobacter calcoaceticus*, *FEMS Microbiol. Lett.*, 22:297.

Fixter, L.M., Nagi, M.N., McCormack, J.G., and Fewson, C.A., 1986, Structure, distribution and function of wax esters in *Acinetobacter calcoaceticus*, *J. Gen. Microbiol.*, 132:3147.

Fuhs, G.W., and Chen, M., 1975, Microbiological basis of phosphate removal in the activated sludge process for treatment of wastewater, *Microb. Ecol.*, 2:119.

Gallagher, I.H.C., 1971, Occurrence of waxes in *Acinetobacter*, *J. Gen. Microbiol.*, 68:245.

Geigert, J., Neidleman, S.L., and DeWitt, S.K., 1984, Further aspects of wax ester biosynthesis by *Acinetobacter* sp. H01-N, *J. Amer. Oil Chem. Soc.*, 61:1747

Halvorson, H.O., Suresh, N., Roberts, M.F., Cocca, M., and Chikarmane, M.M., 1987, Metabolic surface polyphosphate pool in *Acinetobacter lwoffi*, *in* "Phosphate Metabolism and Cellular Regulation in Microorganisms," p.220, A Torriani-Gorini, F.G.

Rothman, S. Silver, A. Wright and E. Yagil, eds., American Society for Microbiology, Washington D.C.

Hao, O.J., and Chang, C.H., 1987, Kinetics of growth and phosphate uptake in pure culture studies of *Acinetobacter* species, *Biotech. Bioeng.*, 29:819.

Hardy, G.A., and Dawes, E.A., 1985, Effect of oxygen concentration on the growth and respiratory efficiency of *Acinetobacter calcoaceticus*, *J. Gen. Microbiol.*, 131:855.

Harold, F.M., 1966, Inorganic polyphosphates in biology: structure, metabolism and functions, *Bacteriol. Rev.*, 30:772.

Haywood, G.W., Anderson, A.J., Chu, L., and Dawes, E.A., 1988, Characterization of two 3-ketothiolases possessing differing substrate specificities in the polyhydroxyalkanoate synthesizing organism *Alcaligenes eutrophus*, *FEMS Microbiol. Lett.*, 52:91.

Heymann, J.B., Eagle, L.M., Greben, H.A., and Potgieter, D.J.J., 1989, The isolation and characterisation of volutin granules as subcellular components involved in biological phosphorus removal, *Water Sci. Tech.*, 21:397.

Hiraishi, A., Masumune, K., and Kitamura, M., 1989, Characterisation of the bacterial population in aerobic-anaerobic activated sludge system on the basis of respiratory quinone profiles, *Appl. Environ. Microbiol.*, 55:897.

Inui, H., Miyatake, K., Nakano, Y., and Kitaoka, S., 1982, Wax ester fermentation in *Euglena gracilis*, *FEBS Lett.*, 150:89.

Jantzen, E., Bryn, K., Bergan, T., and Bovre, K., 1975, Gas chromatography of bacterial whole cell methanolysates VII. Fatty acid composition of *Acinetobacter* in relation to the taxonomy of *Neisseriaceae*, *Acta Path. Microbiol. Scand.*, 83B:569.

Kulaev, I.S., and Vagabov, V.M., 1983, Polyphosphate metabolism in microorganisms, *Adv. Microb. Physiol.*, 4:83.

Lloyd, G.M., and Russell, N.J., 1983, Biosynthesis of wax esters in the psychrophilic bacterium *Micrococcus cryophilus*, *J. Gen. Microbiol.*, 129:2641.

Lotter, L.H., 1987a, Metabolic behaviour of *Acinetobacter* spp. in enhanced biological phosphorus removal - a biochemical model. Reply to comments, *Water SA*, 13:251.

Lotter, L.H., 1987b, Preliminary observations on polyhydroxybutyrate metabolism in the activated sludge process, *Water SA*, 13:189.

Lotter, L.H., and Dubery, I.A., 1989, Metabolic regulation of beta-hydroxy-butyrate dehydrogenase in *Acinetobacter calcoaceticus var lwoffi*, *Water SA*, 15:65.

Lotter, L.H., Wentzel, M.C., Loewenthal, L.E., Ekama, G.A., and Marais, GvR., 1986, A study of selected characteristics of *Acinetobacter* spp. isolated from activated sludge in aerobic/anoxic/aerobic and aerobic systems, *Water SA*, 12:203.

Makula, R.A., Lockwood, P.J., and Finnerty, W.R., 1975, Comparative analysis of lipids of *Acinetobacter* grown on hexadecane, *J. Bacteriol.*, 121:250.

Moss, C.W., Wallace, P.L., Hollis, D.G., and Weaver, R.E., 1988, Cultural and chemical characterization of CDC groups E0-2, M-5, and M-6, *Moraxella* species, *Oligella urethralis*, *Acinetobacter* species, and *Psychrobacter immobilis*, *J. Clin. Microbiol.*, 26:484.

Murphy, M., and Lotter, L.H., 1986, The effect of acetate and succinate on polyphosphate formation and degradation in activated sludge with particular reference to *Acinetobacter calcoaceticus*, *Appl. Microbiol. Biotech.*, 24:512.

Nagi, M.N., 1981, Some studies on wax esters of *Acinetobacter calcoaceticus*, *PhD thesis, University of Glasgow, UK*.

Nagi, M.N., and Fixter, L.M., 1981, Regulation of the wax content of *Acinetobacter calcoaceticus*. NCIB 8250, *Abstr. Ann. Meet. Amer. Soc. Microbiol.*, K11

Neidleman, S.L., 1987, Effects of temperature on lipid unsaturation, *Biotech. Genet. Eng. Rev.*, 5:245.

Nesbitt, J.B., 1969, Phosphorus removal, the state of the art, *J. Water Pollut. Cont. Fed.*, 41:701.

Nicolls, H.A., and Osborn, D.W., 1979, Bacterial stress, a prerequisite for biological removal of phosphorus, *J. Water Pollut. Cont. Fed.*, 51:557.

Odham, G., Tunlid, A., Westerdahl, G., and Marden, P., 1986, Combined determination of poly-ß-hydroxyalkanoic acid and cellular fatty acids in starved marine bacteria and sewage sludge by gas chromatography with flame ionization or mass spectrometry detection, *Appl. Environ. Microbiol.*, 52:905.

Oeding, V., and Schlegel, H.G., 1973, ß-Ketothiolase from *Hydrogenomonas eutropha* H16 and its significance in the regulation of poly-ß-hydroxybutyrate metabolism, *Biochem. J.*, 134:239.

Pedros-Alio, C., Mas, J., and Guerrero, R., 1985, The influence of poly-ß-hydroxybutyrate accumulation on cell volume and buoyant density in *Alcaligenes eutrophus*, *Arch. Microbiol.*, 143:178.

Preiss, J., 1984, Bacterial glycogen synthesis and its regulation, *Ann. Rev. Microbiol.*, 38:419.

Preiss, J., 1989, Chemistry and metabolism of intracellular reserves, *in* "Bacteria in Nature, vol. 3," p.189, J.S. Poindexter and E.R. Leadbetter, eds. Plenum Press, New York.

Rao, N.N., Roberts, M.F., and Torriani, A., 1987 Polyphosphate accumulation and metabolism in *Escherichia coli*, *in* "Phosphate Metabolism and Cellular Regulation in Microorganisms," p.213, A. Torriani-Gorini, F.G. Rothman, S. Silver, A. Wright and E. Yagil, eds., American Society for Microbiology, Washington D.C.

Reusch, R.N., and Sadoff, H.L., 1988, Putative structure and functions of a poly-ß-hydroxybutyrate-calcium polyphosphate channel in bacterial plasma membranes, *Proc. Natl. Acad. Sci. USA.*, 85:4176.

Reusch, R.N., Hiske, T.W., and Sadoff, H.L., 1986, Poly-ß-hydroxybutyrate membrane structure and its relationship to genetic transformation in *Escherichia coli*, *J. Bacteriol.*, 168:553.

Reusch, R., Hiske, T., Sadoff, H., Harris, R., and Beveridge, T., 1987, Cellular incorporation of poly-ß-hydroxybutyrate into plasma membranes of *Escherichia coli* and *Azotobacter vinelandii* alters native membrane structure, *Can. J. Microbiol.*, 33:435.

Roberts, M.R., 1987, Metabolic behaviour of *Acinetobacter* spp. in enhanced biological phosphorus removal - a biochemical model. Comments. *Water SA.*, 13:251.

Rock, C.O., and Cronan, J.E., 1985, Lipid metabolism in prokaryotes, *in* "Biochemistry of Lipids and Membranes," p.73, D.E. Vance and J.E. Vance, eds., Benjamin/Cummings, Menlo Park, California.

Russell, N.J., and Volkman, J.K., 1980, The effect of growth temperature on wax ester composition in psychrophilic bacterium *Micrococcus cryophilus* ATCC 15174, *J. Gen. Microbiol.*, 118:131.

Sargent, J.R., 1978, Marine wax esters, *Sci. Prog.*, 65:437.

Scott, C.C.L., Makula, R.A., and Finnerty, W.R., 1976, Isolation and characterization of membranes from a hydrocarbon-oxidizing *Acinetobacter* sp., *J. Bacteriol.*, 127:469.

Senior, P.J., and Dawes, E.A., 1973, The regulation of poly-ß-hydroxybutyrate metabolism in *Azotobacter beijerinckii*, *Biochem. J.*, 134:225.

Shabtai, Y., and Gutnick, D.L., 1985, Exocellular esterase and emulsan release from the cell surface of *Acinetobacter calcoaceticus*, *J. Bacteriol.*, 161:1176.

Sherwani, M.K., and Fixter, L.M., 1989, Multiple forms of carboxylesterase activity in *Acinetobacter calcoaceticus*, *FEMS Microbiol. Lett.*, 58:75.

Shively, J.M., 1974, Inclusion bodies of prokaryotes, *Ann. Rev. Microbiol.*, 28:167.

Singer, M.E., and Finnerty, W.R., 1985a, Fatty aldehyde dehydrogenases in *Acinetobacter* sp. strain H0-1N: role in hexadecane and hexadecanol metabolism, *J. Bacteriol.*, 164:1011.

Singer, M.E., and Finnerty, W.R., 1985b, Alcohol dehydrogenases in *Acinetobacter* sp. strain H0-1N: role in hexadecane and hexadecanol metabolism, *J. Bacteriol.*, 164:1017.

Singer, M.E., Tyler, S.M., and Finnerty, W.R., 1984, Growth of *Acinetobacter* sp. strain H01-N on hexadecanol: physiological and ultrastructural characteristics, *J. Bacteriol.*, 162:162.

Stephenson, T., 1987, *Acinetobacter*: its role in biological phosphate removal, *in* "Biological Phosphate Removal from Wastewaters," p. 313, R. Ramadori, ed., Pergamon Press, Oxford.

Stewart, J.E., and Kallio, R.E., 1959, Bacterial hydrocarbon oxidation. II Ester formation from alkanes, *J. Bacteriol.*, 78:726.

Streichan, M., Golecki, J.R., and Schon, G., 1990, Polyphosphate removal from sewage plants with different processes for biological phosphorus removal, *FEMS Microbiol. Ecol.*, 73:113.

Suresh, N., Warburg, R., Timmerman, M., Wells, J., Coccia, M., Roberts, M.F., and Halvorson, H.O., 1985, New strategies for the isolation of microorganisms for phosphate accumulation, *Water Sci. Tech.*, 17:99.

Suresh, N., Roberts, M.F., Coccia, M. Chikarmane, H.M., and Halvorson, H.O., 1986, Cadmium-induced loss of surface polyphosphate in *Acinetobacter lwoffi,, FEMS Microbiol. Lett.*, 36:91.

Tal, S., and Okon, Y., 1985, Production of the reserve material, poly-ß-hydroxybutyrate and its function in *Azospirillum brasiliense, Can. J. Microbiol.*, 31:608.

Tempest, D.W., 1969, Quantitative relationship between inorganic cations and anionic polymers in growing bacteria, *in* "Microbial Growth, 19th Symp. Soc. Gen. Microbiol.," p.87, P.M. Meadow and S.J. Pirtm, eds., Cambridge University Press, Cambridge.

T'Seyen, J., Malnou, D., Block, J.C., and Faup, G., 1985, Polyphosphate kinase activity during phosphate uptake by bacteria, *Water Sci. Tech* 17:43.

Van Groenestijn, J.W., Deinema, M.M., and Zehnder, A.J.B., 1987, ATP production from polyphosphate in *Acinetobacter* strain 210A, *Arch. Microbiol.*, 148:14.

Van Groenestijn, J.W., Vlekke, G.F.M., Anink, D.M.E. Deinema, M.M., and Zehnder, A.J.B., 1988, Role of cations in the accumulation and release of phosphate by *Acinetobacter* strain 210A, *Appl. Environ. Microbiol.*, 54:2894.

Van Groenestijn, J.W., Bentvelsen, M.M.A., Deinema, M.M., and Zehnder, A.J.B., 1989a, Polyphosphate-degrading enzymes in *Acinetobacter* spp. and activated sludge, *Appl. Environ. Microbiol.*, 55:219.

Van Groenestijn, J.W., Zuidena, M., Van de Worp, J.J.M., Deinema, M.M., and Zehnder, A.J.B., 1989b, Influence of environmental parameters on polyphosphate accumulation in *Acinetobacter* sp., *Ant. v. Leeuw. J. Microbiol.*, 55:67.

Vasiliadis, G., Duncan, A., Bayly, R.C., and May, J.W., 1990, Polyphosphate production by strains of *Acinetobacter, FEMS Microbiol. Lett.*, 70:37.

Wentzel, M.C., Lotter, L.H., Loewenthal, R.E., and Marais. Gv.R., 1986, Metabolic behaviour of *Acinetobacter* spp. in enhanced biological phosphorus removal. A biochemical model, *Water SA*, 12:209.

Wentzel, M.C., Loewenthal, R.E., Ekama, G.A., and Marais, G.V.R., 1988, Enhanced polyphosphate organism cultures in activated sludge systems - Part I Enhanced culture development, *Water SA*, 14:141.

Wilkinson, J.F., 1959, The problem of energy-storage compounds in bacteria, *Exp. Cell Res., Suppl.* 7:111.

Wood, H.G., and Clark, J.E., 1988, Aspects of inorganic polyphosphates, *Ann. Rev. Biochem.*, 57:235.

Yall, I., Bangwan, W.H., Knudsen, C. and Sinclair, N.A., 1970, Biological uptake of phosphate by activated sludge, *Appl. Microbiol.*, 20:145.

Ye, Q., Ohtake, H., and Toda, K., 1988, Phosphorus removal by pure and mixed cultures of microorganisms, *J. Ferment. Tech.*, 66:207.

ENERGY GENERATION AND THE GLUCOSE DEHYDROGENASE

PATHWAY IN *ACINETOBACTER*

J.A. Duine

Vakgroep Microbiologie & Enzymologie
Technische Universiteit Delft
Julianalaan 67
2628 BC Delft
The Netherlands

INTRODUCTION

Direct, non-phosphorylative, bacterial glucose oxidation was discovered nearly half a century ago (Barron and Friedemann, 1941; Lockwood et al., 1941). The possession of such a property was deduced from the fact that gluconic and ketogluconic acids were formed from glucose. The first step in this pathway is the oxidation of glucose at the C_1 position to form 1,5-gluconolactone, which is hydrolysed by enzyme action or spontaneously to form gluconic acid; direct glucose oxidation can also occur at other positions, but these conversions are not relevant here. The reaction is catalysed by a glucose dehydrogenase, whereas glucose oxidase (EC 1.1.3.4) occurs only in yeasts and fungi. The enzyme belongs either to the category of NAD(P)-dependent dehydrogenases or to the dye-linked dehydrogenases. In the former case, NAD(P) is reduced in the reaction and then re-converted into NAD by the NADH dehydrogenase of the respiratory chain; in the latter case, the reduction equivalents on the cofactor are directly transferred *in vivo* to a component of the respiratory chain, and *in vitro* to an artificial dye. During the 1950's, the latter type of glucose dehydrogenase was discovered in many *Pseudomonas* and *Acetobacter* strains, and also in *Bacterium anitratum*. It should be noted that the bacterial dehydrogenase discussed here is not the flavoprotein glucose dehydrogenase (EC 1.1.99.10) which has only been reported to occur in *Aspergillus oryzae* (Bak, 1967).

The Biology of Acinetobacter, Edited by K. J. Towner *et al.*
Plenum Press, New York, 1991

Bacterium anitratum, nowadays known as *Acinetobacter calcoaceticus*, was used by Hauge and colleagues to obtain the dye-linked glucose dehydrogenase and study its properties. In a series of papers it was demonstrated that the enzyme contained a co-factor whose identity was not known at that time (Hauge, 1960a,b, 1961, 1964, 1966a,b; Hauge and Hallberg, 1964; Hauge and Mürer, 1964). The presence of the co-factor could easily be detected by the ability to reconstitute the glucose dehydrogenase activity in cell-free extracts prepared from anaerobically-grown *Rhodopseudomonas* (now *Rhodobacter*) *sphaeroides* (Niederpruem and Doudoroff, 1965). Based on this assay, it was shown that glucose dehydrogenase containing this co-factor occurs in other aerobically-grown Gram-negative bacteria. In addition, the assay demonstrated the natural occurrence of the glucose dehydrogenase apo-enzyme.

From what is currently known, the choice of the particular *B. anitratum* strain (provided to us by Prof. J.G. Hauge and deposited in our culture collection, first as *A. calcoaceticus* LMD 79.39 and later as LMD 79.41) may not have been fortuitous as it contains a type of glucose dehydrogenase that is most easily purified, as described below. On the other hand, only low levels of this enzyme are produced, which could be the reason why identification of its co-factor was achieved only many years later. Thus, 15 years after the peculiar properties of the co-factor of methanol dehydrogenase had been described (Anthony and Zatman, 1964), the similarity to the co-factor of glucose dehydrogenase was noticed, both by us and by Imanaga et al. (1979). After the elucidation of the co-factor structure of methanol dehydrogenase (Salisbury et al., 1979; Duine and Frank, 1980), that of glucose dehydrogenase was also shown to be 2,7,9-tricarboxyl-1*H*-pyrrolo[2,3-*f*]-quinoline-4,5-dione, or pyrroloquinoline quinone (PQQ) (Duine et al., 1979). In the past ten years, the quinoprotein nature (a quinoprotein is defined as an enzyme with PQQ as a co-factor) of glucose dehydrogenase from many other Gram-negative bacteria has been established. Although it has been reported that the glucose dehydrogenase from the Gram-positive organism *Zymomonas mobilis* is also a quinoprotein (Strohdeicher et al., 1989), the enzyme is probably quite different because the co-factor is covalently bound.

Glucose Dehydrogenase: Different Forms or Different Enzymes?

It was noted quite early (Hauge, 1966a) that glucose dehydrogenase activity present in a cell-free extract could be separated into two fractions by ion exchange chromatography; one fraction being adsorbed, the other not. Another distinction could be made on "solubility", as centrifugation studies suggested that one part of the activity was soluble (i.e. remaining in the supernatant), while the other part remained attached to membranes (in the pellet). Although these fractions showed different substrate and electron acceptor activities, they were considered to derive from one and the same enzyme since the properties of the particle-bound form could be changed into those of the soluble form by detergent treatment (Hauge and Hallberg, 1964; Hauge, 1966a). On the basis of what is currently known, namely that two quite different glucose dehydrogenases occur in this strain (soluble and membrane-bound glucose dehydrogenases, represented by "s-GDH" and "m-GDH" respectively), this observation is difficult to understand. The "transformation" could be explained by assuming that both s-GDH and m-GDH were present in the pellet and that m-GDH activity was selectively destroyed by the treatment; membrane preparations do indeed show activity with disaccharides and artificial electron acceptors (see e.g. Matsushita et al., 1988), and this points to the presence of s-GDH. In studies carried out with several *Acinetobacter* strains, it was concluded that different forms of glucose dehydrogenase exist (Duine et al., 1982). However, the data on the induction of these forms in the original strain (at that time deposited as LMD 79.39) may have been biased by the fact that, as was later discovered, the culture slant was contaminated with a *Pseudomonas putida* strain, leading to the erroneous conclusion that *A. calcoaceticus* contains a quinoprotein alcohol dehydrogenase (Duine and Frank, 1981; Beardmore-Gray and Anthony, 1983; B.W. Groen, R.A. van der Meer and J.A. Duine, unpublished results). On the other hand, it is clear that the glucose dehydrogenases from the *Acinetobacter* strains are dissimilar. Different glucose dehydrogenases have also been reported to occur simultaneously in certain *Pseudomonas* species (Hauge, 1966a; Hayaishi, 1966; Vicente and Canovas, 1973). It remains to be investigated whether these phenomena can be explained by the presence of both an m-GDH and an s-GDH, or by a single glucose dehydrogenase manifesting itself in different forms because it is present in different environments. The curious result that the substrate specificity for aldose sugar oxidation by whole cells is quite different from that observed in cell-free extracts (Dokter et al., 1987) can now be explained from the fact that s-GDH shows no activity *in vivo*, as discussed below.

Scope and Limitations

When reading the literature on the glucose dehydrogenase of the *A. calcoaceticus* strain originally used by Hauge, the conclusions in the early papers should be reinterpreted in the light of what it now known about s-GDH and m-GDH. Since glucose dehydrogenases from other strains have scarcely been studied, it is not known whether this caution should apply in general.

Detection of the glucose oxidation pathway in *Acinetobacter* strains is achieved by measuring growth on glucose or by measuring acid formation under aerobic conditions in special media supplemented with glucose. However, identification of the enzyme responsible for the observed phenomenon has generally been neglected. Moreover, the non-acid forming strains (frequently classed as *A. lwoffii*) should be reinvestigated as they may produce only the protein part of glucose dehydrogenase, which can be detected by screening with and without PQQ (van Schie et al., 1984). Although this approach has recently been followed by Bouvet and Bouvet (1989), it will be already clear that data on the distribution of the enzyme in this genus cannot be fully trusted. Similarly, since most of the work has been carried out with *A. calcoaceticus* strain LMD 79.41, the conclusions drawn in this review apply only to this strain, unless explicitly indicated otherwise, and care should be taken in making generalisations. In this connection, it should be mentioned that this strain does not grow on glucose or gluconic acid. Furthermore, compared with the sub-group which does have these characteristics (Baumann et al., 1968), it shows a somewhat atypical behaviour as it does not grow on pentoses and does not produce lactobionic acid from lactose. A recent identification by the criteria of Bouvet and Grimont (1986) in two separate laboratories has now revealed that strain LMD 79.41 belongs to either *Acinetobacter* genospecies 2 or 3 (Dijkshoorn et al., 1990).

THE SOLUBLE TYPE OF GLUCOSE DEHYDROGENASE (s-GDH)

Although the distinction between s-GDH and m-GDH has only recently been made, from the properties reported it can be deduced that the enzyme investigated by Hauge was s-GDH. Since classification of this glucose dehydrogenase was carried out some years ago, it seems logical to reserve the EC number 1.1.99.17 for s-GDH and to give m-GDH a novel EC number.

General Properties

s-GDH has been purified (Dokter et al., 1986; Geiger and Görisch, 1986) and most of its properties are shown in Table 1. It is a dimeric, strongly basic protein with a broad substrate specificity. Thus, not only pentose and hexose sugars, but also disaccharides are oxidised; in this respect, aldose dehydrogenase would be a better name. There is a preference for the ß-anomer of glucose and the product formed is the 1,5-gluconolactone (Hauge, 1960b). A substantial isotopic effect on the reaction rate was observed with glucose labelled at the C_1 position with ^2H (Hauge et al., 1968). Several artificial electron acceptors exist for the enzyme (Hauge, 1966b). All are cationic or neutral species, as deduced from the fact that 2,6-dichlorophenolindophenol (DCPIP) is active only below pH 6.5, but the pH optimum observed with Wurster's Blue is 9.0 (Dokter et al., 1986). In a search for the natural electron acceptor, only a cytochrome appeared to be active, not ubiquinones. Purification revealed that it is a periplasmic, small, basic haemoprotein, called cytochrome b_{562}, and similar to a cytochrome from *Escherichia coli* with an unknown function (Dokter et al., 1988). *A. calcoaceticus* does not contain cytochromes *c*. A start has been made on unravelling the cytochrome *b* complexes (Geerlof et al., 1989; Dokter et al., 1990).

The mechanism of action seems straightforward, namely reaction of the oxidised form with the substrate to yield the 2-electron reduced form. Stopped flow experiments revealed a fluorescing intermediate (J. Frank and J.A.Duine, unpublished results) which could be a PQQ-substrate complex,

Table 1. Comparison of the properties of s-GDH and m-GDH

	s-GDH	m-GDH
M_r (enzyme)	100,462	86,956
M_r (subunit)	50,231	86,956
Cofactors (per enzyme molecule)	2 PQQ 4 Ca^{2+}	x PQQ x Ca^{2+}/Mg^{2+}?
Metal ions in reconstitution	$Ca^{2+}(Mn^{2+},Cd^{2+})$	Mg^{2+}, Ca^{2+}
Removal of PQQ by chelators	no	yes
Midpoint potential (pH 7)	+50 mV	?
Isoelectric point	9.5	(low)
Substrate specificity		
glucose	+	+
2-deoxyglucose	-	+
lactose	+	-
maltose	+	-
pH optimum	9.0	(8.0,4.6)
Substrate inhibition	yes	not observed
Natural electron acceptor	cytochrome b_{562}	ubiquinone (Q_9)
Location	periplasm	membrane-integrated (outside)

just as has been observed for methanol dehydrogenase (Frank et al., 1989). Ca^{2+} ions are required for the reconstitution of the apo-form of s-GDH to active holo-enzyme (Geiger and Görisch, 1989; B.W. Groen and J.A. Duine, unpublished results). Absorption spectra of the redox forms of the enzyme are depicted in Fig.1. The semiquinone form, which should theoretically occur in the oxidative half reaction in which reduced glucose dehydrogenase reacts with one-electron acceptors of the respiratory chain, has not been studied to date.

Structural Aspects

In reconstitution studies, Ca^{2+} can be replaced by Mn^{2+} and Cd^{2+}, but not by Mg^{2+}. Four Ca^{2+} ions occur per enzyme molecule, and these cannot be removed from the holo-enzyme with chelators. The cofactor and the metal ions also play a structural role since they are required in the refolding process after heat denaturation, and Ca^{2+} stabilises the enzyme against this denaturation (Geiger and Görisch, 1989). Detachment of PQQ requires rather drastic conditions (Hauge and Mürer, 1964; Duine et al., 1979; Kilty et al., 1982) and the effects are not always reproducible (J.A. Duine, unpublished results). As it has only recently become apparent that both the metal ion and PQQ are required for reconstitution, past experiments in which apo-enzyme was prepared to assay PQQ, or to establish

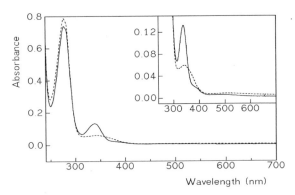

Fig.1. Absorption spectra of s-GDH: ———— , reduced form (maximum 338 nm); ----, oxidised form (maximum 350 nm).

activity of PQQ-analogues, may have suffered from lack of knowledge in this respect. Thus, although it was claimed that synthetic phenanthroline quinones are active in reconstitution (Duine et al., 1980; Conlin et al., 1985), this could not be confirmed in later experiments (J. Frank and J.A. Duine, unpublished results). A plausible explanation could be that the samples tested contained the tiny amount of Ca^{2+} required for reconstitution and that PQQ was not completely removed from the apo-enzyme preparation.

The gene of s-GDH has been cloned, sequenced and expressed in an *E. coli* strain (Cleton-Jansen et al., 1989). It contains a leader sequence of 24 amino acids which has the characteristics of a signal peptide. Growth in the presence of the detergent Triton X-100 had already shown that s-GDH occurs in the periplasm, as deduced from its presence in the culture fluid under these conditions (Dokter et al., 1985). Similar observations have been made recently for strain 69-V where selective detachment of the activity (m-GDH?) depended on the detergent used (Hommel et al., 1989).

Applications

The gene of s-GDH has been combined with a strong promoter and expressed in *E. coli* so that substantial amounts of apo-enzyme could be obtained; there was no holo-enzyme since *E. coli* does not produce free PQQ (Cleton-Jansen et al., 1989; van der Meer et al., 1990). Since the holo-enzyme has an extremely high turnover number of around 400,000 /min (Hauge, 1966b; Dokter et al., 1986), a sensitive assay for PQQ could be developed with these preparations (van der Meer et al., 1990). Production of the enzyme may also be important for other applications as the natural holo-enzyme has been applied in an amperometric biosensor; the high catalytic activity and oxygen-insensitivity of s-GDH are attractive aspects compared with glucose oxidase (D'Costa et al., 1986). Studies on the binding and catalysis of PQQ analogues have been carried out with glucose dehydrogenase from *E. coli* (Shinagawa et al., 1986) and are in progress with apo-s-GDH from *A. calcoaceticus*.

A Physiological Role?

The typical substrate specificity of s-GDH is reflected in that of cell-free extracts, not in the oxidation pattern of whole cells (Dokter et al., 1987). A mutant unable to produce s-GDH showed the same substrate specificity as wild-type cells and no unusual growth behaviour (Cleton-Jansen et al., 1988b). This leads to the conclusions that s-GDH is not involved in cellular sugar oxidation and has no role at all in the organism or is operating in a dispensable function. Since both s-GDH and cytochrome b_{562} are localised in the periplasm, the non-functionality is not related to transport problems for the substrate. Thus, an essential component for electron transfer to the respiratory chain may be lacking which could have been lost in the evolutionary history of this organism; this assumed incompleteness is in accordance with the low reaction rate of reduced s-GDH with cytochrome b_{562} *in vitro* (Dokter et al., 1988). However, this assumption leads to the curious implication that in the past the organism would have possessed the ability to oxidise glucose with two dehydrogenases having the same cofactor, but no structural similarity of the protein components. However, the real conclusion seems to be that, despite all the research effort that has been concentrated on the characterisation of s-GDH from this strain, the physiological role of the enzyme, if any, is completely unknown.

MEMBRANE-BOUND GLUCOSE DEHYDROGENASE (m-GDH)

Since m-GDH was previously thought to be identical with s-GDH, no real efforts were made in the past to characterise it. Another reason for its elusiveness is the hydrophobic nature of the protein, requiring detergents to remove it from the membranes and making purification difficult (Matsushita et al., 1989). Now, 30 years after its discovery (Hauge, 1960), it is realised that this enzyme should have been preferentially studied as it is the one which is responsible for cellular glucose oxidation.

General Properties

From the primary structure deduced from the DNA sequence (Cleton-Jansen et al., 1988a), it appears that the *N*-terminal part of the enzyme (140 amino acids) is very hydrophobic. Five hydrophobic regions, separated from each other by hydrophilic amino acids, are putative membrane spanning segments. There is no signal sequence and it has been suggested that the *N*-terminal part anchors the enzyme to the cytoplasmic membrane (Cleton-Jansen et al., 1988b). Attachment should occur in such a way that the active site is still available for sugars coming from the outside because no transport mechanisms have been found. A domain motif has been found which also occurs in mitochondrial NADH dehydrogenase, coding for interaction with ubiquinone (Friedrich et al., 1990). Although the gene has been cloned and expressed in an *E. coli* strain, production of adequate amounts of the enzyme is still problematic since the cells were killed in attempts to achieve high expression (A-M. Cleton-Jansen and B.W. Groen, unpublished results).

Comparison with data from the literature suggests that m-GDH is widespread among Gram-negative bacteria (van Schie et al., 1987a). This has been further substantiated for a range of bacteria by showing that an immunological cross-reaction occurs between the antibody directed against m-GDH from *A. calcoaceticus* (Matsushita et al., 1986) and the glucose dehydrogenases from *E. coli* and *Pseudomonas aeruginosa*, where the genes for the enzyme showed sequence similarity with that of *A. calcoaceticus* (Cleton-Jansen et al., 1988b). In studies on the partially purified enzyme from *Ps. fluorescens*, selective protein modification indicated several amino acids in the active site which are essential for binding of PQQ or for activity (Imanaga, 1989).

The enzyme has been purified and partly characterised (Table 1; Matsushita et al., 1989). It is a monomeric protein with a narrower substrate specificity and lower catalytic activity than s-GDH. The substrate specificity of m-GDH from other bacteria is variable, ranging from strict specificity for glucose to very broad specificity, including disaccharides. Evidence has been provided that the natural electron acceptor is ubiquinone Q_9 (Matsushita et al., 1989). If it appears that the redox potential of m-GDH is as high as that of s-GDH (+ 50 mV; Dokter et al., 1988), problems may arise if the oxidation state of the respiratory chain is inadequate. Perhaps the decrease of the cellular glucose oxidation rate, but not of the ethanol oxidation rate, at low oxygen tensions (Dokter et al., 1987) is an indication for that. Binding of PQQ also requires a bivalent metal ion (Ca^{2+} or Mg^{2+}). In contrast to s-GDH, the metal ion, and concomitantly PQQ, could be easily removed by a chelator from m-GDH of *Ps. aeruginosa*, enabling easy preparation of an apo-enzyme suited for PQQ determinations (Duine et al., 1983). m-GDH apo-enzyme is present in *A. lwoffii* (van Schie et al., 1984, 1987a) and, of course, in PQQ⁻ mutants (van Kleef et al., 1987). Thus whole cells of these strains can be used for the detection of PQQ. Upon transfer of the genes for PQQ biosynthesis (obtained from *A. calcoaceticus*) to *A. lwoffii*, holo-m-GDH was produced (Goosen et al., 1989).

Function in the Entner-Doudoroff Route

Glycolysis has not been detected in *A. calcoaceticus*, probably because of the absence of hexokinase. A few percent of the strains grow on glucose via the Entner-Doudoroff pathway (Juni, 1978). m-GDH is required in these cases to provide the gluconic acid. Most strains do not grow on glucose, although half of them possess m-GDH, as judged by acid formation from aldose sugars (Baumann et al., 1968). Mutation may occur to produce growth on glucose (Juni, 1972), as has also been reported for strain LMD 79.41 (van Schie et al., 1989). All observations suggest that cryptic genes are present (especially for gluconokinase) which become derepressed upon mutation. The ability to grow on glucose can also be acquired by genetic transformation (Juni, 1972). In principle, acquisition of m-GDH could enable growth on those sugars which can be converted to assimilatable lactones; indeed, all strains which can grow on pentoses contain glucose dehydrogenase (Baumann et al., 1968). However, in this connection it should be mentioned that a lactonase, observed in several *Pseudomonas* strains, has so far not been reported for this bacterium, and spontaneous hydrolysis, especially of an aldonolactone from a pentose, scarcely occurs at pH 7 (van Schie et al., 1989). This could imply that, although an assimilation route for a certain aldonic acid is available, in practice the rapid acidification during the lag phase may prohibit growth if buffering of the medium is not adequate.

Natural and man-made mutants occur which lack the ability to synthesise m-GDH, but show normal growth behaviour. This indicates that the absence of the protein is not crucial for growth on non-sugar substrates. The viability of PQQ$^-$ mutants supports this view. The apo-enzyme functions as a scavenger for PQQ because nearly stoichiometric amounts added to the growth medium are sufficient to achieve the holo-enzyme status. However, other quinoprotein dehydrogenases perform even better. Thus, by applying low levels of PQQ, it was found that the in-vivo affinity of quinate dehydrogenase for PQQ is higher than that of m-GDH (van Kleef et al., 1988).

ENERGY CONSERVATION

Since oxidation of glucose to gluconolactone is achieved by a dehydrogenase, the question can be posed as to whether respiratory activity via this enzyme leads to any useful energy being conserved for the organism. In principle, this process could be most easily studied with a strain such as LMD 79.41 as the enzymes for the following interfering processes are absent: glycolysis; NAD(P)-dependent glucose dehydrogenation; ketogluconate formation; gluconolactone or gluconate conversion.

Growth Experiments

To create optimal conditions for ameasurable effect of m-GDH, growth yield experiments should be performed using a carbon source with a low energy content (i.e. highly oxidised) so that the carbon source is mainly used for assimilation and energy provision derives to a large extent from the m-GDH pathway. These conditions are not easily met in batch growth experiments as it is generally energy spillage, rather than energy limitation, that applies. This may be why negative results were reported in the past for growth yield improvement, although there were some reports with a positive outcome (Bell and Marus, 1966; Kitagawa et al., 1986a). Several research groups have performed growth experiments of this type with continuous cultures. Studies with strain 69-V on an acetate medium supplemented with glucose, added in a gradient form, gave a high yield of biomass (Müller and Babel, 1986). The presence of an unknown pathway for glucose or gluconic acid assimilation in this strain can be excluded since only low amounts of label were incorporated by the cells in the presence of labelled glucose; some intracellular labelled gluconate and 6-phosphogluconate was detected (Kleber et al., 1984), and no glucose uptake, as previously reported by Cook and Fewson (1973). Very high growth yields were obtained for strain LMD 79.41 (De Bont et al., 1984; van Schie et al., 1987c). These increases were found at pH 8.3, not at pH 7 or 5. Yield improvement was correlated with lactone hydrolysis, as was especially clear for the experiments with xylose at pH 7 where no hydrolysis and no yield increase was observed; however, it is a puzzling phenomenon that energisation of the transport system via m-GDH occurred under this condition (van Schie et al., 1985). Checks with labelled sugars also revealed that assimilation cannot account for the high yields; with glucose, incorporation of 2.9% label in biomass and 9.2% in CO_2; with xylose, 0.2% and 2.6%, respectively. Since the increase could not even be explained from coupling of the dehydrogenase at the level of NADH, it was assumed that the glucose oxidation step was able to improve the sub-optimal metabolism of acetate. Unfortunately, the literature reports on the efficiency of acetate metabolism and ratios are not unanimous. Thus, acetate conversion has been reported to occur with variable yields (Du Preez et al., 1981; Fewson, 1985; Müller and Babel, 1986; Gommers et al., 1988); these may be related to strain differences and/or dilution rates, and P/O ratios of one (Fewson, 1985) and three (Meyer and Jones, 1973) have been mentioned. Nevertheless, the conclusions with respect to strain LMD 79.41 may be premature because the experiments have subsequently been repeated with a slightly different experimental arrangement and the very high yield improvements could not be confirmed. This discrepancy could be related to the easy flotation of this organism under certain conditions; cells are in the gas liquid boundary layer and removed with increased efficiency in the effluent (H. Noorman, personal communication). From recent experiments, a more realistic increase was observed, as reflected by the number of moles of ATP generated per mole of O_2, which is twice as large on acetate plus glucose than on acetate (H. Noorman, personal communication). Summarising, it is clear that the pathway via m-GDH can lead to increases in yield under appropriate conditions of growth, but that the quantitative aspects still need attention.

Transport Studies

Transport processes can be driven by glucose oxidation via m-GDH (van Schie et al., 1985, 1987b; Kitagawa et al., 1986b). The proton translocation achieved can be used to energise transport and ATP synthesis.

Since several dehydrogenases deliver their electrons to the respiratory chain at the level of ubiquinone Q_9, one may wonder whether competition occurs. However, studies on the availability of the respiratory chain for different dehydrogenases have not been carried out for this organism. Since the same rates were observed for the oxidation of sugars with strain 69-V

(Asperger et al., 1981), it was concluded that the dehydrogenase step is not rate-limiting and the bottleneck should occur somewhere else in the respiratory chain. However, the conclusion does not apply to strain LMD 79.41, where differences in the affinity constant as well as in the maximal rate were observed with different sugars (Dokter et al., 1987). Thus, the rate-limiting step may vary according to the organism and the growth condition. Kinetically distinct pathways have been suggested for NADH and $PQQH_2$ oxidation (Beardmore-Gray and Anthony, 1986). The dissimilarity may be related to the location of m-GDH at the periplasmic site and of NADH dehydrogenase at the cytosolic site of the cytoplasmic membrane.

Induction Experiments

Studies on overflow metabolism (Neijssel et al., 1989) proceeding via m-GDH in *Klebsiella aerogenes* have shown that energy stress leads to induction of the enzyme. This correlation was found either during growth in the presence of uncoupler (collapse of the proton motive force) or during growth limitation with certain nutrients, leading to futile cycling. Continuous culture experiments with *A. calcoaceticus* strain LMD 79.41 showed that the enzyme is produced constitutively, independently of the growth substrate or the presence of an aldose sugar (van Schie et al., 1988). Since the highest levels were attained at low growth rates, the energy status of the cells may also have at least a partial influence on the induction in this bacterium. The conclusions drawn here may, however, be premature, since on repeating the experiments with acetate plus xylose (instead of glucose), it appeared that both constitutive as well as inducible synthesis occurs (H. Noorman, personal communication).

The synthesis of apo-m-GDH and PQQ can occur independently (van Kleef and Duine, 1989). This independence is very pronounced in the case of the PQQ-less strain *A. lwoffii* LMD 73.1, but is also indicated by the occurrence of non-coordinated synthesis in the case of *A. calcoaceticus* LMD 82.3 (van Schie et al., 1989). A curious phenomenon is the absence of m-GDH activity in cells grown under oxygen limitation (van Schie et al., 1988). This negative result is not due to absence of the cofactor because PQQ addition had no effect and cells resumed their activity upon a short oxygen exposure. Since respiration of other substrates was not impaired, the route for glucose conversion may involve a special factor which has to be kept in its oxidised form.

CONCLUSIONS AND PROSPECTS

Until recently, s-GDH had been found only in *A. calcoaceticus* strain LMD 79.41, but it is now known to exist in other *Acinetobacter* strains (K. Matsushita, personal communication). Although the enzyme oxidises a broad range of aldose sugars *in vitro*, all the data suggest that it does not participate in sugar oxidation or in any other cellular process. Thus one is left with the conclusion that either the enzyme is a relic from the ancient past, for which natural selection had no tools to remove the genetic information, or that it has a still unknown function. The same difficulty is met when one tries to explain the widespread occurrence of m-GDH apo-enzyme in bacteria which do not synthesise free PQQ and which will most probably never see PQQ in their natural environment (Neijssel, 1987). Studies on glucose dehydrogenase remain relevant in order to answer these questions.

s-GDH is an attractive model enzyme to study the mechanism of quinoprotein dehydrogenases: the spectra of its redox forms are similar to those of PQQ, suggesting a rather simple binding to the protein chain; the oxidised form is stable and no activator is required for the substrate conversion step; the availability of stable apo-enzyme gives possibilities for studying the binding and catalytic mechanism of PQQ, labelled PQQ, and PQQ-analogues. Since the enzyme has a high turnover number and is specific for PQQ (only slight modifications of the molecule are allowed), a very sensitive and reliable assay for PQQ has been developed, based on apo-enzyme which is free of holo-enzyme, prepared by heterologous expression in *E. coli*.

m-GDH is responsible for the oxidation of sugars by whole cells. The hydrophobic enzyme has been only partially characterised. Its protein structure is quite different from that of s-GDH, but the global information available suggests mechanistic similarity between the two enzymes. Although m-GDH probably has a high redox potential, ubiquinone is the natural electron acceptor. Perhaps this discrepancy will be explained when more is known about the interplay between the Q-cycle of this organism and the mechanism of one-electron transfer steps between m-GDH and the respiratory chain.

Growth experiments and studies of transport and ATP formation convincingly demonstrate that the glucose oxidation step can provide useful energy to the organism. For some organisms containing m-GDH, but which are unable to metabolise gluconic acid, glucose oxidation via m-GDH provides an auxiliary energy source. Although these organisms form an ideal object for energetic studies, the results of growth experiments with mixed substrates (acetate plus glucose) are still confusing. Efforts to improve experimental arrangements might be rewarding as in principle, related to their simplicity, opportunities are provided to manipulate carbon flow, energy dissipation and redox balances. This could also have practical significance since *Gluconobacter oxydans*, the industrial organism used for gluconic acid production, is much more complicated as it has multiple catabolic routes for glucose degradation.

Finally, what is the significance of m-GDH for *A. calcoaceticus* in its natural environment? The affinity constant for the aldose sugars is rather high: 2 mM for glucose, 4 mM for D-xylose, and much higher for other sugars (Dokter et al., 1987). Cellular oxidation of glucose has a pH optimum of about 6, but as no lactonase is present, the benefit of the oxidation is

306

questionable at this pH. Moreover, the oxidation rates rapidly drop at lower oxygen tensions. Taking these restrictions together, one may wonder whether *A. calcoaceticus* in its natural environment will benefit from this enzyme for energetic purposes. The objections apply even more strongly in the case of *A. lwoffii*, because it not only requires the high concentrations of sugars, but also PQQ to functionalise its enzyme. Although one could be optimistic about the chance that these conditions are met, is m-GDH really designed for energy provision? For instance, the production of gluconolactone or gluconic acid *per se* may be beneficial as it creates an attractive micro-environment for *A. calcoaceticus*, but not for other organisms. Support for this idea can be derived from the situation which exists for those yeasts and fungi that contain glucose oxidase; this enzyme catalyses the same reaction (although additional H_2O_2 formation occurs, this becomes rapidly decomposed by the catalase), but without yielding useful energy. Perhaps the constitutive synthesis of the enzyme serves that purpose, while the gene(s) to make a complete route for glucose catabolism (gluconate kinase) remain silent until conditions are favourable for derepression by mutation. The final conclusion is that, although it is clear that m-GDH is able to function in an energy conserving process in *A. calcoaceticus*, its ecological role and real function are still far from understood.

REFERENCES

Anthony, C., and Zatman, L.J., 1967, The microbial oxidation of methanol. The prosthetic group of the alcohol dehydrogenase of *Pseudomonas* sp. M27, *Biochem. J.*, 104:960.

Asperger, O., Borneleit, P., and Kleber, H.-P., 1981, Untersuchungen zum Elektronentransport in *Acinetobacter calcoaceticus*, *Ab. Akad. Wissen. DDR.*, 3:259.

Bak, T.G., 1967, Studies on glucose dehydrogenase from *Aspergillus oryzae*. II Purification and physical and chemical properties, *Biochim. Biophys. Acta*, 139:277.

Barron, E.S.G., and Friedemann, T.E., 1941, Studies on biological oxidations. XIV. Oxidations by microorganisms which do not ferment glucose, *J. Biol. Chem.*, 137:593.

Baumann, P., Doudoroff, M., and Stanier, R.Y., 1968, A study of the *Moraxella* group. II Oxidative-negative species (genus Acinetobacter), *J. Bacteriol.*, 95:1520.

Beardmore-Gray, M., and Anthony, C., 1983, The absence of quinoprotein alcohol dehydrogenase in *Acinetobacter calcoaceticus*, *J. Gen. Microbiol.*, 129:2979.

Beardmore-Gray, M., and Anthony, C., 1986, The oxidation of glucose by *Acinetobacter calcoaceticus:* interaction of the quinoprotein glucose dehydrogenase with the electron transport chain, *J. Gen. Microbiol.*, 132:1257.

Bell, E.J., and Marus, A., 1966, Carbohydrate metabolism of *Mima polymorpha* I. Supplemental energy from glucose added to a growth medium, *J. Bacteriol.*, 91:2223.

Bouvet, P.J.M., and Bouvet,O.M.M., 1989, Glucose dehydrogenase activity in *Acinetobacter* species, *Res. Microbiol.*, 140:531.

Bouvet, P.J.M., and Grimont, P.A.D., 1986, Taxonomy of the genus *Acinetobacter* with the recognition of *Acinetobacter baumannii* sp. nov., *Acinetobacter haemolyticus* sp. nov., *Acinetobacter johnsonii* sp. nov., and *Acinetobacter junii* sp. nov., and emended escriptions of *Acinetobacter calcoaceticus* and *Acinetobacter lwoffii*, *Int. J. Syst. Bacteriol.*, 36:228.

Cleton-Jansen, A.M., Goosen, N., Odle, G., and van de Putte, P., 1988a, Nucleotide sequence of the gene coding for quinoprotein glucose dehydrogenase from *Acinetobacter calcoaceticus*, *Nucl. Acid Res.*, 16:6228.

Cleton-Jansen, A.M., Goosen, N., Wenzel, T.J., and van de Putte, P., 1988b, Cloning of the gene encoding quinoprotein glucose dehydrogenase from *Acinetobacter calcoaceticus*: evidence for the presence of a second enzyme, *J. Bacteriol.*, 170:2121.

Cleton-Jansen, A.M., Goosen, N., Vink, K., and van de Putte, P., 1989, Cloning, characterization and DNA sequencing of the gene encoding the M_r 50,000 quinoprotein glucose dehydrogenase from *Acinetobacter calcoaceticus*, *Mol. Gen. Genet.*, 217:430.

Conlin, M., Forrest, H.S., and Bruice, T.C., 1985, Replacement of methoxatin by 4,7-phenanthroline-5,6-dione and the inability of other phenanthroline quinones, as well as 7,9-didecarboxymethoxatin, to serve as cofactors for the methoxatin-requiring glucose dehydrogenase of *Acinetobacter calcoaceticus*, *Biochem. Biophys. Res. Comm.*, 131:564.

Cook, A.M., and Fewson, C.A., 1973, Role of carbohydrates in the metabolism of *Acinetobacter calcoaceticus*, *Biochim. Biophys. Acta*, 320:214.

D'Costa, E.J., Higgins, I.J., and Turner, A.P.F., 1986, Quinoprotein glucose dehydrogenase and its application in an amperometric glucose sensor, *Biosensors* 2:71.

De Bont, J.A.M., Dokter, P., van Schie, B.J., van Dijken, J.P., Frank, J., Duine, J.A., and Kuenen, J.G., 1984, Role of quinoprotein glucose dehydrogenase in gluconic acid production by *Acinetobacter calcoaceticus*, *Ant. v. Leeuw.*, 50:76.

Dijkshoorn, L., Tjernberg, I., Pot, B., Ursing, M.J.F., and Kersters, K., 1990, Numerical analysis of cell envelope profiles of *Acinetobacter* strains classified by DNA-DNA hybridization, *Syst. Appl. Microbiol.*, in press.

Dokter, P., van Kleef, M.A.G., Frank, J., and Duine, J.A., 1985, Production of quinoprotein D-glucose dehydrogenase in the culture medium of *Acinetobacter calcoaceticus*, *Enz. Microb. Tech.*, 7:613.

Dokter, P., Frank, J., and Duine, J.A., 1986, Purification and characterization of quinoprotein glucose dehydrogenase from *Acinetobacter calcoaceticus* LMD 79.41, *Biochem. J.*, 239:163.

Dokter, P., Pronk, J.T., van Schie, B.J., van Dijken, J.P., and Duine, J.A., 1987, The *in vivo* and *in vitro* substrate specificity of *Acinetobacter calcoaceticus* LMD 79.41, *FEMS Microbiol. Lett.*, 43:195.

Dokter, P., van Wielink, J.E., van Kleef, M.A.G., and Duine, J.A., 1988, Cytochrome-562 from *Acinetobacter calcoaceticus* LMD 79.41, *Biochem. J.*, 254:131.

Dokter, P., van Wielink, J.E., Geerlof, A., Oltman, F., Stouthamer, A.H., and Duine, J.A., 1990, Purification and partial characterization of the membrane-bound haem-containing proteins from *Acinetobacter calcoaceticus* LMD.41, *J. Gen. Microbiol.*, in press.

Du Preez, J.C., Toerien, D.F., and Lategan, P.M., 1981, Growth parameters of *Acinetobacter calcoaceticus* on acetate and ethanol, *Appl. Microbiol. Biotech.*, 13:45.

Duine, J.A., and Frank, J., 1981, Quinoprotein alcohol dehydrogenase from a non-methylotroph, *Acinetobacter calcoaceticus*, *J. Gen. Microbiol.*, 122:201.

Duine, J.A., Frank, J., and van Zeeland, J.K., 1979, Glucose dehydrogenase from *Acinetobacter calcoaceticus*: a quinoprotein, *FEBS Lett.*, 108:443.

Duine, J.A., Frank, J., and Verwiel, P.E.J., 1980, Structure and activity of the prosthetic group of methanol dehydrogenase, *Eur. J. Biochem.*, 108:187.

Duine, J.A., Frank, J., and van der Meer, R., 1982, Different forms of quinoprotein aldose-(glucose-) dehydrogenase in *Acinetobacter calcoaceticus*, *Arch. Microbiol.*, 131:27.

Duine, J.A., Frank, J., and Jongejan, J.A., 1983, Detection and determination of pyrroloquinoline quinone, the coenzyme of quinoproteins, *Anal. Biochem.*, 133:239.

Fewson, C.A., 1985, Growth yields and respiratory efficiency of *Acinetobacter calcoaceticus*, *J. Gen. Microbiol.*, 131:865.

Frank, J., van Krimpen, S.H., Verwiel, P.E.J., Jongejan, J.A., Mulder, A.C., and Duine, J.A., 1989, On the mechanism of inhibition of methanol dehydrogenase by cyclopropane-derived inhibitors, *Eur. J. Biochem.*, 184:187.

Friedrich, T., Strohdeicher, M., Hofhaus, G., Preis, D., Sahm, H., and Weiss, H., 1990, The same domain motif for ubiquinone reduction in mitochondrial or chloroplast NADH dehydrogenase and bacterial glucose dehydrogenase, *FEBS Lett.*, 265:37.

Geerlof, A., Dokter, P., van Wielink, J.E., and Duine, J.A., 1989, Haem-containing protein complexes of *Acinetobacter calcoaceticus*, in "PQQ and Quinoproteins," p.106, J.A. Jongejan, and J.A. Duine, eds., Kluwer Academic Publishers, Dordrecht.

Geiger, O., and Görisch, H., 1986, Crystalline quinoprotein glucose dehydrogenase from *Acinetobacter calcoaceticus*, *Biochemistry*, 25:6043.

Geiger, O., and Görisch, H., 1989, Reversible thermal inactivation of the quinoprotein glucose dehydrogenase from *Acinetobacter calcoaceticus*, *Biochem. J.*, 261:415.

Gommers, P.J.F., van Schie, B.J., van Dijken, J.P., and Kuenen, J.G., 1988, Biochemical limits to microbial growth yields: an analysis of mixed substrate utilization, *Biotech. Bioeng.*, 32:86.

Goosen, N., Horsman, H.P.A., Huinen, R.G.M., de Groot, A., and van de Putte, P., 1989, Genes involved in the biosynthesis of PQQ from *Acinetobacter calcoaceticus, in* "PQQ and Quinoproteins," p.169, J.A.Jongejan and J.A.Duine, eds., Kluwer Academic Publishers, Dordrecht.

Hauge, J.G., 1960a, Purification and properties of glucose dehydrogenase and cytochrome *b* from *Bacterium anitratum, Biochim. Biophys. Acta,* 45:250.

Hauge, J.G., 1960b, Kinetics and specificity of glucose dehydrogenase from *Bacterium anitratum, Biochim. Biophys. Acta,* 45:263.

Hauge, J.G., 1961, Glucose dehydrogenase in bacteria: a comparative study, *J. Bacteriol.,* 82:609.

Hauge, J.G., 1964, Glucose dehydrogenase of *Bacterium anitratum*: an enzyme with a novel prosthetic group, *J. Biol. Chem.,* 239:3630.

Hauge, J.G., 1966a, Glucose dehydrogenase-particulate. I *Pseudomonas* species and *Bacterium anitratum, Meth. Enzymol.,* 9:92.

Hauge, J.G., 1966b, Glucose dehydrogenase soluble. II *Bacterium anitratum, Meth. Enzymol.,* 9:107.

Hauge, J.G., and Hallberg, P.A., 1964, Solubilization and properties of the structurally-bound glucose dehydrogenase of *Bacterium anitratum, Biochim. Biophys. Acta,* 81:251.

Hauge, J.G., and Mürer, E.H., 1964, Studies on the nature of the prosthetic group of glucose dehydrogenase of *Bacterium anitratum, Biochim. Biophys. Acta,* 81:244.

Hauge, J.G., Foulds, G., and Bentley, R., 1968, Kinetic isotope effects in enzymic oxidations of D-glucose-1-H and D-glucose-1-^{2}H, *Biochim. Biophys. Acta,* 159:398.

Hayaishi, O., 1966, Lactose dehydrogenase, *Meth. Enzymol.,* 9:73.

Hommel, R., Götzrath, M., and Kleber, H-P., 1989, Enzyme production by growing cells of Acinetobacter in presence of sophoroselipid and Triton X-100, *Acta Biotech.,* 5:461.

Imanaga, Y., 1989, Investigations on the active site of glucose dehydrogenase from *Pseudomonas fluorescens, in* "PQQ and Quinoproteins," p.87, J.A. Jongejan and J.A. Duine, eds., Kluwer Academic Publishers, Dordrecht.

Imanaga, Y., Hirano-Sawatake, Y., Arita-Hashimoto, Y., Iton-Shibouta, Y., and Katon-Semba, R., 1979, On the cofactor of glucose dehydrogenase of *Pseudomonas fluorescens, Proc. Jap. Acad.,* 55B:264.

Juni, E., 1972, Interspecies transformation of *Acinetobacter*: genetic evidence for a ubiquitous genus, *J. Bacteriol.,* 122:917.

Juni, E., 1978, Genetics and physiology of *Acinetobacter, Ann. Rev. Microbiol.,* 32:349.

Kilty, C.G., Maruyama, K., and Forrest, H.S., 1982, Reconstitution of glucose dehydrogenase using synthetic methoxatin, *Arch. Biochem. Biophys.,* 218:623.

Kitagawa, K., Tateishi, A., Nakano, F., Matsumoto, T., Morohoshi, T., Tanino, T., and Usui, T., 1986a, Generation of energy coupled with membrane-bound glucose dehydrogenase in *Acinetobacter calcoaceticus*, *Agric. Biol. Chem.*, 50:1453.

Kitagawa, K., Tateishi, A., Nakano, F., Matsumoto, T., Morohoshi, T., Tanino, T., and Usui, T., 1986b, Sources of energy and energy coupling reactions of the active transport systems in *Acinetobacter calcoaceticus*, *Agric. Biol. Chem.*, 50:2939.

Kleber, H.P., Haferburg, D., Asperger, O., Schmidt, M., and Aurich, H., 1984, Aufnahme und Oxidation von monosacchariden bei *Acinetobacter calcoaceticus*, *Zeits. Allg. Mikrobiol.*, 24:691.

Lockwood, L.B., Tabenkin, B., and Ward, G.E., 1941, The production of gluconic acid and 2-ketogluconic acid from glucose by species of *Pseudomonas* and *Phytomonas*, *J. Bacteriol.*, 42:51.

Matsushita, K., Shinagawa, E., Inone, T., Adachi, O., and Ameyama, M., 1986, Immunological evidence for two types of PQQ-dependent D-glucose dehydrogenases in bacterial membranes and the location of the enzyme in *Escherichia coli*, *FEMS Microbiol. Lett.*, 37:141.

Matsushita, K., Shinagawa, E., Adachi, O., and Ameyama, M., 1988, Quinoprotein D-glucose dehydrogenase in *Acinetobacter calcoaceticus* LMD 79.41: the membrane-bound enzyme is distinct from the soluble enzyme, *FEMS Microbiol. Lett.*, 55:53.

Matsushita, K., Shinagawa, E., Adachi, O., and Ameyama, M., 1989, Quinoprotein D-glucose dehydrogenase of the *Acinetobacter calcoaceticus* respiratory chain: membrane-bound and soluble forms are different molecular species, *Biochemistry*, 28:6276.

Meyer, D.J., and Jones, C.W., 1973, Oxidative phosphorylation in bacteria which contain different cytochrome oxidases, *Eur. J. Biochem.*, 36:144.

Müller, R.H., and Babel, W., 1986, Glucose as an energy donor in acetate growing *Acinetobacter calcoaceticus*, *Arch. Microbiol.*, 144:62.

Neijssel, O.M., 1987, PQQ-linked enzymes in enteric bacteria, *Microbiol. Sci.*, 4:87.

Neijssel, O.M., Hommes, R.W.J., Postma, P.W., and Tempest, D.W., 1989, *in* "PQQ and Quinoproteins," p.57, J.A. Jongejan, and J.A. Duine, eds., Kluwer Academic Publishers, Dordrecht.

Niederpruem, D.J., and Doudoroff, M., 1965, Cofactor-dependent aldose dehydrogenase of *Rhodopseudomonas spheroides*, *J. Bacteriol.*, 89:697.

Salisbury, S.A., Forrest, H.S., Cruse, W.B.T., and Kennard, O., 1979, A novel coenzyme from bacterial primary alcohol dehydrogenase, *Nature*, 280:843.

Shinagawa, E., Matsushita, K., Nonobe, M., Adachi, O., Ameyama, M., Ohshiro, Y., Itoh, S., and Kitamura, Y., 1986, The 9-carboxylic acid group of pyrroloquinoline quinone, a novel prosthetic group, is essential in the formation of holo-enzyme of D-glucose dehydrogenase, *Biochem. Biophys. Res. Comm.*, 139:1279.

Strohdeicher, M., Bringer-Meyer, S., Neusz, B., van der Meer, R., Duine, J.A., and Sahm, H., 1989, Glucose dehydrogenase from *Zymomonas mobilis*: evidence for a quinoprotein, *in* "PQQ and Quinoproteins," p.103, J.A. Jongejan and J.A. Duine, eds., Kluwer Academic Publishers, Dordrecht.

van Kleef, M.A.G., and Duine, J.A., 1988, Bacterial NAD(P)-independent quinate dehydrogenase is a quinoprotein, *Arch. Microbiol.*, 150:32.

van Kleef, M.A.G., and Duine, J.A., 1989, Factors relevant in bacterial pyrroloquinoline quinone production, *Appl. Environ. Microbiol.*, 55:1209.

van Kleef, M.A.G., Dokter, P., Mulder, A.C., and Duine, J.A., 1987, Detection of the cofactor pyrroloquinoline quinone, *Anal. Biochem.*, 162:143.

van der Meer, R.A., Groen, B.W., van Kleef, M.A.G., Frank, J., Jongejan, J.A., and Duine, J.A., 1990, Isolation, preparation and assay of pyrroloquinoline quinone (PQQ), *Meth. Enzymol.* in press.

van Schie, B.J., van Dijken, J.P., and Kuenen, J.G., 1984, Non-coordinated synthesis of glucose dehydrogenase and its prosthetic group PQQ in *Acinetobacter* and *Pseudomonas* species, *FEMS Microbiol. Lett.*, 24:133.

van Schie, B.J., Hellingwerf, K.J., van Dijken, J.P., Elferink, M.G.L., van Dijl, J.M., Kuenen, J.G., and Konings, W.N., 1985, Energy transduction by electron transfer via a pyrroloquinoline quinone dependent glucose dehydrogenase in *Escherichia coli, Pseudomonas aeruginosa,* and *Acinetobacter calcoaceticus (var. lwoffi), J. Bacteriol.*, 163:493.

van Schie, B.J., de Mooy, O.H., Linton, J.D., van Dijken, J.P., and Kuenen, J.G., 1987a, PQQ-dependent production of gluconic acid by *Acinetobacter, Agrobacterium* and *Rhizobium* species, *J. Gen. Microbiol.*, 133:867.

van Schie, B.J., Pronk, J.T., Hellingwerf, K.J., van Dijken, J.P., and Kuenen, J.G., 1987b, Glucose-dehydrogenase-mediated solute transport and ATP synthesis in *Acinetobacter calcoaceticus, J. Gen. Microbiol.*, 133:3427.

van Schie, B.J., Rouwenhorst, R.J., de Bont, J.A.M., van Dijken, J.P., and Kuenen, J.G., 1987c, An *in vivo* analysis of the energetics of aldose oxidation by *Acinetobacter calcoaceticus, Appl. Microbiol. Biotech.*, 26:560.

van Schie, B.J., van Dijken, J.P., and Kuenen, J.G., 1988, Effects of growth rate and oxygen tension on glucose dehydrogenase activity in *Acinetobacter calcoaceticus, Ant. v. Leeuw.*, 55:53.

van Schie, B.J., Rouwenhorst, R.J., van Dijken, J.P., and Kuenen, J.G., 1989, Selection of glucose assimilating variants of *Acinetobacter calcoaceticus* LMD 79.41 in chemostat culture, *Ant. v. Leeuw.*, 55:39.

Vicente, M., and Canovas, J.L., 1973, Glucolysis in *Pseudomonas putida*: physiological role of alternative routes from the analysis of defective mutants, *J. Bacteriol.*, 116:908.

ACINETOBACTER - CITRIC ACID CYCLIST WITH

A DIFFERENCE

P.D.J. Weitzman

Faculty of Life Sciences
Cardiff Institute of Higher Education
Cardiff CF5 2YB
UK

INTRODUCTION

My own introduction to the genus *Acinetobacter* was the result of a purely chance encounter - indeed an experimental accident - for which I have reason to be most grateful. I had decided to examine enzyme regulatory mechanisms in the Krebs' citric acid cycle of bacteria. The ubiquitous nature of this pathway, its dual metabolic function (catabolic and anabolic) in the generation of both energy and biosynthetic intermediates, and its cyclic character combine to present an interesting challenge to the investigator of metabolic regulation. The present article summarises work carried out in my laboratory over a number of years. Fig. 1 shows the reaction scheme of the citric acid cycle.

CITRATE SYNTHASE

I started investigations on citrate synthase - the enzyme that condenses acetyl-CoA and oxaloacetate to produce citrate. This enzyme may be considered the "first" or "initiating" enzyme of the citric acid cycle since it catalyses the only step in which carbon atoms enter the cycle by combining with a cycle intermediate. In extracts of *Escherichia coli* I found that the enzyme was strongly inhibited specifically by NADH (Weitzman, 1966). As NADH may be thought of as the product of the cycle, this represented a case of product feedback inhibition in a cyclic metabolic pathway and clearly merited further investigation. The fact that NADH was without effect on citrate synthases from eukaryotic organisms suggested the possibility that the effect observed with the *E. coli* enzyme might be characteristic of other bacteria, perhaps even a general property of prokaryotes.

The Biology of Acinetobacter, Edited by K. J. Towner *et al.*
Plenum Press, New York, 1991

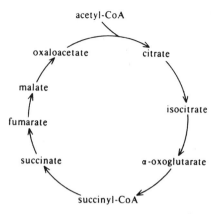

Fig. 1. The citric acid cycle

To examine the *E. coli* citrate synthase in more detail, I embarked on its large-scale purification and it was here that a lack of skill resulted in a fortunate accident. It was suggested that if I used a heavy inoculum of *E.coli*, sterilisation of the bottles containing 15 l of growth medium could probably be dispensed with, particularly as these were made up with freshly-distilled water. I innocently followed this advice, harvested the cells produced and devised an effective procedure to purify the citrate synthase. The pure enzyme retained its sensitivity to inhibition by NADH, though the kinetic characteristics of this inhibition differed from those observed with crude extracts of *E.coli*. Whereas the latter had shown a *hyperbolic* dependence of inhibition on NADH concentration, the pure enzyme displayed a *sigmoidal* dependence. Possible explanations in terms of restricted conformational flexibility in the presence of (association with?) other proteins in the crude extract offered themselves as speculations. However, another marked difference between the behaviour of the crude and pure enzymes was observed. With the pure enzyme, the NADH inhibition could be relieved by AMP (the "reactivation" again displaying a sigmoidal dependence on AMP concentration), whereas no such effect could be seen with the crude enzyme.

The reader will already have jumped to the correct conclusion, though for us, blinkered by our expectations, it took some time to realise and convince ourselves that the large-scale growth of *E.coli* had, in reality, resulted in the production of another bacterium which had dramatically outgrown the *E.coli* inoculum. One organism dominated the large-scale cultures and this was shown to be *Acinetobacter calcoaceticus*. Crude extracts of this bacterium showed the same citrate synthase properties as did the purified enzyme. It was the difference between these properties and those of *E.coli* citrate synthase that suggested it would be worthwhile to examine citrate synthases from a wide range of bacterial species to ascertain both the generality of NADH inhibition and any other regulatory diversity.

Our survey (Weitzman and Jones, 1968) revealed a striking pattern of enzyme behaviour with two clear-cut divisions. First, citrate synthases from Gram-negative bacteria are all inhibited by NADH, whereas those from Gram-positive bacteria are unaffected. Second, within the Gram-negative group, it is only in the case of the citrate synthases from strictly aerobic organisms that AMP reactivation of NADH inhibition occurs; in the facultative anaerobes (such as *E. coli*) AMP exerts no such effect. The subsequent examination of yet more bacterial citrate synthases, both by ourselves and by other investigators, has essentially confirmed these regulatory patterns.

The striking difference between citrate synthases from Gram-negative and Gram-positive bacteria with respect to NADH inhibition is paralleled by the difference between the molecular complexity and sizes of the two classes of enzyme (Weitzman and Dunmore, 1969; Weitzman, 1981). Gram-negative bacterial citrate synthase appears to be a hexameric molecule of identical or similar subunits (M_r 250,000-300,000), whereas the Gram-positive bacterial enzyme is a dimer (M_r approx. 95,000).

Why the phenomenon of NADH inhibition should occur only in the Gram-negative organisms remains a mystery. On the other hand, the incidence of AMP reactivation exclusively in strictly aerobic organisms may be rationalised in terms of the absolute dependence of such organisms on the citric acid cycle (and hence on citrate synthase) for energy production. It thus makes some sense that AMP, a signal of depleted metabolic energy, should serve as a positive effector (stimulator) of a key enzyme in the cycle, thereby increasing the activity of the cycle and overcoming the low-energy condition. In contrast, organisms such as *E. coli* are metabolically capable of producing energy without recourse to the citric acid cycle, i.e. by fermentation, and would thus not require AMP stimulation of citrate synthase.

Although the properties of the *Acinetobacter* citrate synthase are not so different from those of the enzyme produced by other Gram-negative aerobes, e.g. the pseudomonads, it was the difference between the properties of the enzyme from *Acinetobacter* and *E. coli* which initially prompted the whole survey of citrate synthases. *Acinetobacter* may therefore claim to have played a crucial part in the discovery of the remarkable patterns of regulatory diversity displayed by this enzyme. We shall later see again that the distinctive properties of *Acinetobacter* enzymes may prompt broader investigations and lead to other interesting and novel findings.

The *Acinetobacter* citrate synthase possesses several features that facilitate its further study: (i) it may be purified to homogeneity by conventional procedures and is a particularly stable enzyme; (ii) it may be stored cold for several months with little loss of activity or regulatory sensitivity; (iii) unlike the *E. coli* enzyme, the citrate synthase produced by *Acinetobacter* does not appear to have sensitive thiol groups whose chemical modification or oxidation would lead to inactivation or desensitisation and, moreover, its hexameric molecule shows no ready tendency to disaggregate into smaller units. These features enabled us to examine the enzyme by electron microscopy and to look for gross conformational changes accompanying interactions with the regulatory effectors (Rowe and Weitzman, 1969). Measurement of individual enzyme molecules in the electron micrographs clearly demonstrated that the enzyme molecule undergoes an asymmetrical "swelling" on interaction with NADH (i.e. on transition to the inactive state), and that this is completely reversed in the additional presence of AMP (return to the active state). This "visualisation" of an allosteric transition was a gratifying confirmation of other, indirect, probes of conformational alterations.

We have also examined the effects of "cross-linking" the *Acinetobacter* citrate synthase molecule with bifunctional reagents, thereby "freezing" the conformation and restricting its alteration (Mitchell and Weitzman, 1983; Lloyd and Weitzman, 1987). Interestingly, the cross-linked enzyme continues to exhibit sensitivity to both NADH and AMP, but the cooperative interactions between effector sites, which give rise to the sigmoidal responses to both NADH and AMP, are abolished. Cleavage of the cross-links restores these sigmoidal responses, presumably by releasing the conformational constraints. Such cross-linking experiments are a useful aid to further exploration of the molecular mechanisms of the regulatory inhibition and reactivation. The finer details of the molecular structure of *Acinetobacter* citrate synthase have been successfully investigated by Duckworth and co-workers (e.g. Morse and Duckworth, 1980; Donald and Duckworth, 1987) and should lead, in the future, to a fuller molecular understanding of the catalytic and regulatory functions of this enzyme.

ISOCITRATE DEHYDROGENASE

Isocitrate dehydrogenase catalyses the first oxidative decarboxylation reaction of the citric acid cycle, converting isocitrate into oxoglutarate. Although some bacteria contain NAD-linked isocitrate dehydrogenase, the majority appear to contain only the NADP-linked enzyme.

In the course of examining *Acinetobacter* isocitrate dehydrogenase we observed that, with extracts of acetate-grown cells, the enzyme activity appeared to increase during the course of the assay; in contrast, extracts of cells grown in nutrient broth gave normal linear rates in the assay. It was found that the progressive activation of isocitrate dehydrogenase resulted from the accumulation in the assay reaction mixture of glyoxylate, produced by the action of isocitrate lyase on the isocitrate substrate. The absence of isocitrate lyase in cells grown in nutrient broth accounted for the absence of the "activation" effect in extracts of such cells. In the course of further investigation of this glyoxylate effect it was discovered that *Acinetobacter* produces two distinct isoenzymes of NADP-linked isocitrate dehydrogenase - IDH-I, a "small" isoenzyme (M_r approx. 100,000) and IDH-II, a "large" isoenzyme (M_r approx. 300,000) (Self and Weitzman, 1970, 1972). It is the "large" isoenzyme which shows several unusual and interesting properties.

316

The activity of IDH-II is markedly stimulated by AMP and, to a lesser extent, ADP (Parker and Weitzman, 1970). As in the case of citrate synthase, stimulation of IDH-II by AMP may be rationalised in terms of "energy control" of the citric acid cycle. The low-energy signal AMP (or ADP) stimulates metabolite flux through the cycle, thereby overcoming the low-energy condition and constituting a homeostatic mechanism. Confirmatory studies on the nucleotide control of IDH-II have been reported by Kleber and Aurich (1976).

Some activation of *Acinetobacter* isocitrate dehydrogenase by glyoxylate was referred to above. Separation of the two isoenzymes revealed that it is only IDH-II that is stimulated by glyoxylate (Self et al., 1973). A similar stimulatory effect is also exerted by pyruvate, which bears a strong structural similarity to glyoxylate. It is tempting to suggest that the activation of IDH-II by glyoxylate constitutes a mechanism for controlling metabolic flux at the isocitrate branchpoint. Isocitrate may be metabolised either via isocitrate dehydrogenase and the citric acid cycle, or via isocitrate lyase and the glyoxylate cycle. The latter route results in the formation of glyoxylate which, by stimulating IDH-II, promotes metabolism via the dehydrogenase route and hence exerts a balance between the two options. It is also significant that we found the adaptation of *Acinetobacter* to growth on acetate to be accompanied by an increase in the level of the IDH-II isoenzyme (Reeves et al., 1983, 1986).

It is also noteworthy that the two types of activator - AMP and glyoxylate - exert their actions in an apparently independent manner; their joint presence results in a summation of their individual activating effects (Self et al., 1973).

As if these activations do not reflect sufficient functional complexity of IDH-II, this enzyme also displays the phenomenon of "hysteresis" (O'Neil and Weitzman, 1988). Partial purification of IDH-II in low-salt buffer by dye-ligand chromatography results in a preparation which takes around 20 min to display a constant steady rate in the normal spectrophotometric enzyme assay. Such a hysteretic effect suggests the slow transition of the enzyme from an inactive to an active form. Both AMP and glyoxylate abolish this hysteresis. It remains to be determined whether the hysteresis plays any part in the normal physiological functioning of *Acinetobacter* isocitrate dehydrogenase, but it certainly adds yet another facet to the regulatory complexity observed *in vitro*. It should also be emphasised that the unusual properties of isocitrate dehydrogenase described here confer a distinctiveness on *Acinetobacter* and are not general features of the bacterial enzyme.

OXOGLUTARATE DEHYDROGENASE

Following isocitrate dehydrogenase, the next step in the citric acid cycle is the oxidative decarboxylation of oxoglutarate to succinyl-CoA. This step is catalysed by the multi-enzyme complex oxoglutarate dehydrogenase.

Once again, the *Acinetobacter* enzyme displays distinctive features (Weitzman, 1972; Parker and Weitzman, 1973). The *Acinetobacter* oxoglutarate dehydrogenase is inhibited by NADH, and this inhibition is relieved by AMP (or ADP). Moreover, AMP and ADP stimulate enzyme activity even in the absence of NADH. In contrast, ATP produces no stimulation but, instead, a small degree of inhibition. The AMP and ADP stimulatory effects are associated with a marked reduction in the value of K_m for oxoglutarate. Again, this suggests the operation of a physiological regulatory mechanism whereby the low-energy signals AMP or ADP stimulate metabolite flux round the cycle.

PYRUVATE DEHYDROGENASE

Observation of the unusual regulatory properties of the oxoglutarate dehydrogenase from *Acinetobacter* prompted an examination of the analogous multi-enzyme pyruvate dehydrogenase complex, which catalyses the conversion of pyruvate to acetyl-CoA. Broadly similar observations were made (Jaskowska-Hodges, 1975). *Acinetobacter* pyruvate dehydrogenase exhibits sensitivity to inhibition by NADH and to activation by AMP. Although this enzyme is not, strictly speaking, a component of the citric acid cycle, it functions to produce acetyl-CoA, one of the substrates for citrate synthase. Stimulation by AMP may therefore be considered as enhancing flux through the cycle.

SUCCINATE THIOKINASE

Succinate thiokinase catalyses the breakdown of succinyl-CoA to succinate and coenzyme A, with the concomitant phosphorylation of a nucleoside diphosphate to the corresponding triphosphate. The nucleoside diphosphate may be either ADP or GDP. The generally held view (e.g. in textbooks of biochemistry) is that succinate thiokinase of animal origin functions with GDP, whereas bacterial and plant succinate thiokinases utilise ADP.

We found a surprisingly high K_m value for ADP (about 1 mM) with *Acinetobacter* succinate thiokinase compared with the value of the *E.coli* enzyme (about 0.01 mM). This prompted us to test the *Acinetobacter* enzyme with GDP. The K_m value for this nucleotide was around 0.02 mM, but the same V_{max} was displayed with GDP as with ADP. The apparent "preference" of *Acinetobacter* succinate thiokinase for GDP over ADP prompted a survey of a range of bacteria for the nucleotide utilisation of their succinate thiokinase (Weitzman and Jaskowska-Hodges, 1982). The bacteria could be classified into four groups. One group contained *Acinetobacter* and a few other species, all of which had a succinate thiokinase with a very high K_m value for ADP and a low K_m for GDP. The enzyme of a second group had a low K_m value for both ADP and GDP, while that of a third group showed a low K_m value for ADP and a higher K_m value for

GDP. The fourth group of bacteria had a succinate thiokinase which appeared to function only with ADP. It is noteworthy that the grouping together of bacteria on the basis of this nucleotide utilisation of their succinate thiokinases shows some taxonomic rationale.

These findings stem from the initial unexpected observations on *Acinetobacter* succinate thiokinase, in much the same way as the extensive studies on the diversity of citrate synthases have their origin in the initial observations of the distinctive properties of the *Acinetobacter* enzyme. In the case of succinate thiokinase it is interesting to note that the initial prompt by *Acinetobacter* has led to new findings about this enzyme in the animal kingdom. The diversity of *bacterial* succinate thiokinases tempted us to question the accepted view of the nucleotide utilisation of *animal* succinate thiokinase. Although, as mentioned above, succinate thiokinase from animal sources is usually claimed to be specific for GDP, we found a range of animal tissues to show succinate thiokinase activity with both GDP and ADP (Weitzman et al., 1986). Interestingly, the ratio of activity with GDP to that with ADP varies across tissues and this unexpected observation was resolved by the demonstration of the presence of two distinct succinate thiokinases in these tissues - one enzyme specific for GDP and the other for ADP. Further studies have shed light on the distinct physiological roles of the two succinate thiokinases and have suggested that it is the ADP-linked enzyme which functions in the energy-generating citric acid cycle, whereas the GDP-linked enzyme is associated with ketone body utilisation and haem biosynthesis (Jenkins and Weitzman, 1986, 1988; Jenkins et al., 1988).

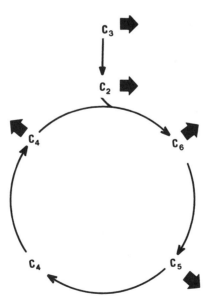

Fig. 2. Biosynthetic branch-points in the citric acid cycle.   Heavy arrows indicate biosynthetic reactions

MULTIPOINT CONTROL

The regulatory properties of the citric acid cycle enzymes of *Acinetobacter*, described above, constitute an unusual system of AMP control of the cycle in this organism, which I have termed "multipoint control" (Weitzman, 1981). Reference to the scheme of the citric acid cycle shown in Fig. 2 emphasises the multi-branchpoint nature of the cycle. Pyruvate (C_3), acetyl-CoA (C_2) oxaloacetate (C_4), isocitrate (C_6) and oxoglutarate (C_5) are branchpoint metabolites which may be metabolised either via the cycle or into other (biosynthetic) pathways. As a prime function of the cycle is in the provision of energy (via the formation of NADH and its oxidation coupled to phosphorylation and ATP production), it is logical that metabolites signalling a low-energy condition, e.g. AMP, should act to channel flux round the cycle rather than allow diversion into competing pathways. *Acinetobacter* displays a coordinated enzymic sensitivity to AMP which may permit multipoint control of its citric acid cycle. By acting as a positive effector of pyruvate dehydrogenase ($C_3 \longrightarrow C_2$), citrate synthase ($C_2 + C_4 \longrightarrow C_6$), isocitrate dehydrogenase ($C_6 \longrightarrow C_5$) and oxoglutarate dehydrogenase ($C_5 \longrightarrow C_4$), AMP can enhance metabolite flux through the cycle and thereby exert a sensitive control over its energy-generating role.

ENVOI

I hope that the brief description given here of the novel features of some citric acid cycle enzymes from *Acinetobacter* has justified the choice of title for this contribution. The study of the *Acinetobacter* enzymes has revealed some very unusual and interesting properties, served as a stimulus to the examination of other organisms, and led to the discovery of extensive enzyme diversity.

At the outset, I confessed to the accidental nature of my introduction to *Acinetobacter*. In conclusion, I must acknowledge that this accidental encounter has proved a most productive and valuable association.

REFERENCES

Donald, L.J., and Duckworth, H.W., 1987, Expression and base sequence of the citrate synthase gene of *Acinetobacter anitratum*, *Biochem. Cell. Biol.*, 65:930.

Jaskowska-Hodges, H., 1975, The control of citric acid cycle enzymes in *Acinetobacter*, *PhD thesis, Univ. Leicester*.

Jenkins, T.M., and Weitzman, P.D.J., 1986, Distinct physiological roles of animal succinate thiokinases: association of guanine nucleotide-linked succinate thiokinase with ketone body utilization, *FEBS Lett.*, 205:215.

Jenkins, T.M., and Weitzman, P.D.J., 1988, Physiological roles of animal succinate thiokinases: specific association of the guanine nucleotide-linked enzyme with haem biosynthesis, *FEBS Lett.*, 230:6.

Jenkins, T.M., Eisenthal, R., and Weitzman, P.D.J., 1988, Two distinct succinate thiokinases in both bloodstream and procyclic forms of *Trypanosoma brucei*, *Biochem. Biophys. Res. Comm.*, 151:257.

Kleber, H.-P., and Aurich, H., 1976, Control of NADP-specific isocitrate dehydrogenase from *Acinetobacter* by nucleotides, *FEBS Lett.*, 61:282.

Lloyd, A.J., and Weitzman, P.D.J., 1987, Modification of the regulatory properties of *Acinetobacter* citrate synthase by cross-linking, *Biochem. Soc. Trans.*, 15:840.

Mitchell, C.G., and Weitzman, P.D.J., 1983, Reversible effects of cross-linking on the regulatory cooperativity of *Acinetobacter* citrate synthase, *FEBS Lett.*, 151:260.

Morse, D., and Duckworth, H.W., 1980, A comparison of the citrate synthases of *Escherichia coli* and *Acinetobacter anitratum*, *Can. J. Biochem.*, 58:696.

O'Neil, S., and Weitzman, P.D.J., 1988, *Acinetobacter calcoaceticus* isocitrate dehydrogenase - fractionation, AMP stimulation and hysteretic behaviour, *Biochem. Soc. Trans.*, 16: 870.

Parker, M.G., and Weitzman, P.D.J., 1970, Regulation of NADP-linked isocitrate dehydrogenase activity in *Acinetobacter*, *FEBS Lett.*, 7:324.

Parker, M.G., and Weitzman, P.D.J., 1973, The purification and regulatory properties of α-oxoglutarate dehydrogenase from *Acinetobacter lwoffi*, *Biochem. J.*, 135:215.

Reeves, H.C., O'Neil, S., and Weitzman, P.D.J., 1983, Modulation of isocitrate dehydrogenase activity in *Acinetobacter calcoaceticus* by acetate, *FEBS Lett*, 163:265.

Reeves, H.C., O'Neil, S., and Weitzman, P.D.J., 1986, Changes in NADP-isocitrate dehydrogenase isoenzyme levels in *Acinetobacter calcoaceticus* in response to acetate, *FEMS Micobiol. Lett.*, 35:209.

Rowe, A.J., and Weitzman, P.D.J., 1969, Allosteric changes in citrate synthase observed by electron microscopy, *J. Mol. Biol.*, 43:345.

Self, C.H., and Weitzman, P.D.J., 1970, Separation of isoenzymes by zonal centrifugation, *Nature*, 225:644.

Self, C.H., and Weitzman, P.D.J., 1972, The isocitrate dehydrogenases of *Acinetobacter lwoffi*: separation and properties of two nicotinamide-adenine dinucleotide phosphate-linked isoenzymes, *Biochem. J.*, 130:211.

Self, C.H., Parker, M.G., and Weitzman, P.D.J., 1973, The isocitrate dehydrogenases of *Acinetobacter lwoffi*: studies on the regulation of a nicotinamide-adenine dinucleotide phosphate-linked isoenzyme, *Biochem. J.*, 132:215.

Weitzman, P.D.J., 1966, Regulation of citrate synthase activity in *Escherichia coli, Biochim. Biophys. Acta*, 128:213.

Weitzman, P.D.J., 1972, Regulation of α-ketoglutarate dehydrogenase activity in *Acinetobacter, FEBS Lett.*, 22:323.

Weitzman, P.D.J., 1981, Unity and diversity in some bacterial citric acid cycle enzymes, *Adv. Microb. Physiol.*, 22:185.

Weitzman, P.D.J., and Dunmore, P., 1969, Citrate synthases: allosteric regulation and molecular size, *Biochim. Biophys. Acta,* 171:198.

Weitzman, P.D.J., and Jaskowska-Hodges, H., 1982, Patterns of nucleotide utilisation in bacterial succinate thiokinases, *FEBS Lett.,* 143:237.

Weitzman, P.D.J., and Jones, D., 1968, Regulation of citrate synthase and microbial taxonomy, *Nature,* 219:270.

Weitzman, P.D.J., Jenkins, T., Else, A.J., and Holt, R.A., 1986, Occurrence of two distinct succinate thiokinases in animal tissues, *FEBS Lett.,* 199:57.

METABOLISM OF ALKANES BY *ACINETOBACTER*

O. Asperger and H.-P. Kleber

Sektion Biowissenschaften
Bereich Biochemie
Karl-Marx-Universität Leipzig
Talstrasse 33
7010 Leipzig
Germany

INTRODUCTION

The ability to oxidise alkanes is widely distributed among prokaryotic and eukaryotic microorganisms. *Acinetobacter* is among the bacterial genera most often found in petroleum-contaminated habitats and has been extensively used in studies of *n*-alkane oxidation (Einsele, 1983). Research with this bacterium has contributed to the exploration of many aspects of microbial *n*-alkane metabolism.

Interest in microbial alkane oxidation is multifarious. It encompasses various practical aspects such as ecology, pollution and waste disposal, problems of deterioration in the petroleum industry, geological prospecting, manufacture of single cell protein or diverse biosynthetic products (amino acids, organic acids, vitamins, lipids, emulsifiers etc.) on the base of petroleum as sole source of carbon and energy, as well as the biotransformation of hydrophobic chemicals to defined products by regio- and stereoselective oxidation. Besides these practical applications, there is a fundamental interest of chemists and biochemists in the biological mechanisms underlying the oxidative functionalisation of such extremely hydrophobic and poorly reactive compounds as *n*-alkanes.

Ever since the description of wax ester formation (Stewart and Kallio, 1959) by a Gram-negative, hexadecane-oxidising coccus (Stewart et al., 1959), originally called *Micrococcus cerificans* HO1-N (Finnerty et al., 1962), research with strains of *Acinetobacter* has contributed substantially to most reviews of microbial *n*-alkane metabolism (e.g. McKenna and Kallio, 1965; van der Linden and Thijsse, 1965; Davis, 1967; Klug and Markovetz,

The Biology of Acinetobacter, Edited by K. J. Towner *et al.*
Plenum Press, New York, 1991

1971; Gutnick and Rosenberg, 1977; Ratledge, 1978; Aurich, 1979; Einsele, 1983; Bühler and Schindler, 1984; Singer and Finnerty, 1984a,b). There are some review articles devoted mainly to aspects of *n*-alkane metabolism by *Acinetobacter* (Finnerty, 1977; Kleber et al., 1983, 1985; Finnerty and Singer, 1985; Rosenberg and Kaplan, 1987), but most consider only special topics or are restricted to individual strains. It is now more than ten years since this subject was covered in a general review on *Acinetobacter* (Juni, 1978). It is therefore the aim of this article, by summarising most of the available results related to *n*-alkane metabolism of *Acinetobacter*, to outline the importance of *Acinetobacter* research for the understanding of microbial hydrocarbon oxidation, and contribute to a comprehensive view of the biochemistry and physiology of this genus.

Generally, the growth of bacteria on oil is thought to involve three specific processes (Erickson et al., 1973): (i) interaction of the cells with, and uptake of, the insoluble substrate via adherence at the oil-water interface, pseudo-solubilisation of the carbon source (Goma et al., 1973), or uptake of hydrocarbon dissolved in the bulk aqueous phase; (ii) the introduction of molecular oxygen into the hydrocarbon and its subsequent biotransformation to a substrate which can then enter the normal metabolic pathway of the bacterial cell (Singer and Finnerty, 1984a); and (iii) the expression of specific regulatory signals governing the primary pathways (Grund et al., 1975) and metabolism as a whole. These aspects will be considered by presenting results on *n*-alkane assimilation and product formation, on alkane uptake and alkane-induced morphological changes, as well as on the enzymology of primary oxidation pathways and other metabolic sequences.

GENERAL FEATURES OF *n*-ALKANE OXIDATION

Aliphatic hydrocarbon assimilation seems to be a common property of most *Acinetobacter* strains. Non-utilisation appears to be the exception. Of 106 strains tested by Baumann et al. (1968) in their fundamental study on the taxonomy of *Acinetobacter*, only 19 strains failed to assimilate and utilise *n*-dodecane, *n*-tridecane and *n*-hexadecane. The alkane utilisers were distributed among all seven phenotypic groups postulated by Baumann et al. (1968) and, of ten representative strains, only one did not assimilate hydrocarbons. Similarly, the alkane-assimilating strains tested by Baumann et al. (1968) were also distributed among all six of the *Acinetobacter* species proposed by Bouvet and Grimont (1986) on the basis of 12 genospecies identified by DNA hybridisation experiments. This means that acinetobacters in general, regardless of whether they originate from soil and water, as it is the case for most strains of the species "*A. calcoaceticus sensu stricto*" (Bouvet and Grimont, 1986), or from animal or human habitats, possess the ability to assimilate aliphatic hydrocarbons.

This conclusion is in good agreement with the assumption that, generally, the genes encoding *n*-alkane assimilation are located on the chromosome. The assumption is supported by the fact that plasmids have not been detected in strains that have been well-studied with respect to *n*-alkane assimilation (*Acinetobacter* sp. HO1-N, Singer & Finnerty, 1984b; *A. calcoaceticus* 69-V, O. Asperger, unpublished results.) Data on the genes encoding alkane assimilation by *Acinetobacter* are so far only preliminary. Attempts by Singer and Finnerty (1984c) to obtain mutants by means of transpositional methods failed, but, as a result of chemically induced point mutations, 120 alkane-negative mutants (C_{12}-C_{16}) that were palmitate-positive and also grew well with alcohols and aldehydes (C_{12}-C_{16}) have been isolated from strains HO1-N and BD413 (Singer and Finnerty, 1984b). The mutants clustered into ten groups on the basis of reciprocal transformation crosses. From intergroup crosses, it was concluded that at least two distinct unlinked loci (*alk x* and *alk y*) were necessary for alkane oxidation by *Acinetobacter*.

In addition, certain strains of *Acinetobacter* may have special capabilities with respect to *n*-alkane metabolism that are encoded by plasmid DNA. Thus, the plasmid-carrying strain *A. calcoaceticus* RA57, which can grow on, and disperse, crude oil, loses this ability concomitantly with the curing of one of its four plasmids (pSR4), although its abilities to assimilate long-chain *n*-alkanes when applied in the vapour phase, to adhere to hydrocarbons, and to produce bioemulsifiers, were retained (Rusansky et al., 1987). Perhaps pSR4 encodes a factor which is tightly associated with the cell surface.

High specific growth rates during cultivation on liquid minimal media with *n*-hexadecane as sole carbon source, up to 1.3 /h in the case of *A. calcoaceticus* EB104, indicate that *Acinetobacter* is highly adapted to the oxidation of aliphatic hydrocarbons. However, *n*-alkanes are oxidised only by cells grown on them or cultivated in the presence of appropriate inducers (Aurich and Eitner, 1973; Asperger and Aurich, 1977; Asperger, 1990), thereby indicating the inducibility of the *n*-alkane assimilation system. Since several non-hydrocarbon substrates did not repress the induction of *n*-alkane oxidation capability (Haferburg et al., 1983; Asperger, 1990), it is clear that catabolite repression does not operate. Thus, induction must occur in direct response to the alkane molecule.

SUBSTRATE SPECIFICITY, PRODUCT PATTERNS AND DEGRADATION PATHWAY

Most *Acinetobacter* strains tested for the ability to utilise *n*-alkanes assimilate only compounds with more than 10 to 12 carbon atoms (Table 1). Hence, *Acinetobacter* strains can be used as model organisms for the assimilation of long-chain *n*-alkanes. Only a few strains, all of them capable of synthesising cytochrome P-450 (cf. Table 7), additionally utilise middle-chain *n*-alkanes (C_6-C_9) as a sole source of carbon, provided that the hydrocarbons are supplied in the vapour phase (Kleber et al., 1983; Eremina et al., 1987).

Table 1. Utilisation of *n*-alkanes as sole carbon source by *Acinetobacter* strains as a function of the chain-length

Strain	Alkane chain length									Ref.
	5	6	7	8	9	10	11	12-18	>18	
Acinetobacter sp. HO1-N	-	-	-	-	-	+	+	+	+	(1)
A.calcoaceticus 69-V	-	-	-	-	-	-	+	+	+	(2,3)
A.calcoaceticus EB106		-	-		-	-	+	+		(3)
A.calcoaceticus EB114		-	-		-	-	+	+		(3)
A.calcoaceticus CCM2355	-	-	-	-	-	+	+			(3)
A.calcoaceticus CCM5594	-	-	-	-	-	+	+			(3)
A.calcoaceticus CCM5595	-	-	-	-	-	+	+			(3)
A.lwoffii CCM2376	-	-	-	-	-	+	+			(3)
A.lwoffii CCM5572	-	-	-	-	-	+	+			(3)
A.lwoffii NCIB8250	-	-	-	+	+	+	+	+		(4)
A.lwoffii NCIB10533	-	-	-	-	-	-	-	+		(5)
A.lwoffii 0-16									+	(6)
A.calcoaceticus EB102			-	-	+	+	+	+		(3)
A.calcoaceticus EB104	-	+	+	+	+	+	+	+	+	(3)
A.calcoaceticus EB113		+		+	-	+	+			(3)

References: (1) Finnerty et al., 1962; (2) Kleber et al., 1973; (3) Kleber et al., 1983; (4) Fewson, 1967; (5) Grant, 1973; (6) Breuil et al., 1978.

Chain-length specificity during assimilation of *n*-alkanes as sole source of carbon does not reside in the substrate specificity of the primary *n*-alkane-oxidising enzymes. Thus, respiratory studies revealed that all *n*-alkanes (from *n*-hexane upwards) were oxidised not only by the middle-chain *n*-alkane-assimilating strain EB104, but also by the strain *A.calcoaceticus* 69-V (Fig.1; Asperger and Aurich, 1977). Inhibition or retardation of growth after addition of middle-chain *n*-alkanes to cultures of both strains growing on *n*-hexadecane, or on a complex medium, indicated that sensitivity of *Acinetobacter* to the toxic effect of middle-chain alkanes is probably the reason for the restricted chain-length growth specificity. This fact has also been reported for other *n*-alkane-assimilating microorganisms (Ratledge, 1978). Why certain strains of *Acinetobacter* (Table 1) are able to utilise middle-chain *n*-alkanes (Eremina et al., 1987) has still to be resolved, but there is a striking correlation between this property and the occurrence of an alkane-inducible cytochrome P-450 in these strains (cf. Table 7).

It is clear from the respiratory studies (Fig. 1) that the substrate spectrum of acinetobacters comprises all liquid and even solid *n*-alkanes. Furthermore, a broad variety of *n*-alkane derivatives are oxidised, as can be

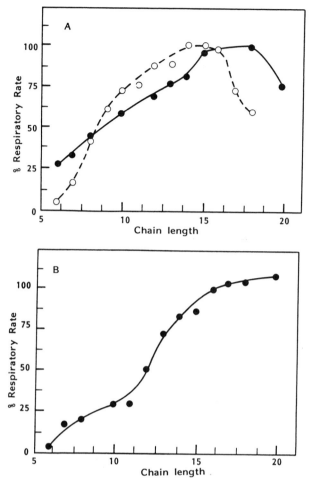

Fig. 1. Dependence of respiratory rates on the chain length of n-alkanes for resting cells of A. calcoaceticus EB104 (A) and A. calcoaceticus 69-V (B) after growth on n-hexadecane (●———●) or nonane (o-- --o). The relative values are given as percentages of the respiration rate with n-hexadecane as substrate, for which the rate was 700 nmol O_2/min/mg protein (data from Asperger and Aurich, 1977; Asperger, 1990).

deduced from growth and respiratory experiments (Table 2) or from studies of product formation (Tables 3 and 4). According to Amund and Higgins (1985), even-numbered phenylalkanes are totally degraded, via phenylacetic acid, by an oil-degrading strain of A. lwoffii, whereas odd-numbered phenylalkanes are transformed into dead-end products such as trans-cinnamic and 3-phenylpropionic acid.

Stoichiometric oxygen/alkane coefficients determined by respiratory measurements with the Clark electrode show that alkyl chains are rapidly oxidised beyond the carboxylic acid stage to intermediary metabolites and CO_2 (Asperger and Aurich, 1977). Although growth on, or respiration with

327

Table 2. Substrates of *n*-alkane monooxygenases of *Acinetobacter* as detected by growth experiments or by respiration studies with *n*-alkane-induced cells

Class of compound	Compounds positively tested with the strains		
	HO1-N[a]	69-V[b]	EB 104[c]
n-Alkanes	C_{10} - C_{20}	C_6 - C_{20}	C_6 - C_{20}
Secondary alkanols		Hexadecan-2-ol	Hexadecan-2-ol Octan-2-ol
Ketones			Heptadecan-9-one
		Hexadecan-3-one	Hexadecan-3-one
		Tridecan-7-one	Tridecan-2-one
Alkenes	1-Octadecene 1-Hexadecene	1-Hexadecene	1-Hexadecene 1-Tetradecene 1-Octene
Iso-alkanes	Methylpentadecanes (2-; 3-; 4-; 5-; 6-; 7- and 8-isomers)		
Phenyl-alkanes	1-Phenyldodecane 2-Phenyldodecane 1-Phenylundecane 1-Phenyldecane	1-Phenyltridecane	1-Phenyltridecane 1-Phenyldodecane
Halo-alkanes	1-Chlorohexadecane	1-Bromotetradecane	1-Bromotetradecane 1-Bromodecane
Ethers	Didecyl ether Dinonyl ether Dioctyl ether Diheptyl ether		Dioctyl ether Dihexyl ether
Thioether			Dioctylsulphide

References: [a]Finnerty et al., 1962; Hankin and Kollatukudy, 1968; Stewart et al., 1960; Kollatukudy and Hankin, 1968; Modrzakowski et al., 1977; Modrzakowski and Finnerty, 1980; [b]Asperger and Aurich, 1977; Aurich and Eitner, 1973; Kleber et al., 1973; [c]Asperger, 1990.

Table 3. Lipid accumulation by *Acinetobacter* sp. HO1-N in (A) cells and (B) culture broth during late exponential phase growth on *n*-hexadecane or *n*-hexadecanol. Cell accumulation is expressed in terms of μmol/mg dry weight, and broth accumulation in terms of μmol/l

Lipid	Nutrient broth		Hexadecane[a]		Hexadecanol[b]	
	A	B	A	B	A	B
Phospholipids	46	-	129	-	100	-
Glycerides	2.2	2.4	9.3	436	29	-
Wax esters	11	-	18	280	306	320
Free fatty acids	7.5	4	8.2	60	0.3	-

References: [a]Makula et al., 1975; [b]Singer et al., 1985.

di-terminally branched alkanes was not observed, the oxidation of the terminal methyl group in such compounds may be possible since the formation of pristanol and pristanediol from pristane by *Acinetobacter* has been reported in the patent literature (Hiratsuka et al., 1980).

A rather unusual method of degradation of the carbon skeleton has been proposed with respect to the oxidation of symmetrical dialkyl ethers by *Acinetobacter* sp. HO1-N (Modrzakowski et al., 1977; Modrzakowski and Finnerty, 1980). In cells grown on these substrates as sole source of carbon, alkoxycarboxylic acids having the same carbon number as the substrates were detected in the cellular phospholipid fraction, and alkoxyacetic acids were also found in the neutral lipid fraction as well as in the spent culture broth. Only acetate derivatives, but not propionate derivatives (even in the case of odd-numbered alkyl residues of the ethers) were described. On the basis that dicarboxylic acids with a carbon number of n-2 (n = carbon number of ether alkyl residue) were detected in the cells, the authors postulated a scission of the initial alkoxycarboxylic acid product in the position ß to the ether oxygen. More recently, Finnerty (1984) reported the formation of short-chain *n*-alkanes, together with a dicarboxylic acid, from long-chain *n*-alkanes or long-chain fatty acids, but did not give experimental details and yields of products.

In addition to the different types of bioemulsifiers produced in enlarged amounts during growth on *n*-alkanes (cf. Table 6), various cellular constitutents and products are accumulated in which metabolites of primary alkane oxidation are incorporated. These include wax esters, glycerides, phospholipids, and small amounts of free fatty acids (cf. Table 3; Vachon et al., 1982; Müller and Voigt, 1982). Phospholipids accumulate mainly inside the cell, reflecting the generation of intracytoplasmic membranes during growth on hydrophobic substrates (Kennedy and Finnerty, 1975; Müller et

Table 4. Components of wax esters produced by *Acinetobacter* sp. HO1-N after growth on different *n*-alkanes and alkane derivatives

Growth substrate	Alcohol	Fatty acid	Ref.
n-Tetradecane	Tetradecyl	Hexadecanoic	(1)
n-Hexadecane	Hexadecyl	Hexadecanoic	(1, 2)
n-Heptadecane	Heptadecyl	Heptadecanoic	(3)
		Pentadecanoic	
		Hexadecanoic	
n-Octadecane	Octadecyl	Octadecanoic	(1)
		Hexadecanoic	
1-Hexadecene	15-Hexadecenyl	Hexadecanoic	(4)
1-Octadecene	17-Octadecenyl	Hexadecanoic	(4)
		Heptadecanoic	
		Octadecanoic	
1-Chlorohexadecane	16-Chlorohexa-decyl	16-Chlorohexa-decanoic	(5)
		Hexadecanoic	

References: (1) Stewart and Kallio, 1959; (2) Stewart et al., 1959; (3) Stevenson et al., 1962; (4) Stewart et al., 1960; (5) Kollatukudy and Hankin, 1968.

al., 1983). Wax ester accumulation by *Acinetobacter* is not restricted to *n*-alkane assimilation. It is found in cells growing on *n*-hexadecanol (Table 3), and has also been reported for non-hydrocarbon-grown acinetobacters (Gallagher, 1971; Fixter and Fewson, 1974; Bryn et al., 1977; Neidleman and Geigert, 1983; Fixter et al., 1986). All 19 strains of *Acinetobacter* tested by Fixter et al. (1986) contained wax esters and 16 of these strains had increased wax contents when harvested from stationary phase of N-limited batch cultures. It appears that wax esters are widespread energy storage components in the genus *Acinetobacter* and their mobilisation under conditions of C-starvation has been shown (Fixter et al. 1986).

The chemical structures of the fatty alcohol moiety and, to a certain degree, those of the fatty acids in the waxes of *n*-alkane-grown cells correspond to those of the hydrocarbon substrates (Table 4). Other cellular lipids are characterised by a significant increase in the percentage of fatty acids having the same carbon number as the alkane that has been used as carbon source (Table 5). This shows that fatty acids originating from the primary oxidation of the alkanes are directly incorporated into the cellular lipids. Furthermore, a terminal oxidation pathway of the *n*-alkanes can be deduced from these results. Additional support for this suggestion is given by the failure of diterminal substituted alkane derivatives, such as

Table 5. Percentage fatty acid composition of *Acinetobacter* sp. HO1-N after growth on different *n*-alkanes (data from Makula and Finnerty, 1968)

Fatty acid	Nutrient broth	Acetate	*n*-Alkanes C13	C14	C15	C16	C17
C10	0.3	0.1	1.6	1.7	1.1	0.8	-
C11	-	-	1.7	-	1.2	-	2.0
C12	3.2	5.0	7.2	-	1.3	7.8	1.4
C13	-	-	7.7	-	1.1	-	1.8
C14	0.8	0.5	2.2	28.4	0.2	2.0	0.4
C15[a]	-	-	1.2	-	68.8	-	10.2
C16[a]	36.6	41.1	28.8	27.8	4.5	72.9	4.4
C17[a]	-	-	15.4	-	7.0	-	0.9
C18[a]	48.8	53.3	34.0	34.8	14.9	9.1	9.1

[a] The sum of saturated and unsaturated fatty acids.

hexadecane-1,15-dion or 1,12-dibromododecane (Asperger and Aurich, 1977; Asperger, 1990), to support respiration by alkane-grown cells, whilst the corresponding monoterminal substituted derivatives are oxidised (cf. Table 2). Still more stringent evidence was provided by the identification of a fatty acid of the same substrate chain-length as the main product of alkane oxidation catalysed by cell-free extracts of *A. calcoaceticus* strain 69-V (Asperger et al., 1978a).

ALKANE UPTAKE AND MORPHOLOGICAL IMPLICATIONS OF ALKANE ASSIMILATION

A serious problem in the biological oxidation of hydrocarbon compounds arises from their extremely low solubility in water. Therefore, in addition to enzymes for the initial oxidative attack on the hydrocarbon skeleton, alkane-oxidising microorganisms also require appropriate mechanisms to facilitate translocation of the alkane molecule from the bulk hydrocarbon phase to the active site of the enzyme. With respect to the intact cell, the interaction with alkanes seems to be facilitated by *Acinetobacter* in two ways: emulsification and adherence.

Surface-active compounds of low molecular mass have, to date, not been described for *Acinetobacter*. The bioemulsifiers produced by *Acinetobacter* are either high molecular weight soluble compounds such as "Emulsan" (Table 6; Rosenberg et al., 1979; Zuckerberg et al., 1979; Rosenberg, 1986) or they are particulate and form vesicles constituted out of lipopolysaccharides, phospholipids and proteins (Table 6; Käppeli and Finnerty, 1979; Claus et al., 1984; Borneleit et al., 1988). While the former

Table 6. Formation of surface-active compounds by *Acinetobacter*

Strain	Product designation	Chemical composition	Ref.
RAG-1/	Emulsan	Polysaccharide-protein	(1)
BD413		complex (covalently bound	(2)
		fatty acids)	(3)
			(4)
2CA2	Emulsifier 2CA2	Protein-lipid-carbohydrate complex	(5)
HO1-N	Extracellular vesicles	Lipopolysaccharide-phospholipid-protein vesicles	(6)
	Soluble Emulsifier	Lipoprotein	(7)
69-V	Extracellular vesicles	Lipopolysaccharide-phospholipid-protein vesicles	(8)

References: (1) Rosenberg et al., 1979; (2) Zuckerberg et al., 1979; (3) Kaplan and Rosenberg, 1982; (4) Rosenberg and Kaplan, 1987; (5) Neufeld and Zajic, 1984; (6) Käppeli and Finnerty, 1979; (7) Käppeli and Finnerty, 1980; (8) Claus et al., 1984.

seem to represent partially degraded capsular constituents (Rosenberg and Kaplan, 1987), the latter appear to derive from the outer membrane (Käppeli and Finnerty, 1979; Borneleit et al., 1988). They all affect *n*-alkane accommodation, mainly by the stabilisation of emulsions (Rosenberg and Kaplan, 1987).

However, since the enhancement of alkane uptake rates in the presence of extracellular vesicles (Claus et al., 1984) did not exceed the rates of alkane respiration measured with ultrasonically dispersed alkanes (Asperger and Aurich, 1977; Haferburg et al., 1983), it appears unlikely that the extracellular surface-active agents are directly involved either in the mechanism of alkane transport across the cellular membranes or in the interaction of the alkanes with the alkane-oxidising enzymes. In agreement with this, vesicle phospholipids were found not to be imported into the cell of *Acinetobacter* sp. HO1-N (Finnerty and Singer, 1985).

The other mechanism - adherence to alkane droplets, as demonstrated by ultramicrographs for various strains of *Acinetobacter* (Kennedy et al., 1975; Rosenberg and Kaplan, 1987) - seems to be achieved by thin fimbriae. These have been observed for *A. calcoaceticus* RAG-1 (Rosenberg et al., 1982) and reported for another alkane-grown strain formerly designated as *Moraxella glucidolytica* (Horisberger, 1977). Furthermore, Neufeld and Zajic (1984) demonstrated the reduction of surface tension by hexadecane-grown cells themselves.

One of the exciting discoveries arising from studies of *n*-alkane assimilation by *Acinetobacter* has been the detection of intracellular membrane-surrounded alkane inclusions and intracytoplasmic membranes in alkane-grown *Acinetobacter* sp. HO1-N (Kennedy et al., 1975; Kennedy and Finnerty, 1975; Scott and Finnerty, 1976a,b) and *A. calcoaceticus* 69-V (Müller et al., 1983). Inclusion membranes of strain HO1-N have been described as monolayers (Scott and Finnerty, 1976b), whereas bilayers have been claimed for strain 69-V (Müller et al., 1983). These intracellular membranes have been discussed as possible sites for the location of the initial enzymes of alkane oxidation, but firm evidence in support of this view has still to be provided.

ENZYMOLOGY RELATED TO ALKANE ASSIMILATION AND ITS REGULATION

In most organisms, the initial metabolism of *n*-alkanes takes the form of monoterminal oxidation:

$$R\text{-}CH_3 \longrightarrow R\text{-}CH_2OH \longrightarrow R\text{-}CHO \longrightarrow R\text{-}COOH$$

A similar pathway of *n*-alkane oxidation by *Acinetobacter* can be deduced from studies with whole cells and cell-free extracts as described above. Enzymes required for such a pathway are: alkane monooxygenase, alcohol dehydrogenase or oxidase, and aldehyde dehydrogenase or oxidase. Furthermore, the regulation of enzymes controlling intermediary metabolism and lipid metabolism may be substantially affected in *n*-alkane-assimilating cells.

Initial Step of n-Alkane Oxidation

Evidence in support of the oxygenative nature of the initial enzyme in alkane oxidation by *Acinetobacter* sp. HO1-N was first provided by Stewart et al. (1959), who demonstrated the incorporation of $^{18}O_2$ into the alkane-derived constituents of wax esters. Based on the fact that *n*-alkyl hydroperoxides are rapidly oxidised by *Acinetobacter* sp. HO1-N (Stewart et al., 1959), a pathway has been suggested that involves *n*-alkyl hydroperoxide as the initial product of alkane oxidation and, consequently, requires a dioxygenase as the first enzyme, followed by an alkyl hydroperoxide reductase (Finnerty, 1977; Singer and Finnerty, 1984a). However, this hypothesis lacks enzymological proof, since alkane-oxidising activity has not, to date, been demonstrated in cell-free extracts of this strain.

Cell-free *n*-alkane oxidation - generally until now demonstrated only for a few alkane-assimilating bacteria that had mostly been grown on middle-chain *n*-alkanes and tested with such substrates - was achieved in our laboratory with strain *A. calcoaceticus* 69-V (Aurich et al., 1977; Asperger et al., 1978a) and strain EB104 (Asperger et al., 1985a). In both cases,

Table 7. Cytochrome composition of various strains of *Acinetobacter* grown on *n*-hexadecane

Strain	Cytochromes						Ref.
	a	b	c	d	o	P-450	
69-V	-	+	-	-	+	-	(1)
CCM2355	-	+	-	-	+	-	(2)
CCM5594	-	+	-	-	+	-	(2)
CCM5595	-	+	-	-	+	-	(2)
CCM5572	-	+	-	-	+	-	(2)
CCM2376	-	+	-	-	+	-	(2)
EB114	-	+	-	-	+	-	(2)
EB102	-	+	-	-	+	+	(2)
EB103	-	+	-	-	+	+	(2)
EB104	-	+	-	-	+	+	(2)
EB10C	-	+	-	-	+	+	(2)
EB10D	-	+	-	-	+	+	(2)
EB113	-	+	-	-	+	+	(2)
A6E						+	(3)
ATCC33305						-	(3)
AH60						-	(3)
DON2						-	(3)
HO1-N	-	+	-	+	+	-	(4)

References: (1) Asperger et al., 1978b; (2) Asperger et al., 1981b; (3) Wyndham, 1987; (4) Ensley and Finnerty, 1980.

alcohols and fatty acids were identified as the main products of alkane oxidation, indicating the functioning of alkane monooxygenases, alcohol dehydrogenases and aldehyde dehydrogenases in cell-free extracts. Alkane oxidation by cell-free extracts of strain 69-V and EB104 was oxygen- and NADH-dependent, providing evidence for a monooxygenase as the initial enzyme of alkane oxidation.

Troublesome assay procedures and enormous loss of activity after cell disruption render the purification of the oxygenases based on activity measurements highly difficult. This is reflected by the fact that to date, apart from methane monooxygenases (Dalton, 1980), the rubredoxin-dependent system of the middle-chain *n*-alkane-utilising bacterium *Pseudomonas oleovorans* (Coon et al., 1973) represents the only alkane monooxygenase isolated from bacteria other than *Acinetobacter*. For *Acinetobacter*, a partially successful approach to the characterisation of the alkane monooxygenases was to search for, and isolate, electron transport proteins that were potential components of monooxygenases. Cytochrome P-450 - the alkane monooxygenase of yeasts - is absent from most of the strains of *Acinetobacter* that have been traditionally investigated, but occurs in certain strains (Table 7; Asperger et al., 1978b, 1981b). It is induced in strain EB104 by a broad variety of aliphatic and aromatic hydrocarbons (Asperger et al., 1984, 1985b) and induction can be enhanced by oxygen limitation (Asperger et al., 1986).

Cytochrome P-450 was isolated from *A. calcoaceticus* EB104 - either grown on *n*-hexadecane (Müller et al., 1989) or in nutrient broth with the addition of *n*-hexane as inducer (Asperger, 1990) - as an electrophoretically homogeneous protein with a M_r of 52,000 and spectral properties typical of other cytochromes P-450 (Müller et al., 1989). Purified cytochrome P-450 is enzymically reduced by NADH in the presence of a ferredoxin and a ferredoxin reductase, both isolated from the same bacterium (Asperger et al., 1983, 1985a). The ferredoxin has a M_r of about 9,000 and absorption maxima of the oxidised form at 415 and 460 nm, characteristic for a (2Fe-2S)-ferredoxin. Identification of the cytochrome P-450 system (Fig.2) as an alkane monooxygenase was based on the following evidence: (a) induction of cytochrome P-450 by the unoxidised *n*-alkane, but not by an intermediate oxidation product, as shown by regulatory studies with a broad variety of different inducers (Kleber et al., 1985); (b) inhibition of cell-free alkane-oxidising activity by CO, typical of cytochrome P-450-dependent reactions (Asperger et al., 1985a); (c) recovery of cytochrome P-450, ferredoxin and ferredoxin reductase as well as alkane-oxidising activity from the cytosolic protein fraction (Asperger et al., 1985a); (d) concomitant induction of cytochrome P-450 and alkane monooxygenase activity by the non-aliphatic hydrocarbon biphenyl (Asperger et al., 1985a); (e) reconstitution of alkane monooxygenase activity by the isolated proteins cytochrome P-450, ferredoxin and ferredoxin reductase (Asperger et al., 1985a).

Hitherto, the detection of alkane-inducible cytochrome P-450 in partially purified alkane monooxygenase preparations of *Rhodococcus rhodochrous* ATCC 19067 (formerly designated as *Corynebacterium* sp. 7E1C) (Cardini and Jurtshuk, 1970) had been the only indication for the existence of cytochrome P-450-dependent alkane monooxygenases in bacteria. The isolation of such a system from *Acinetobacter* provides more convincing evidence and it may be supposed that still more cytochrome P-450-dependent alkane monooxygenases will be found in bacteria.

There are, however, other *n*-alkane-assimilating bacteria in which cytochromes P-450 have definitely not been found, including several strains of *Acinetobacter* (Table 7). The terminal oxidase of the alkane monooxygenase of these cytochrome P-450-negative *Acinetobacter* strains remains unknown, but a rubredoxin (Aurich et al., 1976) and NADH-dependent rubredoxin reductase (Claus et al., 1978, 1979) have been isolated from alkane-grown cells of *A. calcoaceticus* 69-V. The rubredoxin exhibited an optical absorption spectrum typical of this class of iron-sulphur proteins with maxima at 380 and 490 nm in the oxidised state, a M_r of 6,000 and an iron content of 1 mol/mol (Aurich et al., 1976). Since the rubredoxin was induced by growth on *n*-alkanes, as determined by means of a radioimmunoassay (Claus et al., 1980a,b), it can be supposed that the alkane monooxygenase of *A.calcoaceticus* 69-V is a rubredoxin-dependent enzyme comparable to that of *P.oleovorans* (Fig.2; Coon et al., 1973).

While alkanes are oxidised only by alkane-grown cells or by induced cells, fatty alcohols, fatty aldehydes and fatty acids are also respired by cells grown on non-hydrocarbons (Aurich and Eitner, 1973; Asperger and Aurich, 1977). Hence, the alkane monooxygenases of *Acinetobacter* are inducible enzymes, which is confirmed by the inducibility of cytochrome P-450 (Asperger et al., 1984) and rubredoxin (Claus et al., 1980b), whereas constitutive alcohol and aldehyde dehydrogenases, as well as a constitutive ß-oxidation pathway, probably also exist. So far as the cellular location of the alkane monooxygenases of *Acinetobacter* is concerned, there is no convincing evidence for them being integral proteins of the cytoplasmic membranes. Activity of the enzymes from strains 69-V (Asperger et al., 1978a) and EB104 (Asperger, 1990) is stimulated by particulate fractions, and also cytochrome P-450 can, in part, be recovered from particulate fractions (Müller et al., 1989). However, it could be unequivocally demonstrated that cytochrome P-450 is not an integral protein of the cytoplasmic membrane (Müller et al., 1989) and that, under conditions of low intracellular alkane concentration, it is localised in the cytosolic protein

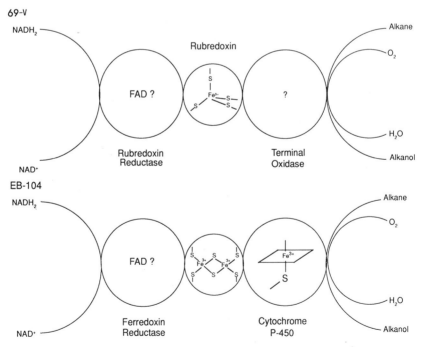

Fig. 2. Schematic representation of the proposed alkane monooxygenases in *Acinetobacter* strains 69-V and EB104.

fraction (Asperger et al., 1983). It may therefore be proposed that the enhancement of activity by particulate cell fractions occurs because alkanes can be accommodated by the help of membranous phospholipid-rich structures. Reconstitution experiments with cell-free components of the alkane monooxygenase system should help to test the validity of this hypothesis. With the cytochrome P-450 system from *Acinetobacter*, a cell-free alkane monooxygenase from long-chain *n*-alkane-assimilating bacteria is available for the first time. Because of its soluble nature, it should be a suitable model system for exploring the mechanisms that achieve high intracellular conversion rates of such extremely hydrophobic substrates as *n*-alkanes.

The substrate specificity of the alkane monooxygenases of *A.calcoaceticus* 69-V and EB104 is reflected by the substrate pattern listed in Table 2, considering that these substrates were oxidised only by *n*-alkane-induced cells and that alcohol and aldehyde dehydrogenases are present constitutively.

Alcohol and Aldehyde Dehydrogenases

Some data on alcohol and aldehyde dehydrogenases of strains 69-V, EB104 and HO1-N are summarised in Tables 8 and 9. Obviously there are multiple enzymes, some of which are induced by growth on *n*-alkanes. Of the alcohol dehydrogenases, only the constitutive, cytosolic $NADP^+$-dependent enzyme from *A.calcoaceticus* 69-V has been purified (Tauchert et al., 1976). Firm conclusions about the importance of inducible alcohol dehydrogenases for the degradation of *n*-alkanes by *Acinetobacter* therefore cannot yet be drawn. Induction of a membrane-bound, $NADP^+$-dependent aldehyde dehydrogenase is well established (Aurich and Eitner, 1977; Aurich et al., 1983; Singer and Finnerty, 1985a). This enzyme has been purified (Sorger and Aurich, 1978), reconstituted into liposomes (Aurich et al., 1985) and kinetically characterised (Aurich et al., 1987).

There is controversy with respect to the location of the aldehyde dehydrogenase of *A.calcoaceticus* 69-V. Sorger et al. (1986) interpreted the ultracytochemically-demonstrated aldehyde-dependent accumulation of NADPH around the inclusions of hexadecane-grown cells as the location in an inclusion-surrounding membrane. However, Fischer et al. (1984) postulated cytoplasmic membrane-bound enzyme in the same strain.

Enzymes of Intermediary Metabolism

Fatty acids produced by the initial steps of alkane oxidation are either directly incorporated into cellular lipids or waxes, or they are rapidly further degraded by ß-oxidation. Nothing is known about the specificity or regulation of the enzymes involved in these processes. There has been intensive study of the tricarboxylic acid cycle enzymes and the anaplerotic sequences involved in alkane assimilation by *A.calcoaceticus* 69-V (Kleber and Aurich, 1973, 1974; Kleber, 1978). Isocitrate lyase and malate

Table 8. Alcohol dehydrogenases of alkane-cultivated strains of *Acinetobacter*

Strain	Enzyme	Cosubstrate	Localisation	Induction	Maximum Sp.Act.[a]	Ref.
69-V	I	$NADP^+$	Cytosol	Constitutive	$7(C_8)$[b]	(1)
	II	DCPIP	Membranes	?	$4(C_8)$	(2)
EB104	I	$NADP^+$	Cytosol	Constitutive	$24(C_8)$	(3)
	II	DCPIP	Membranes	Constitutive	$3(C_8)$	(3)
HO1-N	ADH-A[c]	NAD^+	Cytosol	Ethanol	$92.6(C_2)$	(4)
	ADH-B	$NADP^+$	Cytosol	Constitutive	$11(C_2)$	(4)
	HDH	NAD^+	Cytosol	Hexadecanol	$5.9(C_{16})$	(4)
	HDH		Membranes	Hexadecane		(4)

[a] nmol/min/mg protein; [b] Chain-length of the substrate; [c] ADH: Alcohol dehydrogenase; HDH: Hexadecanol dehydrogenase.

References: (1) Tauchert et al., 1976; (2) Tauchert et al., 1975; (3) Jirausch et al., 1986; (4) Singer and Finnerty, 1985b.

Table 9. Aldehyde dehydrogenases of alkane-cultivated strains of *Acinetobacter*

Strain	Enzyme	Co-substrate	Localisation	Induction	Maximum Sp.Act.[a]	Ref.
69-V		$NADP^+$	Membranes	Alkane	$600(C_8)$[b]	(1)
				Alkanol		(2)
EB104		$NADP^+$	Membranes	?	?	(3)
H01-N	FALDH-a[c]	$NADP^+$	Membranes	Hexadecane	$147(C_{10})$	(4)
				Hexadecanol		(4)
	FALDH-b[c]	$NADP^+$	Membranes	Constitutive	?	(4)
	ALDH[c]	NAD^+	Cytosol	Ethanol	$165(C_2)$	(4)

[a] nmol/min/mg protein; [b] Chain-length of the substrate; [c] FALDH: Fatty aldehyde dehydrogenase; ALDH: aldehyde dehydrogenase.

References: (1) Aurich and Eitner, 1977; (2) Aurich et al., 1983; (3) Klossek et al., 1985; (4) Singer and Finnerty, 1985a.

dehydrogenase, key enzymes for the metabolism of excess C_2-substrates, are repressed in the presence of carbohydrates, phosphoenolpyruvate, or during growth on C_4-dicarboxylic acids. However, they are induced by n-alkanes, fatty acids, and acetate, whereas malic enzyme is repressed under these conditions (Kleber, 1973, 1978), avoiding the loss of C_4-units of the citric acid cycle. In addition, fatty acids and acetyl-CoA inhibit the latter enzyme (Kleber, 1975). Thus, this *Acinetobacter* is well-equipped for the metabolism of n-alkanes via oxidation to fatty acids followed by ß-oxidation. Induced levels of isocitrate lyase and malate dehydrogenase, as well as the repression of malic enzyme in n-alkane-grown cells, contribute to such a catabolic pathway.

Enzymes of Lipid Metabolism

Alkane metabolism leads to significant changes in cellular lipid composition and to the formation of several lipid products such as waxes, glycerides or glycolipids. However, little is known about how enzymes involved in lipid biosynthesis or catabolism are regulated during n-alkane assimilation. Scott et al. (1976) found that cytoplasmic membranes of *Acinetobacter* sp.HO1-N contain an unusually high specific activity of phosphatidic acid cytidyl transferase (McCaman and Finnerty, 1968) which is increased threefold in n-alkane-grown cells (Finnerty, 1980). Elevated activities of α-glycerol-CMP phosphatidyl transferase, L-serine-CMP phosphatidyl transferase and phosphatidyl serine decarboxylase in the cytoplasmic membrane have also been measured, whereas phosphatidic acid phosphatase was decreased during growth on n-hexadecane compared with nutrient broth-grown cells (Finnerty and Singer, 1985). Since neither phospholipid content nor the composition of cytoplasmic membranes depended on the carbon source (Makula and Finnerty, 1970,1971; Makula et al., 1975), it is possible that the elevated enzyme activities may reflect the formation of intracellular membranes during growth on n-alkanes.

A phospholipase A, detected in the outer membrane of *Acinetobacter* sp. HO1-N (Torregrossa et al., 1977b) by studying the occurrence of lysocardiolipin (Torregrossa et al., 1977a), was two to three times more active in hexadecane-grown than in nutrient broth-grown cells (Torregrossa et al., 1977b). In addition to cardiolipin, it also hydrolyses phosphatidylglycerol and phosphatidylethanolamine (Torregrossa et al., 1978). An esterase with high activity against palmitoyl-CoA has been described as a marker enzyme of the outer membrane of *A.calcoaceticus* 69-V (Claus et al., 1985; Fischer, 1986); however, the ability of this enzyme to serve as a phospholipase and its regulation by the carbon source have not been studied. It would be of interest to discover whether these outer membrane hydrolase activities are functionally linked with the formation of vesicles and/or capsular materials and therefore with the capacity of acinetobacters to deal with n-alkanes. *A.calcoaceticus* RAG-1 produces an extracellular esterase for which the bioemulsifier emulsan is a substrate (Shabtai and Gutnick, 1985), and it has been proposed that it has a role in emulsan release from the cell surface.

It remains an open question whether or not formation of lipase by *Acinetobacter* plays a functional role in *n*-alkane metabolism. Production of lipase during growth on *n*-alkanes has been reported for strain *A.lwoffii* O_{16} (Breuil et al., 1978) and for strain *A.calcoaceticus* 69-V (Haferburg and Kleber, 1982, 1983). Little or no activity is observed during growth on defined media with non-hydrocarbons as carbon sources. Lipase production occurs in complex media and is largely enhanced by fatty acid esters and certain detergents (Breuil et al., 1978; Haferburg and Kleber, 1982). Thus, lipase production may be more a consequence of, than a prerequisite for, alkane metabolism. There are no other extracellular enzymes known so far whose production is linked to alkane utilisation.

Enzymes of fatty acid biosynthesis have not yet been studied in detail, although Sampson and Finnerty (1974) found evidence for their repression in *Acinetobacter* sp. HO1-N growing at the expense of *n*-hexadecane.

Respiratory Chain and Electron Transport Systems

Alkane assimilation does not seem to influence energy transducing systems. Thus, studies with *A.calcoaceticus* 69-V (Asperger et al., 1978b, 1981a) and *Acinetobacter* sp. HO1-N (Ensley and Finnerty, 1980) did not reveal significant changes of respiratory enzyme activities in response to *n*-alkanes as carbon sources. In hexadecane-grown *A.calcoaceticus* 69-V, membrane-bound oxidase systems for L-malate, D-glucose and L-glutamate (as well as the corresponding substrate:acceptor oxidoreductases) occur in addition to the classical oxidase systems for NADH and succinate (Asperger et al., 1981a). The same enzymes with similar specific activities are also found in cells grown at the expense of other carbon sources. Both strains contain cytochromes b and o, but no cytochrome c; under conditions of low aeration cytochrome d is also found. Compared with non-hydrocarbon carbon sources, growth on *n*-hexadecane did not significantly change the specific cytochrome content of *A.calcoaceticus* cells (Asperger et al., 1978b). In the case of *Acinetobacter* sp. HO1-N, only moderate changes of the cytochrome o content of cytoplasmic membranes have been observed in response to the carbon source (Ensley and Finnerty, 1980).

OUTLOOK

This review demonstrates that numerous insights into bacterial *n*-alkane metabolism have been gained by investigating several strains of the genus *Acinetobacter*. However, various questions remain to be resolved before the pathway taken by an alkane molecule from the bulk hydrocarbon phase to the active site of enzymes inside the cell and its subsequent conversion to a fatty acid can be depicted in an even more convincing fashion. One of the main intellectual challenges concerns the penetration of alkanes across the cell envelope to the initial oxygenative enzyme system, wherever it is located. All the hypotheses about possible mechanisms of this process still lack unequivocal evidence. Location of the alkane monooxygenases in cellular membranes is not very probable. Nevertheless, the necessity of superstructural assemblies for *n*-alkane oxidation becomes evident when rates of *n*-alkane conversion by intact cells (up to 100 nmol/min/mg protein), as calculated from the growth rates, are compared with maximum activities of cell-free extracts or reconstituted alkane monooxygenase systems (about 1 nmol/min/mg protein). It is to be hoped that the isolated cytochrome P-450 system from *A. calcoaceticus* EB104 will provide an appropriate instrument for studying these questions and for gaining more insights into the cellular organisation of bacterial systems that can oxidise strongly hydrophobic compounds.

Furthermore, several practical aspects of the capability of *Acinetobacter* for alkane oxidation should be considered. Principally, the high growth rate on long-chain *n*-alkanes makes it appear that *Acinetobacter* might be an appropriate organism for the manufacture of biosynthetic products, the production of single cell protein and the bioremediation of oil-contaminated soil or water. The latter field of application is favoured by the generation of surface-active compounds for which a broad spectrum of other applications exists (Haferburg et al., 1986; Rosenberg, 1986). In addition, the surface-active compounds, wax esters or glycerides produced during growth on *n*-alkanes or derivatives of them may be exploited as sources of special carbohydrates, fatty acids or fatty alcohols.

Another aspect to be considered is the use of alkane-induced *Acinetobacter* cells or enzymes for the biotransformation of selected substrates to defined products. Practical use of the alkane monooxygenases of *Acinetobacter* may be favoured by the following features: (a) high regioselectivity for the hydroxylation of the alkyl terminus; (b) high specific activities for the substrate conversion, which can be deduced from the rapid growth rates on *n*-hexadecane or from respiratory measurements with resting cells; and (c) broad specificity against chain-length or different substituted derivatives, which allows the conversion of a broad variety of alkyl chain-containing compounds. Biotransformation processes may be achieved with whole cells or with isolated enzymes. A serious disadvantage of whole cells is the rapid consecutive degradation of the primary oxidation products. This could be circumvented by product recovery in wax esters under appropriate fermentation conditions, by the selection of mutants prepared by classical methods, or by transfer of the alkane monooxygenase genes into other hosts via appropriate vectors. The isolation of cytochrome P-450 from *A. calcoaceticus* EB104 and the start that has been made on its cloning provides the basis for such strategies.

REFERENCES

Amund, O.O., and Higgins, I.J., 1985, The degradation of 1-phenylalkanes by an oil-degrading strain of *Acinetobacter lwoffi, Ant. van Leeuw. J. Microbiol. Serol.*, 51:45.

Asperger, O., 1990, Alkanoxidation und Elektronentransportsysteme bei *Acinetobacter calcoaceticus* unter besonderer Berücksichtigung des Vorkommens von Eisenschwefelproteinen und Cytochrom P-450, *Dissertation B. Leipzig: Karl-Marx-Universität*.

Asperger, O., and Aurich, H., 1977, Anwendung polarographischer O_2-Messungen auf die Veratmung von *n*-Alkanen und deren Derivaten durch *Acinetobacter calcoaceticus, Zeits. Allg. Mikrobiol.*, 17:419.

Asperger, O., Futtig, A., Behrends, B., Kleber, H.-P., and Aurich, H., 1978a, Oxidation langkettiger *n*-Alkane durch zellfreie Extrakte von *Acinetobacter calcoaceticus*. Identifizierung und Bestimmung von *n*-Tetradekansäure nach Inkubation mit *n*-Tetradekan, *Wiss. Zeits. Karl-Marx-Univ. Leipzig, Math. Naturwiss. Reihe*, 27:3.

Asperger, O., Kleber, H.-P., and Aurich, H., 1978b, Zytochrom-Zusammensetzung von *Acinetobacter calcoaceticus, Acta Biol. Med. German.*, 37:191.

Asperger, O., Borneleit, P., and Kleber, H.-P., 1981a, Untersuchungen zum Elektronentransportsystem in *Acinetobacter calcoaceticus, Abh. Akad. Wiss. DDR*, 3:259.

Asperger, O., Naumann, A., and Kleber, H.-P. 1981b, Occurrence of cytochrome P-450 in *Acinetobacter* strains after growth on *n*-hexadecane, *FEMS Microbiol. Lett.*, 11:309.

Asperger, O., Müller, R., and Kleber, H.-P., 1983, Isolierung von Cytochrom P-450 und des entsprechenden Reductasesystems aus *Acinetobacter calcoacticus, Acta Biotech.*, 3:319.

Asperger, O., Naumann, A., and Kleber, H.-P., 1984, Inducibility of cytochrome P-450 in *Acinetobacter calcoaceticus* by *n*-alkanes, *Appl. Microbiol. Biotech.*, 19:398.

Asperger, O., Müller, R., and Kleber, H.-P., 1985a, Reconstitution of *n*-alkane-oxidising activity of a purified cytochrome P-450 from *Acinetobacter calcoaceticus* strain EB104, *in*: "Cytochrome P-450, Biochemistry, Biophysics and Induction," p.447, L. Vereczkey, and K. Magyar, eds., Elsevier Science Publishers, Amsterdam.

Asperger, O., Stüwer, B., and Kleber, H.-P., 1985b, Aryl hydrocarbons as inducers of cytochrome P-450 in *Acinetobacter calcoaceticus, Appl. Microbiol. Biotech.*, 21:309.

Asperger, O., Sharychev, A.A., Matyashova, R.N., Losinov, A.B., and Kleber, H.-P., 1986, Effect of oxygen limitation on the content of *n*-hexadecane-inducible cytochrome P-450 in *Acinetobacter calcoaceticus* strain EB104, *J. Basic Microbiol.*, 26:571.

Aurich, H., 1979, Die Oxidation aliphatischer Kohlenwasserstoffe durch Bakterien, *Sitz. Akad. Wiss. DDR, Math. Naturwiss. Tech.*, 16(N):3.

Aurich, H., and Eitner, G., 1973, Oxidation von *n*-Hexadecan durch *Acinetobacter calcoaceticus*. Bedingungen und Induktion beteiligter Enzyme, *Zeits. Allg. Mikrobiol.*, 13:539.

Aurich, H., and Eitner, G., 1977, Induktion der NADP$^+$-abhangigen Aldehyddehydrogenase durch Kohlenwasserstoffe bei *Acinetobacter calcoaceticus, Zeits. Allg. Mikrobiol.,* 17:263.

Aurich, H., Sorger, D., and Asperger, O., 1976, Isolierung und Charakterisierung eines Rubredoxins aus *Acinetobacter calcoaceticus, Acta Biol.Med.German.,* 35:443.

Aurich, H., Brückner, A., Asperger, O., Behrends, B., and Futtig, A., 1977, Oxidation von *n*-Tetradecan-1-^{14}C durch zellfreie Extrakte aus *Acinetobacter calcoaceticus, Zeits. Allg. Mikrobiol.,* 17:249.

Aurich, H., Sorger, H., Bergmann, R., Lasch, J., and Koelsch, R., 1983, Purification and characterization of membrane-bound aldehyde dehydrogenase from *Acinetobacter calcoaceticus* grown on long-chain alkanes, *in* "Enzyme Technology," p.37, R.M. Lafferty, ed., Springer Verlag, Berlin.

Aurich, H., Bergmann, R., Lasch, J., Koelsch, R. and Sorger, H., 1985, Wechselwirkung der Aldehyddehydrogenase aus *Acinetobacter calcoaceticus* mit Membranlipiden. (II), *J. Basic Microbiol.,* 25:631.

Aurich, H., Sorger, H., Bergmann, R., and Lasch, J., 1987, Zur Kinetik der membrangebundenen Aldehyd-Dehydrogenase aus *Acinetobacter calcoaceticus, Biol. Chem. Hoppe-Seyler,* 368:101.

Baumann, P., Doudoroff, M., and Stanier, R.Y. 1968, A study of the *Moraxella* group. II. Oxidase-negative species (genus *Acinetobacter), J. Bacteriol.,* 95:1520.

Borneleit, P., Hermsdorf, T., Claus, R., Walther, P., and Kleber, H.-P., 1988, Effect of hexadecane-induced vesiculation on the outer membrane of *Acinetobacter calcoaceticus, J. Gen. Microbiol.,* 134:1983.

Bouvet, P.J.M., and Grimont, P.A.D., 1986, Taxonomy of the genus *Acinetobacter* with the recognition of *Acinetobacter baumannii* sp. nov., *Acinetobacter haemolyticus* sp. nov., *Acinetobacter johnsonii* sp. nov., *Acinetobacter junii* sp. nov. and emended description of *Acinetobacter calcoaceticus* and *Acinetobacter lwoffii, Int. J. Syst. Bacteriol.,* 3:228.

Breuil, C., Shindler, B.D., Sijher, J.S., and Kushner, D.J., 1978, Stimulation of lipase production during bacterial growth on alkanes, *J. Bacteriol.,* 133:601.

Bryn, K., Jantzen, E., and Bovre, K., 1977, Occurrence and patterns of waxes in *Neisseriaceae, J. Gen. Microbiol.,* 102:33.

Bühler, M., and Schindler, J., 1984, Aliphatic hydrocarbons, *in* "Biotechnology, vol.6a," p.331, H.-J. Rehm, and G. Reed, eds., Verlag Chemie, Weinheim.

Cardini, G., and Jurtschuk P., 1970, The enzymatic hydroxylation of *n*-octane by *Corynebacterium* sp. strain 7EIC, *J. Biol. Chem.,* 245:2789.

Claus, R., Asperger, O., and Kleber, H.-P., 1978, Nachweis und partielle Anreicherung von Rubredoxin-Reductase aus *Acinetobacter calcoaceticus, Wiss. Zeits. Karl-Marx-Univ. Leipzig, Math. Naturwiss. Reihe,* 27:17.

Claus, R., Asperger, O., and Kleber, H.-P., 1979, Eigenschaften der Rubredoxin-Reductase aus dem alkanassimilierenden Bakterienstamm *Acinetobacter calcoaceticus, Zeits. Allg. Mikrobiol.,* 19:695.

Claus, R., Hädge, D., Asperger, O., Fiebig, H., and Kleber, H.-P., 1980a, Quantitative, immunologische Bestimmungsmethode fur Rubredoxin in Bakterienrohextrakten von *Acinetobacter calcoaceticus, Zeits. Allg. Mikrobiol.,* 20:95.

Claus, R., Asperger, O., and Kleber H.-P., 1980b, Influence of growth phase and carbon source on the content of rubredoxin in *Acinetobacter calcoaceticus, Arch. Microbiol.,* 128:263.

Claus, R., Käppeli, O., and Fiechter, A., 1984, Possible role of extracellular membrane particles in hydrocarbon utilization by *Acinetobacter calcoaceticus* 69-V, *J. Gen. Microbiol.,* 130:1035.

Claus, R., Fischer, B.E., and Kleber, H.-P., 1985, An esterase as marker enzyme of the outer membrane of *Acinetobacter calcoaceticus, J. Basic Microbiol.,* 25:299.

Coon, M.J., Autor, A.P., Boyer, R.F., Lode, E.T., and Strobel, H.W., 1973, On the mechanism of fatty acid, hydrocarbon, and drug hydroxylation in liver microsomal and bacterial enzymes systems, *in* "Oxidase and Related Redox Systems," p.529, T.E. King, H.S. Mason, and M. Morrison, eds., University Park Press, Baltimore.

Dalton, H., 1980, Oxidation of hydrocarbons by methane monooxygenase from a variety of microbes, *Adv. Appl. Microbiol.,* 262:71.

Davis, J. B., 1967, "Petroleum Microbiology," Elsevier Publishing Company, Amsterdam.

Einsele, A., 1983, Biomass from higher *n*-alkanes, *in* "Biotechnology, vol.3," p.44, H.J. Rehm, and G. Reed, eds., Verlag Chemie, Weinheim.

Ensley, B.D., and Finnerty, W.R., 1980, Influence of growth substrates and oxygen on the electron transport system in *Acinetobacter species* HO1-N, *J. Bacteriol.,* 142:859.

Eremina, S.S., Asperger, O., and Kleber, H.-P., 1987, Cytochrome P-450 and the respiration activity of *Acinetobacter calcoaceticus* growing on n-nonane, *Mikrobiologija,* 5:764.

Erickson, L.E., Nakahara, T., and Prokop, A., 1973, Growth in cultures with two liquid phases: hydrocarbon uptake and transport, *Process Biochem.,* 10:9.

Fewson, C.A., 1967, The growth and metabolic versatility of the Gram-negative bacterium NCIB 8250 ('*Vibrio 01*'), *J. Gen. Microbiol.,* 46:255.

Finnerty, W.R., 1977, The biochemistry of microbial alkane oxidation: new insights and perspectives, *Trends Biochem. Sci.,* 2:73.

Finnerty, W.R., 1980, Physiology and biochemistry of bacterial phospholipid metabolism, *Adv. Bacterial Physiol.,* 18:177.

Finnerty, W.R., 1984, The application of hydrocarbon utilizing microorganisms for lipid production, *in* "Biotechnology for the oils and fats industry," *AOCS Monograph*, 11:199.

Finnerty, W.R., and Kallio, R.E., 1964, Origin of palmitic acid carbon in palmitate formed from hexadecane-1-^{14}C and tetradecane-1-^{14}C by *Micrococcus cerificans*, *J. Bacteriol.*, 87:1261.

Finnerty, W.R., and Singer, M.E., 1985, Membranes of hydrocarbon-utilizing microorganisms, *in* "Organization of Procaryotic Cell Membranes," p.1, B.K. Ghosh, ed., CRC-Press, Boca Raton.

Finnerty, W.R., Hawtrey, E., and Kallio, R.E., 1962, Alkane-oxidizing *Micrococci, Zeits. Allg. Mikrobiol.*, 2:169.

Fischer, B.E., 1986, Localization of hydrolytic enzymes in *Acinetobacter calcoaceticus*, *J. Basic Microbiol.*, 26:9.

Fischer, B., Claus, R., and Kleber, H.-P., 1984, Isolation and characterization of the outer membrane of *Acinetobacter calcoaceticus*, *J. Biotech.*, 1:111.

Fixter, L.M., and Fewson, C.A., 1974, The accumulation of waxes by *Acinetobacter calcoaceticus* NCIB 8250, *Biochem. Soc. Trans.*, 2:944.

Fixter, L.M., Nagi, M.N., McCormach, J.G., and Fewson, C.A., 1986, Structure, distribution and function of wax esters in *Acinetobacter calcoaceticus*, *J. Gen. Microbiol.*, 132:3147.

Gallagher, I.H.C. 1971, Occurrence of waxes in *Acinetobacter*, *J. Gen. Microbiol.*, 68:245.

Goma, G., Pareilleux, A., and Durand, G., 1973, Cinetique de degradation des hydrocarbures par *Candida lipolytica*, *Arch. Microbiol.*, 88:97.

Grant, D.J.W., 1973, The degradative versatility, aryl-esterase activity and hydroxylation reactions of *Acinetobacter lwoffii* NCIB 10553, *J. Appl. Microbiol.*, 36:47.

Grund, A., Shapiro, J., Fennewald, M., Bacha, P., Leahy, J., Markbreiter, K., Nieder, M., and Toepfer, M., 1975, Regulation of alkane oxidation in *Pseudomonas putida*, *J. Bacteriol.*, 123:546.

Haferburg, D., and Kleber, H.-P., 1982, Extrazellulare Lipase aus *Acinetobacter calcoaceticus*, *Acta Biotech.*, 2:337.

Haferburg, D., and Kleber, H.-P., 1983, Regulation der extrazellulären Lipase aus einem alkanverwertenden Stamm von *Acinetobacter*, *Acta Biotech.*, 3:185.

Haferburg, D., Asperger, O., Lohs, U., and Kleber, H.-P., 1983, Regulation der Alkanverwertung bei *Acinetobacter calcoaceticus*, *Acta Biotech.*, 3:371.

Haferburg, D., Hommel, R., Claus, R., and Kleber, H.-P., 1986, Extracellular microbial lipids as biosurfactants, *Adv. Biochem. Eng. Biotech.*, 33:53.

Hankin, C., and Kollatukudy, P.E., 1968, Metabolism of a plant wax paraffin (n-nonacosane) by a soil bacterium (*Micrococcus cerificans*), *J. Gen. Microbiol.*, 51:457.

Hiratsuka, J., Furukawa, T., and Noda, M., 1980, Microbial production of pristanol and pristanediol, *Japan Kokai Koho*, 80 00:065.

Horisberger, M., 1977, Structure of the peptidoglycans of *Moraxella glucidolytica* and *Moraxella lwoffi* grown on hydrocarbons, *Arch. Mikrobiol.*, 112:297.

Jirausch, M., Asperger, O., and Kleber, H.-P., 1986, Alcohol oxidation by *Acinetobacter calcoaceticus* EB 104 - a *n*-alkane-utilizing and cytochrome P-450-producing strain, *J. Basic Microbiol.*, 26:351.

Juni, E., 1978, Genetics and physiology of *Acinetobacter*, *Ann. Rev. Microbiol.*, 32:349.

Kaplan, N., and Rosenberg, E., 1982, Exopolysaccharide distribution of and bioemulsifier production by *Acinetobacter calcoaceticus* BD4 and BD413, *Appl. Environ. Microbiol.*, 44:1335.

Käppeli, O., and Finnerty, W.R., 1979, Partition of alkane by an extracellular vesicle derived from hexadecane-grown *Acinetobacter*, *J. Bacteriol.*, 140:707.

Käppeli, O., and Finnerty, W.R., 1980, Characteristics of hexadecane partition by the growth medium of *Acinetobacter* sp., *Biotech. Bioeng.*, 22:495.

Kennedy, R.S., and Finnerty, W.R., 1975, Microbial assimilation of hydrocarbons. II. Intracytoplasmic membrane induction in *Acinetobacter* sp., *Arch. Microbiol.*, 102:83.

Kennedy, R.S., Finnerty, W.R., Sudarsanan, K., and Young, R.A., 1975, Microbial assimilation of hydrocarbons. I. The fine structure of a hydrocarbon oxidizing *Acinetobacter* sp., *Arch. Microbiol.*, 102:75.

Kleber, H.-P., 1973, Repression des Malatenzyms durch n-Alkane in *Acinetobacter calcoaceticus*, *Zeits. Allg. Mikrobiol.*, 13:467.

Kleber, H.-P., 1975, Hemmung des Malatenzyms aus *Acinetobacter calcoaceticus* durch Acetyl-CoA, *Zeits. Allg. Mikrobiol.*, 15:19.

Kleber, H.-P., 1978, Regulation der Synthese und Aktivitat von Enzymen des Citrat- und Glyoxylatzyklus in *Acinetobacter calcoaceticus* unter dem Einfluss der n-Alkanassimilation, *Wiss. Zeits. Karl-Marx-Univ. Leipzig, Math. Naturwiss. Reihe*, 27:55.

Kleber, H.-P., and Aurich, H., 1973, Einfluss von n-Alkanen auf die Synthese der Enzyme des Glyoxylatzyklus in *Acinetobacter calcoaceticus*, *Zeits. Allg. Mikrobiol.*, 13:473.

Kleber, H.-P., and Aurich, H., 1974, Verhalten der Enzyme des Citratzyklus wahrend der n-Alkan-Assimilation bei *Acinetobacter calcoaceticus*, *Zeits. Allg. Mikrobiol.*, 14:575.

Kleber, H.-P., Schöpp, W., and Aurich, H., 1973, Verwertung von n-Alkanen durch einen Stamm von *Acinetobacter calcoaceticus*, *Zeits. Allg. Mikrobiol.*, 13:445.

Kleber, H.-P., Claus, R., and Asperger, O., 1983, Enzymologie der n-Alkanoxidation bei *Acinetobacter*, *Acta Biotech.*, 3:251.

Kleber, H.-P., Müller, R., and Asperger, O., 1985, Cytochrome P-450 in *Acinetobacter*: Occurrence, isolation and regulation, *in* "Environmental Regulation of Microbial Metabolism," p.89, I.S. Kulaev, E.A. Dawes, and D.W. Tempest, eds., Academic Press, London.

Klossek, P., Kirchner, M., and Kurth, J., 1985, Immobilisierung partikel-gebundener Aldehyddehydrogenase aus *Acinetobacter calcoaceticus* EB 104, *J. Basic Microbiol.*, 25:429.

Klug, M.J., and Markovetz, A.J., 1971, Utilization of aliphatic hydrocarbons by microorganisms, *Adv. Microbial Physiol.*, 5:1.

Kollatukudy, P.E., and Hankin, C., 1968, Production of -haloesters from alkylhalides by *Micrococcus cerificans, J. Bacteriol.*, 54:145..

McCaman, R.E., and Finnerty, W.R., 1968, Biosynthesis of cytidine diphosphate-diglyceride by a particulate fraction from *Micrococcus cerificans, J. Biol. Chem.*, 243:5074.

McKenna, E.J., and Kallio, R.E., 1965, The biology of hydrocarbons, *Ann. Rev. Microbiol.*, 19:183.

Makula, R., and Finnerty, W.R., 1968, Microbial assimilation of hydrocarbons. I. Fatty acid derived from normal alkanes, *J. Bacteriol.*, 95:2102.

Makula, R.A., and Finnerty, W.R., 1970, Microbial assimilation of hydrocarbons: Identification of phospholipids, *J. Bacteriol.*, 103:348.

Makula, R.A., and Finnerty, W.R., 1971, Microbial assimilation of hydrocarbons: Phospholipid metabolism, *J. Bacteriol.*, 107:806.

Makula, R.A., Lockwood, P.J., and Finnerty, W.R., 1975, Comparative analysis of the lipids of *Acinetobacter species* grown on hexadecane, *J. Bacteriol.*, 121:250.

Modrzakowski, M.C., and Finnerty, W.R., 1980, Metabolism of symmetrical dialkyl ethers by *Acinetobacter species* HO1-N, *Arch. Microbiol.*, 126:285.

Modrzakowski, M.C., Makula, R.A., and Finnerty, W.R., 1977, Metabolism of the alkane analogue *n*-dioctyl ether by *Acinetobacter species, J. Bacteriol.*, 131:92.

Müller, H., and Voigt, B., 1982, Untersuchungen zur chemischen Zusammensetzung der Lipidfraktion von *Acinetobacter calcoaceticus, Acta Biotech.*, 2:155.

Müller, H., Naumann, A., Claus, R., and Kleber, H.-P., 1983, Intracyto-plasmic membrane induction by hexadecane in *Acinetobacter calcoaceticus, Zeits. Allg. Mikrobiol.*, 23:645.

Müller, R., Asperger, O., and Kleber, H.-P., 1989, Purification of cytochrome P-450 from n-hexadecane-grown *Acinetobacter calcoaceticus, Biomed. Biochim. Acta,* 48:243.

Neidleman, S.L., and Geigert, J., 1983, Long-chain wax esters, *US Patent* 4, 404, 283.

Neufeld, R.J., and Zajic, J.E., 1984, The surface activity of *Acinetobacter calcoaceticus* sp. 2CA2, *Biotech. Bioeng.*, 26:1108.

Ratledge, C., 1978, Degradation of aliphatic hydrocarbons, *in* "Developments in Biodegradation of Hydrocarbons," p.1, R.J. Watkinson, ed., Applied Science Publishers, London.

Rosenberg, E., 1986, Microbial surfactants, *CRC Crit. Rev. Microbiol.*, 3:109.

Rosenberg, E., and Kaplan, N., 1987, Surface-active properties of *Acineto-bacter* exopolysaccharides, *in* "Bacterial Outer Membranes as Model Systems," p.311, M.Inouye, ed., Wiley, New York.

Rosenberg, E., Zuckerberg, A., Rubinovitz, C., and Gutnick, D.L., 1979, Emulsifier of *Arthrobacter* RAG-1: isolation and emulsifying properties, *Appl. Environ. Microbiol.*, 37:409.

Rosenberg, M., Bayer, E.A., DeLarea, J., and Rosenberg, E., 1982, Role of thin fimbriae in adherence and growth of *Acinetobacter calcoaceticus* RAG-1 on hexadecane, *Appl. Environ. Microbiol.*, 44:929.

Rusansky, S., Avigad, R., Micheali, S., and Gutnick, D.L., 1987, Involvement of a plasmid in growth on and dispersion of crude oil by *Acinetobacter calcoaceticus* RA57, *Appl. Environ. Microbiol.*, 53:1918.

Sampson, K.L., and Finnerty, W.R., 1974, Regulation of fatty acid biosynthesis in the hydrocarbon-oxidizing microorganism, *Acinetobacter* sp., *Arch. Microbiol.*, 99:203.

Scott, C.C.L., and Finnerty, W.R., 1976a, A comparative analysis of the ultrastructure of hydrocarbon-oxidizing microorganisms, *J. Gen. Microbiol.*, 94:342.

Scott, C.C.L., and Finnerty, W.R., 1976b, Characterization of intracytoplasmic hydrocarbon inclusions from the hydrocarbon-oxidizing *Acinetobacter species* HO1-N, *J. Bacteriol.*, 127:481.

Scott, C.C.L., Makula, R.A., and Finnerty, W.R., 1976, Isolation and characterization of membranes from a hydrocarbon-oxidizing *Acinetobacter* sp., *J. Bacteriol.*, 127:469.

Shabtai, Y., and Gutnick, D.L., 1985, Exocellular esterase and emulsan release from the cell surface of *Acinetobacter calcoaceticus*, *J. Bacteriol.*, 161:1176.

Singer, M.E., and Finnerty, W.R., 1984a, Microbial metabolism of straight-chain and branched alkanes, *in* "Petroleum microbiology," p.1, R.M. Atlas, ed., Macmillan, New York.

Singer, J.T., and Finnerty, W.R., 1984b, The genetics of hydrocarbon-utilizing microorganisms, *in*: "Petroleum Microbiology," p.299, R.M. Atlas, ed., Macmillan, New York.

Singer, J.T., and Finnerty, W.R., 1984c, Insertional specificity of transposon Tn5 in *Acinetobacter* sp., *J. Bacteriol.*, 157:607.

Singer, M.E., and Finnerty, W.R., 1985a, Fatty aldehyde dehydrogenases in *Acinetobacter species* strain HO1-N: Role in hexadecane and hexadecanol metabolism, *J. Bacteriol.*, 164:1011.

Singer, M.E., and Finnerty, W.R., 1985b, Alcohol dehydrogenases in *Acinetobacter species* strain HO1-N: Role in hexadecane and hexadecanol metabolism, *J. Bacteriol.*, 164:1017.

Singer, M.E., Tyler, S.M., and Finnerty, W.R., 1985, Growth of *Acinetobacter species* HO1-N on n-hexadecanol: Physiological and ultrastructural characteristics, *J. Bacteriol.*, 162:162.

Sorger, H., and Aurich, H., 1978, Mikrobielle Aldehyddehydrogenasen und ihre Bedeutung für die Assimilation aliphatischer Kohlenwasserstoffe. *Wiss. Zeits. Karl-Marx-Univ. Leipzig, Math. Naturwiss. Reihe*, 27:35.

Sorger, H., Aurich, H., Fricke, B., and Vorisek, J., 1986, Ultracytochemical localization of aldehyde dehydrogenase in *Acinetobacter calcoaceticus*, *J. Basic Microbiol.*, 26:541.

Stevenson, D.P., Finnerty, W.R., and Kallio, R.E., 1962, Esters produced from n-heptadecane by *Micrococcus cerificans, Biochem. Biophys. Res. Comm.*, 9:426.

Stewart, J.E., and Kallio, R.E., 1959, Bacterial hydrocarbon oxidation. II. Ester formation from alkanes, *J. Bacteriol.*, 78:726.

Stewart, J.E., Kallio, R.E., Stevenson, D.P., Jones, A.C., and Schissler, D.O., 1959, Bacterial hydrocarbon oxidation. I. Oxidation of n-hexadecane by a gram-negative coccus, *J. Bacteriol.*, 78:441.

Stewart, J.E., Finnerty, W.R., Kallio, R.E., and Stevenson, D.P., 1960, Esters from bacterial oxidation of olefins, *Science*, 132:1254.

Tauchert, H., Roy, M., Schopp, W., and Aurich, H., 1975, Pyridinnucleotidunabhangige Oxidation von langkettigen aliphatischen Alkoholen durch ein Enzym aus *Acinetobacter calcoaceticus, Zeits. Allg. Mikrobiol.*, 15:457.

Tauchert, H., Grunow, M., Harnisch, H., and Aurich, H., 1976, Reinigung und einige Eigenschaften der NADP$^+$-abhangigen Alkoholdehydrogenase aus *Acinetobacter calcoaceticus, Acta Biol. Med. German.*, 35:1267.

Torregrossa, R.E., Makula, R.A., and Finnerty, W.R., 1977a, Characterization of lysocaroliolipin from *Acinetobacter species* HO1-N, *J. Bacteriol.*, 131:486.

Torregrossa, R.E., Makula, R.E., and Finnerty, W.R., 1977b, Outer membrane phospholipase A from *Acinetobacter species* HO1-N, *J. Bacteriol.*, 131:493.

Torregrossa, R.E., Makula, R.E., and Finnerty, W.R., 1978, Hydrolysis of phosphatidylglycerol by outer membrane phospholipase A from *Acinetobacter species* HO1-N, *J. Bacteriol.*, 136:803.

Vachon, V., McGarrity, J.T., Breuil, C., Armstrong, J.B., and Kushner, D.J., 1982, Cellular and extracellular lipids of *Acinetobacter lwoffi* during growth on hexadecane, *Can. J. Microbiol.*, 28:660.

van der Linden, A.C., and Thijsse, G.J.E., 1965, The mechanism of microbial oxidation of hydrocarbons, *Adv. Enzymol.*, 27:469.

Wyndham, R.C., 1987, A screening method for cytochrome P-450 organic peroxidase activity and application to hydrocarbon-degrading bacterial populations, *Can. J. Microbiol.*, 33:1.

Zuckerberg, A., Diver, A., Peeri, Z., Gutnick, D.L., and Rosenberg, E., 1979, Emulsifier of *Arthrobacter* RAG-1: chemical and physical properties, *Appl. Environ. Microbiol.*, 37:414.

METABOLISM OF AROMATIC COMPOUNDS BY *ACINETOBACTER*

C.A. Fewson

Department of Biochemistry
University of Glasgow
Glasgow G12 8QQ
UK

INTRODUCTION

One of the most notable features of strains of *Acinetobacter* is their ability to grow on a wide range of aromatic compounds. In addition, acinetobacters isolated from natural environments can grow in simple defined mineral media with no requirement for growth factors, so they are clearly synthesising all their own aromatic amino acids and related compounds. This review contains a description of the mechanisms used for the dissimilation of aromatic compounds by *Acinetobacter* strains; it concentrates on the best-characterised of these pathways, the mandelate pathway, but includes at least outline descriptions of the ways in which many other aromatic compounds are catabolised. In addition, there is a brief summary of the pathways for the biosynthesis of aromatic amino acids and an indication of the potential competition between these pathways and those for the degradation of aromatic compounds.

UTILISATION OF AROMATIC COMPOUNDS TO SUPPORT GROWTH

It is easy to isolate acinetobacters which can use aromatic or hydroaromatic compounds as sole sources of carbon and energy for growth, or which can partially metabolise them. Such strains are found in many natural environments and in industrial effluents and treatment plants. One of the first strains to be isolated was the so-called 'Vibrio strain 01', which was obtained by Happold and Key (1932) in the course of their work on the destruction of phenols when spent gasworks' liquor was mixed with sewage and then passed through filter beds. This organism, which is probably *A. calcoaceticus* NCIB8250 (Fewson, 1967b; Baumann et al., 1968), can

The Biology of Acinetobacter, Edited by K. J. Towner *et al.*
Plenum Press, New York, 1991

degrade a great range of aromatic compounds (Fewson, 1967a) and was used in some of the earliest studies on the elucidation of the metabolic pathways by which dihydric phenols are subjected to intra-diol ring-cleavage and then converted into ß-ketoadipate (Evans and Happold, 1939; Evans, 1947; Kilby 1948, 1951; reviewed by Stanier and Ornston, 1973).

In their monumental study of the nutritional and biochemical properties of *Acinetobacter*, Baumann et al. (1968) found that many of their strains could grow on aromatic or hydroaromatic compounds. Of 106 strains tested, 86 grew on benzoate, 80 on quinate, 79 on L-tyrosine, 76 on 4-hydroxybenzoate, 66 on phenylacetate, 63 on kynurenate, 60 on anthranilate, 58 on L-phenylalanine, 53 on L-tryptophan, 44 on L-kynurenine, 28 on phenol, 17 on 2-hydroxybenzoate (salicylate), 14 on 3-hydroxybenzoate, five on phenylglyoxylate (benzoylformate), four on L-mandelate and three on D-mandelate. Even these proportions may be underestimates because of possible false-negative results caused by toxicity of compounds such as phenol and salicylate. Many of these strains were obtained from clinical sources or from enrichments on non-aromatic substrates, so clearly the genetic potential to metabolise aromatic compounds is very widespread within the genus. Individual strains can sometimes use more than 30 different aromatic compounds as sole sources of carbon (e.g. Fewson, 1967a; Hasegawa and Suzuki, 1983). All these substrates are used aerobically, but there is a report in the literature of a '*Moraxella* species' metabolising benzoate anaerobically, supported by nitrate respiration (Williams and Evans, 1975). Although there seems no intrinsic reason why acinetobacters should not carry out this type of reaction, this particular organism seems to have been wrongly identified and was actually a *Paracoccus* sp. (Evans, 1988).

EFFICIENCY OF UTILISATION OF AROMATIC COMPOUNDS

Many observations on the utilisation of aromatic compounds by *Acinetobacter* are rather qualitative, but a few attempts have been made to quantify the efficiency of metabolism and growth. These experiments have been done by microcalorimetry (e.g. Lovrien et al., 1980) and by determination of molar growth yields (e.g. Fewson, 1985). Growth yields on both aromatic and non-aromatic substrates are low. There are various possible explanations for this. The most likely is that the effective P/O ratio is only about one under most conditions, even though there may be two sites of oxidative phosphorylation. The values for oxygen uptake used in these calculations were those for respiratory oxygen and excluded the large amounts needed for oxygenative reactions (Fewson, 1985). It is perhaps disappointing that an organism which is so metabolically versatile is so apparently inefficient in this respect. However, this may simply be further evidence that the principal selective pressures operating on bacteria have not been those which have tended to maximise yield (Tempest, 1978). Acinetobacters may be better advertisements for the success of metabolic versatility and rapid growth rate than for high molar growth yields.

GENERAL PATHWAYS FOR THE CATABOLISM OF AROMATIC COMPOUNDS

The soluble and insoluble aromatic compounds which can be metabolised by *Acinetobacter* and other microorganisms may consist of one or more rings that often carry various types of substituent groups. The complex and apparently bewildering variety of catabolic pathways can best be thought of as parts of a systematic attack (Fig. 1): (a) chemotaxis or directed growth towards the substrates; (b) adsorption to the surface of insoluble substrates; (c) excretion of extracellular enzymes or emulsifying agents in order to obtain diffusible products of small molecular mass; (d) entry into the cell; (e) metabolism by peripheral pathways to manipulate the substituent groups and form substrates for ring-cleavage; (f) ring-cleavage; (g) conversion of the products of ring-cleavage into amphibolic intermediates; (h) utilisation of the amphibolic intermediates for growth and energy production.

So far as the initial stages are concerned, *Acinetobacter* is not motile and has to make do with substrates that are present in its immediate environment. The possibility of it using emulsifying agents to solubilise potential substrates in natural environments is only hinted at, despite the

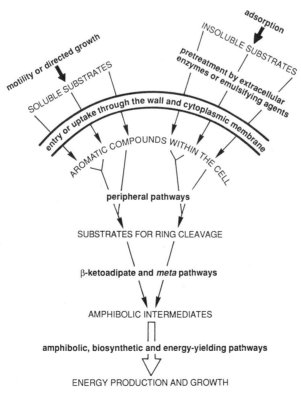

Fig. 1. Microbial utilisation of aromatic compounds

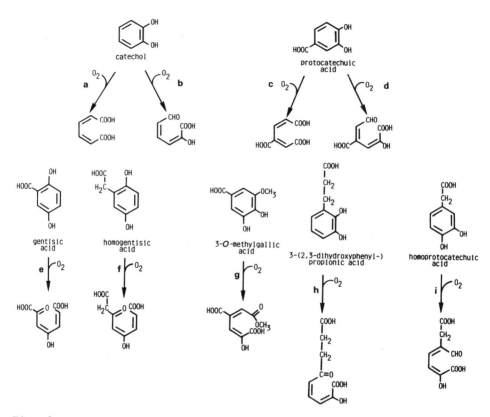

Fig. 2. Intra-diol (*ortho*) (*a* and *c*) and extra-diol (*meta*) (*b,d-i*) oxygenative ring-cleavage reactions found in various strains of *Acinetobacter*. Based on Baumann et al. (1968), Stanier and Ornston (1973), Dagley (1978), Keil et al. (1983), Sze and Dagley (1987).

great deal of work that has been done on the emulsifiers which it produces (Gutnick et al., this volume). Entry into the cell is often assumed to be by free diffusion, but there is evidence that *A. calcoaceticus* possesses specific transport mechanisms for even such simple aromatic compounds as benzoate and mandelate (Cook and Fewson, 1972; Fewson, 1988).

The great majority of work that has been done on the utilisation of aromatic compounds by *Acinetobacter* and other microorganisms has been concerned with elucidating the metabolic pathways which are involved in the breakdown of these compounds and how these pathways are regulated. All the dozens, probably hundreds, of aromatic compounds that can be degraded by microorganisms are converted by converging, peripheral metabolic pathways into a small handful of dihydric phenols. These key intermediates are then subjected to intra-diol (*ortho*) or extra-diol (*meta*) ring-cleavage (e.g. Dagley, 1978; Fewson, 1981). In *Acinetobacter* spp., just seven such compounds have been identified (Fig. 2). Of these, catechol and protocatechuate serve as the points of convergence of a particularly wide range of peripheral degradative pathways (Figs. 3 and 4). In the great

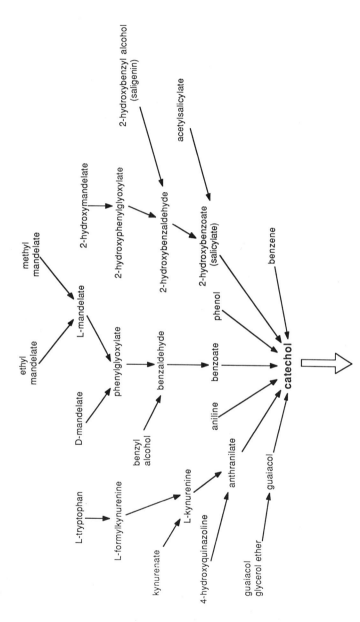

Fig. 3. Conversion of aromatic compounds into catechol by various strains of *Acinetobacter*. Based on Kennedy and Fewson (1968), Grant (1971), Stanier and Ornston (1973), Grant et al. (1975), Wyndham (1986), Winstanley et al. (1987), Vasudevan and Mahadevan (1989).

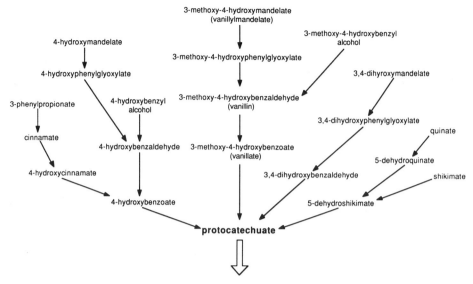

Fig. 4. Conversion of aromatic compounds into protocatechuate by various strains of *Acinetobacter*. Based on Kennedy and Fewson (1968), Stanier and Ornston (1973), Keil et al., (1983).

majority of cases, *ortho* fission of catechol by catechol 1,2-dioxygenase (Fig. 2a) and of protocatechuate (3,4-dihydroxybenzoate) by protocatechuate 3,4-dioxygenase (Fig. 2c) have been demonstrated or assumed to take place (Baumann et al., 1968; Stanier and Ornston, 1973). However, *meta*-cleavage of protocatechuate by protocatechuate 4,5-dioxygenase (Fig. 2d) occurs in a strain of *A. lwoffii* after growth on 4-hydroxymandelate (Sze and Dagley, 1987), while strain 20 of Baumann et al. (1968) was reported to catalyse both *ortho* and *meta* fission of catechol after growth on phenol.

The ß-ketoadipate pathway (Fig. 1 in Ornston and Neidle, this volume) converts products of *ortho*-cleavage into succinate and acetyl-CoA, and the various *meta*-cleavage pathways produce a range of compounds such as pyruvate, acetaldehyde and formate (Stanier and Ornston, 1973; Dagley, 1978). The small molecular mass compounds then enter the amphibolic and anabolic pathways and support growth (Fig. 1). Rather little energy and few, if any, biosynthetically useful intermediates are produced in the early stages of these aerobic pathways for the degradation of aromatic compounds. Indeed, in many cases it is only the very last products of degradation that can be used for either energy production or synthesis. Some of the early catabolic steps may actually be energy-dependent or lower the potential amount of energy that is available for growth; e.g. monooxygenases divert reducing power from the electron transfer system, and therefore from oxidative phosphorylation, into the non-energy producing reduction of one of the oxygen atoms to water. In a few cases, however, there is evidence that early oxidative steps may give rise to useful energy by oxidative phosphorylation; e.g. mandelate supports a slightly higher molar growth yield of *A. calcoaceticus* NCIB8250 than does phenylglyoxylate, presumably because the flavin-linked mandelate dehydrogenase leads to additional production of ATP (Fewson, 1985).

One of the problems that dogs much of the research in this area is the lack of commercial availability and/or instability of many of the potential metabolic intermediates. In many cases they have to be synthesised enzymically as required and acinetobacters can be used for this purpose (e.g. Darrah and Cain, 1969).

INVOLVEMENT OF PLASMIDS IN THE DEGRADATION OF AROMATICS

The widespread presence of plasmids in strains of *Acinetobacter* isolated from natural environments (Towner, this volume) has invited the speculation that some of them might be involved in encoding catabolic pathways, as is quite often the case in pseudomonads. However, so far as is known, the vast majority of aromatic pathways in acinetobacters are chromosomally-encoded. Nevertheless, there are a few well-documented cases of aromatics being degraded by acinetobacters using plasmid-encoded enzymes; examples include the utilisation of chloro-substituted compounds (Furukawa and Chakrabarty, 1982; Shields et al., 1985; see below), resorcin (Barkovsky and Shub, 1986) and benzene (Winstanley et al., 1987).

METABOLISM AFTER RING-CLEAVAGE

The present chapter concentrates on the peripheral pathways used by acinetobacters for the conversion of aromatic compounds into one or other of the substrates for ring-cleavage. The pathways found in *Acinetobacter* spp., *Pseudomonas* spp. and other organisms for the *ortho*-cleavage of catechol and protocatechuate, and the subsequent formation of ß-ketoadipate, have been extensively reviewed (e.g. Kilby, 1951; Cain, 1961; Ornston and Stanier, 1964; Canovas et al., 1967; Stanier and Ornston, 1973; Ornston and Parke, 1977; Kozarich, 1988; Ornston and Neidle, this volume). The regulation of expression of the enzymes of the ß-ketoadipate pathway has been thoroughly studied, especially in strain 73 of *A. calcoaceticus* ('*Moraxella calcoacetica*') (Canovas and Stanier, 1967; Canovas et al., 1967, 1968; Canovas and Johnson, 1968; Neidle and Ornston, 1987; Doten et al., 1987b). Many of the enzymes have been purified, mostly from the transformable strain BD413 (ATCC33305 = ADPI; Juni and Janik, 1969), but in some cases from strains such as 80-1. The enzymes have been characterised and compared with respect to mechanisms of action, *N*-terminal amino acid sequences, amino acid compositions and immunological cross-reactivity (Katagiri and Wheelis, 1971; Patel et al., 1974, 1975, 1976; Zaborsky and Schwartz, 1974; Hou et al., 1976, 1977, 1978; Patel and Ornston, 1976; Yeh et al., 1978, 1980a,b, 1981, 1982; Durham et al., 1980; McCorkle et al., 1980; Yeh and Ornston, 1980, 1981, 1984; Chari et al.,

1987a,b). Many of the relevant genes have been cloned and sequenced (Neidle and Ornston, 1986; Shanley et al., 1986, 1989; Doten et al., 1987a; Neidle et al., 1987, 1988; Hughes et al., 1988; Ornston and Neidle, this volume). The combined results from this sustained attack have led to some of the most original and intriguing speculations concerning the evolution of any metabolic pathway (e.g. Canovas et al., 1967; Stanier and Ornston, 1973; Ornston and Parke, 1977; Ornston and Yeh, 1981). Current thinking is summarised by Ornston and Neidle (this volume).

Meta-cleavage pathways in *Acinetobacter* are much less well understood. A few of the enzymes have been purified (Seltzer, 1973; Leung et al., 1974; Burlingame and Chapman, 1983) and the regulation of the 4-hydroxyphenylacetate *meta*-cleavage pathway in *Acinetobacter* strain 3B-1 has been reviewed by Bayly and Barbour (1984).

THE MANDELATE AND BENZYL ALCOHOL PATHWAYS

Of all the aromatic compounds that are degraded by *Acinetobacter* spp., the converging pathway for the catabolism of mandelate and benzyl alcohol is by far the best characterised (Fig. 5). The information that has been obtained parallels that available for various *Pseudomonas* spp., which in turn builds on the pioneering work of R.Y. Stanier and his colleagues (references in Fewson, 1988). The term 'mandelate pathway' generally refers to the pathway down to the point of ring cleavage.

Growth on Mandelate and Benzyl Alcohol

Of 141 strains of *A. calcoaceticus* that have been tested, 11 can grow on both L- and D-mandelate. One strain (NCIB8250) can grow on only L-mandelate and a further one (EBF65/65) on only D-mandelate (Fewson, 1988). So far as is known, none of these strains was isolated by procedures involving enrichment on mandelate media; indeed, the majority of the acinetobacters which can grow on mandelate were originally isolated from frozen chicken carcasses by E.M. Barnes (Fewson et al., 1988). There seems to have been no systematic survey of the ability of acinetobacters to grow on benzyl alcohol, but it appears to be quite a common feature.

The Mandelate Pathway and its Enzymes

It is not known how mandelate and benzyl alcohol enter the cell. In very preliminary experiments it appeared that there was a barrier against the entry of mandelate into *A. calcoaceticus* NCIB8250 unless the bacteria had been grown under conditions where the mandelate-utilising enzymes were induced (Cook and Fewson, 1972). The implication was that there is a specific and inducible uptake system for mandelate. If so, it may be required only at low extracellular concentrations of mandelate when the lipid-solubility of the substrate will not be sufficient to maintain adequate intracellular concentrations. Transport systems are usually stereospecific,

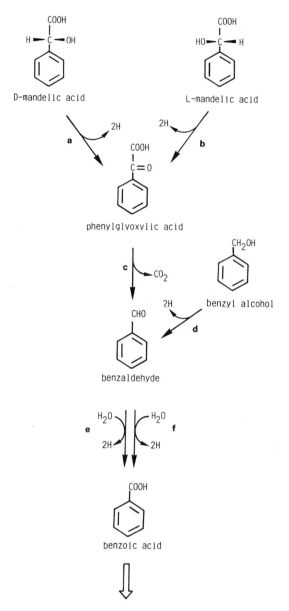

Fig. 5. The mandelate pathway in *Acinetobacter*. Based on Kennedy and Fewson (1968a,b), Fewson (1988).

so presumably there must be a separate system for each of the mandelate enantiomers.

The pathway for the conversion of mandelate and benzyl alcohol into benzoate by *A. calcoaceticus* (Fig. 5) is the same as that found in some pseudomonads, and was elucidated by: (a) experiments on simultaneous adaption; (b) measurement of the relevant enzyme activities; and (c) the detection and utilisation of intermediates (Kennedy and Fewson, 1968a,b;

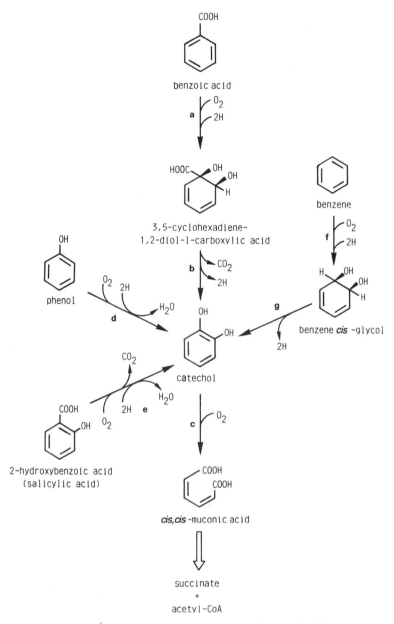

Fig. 6. Oxygenative formation and metabolism of catechol by various strains of *Acinetobacter*. Based on Kennedy and Fewson (1968a,b), Reiner (1971), Winstanley et al. (1987), Fewson (1988).

Fewson, 1988). Benzoate is oxygenatively converted into catechol (Fig. 6; Reiner, 1971), which is then *ortho*-cleaved and metabolised to ß-ketoadipate (Fig. 1 in Ornston and Neidle, this volume).

Those strains of *A. calcoaceticus* (e.g. EBF65/174) which can grow on both enantiomers of mandelate have two stereospecific mandelate dehydrogenases (Fig. 5a,b). Strain NCIB8250 can grow on only the L-enantiomer and has an L-mandelate dehydrogenase (Fig. 5b), but no D-mandelate dehydrogenase, whereas strain EBF65/65 can grow on only the D-enantiomer and has just the D-mandelate dehydrogenase (Fig. 5a). Both of these strains give rise to mutants that can grow on the second enantiomer. In every case this is because of the appearance of an extra dehydrogenase, specific for the other enantiomer (Hills and Fewson, 1983a). All of the enzymes that convert mandelate and benzyl alcohol into benzoate have now been purified from *A. calcoaceticus* NCIB8250, EBF65/65 or their mutants.

The mandelate dehydrogenases (Fig. 5a,b) are membrane-bound flavoproteins and can be solubilised in an active form by careful treatment with appropriate detergents (Allison et al., 1985a). L-Mandelate dehydrogenase contains FMN and has a sub-unit M_r of 44,000 (Hoey et al., 1987), whereas D-mandelate dehydrogenase contains FAD and has a sub-unit M_r of 60,000 (Allison et al., 1985b). *A. calcoaceticus* also contains a pair of membrane-bound flavin-linked lactate dehydrogenases which have been purified (Allison et al., 1985b; Allison and Fewson, 1986). Comparison of the properties of the various enzymes has led to the following conclusions:- (a) The two enzymes specific for the L-enantiomers of mandelate and lactate strongly resemble each other, as do the two enzymes specific for the D-enantiomers, with respect to ease of extraction from the membrane by detergents, sub-unit M_r values, pH optima, pI values, and susceptibility to thiol reagents. Each of these pairs of enzymes may therefore have had a common evolutionary origin. If so, the ability of D-mandelate dehydrogenase to oxidise D-lactate (Allison et al., 1985b) may be a relic of the recruitment of a lactate dehydrogenase to oxidise mandelate, or it may simply be a consequence of the similarity in chemical structures. (b) The two pairs of enzymes, although similar to each other in a general way, are sufficiently different (e.g. in flavin specificity and sub-unit M_r) that a common origin for all four enzymes is either unlikely or was much more remote. (c) The 'evolved' dehydrogenases in mutants of strains NCIB8250 and EBF65/65 that were selected on the basis of ability to grow on the second enantiomer of mandelate are almost identical with the 'original' dehydrogenases in the other strains with respect to immunological cross-reactivity, solubility in detergents, net charge at pH 7.5, pI values, salting-out properties, sub-unit M_r values, apparent K_m values, heat-stability and sensitivity to 4-chloromercuribenzoate (Fewson et al., 1988). This suggests that the most likely explanation for the appearance of the 'evolved' dehydrogenases is some event(s) that allowed the expression of previously silent or cryptic genes. This idea is strengthened by the observation that, in each of the mutants tested, the 'evolved' dehydrogenases were found to be regulated co-ordinately with the pre-existing dehydrogenases, whether tested by the isolation of constitutive mutants or by measuring the effects of various inducers, anti-inducers and compounds that cause catabolite repression (Hills and Fewson, 1983b).

Phenylglyoxylate decarboxylase (Fig. 5c) is a tetrameric, TPP-dependent enzyme with a sub-unit M_r of 58,000 (Barrowman and Fewson, 1985) which is at least superficially similar to the familiar yeast pyruvate decarboxylase (Barrowman et al., 1986).

There are two benzaldehyde dehydrogenases in *A. calcoaceticus* (Livingstone et al., 1972). Benzaldehyde dehydrogenase I (Fig. 5e) is a heat-stable, K^+-activated enzyme and is associated with mandelate oxidation. Benzaldehyde dehydrogenase II (Fig. 5f) is a heat-labile, K^+-independent enzyme and is associated with benzyl alcohol metabolism. Both enzymes have been purified and shown to be NAD-dependent, soluble and tetrameric (MacKintosh and Fewson, 1987, 1988a,b; Chalmers and Fewson, 1989b). Benzyl alcohol dehydrogenase (Fig. 5d) is also a soluble, NAD-dependent tetrameric enzyme (MacKintosh and Fewson, 1987, 1988a,b).

A study of the two benzaldehyde dehydrogenases and benzyl alcohol dehydrogenase affords a good opportunity to test the hypothesis of retrograde evolution of enzymes as well as ideas about gene duplication. All three enzymes share two common substrates (NAD and benzaldehyde) and participate in a peripheral pathway that may be more likely to undergo rapid evolution than central pathways which are constrained by the necessity of preserving features concerned with regulation. The evidence obtained so far from *N*-terminal amino acid sequencing, amino acid compositions, immunological cross-reactivity and kinetic studies is that benzyl alcohol dehydrogenase is probably not homologous with the other two enzymes, and that there is therefore no evidence for retrograde evolution in this pathway. The two benzaldehyde dehydrogenases may, however, be homologous with each other and, if so, could have arisen as a result of a gene doubling event (Chalmers and Fewson, 1989a, unpublished observations).

The two enzymes that convert benzoate into catechol (Fig. 6a,b) have not been purified from *A. calcoaceticus*, but they have been well-characterised in other organisms (Reiner, 1971; references in Fewson, 1988).

Regulation of the Mandelate Pathway

The mandelate pathway enzymes are expressed co-ordinately and are induced by phenylglyoxylate, but not by mandelate. The evidence for this rests on a variety of approaches, including: (a) estimation of the degree of correlation amongst differential rates of synthesis of the enzymes under various conditions of induction and repression; (b) characterisation of blocked mutants; (c) gratuitous induction by thiophenoxyacetate and phenoxyacetate; (d) co-ordinate anti-induction by 2-phenylpropionate; and (e) characterisation of constitutive mutants (Livingstone and Fewson, 1972; Fewson et al., 1978). The enzymes converting benzoate into catechol are regulated independently of the other mandelate enzymes, as are those of the ß-ketoadipate pathway (Stanier and Ornston, 1973).

The mandelate enzymes of *A. calcoaceticus* are subject to some sort of catabolite repression (e.g. by succinate) and to feedback repression (e.g. by benzoate), but the mechanisms of this are unknown (Cook et al., 1975; Fewson, 1988).   In addition, mandelate metabolism dominates benzyl alcohol metabolism even though benzyl alcohol supports a faster growth rate and gives a higher molar growth yield than does mandelate (Beggs et al., 1976; Beggs and Fewson, 1977; Fewson, 1985).   Batch cultures of *A. calcoaceticus* growing on mandelate or phenylglyoxylate show an unusual non-exponential pattern of growth.   There are transient accumulations of benzyl alcohol and benzaldehyde in the medium which are caused by the limitation of mandelate oxidation by low activities of benzaldehyde dehydrogenase and by the diversion of reducing power to the formation of benzyl alcohol (Cook et al., 1975).   The basis for the apparent inability to induce the two enzymes specific for benzyl alcohol oxidation whenever phenylglyoxylate decarboxylase is being induced is not clear (Beggs et al., 1976; Beggs and Fewson, 1977).

There is no evidence for feedback inhibition of any of the mandelate enzymes by subsequent metabolites, and there is no substantial evidence that they exist in any sort of ordered complex within the cell (Hoey and Fewson, 1990).   The general presumption seems to be that peripheral pathways of this sort are regulated only at the level of gene expression, and not at all by regulation of enzyme activity by allosteric or other mechanisms (except, of course, by the availability of substrates and cofactors).   If this is correct, it may explain why intermediates often accumulate while aromatic and similar substrates are being metabolised, and why growth rates are less than maximum.

Genetic studies of the mandelate pathway in *A. calcoaceticus* are at an elementary stage.   The mandelate genes in strain EBF65/65 appear to be clustered near the auxotrophic marker *phe-1*, but are not all contiguous with each other.   A gene responsible for the expression of the cryptic L-mandelate dehydrogenase appears to be close to a gene required for the activity of D-mandelate dehydrogenase (Vakeria et al., 1984).

Metabolism of Ring-Substituted Mandelates and Benzyl Alcohols

A. calcoaceticus can grow on several ring-substituted mandelates and benzyl alcohols, and can partially metabolise others.   This is possible because the enzymes involved in the conversion of mandelate into benzoate can tolerate extensive substitution on the benzene ring, especially at the 3-, 4- and 5-positions, and the enzymes are also induced by the substituted compounds (Kennedy and Fewson, 1968a,b; Fewson, 1988).   Strain NCIB8250 can convert L-4-hydroxymandelate and 4-hydroxybenzyl alcohol into 4-hydroxybenzoate (Fig. 7b-d), and there is an inducible 3-hydroxylase (Fig. 7e) to form protocatechuate prior to *ortho-* cleavage (Fig. 7).   This strain also forms protocatechuate from 3,4-dihydroxymandelate by means of the mandelate enzymes, and from 4-hydroxy-3-methoxymandelate (vanillylmandelate) and 4-hydroxy-3-methoxybenzyl alcohol (vanillyl

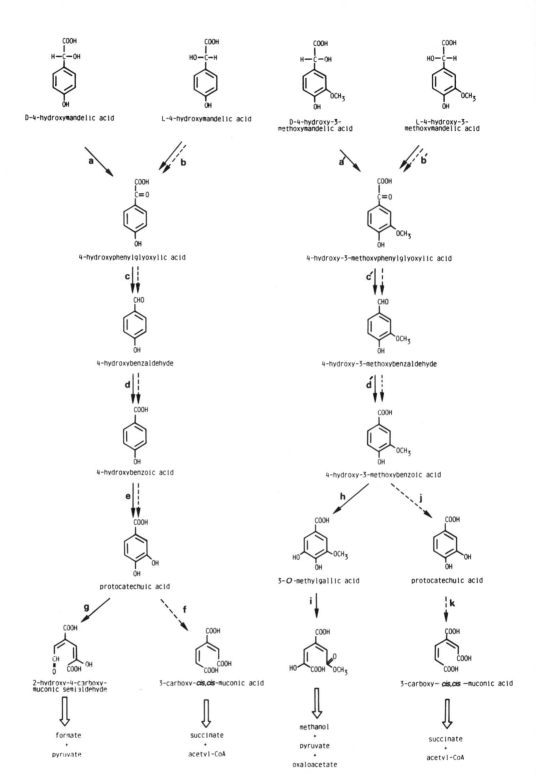

Fig. 7. Utilisation of 4-hydroxymandelate and 4-hydroxy-3-methoxy-mandelate (vanillylmandelate) by *A.calcoaceticus* NCIB8250 (--->) (Kennedy and Fewson, 1968a,b) and *A.lwoffii* (⟶) (Sze and Dagley, 1987).

alcohol) by the additional involvement of an inducible *O*-demethylase (Fig. 7j). 2-Hydroxymandelate and 2-hydroxybenzyl alcohol (saligenin) support growth by oxidation to salicylate, and then conversion to catechol, by means of salicylate hydroxylase (Figs. 3,6e). The much more restricted availability and substrate specificities of the ring-cleavage enzymes means that compounds can be completely oxidised only if they can be converted into an appropriate substrate for one of the ring-cleavage enzymes. The consequence is that many of the substrates of the mandelate enzymes can be only partially oxidised. For example, 3-hydroxymandelate and 3-hydroxybenzyl alcohol are converted into 3-hydroxybenzoate, and this accumulates because there are no hydroxylases which might convert it into protocatechuate or gentisate (2,5-dihydroxybenzoate). Monofluoro-mandelates are oxidised beyond the level of the fluorobenzoates because the small fluorine substitutions can be tolerated to some extent by the later enzymes of the pathway (Clarke et al., 1975).

Sze and Dagley (1987) isolated a strain of *Acinetobacter* which could grow on D,L-4-hydroxy-3-methoxymandelate or D,L-4-hydroxymandelate, but not on either D- or L-mandelate. They isolated the strain from a cattle yard that had presumably been exposed to large amounts of urinary 4-hydroxy-3-methoxymandelate derived from adrenaline. As in strain NCIB8250, the initial enzymes of the mandelate pathway are fairly non-specific in this strain and they convert 4-hydroxy-3-methoxymandelate and 4-hydroxymandelate into 4-hydroxy-3-methoxybenzoate and 4-hydroxybenzoate, respectively, and these are then hydroxylated to form 3-*O*-methylgallate and protocatechuate, which in this strain undergo *meta* cleavage (Fig. 7). Mandelate can be converted into benzoate, but this strain has no benzoate oxygenase, which explains why unsubstituted mandelate cannot support its growth. It would clearly be worthwhile trying to isolate other acinetobacters which can be grown on a variety of substituted mandelates.

METABOLISM OF A WIDE RANGE OF SIMPLE AROMATIC COMPOUNDS

This section summarises some of the work that has been done on the degradation of a great range of aromatic compounds by acinetobacters. Three provisos must be borne in mind: (i) a few of the organisms identified as *Acinetobacter* may in fact have belonged to other genera; however, whenever possible papers have been cited only if the evidence for identification appeared to be adequate; (ii) the literature almost certainly contains descriptions of aromatic degradations carried out by acinetobacters, but where the organisms were not identified or were wrongly assigned to other genera; (iii) there is an unfortunate tendency to postulate catabolic pathways on the basis of inadequate evidence. Whatever the shortcomings of some of the work, the accelerating rate of appearance of publications dealing with aromatic degradation by acinetobacters means that this selection can represent only a very small part of the total repertoire of this versatile organism.

Benzene: a Plasmid-Encoded Pathway in Some Strains

Högn and Jaenicke (1972) isolated a benzene-utilising bacterium from Rhine mud and named it *Moraxella* B; it was, however, oxidase-negative and probably an *Acinetobacter*. Extracts required catalytic amounts of NADH for the conversion of benzene into catechol. This is consistent with the dioxygenation of benzene to benzene *cis*-glycol (cis-1,2-dihydroxycyclohexa-3,5-diene) which is then dehydrogenated to catechol, as occurs in other bacteria (Fig. 6f,g; Fewson, 1988). *A. calcoaceticus* RJE 74, which was isolated from soil, contains the large (about 200 kb), transmissible catabolic plasmid pWW174, and this encodes the enzymes for the catabolism of benzene via the ß-ketoadipate pathway (Winstanley et al., 1987). This was the first report of a plasmid-encoded catechol pathway, and the catechol 1,2-dioxygenase (Fig. 2a) appeared to be significantly more thermostable than the chromosomally-encoded enzyme.

Phenol and Salicylate: Biodeterioration and Bioremediation

Phenol and salicylate are presumably converted into catechol by specific mono-oxygenases, although the evidence rests more on results from pseudomonads and other organisms than from acinetobacters (Figs. 3 and 6d,e). Both compounds are rather toxic to microorganisms, but can support growth of various acinetobacters provided that the concentration is kept low enough (Baumann et al., 1968; Fewson et al., 1970; Jones and Carrington, 1972).

Acinetobacters may be responsible for the biodeterioration of certain pharmaceutical preparations based on salicylate. Acetylsalicylate can be subjected to esterase attack by some acinetobacters, with the salicylate being converted into catechol prior to *ortho*-cleavage (Fig. 3; Grant et al., 1970; Grant, 1971, 1973). Salicylamide degradation by an acinetobacter has also been reported, but in this case the salicylate formed by hydrolysis appeared to be hydroxylated to form gentisate (2,5-dihydroxybenzoate) before ring-cleavage (Grant et al., 1975b).

Kim et al. (1986) reported that *Acinetobacter* strain DY-1 could utilise phenanthrene as sole source of carbon and energy by degrading it to salicylate. There was some evidence that the pathway might be plasmid-encoded.

Some of the classical experiments on the bacterial degradation of aromatics concerned the oxidation of phenol in gasworks' liquor by 'Vibrio 01' (*A. calcoaceticus* NCIB8250) (Evans and Happold, 1939; Evans, 1947; Kilby, 1948). More recently, a great deal of work has been done to clarify the role of acinetobacters in the degradation of phenol in various sorts of industrial and agricultural waste waters. Studies have included examination of the effects of acclimation in activated sludge and various modelling procedures especially aimed at bioremediation of high phenol concentrations, often in the presence of cyanide, thiocyanate or other noxious substances (Carbidenc and Zakharchenko, 1969; Jones and Carrington, 1972; Suzuki and Fujii, 1980, 1985; Borighem and Vereecken, 1981; Anselmo and Novais, 1984; Bourque et al., 1987; Liang et al., 1987; D'Aquino et al., 1988; Korol et al., 1989; Tibbles and Baecker, 1989).

Aniline: Evolution in Action

Various strains of *A. calcoaceticus* can use aniline as sole source of carbon, nitrogen and energy for growth (Wyndham, 1986; Cho et al., 1988). Aniline is converted into catechol (Fig. 3), presumably by a dioxygenase which releases ammonia, and this is followed by *ortho*-cleavage. Most aniline-utilising strains of *A. calcoaceticus* isolated from river water were Ani°, with a low half-saturation constant and a high specific affinity for aniline. A much smaller proportion of the population exhibited an Ani$^+$ phenotype and could grow on relatively high concentrations of aniline (up to 16 mM). Adaptation of mixed river water communities to aniline appeared to depend on the dynamics of parent and mutant populations. In this study there was no evidence that plasmids were involved in aniline degradation, but the frequencies of acquisition and loss of the Ani$^+$ phenotype were thought to be commensurate with repair and subsequent excision of genes by homologous recombination (Wyndham, 1986); however, there is some evidence that aniline degradation may be plasmid-encoded in some acinetobacters (Cho et al., 1988).

3-Hydroxybenzoate and the Gentisate Pathway; 4-Hydroxy-benzoate and the Protocatechuate ortho-Cleavage Pathway

All the *Acinetobacter* strains found by Baumann et al. (1968) to utilise 3-hydroxybenzoate appeared to carry out a monohydroxylation to form gentisate, and this was then ring-cleaved (Fig. 2e). 4-Hydroxybenzoate is usually hydroxylated at the 3-position and the protocatechuate is generally *ortho*-cleaved (Fig. 2c), giving rise to ß-ketoadipate (Cain, 1961), although in some strains it is *meta*-cleaved (Fig. 2d; Sze and Dagley, 1987).

Phenylalanine, Tyrosine, Phenylacetate, Phenylalkanes and 4-Hydroxyphenylacetate: Inter-Related Pathways

Phenylacetate metabolism is common amongst acinetobacters and appears to involve the homogentisate (2,5-dihydroxyphenylacetate) pathway (Figs. 2f and 8d,e; Dagley et al., 1952; Chapman and Dagley, 1962; Baumann et al., 1968; Keil et al., 1983). 4-Hydroxyphenylacetate is hydroxylated to give homoprotocatechuate (3,4-dihydroxyphenylacetate; Fig. 8j), and this is then subjected to *meta*-cleavage by homoprotocatechuate 2,3-dioxygenase (Fig. 2i; Sparnins et al., 1974; Barbour and Bayly, 1977). The regulation of this *meta* pathway has been studied in some detail (Bayly and Barbour, 1984).

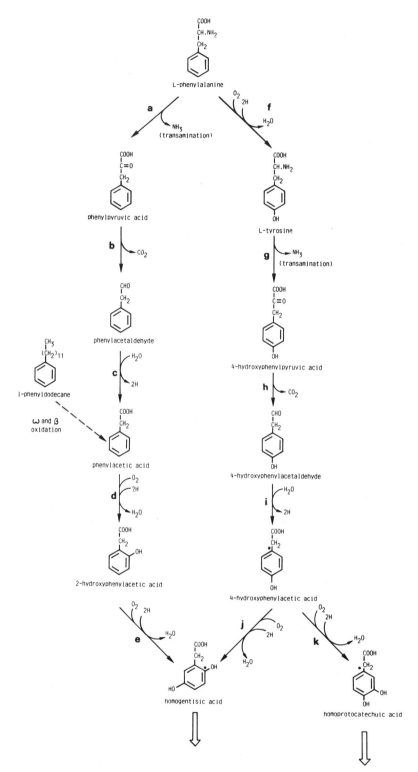

Fig. 8. Degradation via homogentisate and homoprotocatechuate in various strains of *Acinetobacter*. Based on Dagley et al. (1952, 1953), Chapman and Dagley (1962), Keil et al. (1983), Sparnins et al. (1984), Amund and Higgins (1985).

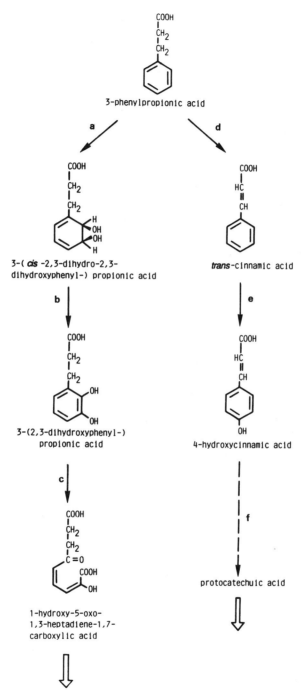

Fig. 9. Metabolism of 3-phenylpropionate and *trans*-cinnamate in various strains of *Acinetobacter*. Based on Keil et al. (1983).

Experiments with 'Vibrio 01' (*A. calcoaceticus* NCIB8250; Dagley et al., 1953), *A. calcoaceticus* EBF65/61 (Barrowman and Fewson, 1985), various strains of *A. calcoaceticus* isolated from soil (Keil et al., 1983) and *Achromobacter eurydice* (Asakawa et al., 1968; - Brisou and Prévot, 1954, have suggested that this organism is probably an acinetobacter) all indicated that L-phenylalanine is degraded via phenylacetate and homogentisate (Fig. 8a-e). Phenylpyruvate decarboxylase (Fig. 8b) has been purified and is a TPP-dependent tetramer with a subunit M_r value of 56,800 (Asakawa et al., 1968; Barrowman and Fewson, 1985).

Amund and Higgins (1985) have suggested, although with little supporting evidence, that *A. lwoffii* AS1 (isolated by elective culture on North Sea Forties crude oil from an activated sludge sample) converts L-phenylalanine to homogentisate via L-tyrosine (Fig. 8f-j). Dagley et al. (1953) had earlier obtained a few results suggesting a similar pathway in 'Vibrio 01'. If this is correct, then it must involve the migration of the acetyl substituent around the aromatic ring in the final hydroxylation step (Fig. 8j[*]; see Hareland et al., 1975). It remains to be seen if this is the pathway followed by the many acinetobacters which can grow on L-tyrosine (Baumann et al., 1968).

A. lwoffii AS1 can completely degrade 1-phenyldodecane. Amund and Higgins (1985) suggested that this was achieved by initial ω-oxidation, followed by ß-oxidation of coenzyme A derivatives, and then production of phenylacetate which was in turn metabolised via homogentisate (Fig. 8d,e). This organism converted 1-phenyltridecane into 3-phenylpropionate and *trans*-cinnamate, again presumably via ω and ß-oxidation. However, in this case these two compounds accumulated, apparently because of the absence of any enzymes for their further metabolism.

Cinnamate and 3-Phenylpropionate: Alternative Pathways

Keil et al. (1983) isolated several strains of *A. calcoaceticus* which could grow on 3-phenylpropionate or *trans*-cinnamate. Most of the 3-phenylpropionate was subjected to attack by a dioxygenase which added two hydroxyl groups to the ring (Fig. 9a) and ring-cleavage was catalysed by 3-(2,3-dihydroxyphenyl-)propionate 1,2-dioxygenase (Figs. 2h and 9c). However, some of the 3-phenylpropionate was oxidised to cinnamate, and this appeared to be converted into protocatechuate (Fig. 9d-f).

DEGRADATION OF LIGNIN-RELATED COMPOUNDS

Many acinetobacters can grow on lignin monomers such as vanillin, vanillate and anisate (e.g. Fewson, 1967a; Golovleva et al., 1983; Hasegawa and Suzuki, 1983). However, this is a common property, shared with many soil microorganisms. The ability of bacteria to grow on larger molecules appears to be a more specialised feature, and it may well be that the primary attack on lignin is generally made by fungi and that bacteria then utilise the

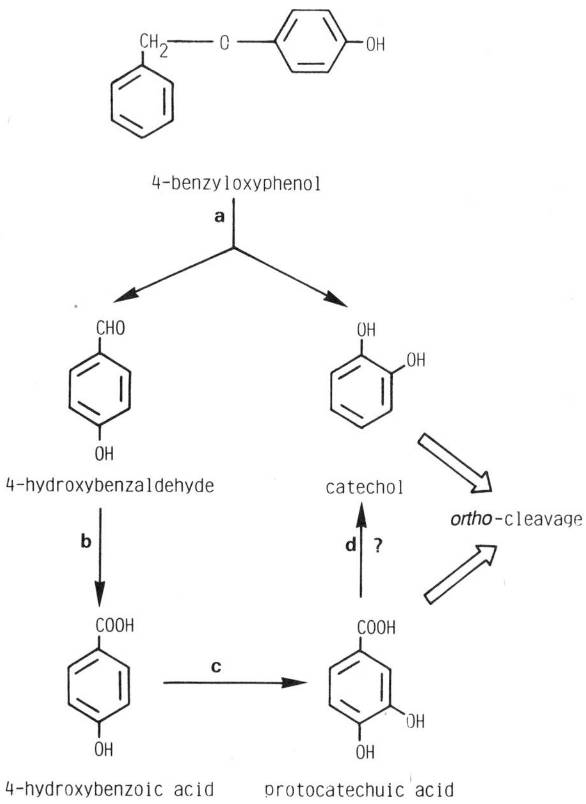

Fig. 10. Metabolism of 4-benzyloxyphenol by an acinetobacter isolated from forest soil. Tentative scheme based on Vasudevan and Mahadevan (1988).

products of fungal metabolism. Nevertheless, it is clear that some bacteria can use substrates more complex than the monomeric products of lignin degradation. For instance, a strain of *A. calcoaceticus* isolated from forest soil could use 4-benzyloxyphenol as sole source of carbon and energy for growth. This compound contains an α-o-4 linkage of the type found in lignin. The evidence for the degradative pathway involved after cleavage of the ether bond (Fig. 10a) is extremely scanty, but is unusual because two ring-cleavage substrates, catechol and protocatechuate, may be formed (Fig. 10; Vasudevan and Mahadevan, 1988). The lignin model compound guaiacol glyceryl ether [3-(2-methoxyphenoxy)1,2-propanediol] appeared to be degraded by another strain of *Acinetobacter* in which the side chain was removed to form guaiacol (*o*-methoxyphenol) and this was converted into catechol prior to *ortho*-cleavage (Vasudevan and Mahadevan, 1989). Another lignin model compound, veratrylglycerol-(*o*-methoxyphenyl) ether was degraded by the mutualistic action of two bacteria in which an acinetobacter first formed guaiacol and this was then used by *Nocardia corallina* (Crawford, 1975).

DEGRADATION OF MULTI-RING COMPOUNDS CONTAINING NITROGEN OR SULPHUR

4-Hydroxyquinazoline was degraded by an acinetobacter isolated from garden soil. There was an initial attack on the heterocyclic ring (Fig. 11a-c) to give anthranilate, which was then deaminated to catechol (Fig. 11d), followed by *meta*-cleavage (Fig. 11; Grant and Al-Najjar, 1975; Grant et al., 1975a).

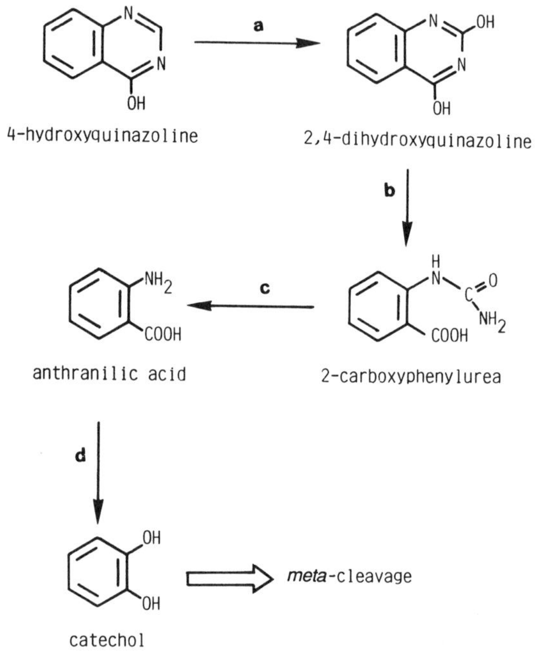

4-hydroxyquinazoline   2,4-dihydroxyquinazoline

anthranilic acid   2-carboxyphenylurea

catechol   *meta*-cleavage

Fig. 11. Metabolism of 4-hydroxyquinazoline by an acinetobacter isolated from garden soil. Based on Grant et al. (1975).

Organisms identified as *Acinetobacter*, or related to *Acinetobacter*, have been isolated which can degrade quinoline (Grant and Al-Najjar, 1976) and isoquinoline (Aislabie et al., 1989).

There is a preliminary report that an *Acinetobacter* strain can use dibenzothiophene as sole source of carbon and energy for growth, but the metabolic pathway involved is not clear (reviewed by Ensley, 1984).

HALOGENATED COMPOUNDS, INCLUDING PCBs AND VARIOUS PESTICIDES

The general topic of biodegradation of halogenated compounds has been well-reviewed (Ghosal et al., 1985; Müller and Lingens, 1986; Rochkind et al., 1986; Reineke and Knackmuss, 1988).

Monohalogenated phenols and benzoates can be mineralised, partially metabolised, or co-metabolised by various strains of *Acinetobacter*. There have been detailed studies of the ability of halogenated compounds to act as inducers and substrates for the relevant enzymes (Beveridge and Tall, 1969; Clark et al., 1975; Reber and Thierbach, 1980; Reber, 1982). Strains of *A. calcoaceticus* able to use 3- or 4-chlorobenzoate as sole sources of carbon and energy for growth have been isolated (Zaitsev and Baskunov, 1985; Adriaens et al., 1989) and 3-chlorobenzoate can be completely oxidised by mixed cultures containing *A. calcoaceticus* and *Alcaligenes faecalis* (Zaitsev, 1988).

Acinetobacters seem to be common members of soil microflora following the application of pentachlorophenol, although this seems to be more a question of tolerance rather than of ability to degrade it (Kato et al., 1980, 1981; Sato et al., 1987).

The herbicide 2,4-dichlorophenoxyacetate can be degraded by various bacteria, including a strain of *Acinetobacter*, and plasmids encoding the enzymes involved in degrading 2,4-dichlorophenoxyacetate can be transferred to *Acinetobacter* strains (Don and Pemberton, 1981). The pathway involves the intermediate formation of 2,4-dichlorophenol, and the enzyme 2,4-dichlorophenol hydroxylase has been purified and partially characterised (Beadle and Smith, 1982; Beadle et al., 1984).

4-Chlorobiphenyl can be completely mineralised by bacteria, including *Acinetobacter* spp., that contain the catabolic plasmid pSS50 (Shields et al., 1985; Hooper et al., 1989). 2,4,4'-Trichlorobiphenyl can be partially metabolised by *Acinetobacter* strain P6, and the pathway of degradation is a useful model for the biodegradation of polychlorinated biphenyls (PCBs) (Fig. 12; Furukawa et al., 1979b). Various acinetobacters are capable of partial metabolism of a range of PCBs, and this has led to systematic studies of the action of acinetobacters on pure PCBs and on commercial mixtures of PCBs such as Kaneclors and Araclor. By and large, the greater the degree of chlorination, the less readily are the PCBs degraded. In addition, isomers with chlorine atoms on only one ring are more readily metabolised than those with substituents on both rings. Isomers with two *ortho*-substitutions (e.g. 2,2' or 2,6) are very recalcitrant to degradation (Furukawa et al., 1978a,b, 1979a, 1982, 1983, 1989; Focht and Brunner, 1985; Kohler et al., 1988).

Several bacteria, including *Acinetobacter*, can carry out the complete dechlorination, demethylation and cleavage of one of the aromatic rings of Methoxychlor [1,1,1-trichloro-2,2-bis(*p*-methoxyphenyl)ethane]

(Golovleva et al., 1984). However, it is likely that the synergistic action of microbial communities is important in the degradation of many pesticides and other residues in Nature. Acinetobacters may well be important constituents of such communities. Thus, co-cultures of *A. calcoaceticus* 4CB1 and P6 were thought to have potential applications for dehalogenation and mineralisation of specific polychlorobiphenyl cogeners because of their complementary metabolic activities (Adriaens et al., 1989). Biodegradative

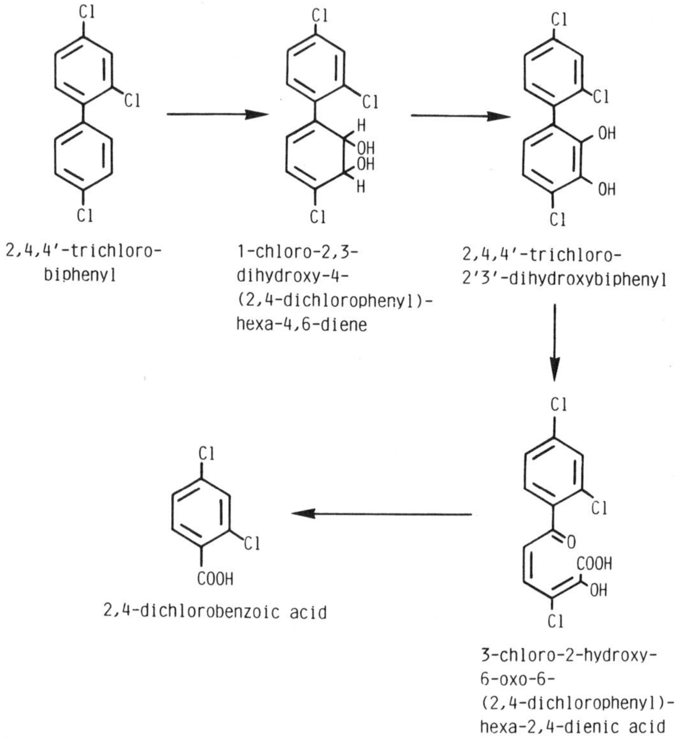

Fig. 12. Primary metabolism of 2,4,4'-trichlorobiphenyl by *Acinetobacter* P6. Based on Furukawa et al. (1979).

studies on Aroclor 1242 indicated that analogue enrichment with biphenyl is the most important factor affecting PCB degradation in soil, but additional enhancement can be achieved by inoculation with *Acinetobacter* (Brunner et al., 1988). A microbial community isolated from wheat root systems was capable of growth on Mecoprop [2-(2-methyl-4-chlorophenoxy)propionate] and contained two *Pseudomonas* species, an *Alcaligenes* species, a *Flavobacterium* species and *A. calcoaceticus* (Lappin et al., 1985).

METHYLATION OF HALOGENATED AROMATICS: A DANGEROUS ALTERNATIVE TO BIODEGRADATION

Microbial *O*-methylation of halogenated phenols can give neutral products that have a high potential for bioconcentration and may be more toxic than the starting compounds. There have been reports of acinetobacters that can *O*-methylate halogenated phenols and thiophenols such as 2,6-dibromophenol, 4,5,6-trichloroguaiacol, pentachlorothiophenol and high molecular mass chlorinated lignin (Neilson et al., 1984, 1988; Allard et al., 1987).

POSSIBLE APPLICATIONS OF ACINETOBACTERS IN AROMATIC WASTE TREATMENT AND THEIR ROLE IN BIODETERIORATION

There are several potential or actual applications of acinetobacters in the degradation of hazardous and unpleasant waste and residue aromatic compounds in addition to those described in previous sections. Examples include the degradation of cresols (Suzuki and Fujii, 1982), phthalic acid esters (Taylor et al., 1981), diphenylethylene (Focht and Joseph, 1984) and anisate (Ventosa et al., 1984; Ferrer et al., 1985); the utilisation of acinetobacters with the potential to degrade aromatic fractions of petroleum (Walker et al., 1975, 1976; Yall, 1979; Ling et al., 1986) and crude oil (Fedorek and Westlake, 1983); fluoranthene-utilisation by bacterial communities growing on the polycyclic aromatic hydrocarbon components of creosote (Mueller et al., 1989); degradation of aromatic hydrocarbons in nuclear reactor coolants (Catelani et al., 1970); degradation of linear alkylbenzene sulphonates (Hrsak et al., 1982).

The possible role of acinetobacters as agents of biodeterioration is exemplified by strain NAV2 which excretes an emulsifier that is capable of emulsifying the saturate and naphthene aromatic fractions of asphalt (Pendrys, 1989).

BIOSYNTHESIS OF AROMATIC AMINO ACIDS AND POSSIBLE INTERACTIONS WITH DEGRADATIVE PATHWAYS

Competition for 5-Dehydroshikimate Between the Route for the Biosynthesis of Aromatic Amino Acids and that for the Catabolism of Hydroaromatic Compounds

Many strains of *A. calcoaceticus* can grow on quinate or shikimate and they all seem to use the shikimate pathway for the biosynthesis of aromatic amino acids and related compounds. Chorismic acid, the key intermediate in the synthesis of aromatic compounds, is formed from erythrose 4-phosphate and phospho*enol*pyruvate by seven enzymes (Fig. 13a,b,f-j). Shikimate and quinate are dehydrogenated by a single, NAD(P)-independent, enzyme (Fig. 13c; Tresguerres et al., 1970a,b). There are two

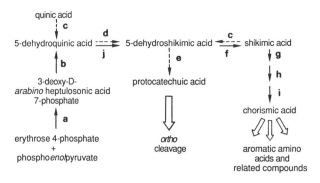

Fig. 13. Competition in *A. calcoaceticus* between the shikimate pathway for the biosynthesis of aromatic amino acids and related compounds (——>) and the pathway for the degradation of quinate and shikimate (--->). Based on Tresguerres et al. (1972), Berlyn and Giles (1973).

dehydroquinase isoenzymes, one of which is constitutive and is involved in biosynthesis (Fig. 13j), and one which is inducible and is involved in the catabolic pathway (Fig. 13d; Ingledew et al., 1971). The biosynthetic and catabolic pathways therefore share common intermediates. It was thought at one time that this problem was overcome by channeling of intermediates within multi-enzyme complexes (Tresguerres et al., 1972), but this does not seem to be the case. Rather, end-product induction of the degradative enzymes by protocatechuate (Canovas et al., 1968; Ingledew et al., 1971) ensures that the biosynthetic intermediates do not induce enzymes for their own degradation, since only at high levels of exogenously supplied quinate or shikimate is enough protocatechuate presumed to be present to induce the catabolic enzymes (Berlyn and Giles, 1973; Ornston and Neidle, this volume). This inclusion of the shikimate/quinate enzymes within the degradative group induced by protocatechuate contrasts with the fact that 4-hydroxybenzoate 3-hydroxylase (Fig. 7e), which also feeds into the protocatechuate pathway, is substrate-induced (Canovas and Stanier, 1967), but this probably causes no difficulty since 4-hydroxybenzoate is not an important anabolic intermediate.

Metabolism of Anthranilate and Tryptophan

Strains of *A.calcoaceticus* that can grow on tryptophan appear to use the kynurenine pathway for catechol formation (Fig. 3). Tryptophan induces the first enzyme, tryptophan oxygenase. The next two enzymes, formamidase and kynureninase, are induced by kynurenine. The last enzyme, anthranilate oxidase, is substrate-induced (Wheelis, 1972), and exogenous anthranilate is a good growth substrate for these strains. Tryptophan is synthesised from anthranilate, which is in turn formed from chorismate produced by the shikimate pathway (Fig. 13; Cohn and Crawford, 1976). Again, the patterns of induction of the catabolic enzymes appear to minimise problems that might arise if the pool of anthranilate required for anabolic purposes was depleted by unregulated catabolism.

Evolutionary Aspects of the Biosynthesis of Tyrosine, Phenylalanine and Tryptophan

Phenylalanine and tyrosine are synthesised from products of the shikimate pathway (see Fig. 13), but are degraded by pathways, described elsewhere in this chapter (see Fig. 8), which do not overlap significantly with the synthetic pathways, and which therefore introduce no particular problems in their regulation. Several enzymes involved in the biosynthesis of aromatic amino acids in *A. calcoaceticus* have been purified and characterised: these include anthranilate synthetase (Sawula and Crawford, 1973), the bifunctional chorismate mutase-prephenate dehydratase (Berry et al., 1985; Ahmad et al., 1988) and two 3-deoxy-D-*arabino*-heptulosonate-7-phosphate synthases (Byng et al., 1985). In addition, some of the genes have been mapped, cloned and sequenced (Sawula and Crawford, 1972; Kaplan et al., 1984; Ross et al., 1990; Haspel et al., this volume). The results obtained from examination of enzyme properties and DNA sequences have led to some intriguing speculations about the evolution of these biosynthetic pathways and about the *Acinetobacter* line of bacterial descent; unfortunately these are beyond the scope of the present chapter, but they have been comprehensively discussed elsewhere (Byng et al., 1985; Ahmad et al., 1987; Ahmad and Jensen, 1988; Jensen and Ahmad, 1988; Crawford, 1989; Ross et al., 1990; Haspel et al., this volume).

ENVOI

The very plethora of aromatic compounds that can be utilised by acinetobacters, together with the scattered and incomplete nature of many of the observations, makes it difficult to draw any significant generalisations, beyond those of the sort illustrated in Fig. 1. Nevertheless, it is clear that detailed studies of all aspects of aromatic metabolism will shed light on the ways in which peripheral metabolic pathways have evolved and are regulated, and will underpin attempts to exploit acinetobacters and other organisms for the biodegradation and bioremediation of hazardous and unpleasant aromatic wastes, spillages and residues.

REFERENCES

Adriaens, P., Kohler, H.P.E., Kohler-Staub, D., and Focht, D.D., 1989, Bacterial dehalogenation of chlorobenzoates and coculture biodegradation of 4,4'-dichlorobiphenyl, *Appl. Environ. Microbiol.*, 55:887.

Ahmad, S., and Jensen, R.A., 1988, The phylogenetic origin of the bifunctional tyrosine-pathway protein in the enteric lineage of bacteria, *Mol. Biol. Evol.*, 5:282.

Ahmad, S., Johnson, J.L., and Jensen, R.A., 1987, The recent evolutionary origin of the phenylalanine-sensitive isozyme of 3-deoxy-D-*arabino*-heptulosonate-7-phosphate synthase in the enteric lineage of bacteria, *J. Mol. Evol.*, 25:159.

Ahmad, S., Wilson, A.T., and Jensen, R.A., 1988, Chorismate mutase: prephenate dehydratase from *Acinetobacter calcoaceticus*. Purification, properties and immunological cross-reactivity, *Eur. J. Biochem.*, 176:69.

Aislabie, J., Rothenburger, S., and Atlas, R.M., 1989, Isolation of microorganisms capable of degrading isoquinoline under aerobic conditions, *Appl. Environ. Microbiol.*, 55:3247.

Allard, A.S., Remberger, M., and Neilson, A.H., 1987, Bacterial *O*-methylation of halogen-substituted phenols, *Appl. Environ. Microbiol.*, 53:839.

Allison, N., and Fewson, C.A., 1986, Purification and properties of L-lactate dehydrogenase from *Acinetobacter calcoaceticus*, *FEMS Microbiol. Lett.*, 36:183.

Allison, N., O'Donnell, M.J., Hoey, M.E., and Fewson, C.A., 1985a, Membrane-bound lactate dehydrogenases and mandelate dehydrogenases of *Acinetobacter calcoaceticus*. Location and regulation of expression, *Biochem. J.*, 227:753.

Allison, N., O'Donnell, M.J., and Fewson, C.A., 1985b, Membrane-bound lactate and mandelate dehydrogenases of *Acinetobacter calcoaceticus*: purification and properties, *Biochem. J.*, 231:407.

Amund, O.O., and Higgins, I.J., 1985, The degradation of 1-phenylalkanes by an oil-degrading strain of *Acinetobacter lwoffi*, *Ant.v.Leeuw.*, 51:45.

Anselmo, A.M., and Novais, J.M., 1984, Isolation and selection of phenol-degrading microorganisms from an industrial effluent, *Biotech. Lett.*, 6:601.

Asakawa, T., Wada, H., and Yamano, T., 1968, Enzymatic conversion of phenylpyruvate to phenylacetate, *Biochim. Biophys. Acta*, 170:375.

Barbour, M.G., and Bayly, R.C., 1977, Regulation of the 4-hydroxy-phenylacetic and meta-cleavage pathway in an *Acinetobacter* sp., *Biochem. Biophys. Res. Comm.*, 79:663.

Barkovsky, A.L., and Shub, G.M., 1986, An *Acinetobacter calcoaceticus* strain utilising various aromatic compounds and carrying a plasmid for resorcin degradation, *Mikrobiologiya*, 55:237.

Barrowman, M.M., and Fewson, C.A., 1985, Phenylglyoxylate decarboxylase and phenylpyruvate decarboxylase from *Acinetobacter calcoaceticus*, *Curr. Microbiol.*, 12:235.

Barrowman, M.M., Harnett, W., Scott, A.J., Fewson, C.A., and Kusel, J.R., 1986, Immunological comparison of microbial TPP-dependent non-oxidative α-keto acid decarboxylases, *FEMS Microbiol. Lett.*, 34:57.

Baumann, P., Doudoroff, M., and Stanier, R.Y., 1968, A study of the *Moraxella* group. II. Oxidase-negative species (genus *Acinetobacter*), *J. Bacteriol.*, 95:1520.

Bayly, R.C., and Barbour, N.G., 1984, The degradation of aromatic compounds by the meta and gentisate pathways, *in* "Microbial Degradation of Aromatic Compounds," p.253, D.T. Gibson, ed., Marcel Dekker, New York.

Beadle, C.A., and Smith, A.R.W., 1982, The purification and properties of 2,4-dichlorophenol hydroxylase from a strain of *Acinetobacter* species, *Eur. J. Biochem.*, 123:323.

Beadle, C.A., Kyprianou, P., Smith, A.R.W., Weight, M.L., and Yon, R.J., 1984, Rapid purification of 2,4-dichlorophenol hydroxylase by biospecific desorption from 10-carboxydecylamino Sepharose, *Biochem. Internat.*, 9:587.

Beggs, J.D., and Fewson, C.A., 1977, Regulation of synthesis of benzyl alcohol dehydrogenase in *Acinetobacter calcoaceticus* NCIB8250, *J. Gen. Microbiol.*, 103:127.

Beggs, J.D., Cook, A.M., and Fewson, C.A., 1976, Regulation of growth of *Acinetobacter calcoaceticus* NCIB8250 on benzyl alcohol in batch culture, *J. Gen. Microbiol.*, 96:365.

Berlyn, M.B., and Giles, N.H., 1973, Organisation of interrelated aromatic metabolic pathway enzymes in *Acinetobacter calco-aceticus*, *J. Gen. Microbiol.*, 74:337.

Berry, A., Byng, G.S., and Jensen, R.A., 1985, Interconvertible molecular-weight forms of the bifunctional chorismate mutase-prephenate dehydratase from *Acinetobacter calcoaceticus*, *Arch. Biochem. Biophys.*, 243:470.

Beveridge, E.G., and Tall, D., 1969, The metabolic availability of phenol analogues to bacterium NCIB8250, *J. Appl. Bacteriol.*, 32:304.

Borighem, G., and Vereecken, J., 1981, Model of a chemostat utilising phenol as inhibitory substrate, *Ecolog. Modelling*, 12:231.

Bourque, D., Bisaillon, J.G., Beaudet, R., Sylvestre, M., Ishaque, M., and Morin, A., 1987, Microbiological degradation of malodorous substances of swine waste under aerobic conditions, *Appl. Environ. Microbiol.*, 53:137.

Brisou, J., and Prévot, A-R., 1954, Études de systématique bactérienne. X. Revision des espèces réunies dans le genre *Achromobacter*, *Ann. Inst. Past.*, 86:722.

Brunner, W., Sutherland, F.H., and Focht, D.D., 1988, Enhanced biodegradation of polychlorinated biphenyls by analog enrichment and bacterial inoculation, *J. Environ. Qual.*, 14:324.

Burlingame, R., and Chapman, P.J., 1983, Stereospecificity in meta-fission catabolic pathways, *J. Bacteriol.* 155:424.

Byng, G.S., Berry, A., and Jensen, R.A., 1985, Evolutionary implications of features of aromatic amino acid biosynthesis in the genus *Acinetobacter*, *Arch. Microbiol.*, 143:122.

Cain, R.B., 1961, The metabolism of protocatechuic acid by a Vibrio, *Biochem. J.*, 79:298.

Canovas, J.L., and Johnson, B.F., 1968, Regulation of the enzymes of the ß-ketoadipate pathway in *Moraxella calcoacetica*. 4. Constitutive synthesis of ß-ketoadipate succinyl-CoA transferases II and III, *Eur. J. Biochem.*, 3:312.

Canovas, J.L., and Stanier, R.Y., 1967, Regulation of the enzymes of the ß-ketoadipate pathway in *Moraxella calcoacetica*. 1. General aspects, *Eur. J. Biochem.*, 1:289.

Canovas, J.L., Ornston, L.N., and Stanier, R.Y., 1967, Evolutionary significance of metabolic control systems, *Science*, 156:1695.

Canovas, J.L., Johnson, B.F., and Wheelis, M.L., 1968, Regulation of the enzymes of the ß-ketoadipate pathway in *Moraxella calcoacetica*. 3. Effects of 3-hydroxy-4-methylbenzoate on the synthesis of enzymes of the protocatechuate branch, *Eur. J. Biochem.*, 3:305.

Canovas, J.L., Wheelis, M.L., and Stanier, R.Y., 1968, Regulation of the enzymes of the ß-ketoadipate pathway in *Moraxella calcoacetica*. 2. The role of protocatechuate as inducer, *Eur. J. Biochem.*, 3:293.

Carbidenc, R., and Zakharchenko, V., 1969, Possibilities of biological decomposition of an effluent incorporating formol and phenol, *Terres Eaux*, 61:8.

Catelani, D., Mosselmans, G., Nienhaus, J., Sorlini, C., and Treccani, V., 1970, Microbial degradation of aromatic hydrocarbons used in reactor coolants, *Experientia*, 26:922.

Chalmers, R.M., and Fewson, C.A., 1989a, The evolution of metabolic pathways: an immunological approach to the evolution of aromatic and aldehyde dehydrogenases, *in* "Enzymology and Molecular Biology of Carbonyl Metabolism", p.193, H.Weiner and T.G. Flynn, eds., Arthur Liss, New York.

Chalmers, R.M., and Fewson, C.A., 1989b, Purification and characterization of benzaldehyde dehydrogenase I from *Acinetobacter calcoaceticus*, *Biochem. J.*, 263:913.

Chapman, P.J., and Dagley, S., 1962, Oxidation of homogentisic acid by cell-free extracts of a vibrio, *J. Gen. Microbiol.*, 28:251.

Chari, R.V.J., Whitman, C.P., Kozarich, J.W., Ngai, K.L., and Ornston, L.N., 1987a, Absolute stereochemical course of the 3-carboxymuconate cycloisomerase from *Pseudomonas putida* and *Acinetobacter calcaceticus:* analysis and implications, *J. Amer. Chem. Soc.*, 109:5514.

Chari, R.V.J., Whitman, C.P., Kozarich, J.W., Ngai, K.L., and Ornston, L.N., 1987b, Absolute stereochemical course of muconolactone δ-lactoneisomerase and of 4-carboxymuconolactone decarboxylase: proton NMR 'ricochet' analysis, *J. Amer. Chem. Soc.*, 109:5520.

Cho, K.Y., Ha, I.H., Baf, K.S., and Kho, Y.H., 1988, Isolation and characterization of aniline degrading bacteria, *Sanop Misaengmul Hakhoechi*, 16:421.

Clarke, K.F., Callely, A.G., Livingstone, A., and Fewson, C.A., 1975, Metabolism of monofluorobenzoates by *Acinetobacter calcoaceticus* NCIB8250, *Biochim. Biophys. Acta*, 404:169.

Cohn, W., and Crawford, I.P., 1976, Regulation of enzyme synthesis in the tryptophan pathway of *Acinetobacter calcoaceticus*, *J. Bacteriol.*, 127:367.

Cook, A.M., and Fewson, C.A., 1972, Evidence for specific transport mechanisms for aromatic compounds in bacterium NCIB8250, *Biochim. Biophys. Acta*, 290:384.

Cook, A.M., Beggs, J.D., and Fewson, C.A., 1975, Regulation of growth of *Acinetobacter calcoaceticus* NCIB8250 on L-mandelate in batch culture, *J. Gen. Microbiol.*, 91:325.

Crawford, I.P., 1989, Evolution of a biosynthetic pathway: the tryptophan paradigm, *Ann. Rev. Microbiol.*, 43:567.

Crawford, R.L., 1975, Mutualistic degradation of the lignin model compound veratrylglycerol (*o*-methoxyphenyl)ether by bacteria, *Can. J. Microbiol.*, 21:1654.

Dagley, S., 1978, Pathways for the utilization of organic growth substrates, *in* "The Bacteria, vol.VI, Bacterial Diversity," p.305, L.N. Ornston and J.R. Sokatch, eds., Academic Press, New York.

Dagley, S., Fewster, M.E., and Happold, F.C., 1952, The bacterial oxidation of phenylacetic acid, *J. Bacteriol.*, 63:327.

Dagley, S., Fewster, M.E., and Happold, F.C., 1953, The bacterial oxidation of aromatic compounds, *J. Gen. Microbiol.*, 8:1.

D'Aquino, M., Korol, S., Santini, P., and Moretton, J., 1988, Biodegradation of phenolic compounds. I. Improved degradation of phenol and benzoate by indigenous strains of *Acinetobacter* and *Pseudomonas, Revista Latinoamer.Microbiologia*, 30:283.

Darrah, J.A., and Cain, R.B., 1969, A convenient biological method for preparing ß-ketoadipic acid, *Lab. Pract.*, 16:989.

Don, R.H., and Pemberton, J.M., 1981, Properties of six pesticide degradation plasmids isolated from *Alcaligenes paradoxus* and *Alcaligenes eutrophus, J. Bacteriol.*, 145:681.

Doten, R.C., Ngai, K.L., Mitchell, D.J., and Ornston, L.N., 1987a, Cloning and genetic organization of the *pca* cluster from *Acinetobacter calcoaceticus, J. Bacteriol.*, 169:3168.

Doten, R.C., Gregg, L.A., and Ornston, L.N., 1987b, Influence of the *cat*BCE sequence on the phenotypic reversion of a *pca*E mutation in *Acinetobacter calcoaceticus, J. Bacteriol.*, 169:3175.

Durham, D.R., Stirling, L.A., Ornston, L.N., and Perry, J.J., 1980, Intergeneric evolutionary homology revealed by the study of protocatechuate 3,4-dioxygenase from *Azotobacter vinelandii, Biochemistry*, 19:149.

Ensley, B.D., 1984, Microbial metabolism of condensed thiophenes, *in* "Microbial Degradation of Aromatic Compounds," p.309, D.T. Gibson, ed., Marcel Dekker, New York.

Evans, W.C., 1947, Oxidation of phenol and benzoic acid by some soil bacteria, *Biochem. J.*, 41:373.

Evans, W.C., 1988, Anaerobic degradation of aromatic compounds, *Ann. Rev. Microbiol.*, 42:289.

Evans, W.C., and Happold, F.C., 1939, The utilization of phenol by bacteria, *J. Chem. Soc.*, 58:55.

Fedorak, P.M., and Westlake, D.W.S., 1983, Selective degradation of biphenyl and methylbiphenyls in crude oil by two strains of marine bacteria, *Can. J. Microbiol.*, 29:497.

Ferrer, M.R., Del Moral, A., Ruiz-Berraquero, F., and Ramos-Cormenzano, A., 1985, Ability of *o*-anisate degrading microorganisms to cometabolize DICAMBA and other related pesticides, *Chemosphere*, 14: 1645.

Fewson, C.A., 1967a, The growth and metabolic versatility of the gram-negative bacterium NCIB8250 ('Vibrio 0l'), *J. Gen. Microbiol.*, 46:255.

Fewson, C.A., 1967b, The identity of the Gram-negative bacterium NCIB8250 ('Vibrio 0l'), *J. Gen. Microbiol.*, 48:107.

Fewson, C.A., 1981, Biodegradation of aromatics with industrial relevance, *in* "Microbial Degradation of Xenobiotics and Recalcitrant Compounds," p.143, T.Leisinger, R. Hutter, A.M. Cook and J. Nuesch, eds., Academic Press, London.

Fewson, C.A., 1985, Growth yields and respiratory efficiency of *Acinetobacter calcoaceticus*, *J. Gen. Microbiol.*, 131:865.

Fewson, C.A., 1988, Microbial metabolism of mandelate: a microcosm of diversity, *FEMS Microbiol. Rev.*, 54:85.

Fewson, C.A., Barclay, A., and Eason, R., 1970, Batch culture and computer simulation of growth of bacterium NCIB8250 using inhibitory substrates, *J. Gen. Microbiol.*, 61:i.

Fewson, C.A., Livingstone, A., and Moyes, H.M., 1978, Mutants of *Acinetobacter calcoaceticus* NCIB8250 constitutive for the mandelate enzymes, *J. Gen. Microbiol.*, 106:233.

Fewson, C.A., Allison, N., Hamilton, I.D., Jardine, J., and Scott, A.J., 1988, Comparison of mandelate dehydrogenase from various strains of *Acinetobacter calcoaceticus*: similarity of natural and evolved forms, *J. Gen. Microbiol.*, 134:967.

Focht, D.D., and Brunner, W., 1985, Kinetics of biphenyl and polychlorinated biphenyl metabolism in soil, *Appl. Environ. Microbiol.*, 50:1058.

Focht, D.D., and Joseph, H., 1984, Degradation of 1,1-diphenylethylene by mixed cultures, *Can. J. Microbiol.*, 20:631.

Furukawa, K., and Chakrabarty, A.M., 1982, Involvement of plasmids in total degradation of chlorinated biphenyls, *Appl. Environ. Microbiol.*, 44:619.

Furukawa, K., Matsumura, F., and Tonomura, K., 1978a, *Alcaligenes* and *Acinetobacter* strains capable of degrading polychlorinated biphenyls, *Agric. Biol. Chem.*, 42:543.

Furukawa, K., Tonomura, K., and Kamibayashi, A., 1978b, Effect of chlorine substitution on the biodegradability of polychlorinated biphenyls, *Appl. Environ. Microbiol.*, 35:223.

Furukawa, K., Tomizuka, N., and Kamibayashi, A., 1979a, Effect of chlorine substitution on the bacterial metabolism of various polychlorinated biphenyls, *Appl. Environ. Micrbiol.*, 38:301.

Furukawa, K., Tonomura, K., and Kamibayashi, A., 1979b, Metabolism of 2,4,4'-trichlorobiphenyl by *Acinetobacter* sp. P6, *Agric. Biol. Chem.*, 43:1577.

Furukawa, K., Tomizuka, N., and Kamibayashi, A., 1982, Bacterial degradation of polychlorinated biophenyls (PCB) and their metabolites, *Adv. Exp. Med. Biol.*, 136A:407.

Furukawa, K., Tomizuka, N., and Kamibayashi, A., 1983, Metabolic breakdown of Kenoclors (polychlorobiphenyls) and their products by *Acinetobacter* sp., *Appl. Environ. Microbiol.*, 46:140.

Furukawa, K., Hayase, N., Taira, K., and Tomizuka, N., 1989, Molecular relationship of chromosomal genes encoding biphenyl/polychlorinated biphenyl catabolism: some soil bacteria possess a highly conserved *bph* operon, *J. Bacteriol.*, 171:5467.

Ghosal, D., You, I.-S., Chatterjee, D.K., and Chakrabarty, A.M., 1985, Microbial degradation of halogenated compounds, *Science*, 228:135.

Golovleva, L.A., Pertsova, R.N., Fedechkina, I.E., Ganbarov, K.G., and Baskunov, B.P., 1983, Microbial decomposition of lignin and a possible role of dioxygenases in its destruction, *Prikladnaya Biokhimiya i Mikrobiologiya*, 19:709.

Golovleva, L.A., Polyakova, A.B., Pertsova, R.N., and Finkel'shtein, Z.I., 1984, The fate of methoxychlor in soils and transformation by soil microorganisms, *J. Environ. Sci. Health*, B19:523.

Grant, D.J.W., 1971, Degradation of acetylsalicylic acid by a strain of *Acinetobacter lwoffii*, *J. Appl. Bacteriol.*, 34:689.

Grant, D.J.W., 1973, The degradative versatility, arylesterase activity and hydroxylation reactions of *Acinetobacter lwoffii* NCIB10553, *J. Appl. Bacteriol.*, 36:47.

Grant, D.J.W., and Al-Najjar, T.R., 1975, Degradation of 4-hydroxy-quinazoline by an *Acinetobacter* and a *Pseudomonas* isolated from soil, *Microbios*, 14:55.

Grant, D.J.W., and Al-Najjar, T.R., 1976, Degradation of quinoline by a soil bacterium, *Microbios*, 15:177.

Grant, D.J.W., de Szecs, J., and Wilson, J.V., 1970, Utilization of acetylsalicylic acid as sole carbon source and the induction of its enzymatic hydrolysis by an isolated strain of *Acinetobacter lwoffii*, *J. Pharmaceut. Pharmacol.*, 22:461.

Grant, D.J.W., Al-Najjar, T.R., Barber, H.L., and Harbour, M.J., 1975a, Tangential pathways of catechol cleavage in an *Acinetobacter* which degrades 4-hydroxyquinazoline and in a *Micrococcus* which degrades 2,4-dihydroxyquinazoline, *Microbios*, 14:195.

Grant, D.J.W., Burgess, S.M., Rambow, E.J., and Yelling, P.W., 1975b, Bacterial degradation of salicylamide, *J. Pharmaceut. Pharmacol.*, 27:49P.

Happold, F.C., and Key, A., 1932, The bacterial purification of gasworks' liquors. The action of the liquors on the bacterial flora of sewage, *J. Hyg.*, 32:573.

Hareland, W.A., Crawford, R.L., Chapman, P.J., and Dagley, S., 1975, Metabolic function and properties of 4-hydroxyphenylacetic acid 1-hydroxylase from *Pseudomonas acidovorans*, *J. Bacteriol.*, 121:272.

Hasegawa, S., and Suzuki, M., 1983, Microbiological decomposition of aromatic compounds by *Acinetobacter calcoaceticus*, *Hokkaidoritsu Eisei Kenkynshoho*, 33:134.

Hills, C.A., and Fewson, C.A., 1983a, Mutant strains of *Acinetobacter calcoaceticus* possessing additional mandelate dehydrogenases. Identification and preliminary characterization of the enzymes, *Biochem. J.*, 209:379.

Hills, C.A., and Fewson, C.A., 1983b, Regulation of expression of novel mandelate dehydrogenases in mutants of *Acinetobacter calcoaceticus, J. Gen. Microbiol.,* 129:2009.

Hoey, M.E., Allison, N., Scott, A.J., and Fewson, C.A., 1987, Purification and properties of L-mandelate dehydrogenase and comparison with other membrane-bound dehydrogenases from *Acinetobacter calcoaceticus, Biochem. J.,* 248:871.

Hoey, M.E., and Fewson, C.A., 1990, Is there a mandelate complex in *Acinetobacter calcoaceticus* or *Pseudomonas putida? J. Gen. Microbiol.,* 136:219.

Högn, T., and Jaenicke, L., 1972, Benzene metabolism of *Moraxella* species, *Eur. J. Biochem.,* 30:369.

Hooper, S.W., Dockendorff, T.C., and Sayler, G.S., 1989, Characteristics and restriction analysis of the 4-chlorobiphenyl catabolic plasmid pSS50, *Appl. Environ. Microbiol.,* 55:1286.

Hou, C.T., Lillard, M.O., and Schwartz, R.D., 1976, Protocatechuate 3,4-dioxygenase from *Acinetobacter calcoaceticus, Biochemistry,* 15: 582.

Hou, C.T., Patel, R., and Lillard, M.O., 1977, Extradiol cleavage of 3-methylcatechol by catechol 1,2-dioxygenase from various microorganisms, *Appl. Environ. Microbiol.,* 33:725.

Hou, C.T., Patel, R.N., Lillard, M.O., Felix, A., and Florance, J., 1978, Circular dichroism of catechol 1,2-dioxygenase from *Acinetobacter calcoaceticus, Bioorgan. Chem.,* 7:115.

Hrsak, D., Busnjak, M., and Johanides, V., 1982, Enrichment of linear alkylbenzene sulphonate (LAS) degrading bacteria in continuous culture, *J. Appl. Bacteriol.,* 53:413.

Hughes, E.J., Shapiro, M.K., Houghton, J.E., and Ornston, L.N., 1988, Cloning and expression of *pca* genes from *Pseudomonas putida* in *Escherichia coli, J. Gen. Microbiol.,* 134:2877.

Ingledew, W.M., Tresguerres, M.E.F., and Canovas, J.L., 1971, Regulation of the enzymes of the hydroaromatic pathway in *Acinetobacter calco-aceticus, J. Gen. Microbiol.,* 68:273.

Jensen, R.A., and Ahmad, S., 1988, Evolution and phylogenetic distribution of the specialized isozymes of 3-deoxy-D-arabino-heptulosonate 7-phosphate synthase in Superfamily-B prokaryotes, *Microbiol. Sci.,* 5:316.

Jones, G.L., and Carrington, E.G., 1972, Growth of pure and mixed cultures of microorganisms concerned in the treatment of carbonization waste liquors, *J. Appl. Bacteriol.,* 35:395.

Juni, E., and Janik, A., 1969, Transformation of *Acinetobacter calco-aceticus (Bacterium anitratum), J. Bacteriol.,* 98:281.

Kaplan, J.B., Goncharoff, P., Seibold, A.M., and Nichols, B.P., 1984, Nucleotide sequence of the *Acinetobacter calcoaceticus trpGDC* gene cluster, *Mol. Biol. Evol.,* 1:456.

Katagiri, M., and Wheelis, M.L., 1971, Comparison of the two isofunctional enol-lactone hydrolases from *Acinetobacter calcoaceticus, J. Bacteriol.,* 106:369.

Kato, H., Sato, K., and Furusaka, C., 1980, Effect of pentachlorophenol (PCP) on soil bacterial flora in soil suspension, *Tohoka Daigaku Nagaku Kenkyuosho Hokoku*, 32:1.

Kato, H., Sato, K., and Furusaka, C., 1981, Effects of PCP on bacterial flora in water-logged soil, *Nippon Noyaku Gakkaishi*, 6:163.

Keil, H., Tittmann, U., and Lingens, F., 1983, Degradation of aromatic carboxylic acids by *Acinetobacter*, *Syst. Appl. Microbiol.*, 4:313.

Kennedy, S.I.T., and Fewson, C.A., 1968a, Metabolism of mandelate and related compounds by bacterium NCIB8250, *J. Gen. Microbiol.*, 53:259.

Kennedy, S.I.T., and Fewson, C.A., 1968b, Enzymes of the mandelate pathway in bacterium NCIB8250, *Biochem. J.*, 107:497.

Kilby, B.A., 1948, The bacterial oxidation of phenol to ß-ketoadipic acid, *Biochem. J.*, 43:v.

Kilby, B.A., 1951, The formation of ß-ketoadipic acid by bacterial fission of aromatic rings, *Biochem. J.*, 49:671.

Kim, C.K., Kim, J.W., Kim, W.C., and Mheen, T.I., 1986, Isolation of aromatic hydrocarbon-degrading bacteria and genetic characterization of their plasmid genes, *Misaengmal Hakhoechi*, 24:67.

Kohler, H.-P. E., Kohler-Staub, D., and Focht, D.D., 1988, Cometabolism of polychlorinated biphenyls: enhanced transformation of Aroclor 1254 by growing bacterial cells, *Appl. Environ. Microbiol.*, 54:1940.

Korol, S., Orsingher, M., Santani, P., Moretton, J., and D'Aquino, M., 1989, Biodegradation of phenolic compounds. II. Effects of inoculum, xenobiotic concentration and adaptation on *Acinetobacter* and *Pseudomonas* phenol degradation, *Revista Latinoamer. Microbiol.*, 31:117.

Kozarich, J.W., 1988, Enzyme chemistry and evolution in the ß-ketoadipate pathway, *in* "Microbial Metabolism and the Carbon Cycle," p.283, S.R. Hagedorn, R.S. Hanson and D.A. Kunz, eds., Harwood, London.

Lappin, H.M., Greaves, M.P., and Slater, J.H., 1985, Degradation of the herbicide Mecoprop [2-(2-methyl-4-chlorophenoxy)propionic acid] by a synergistic microbial community, *Appl. Environ. Microbiol.*, 49:429.

Leung, P.-T., Chapman, P.J. and Dagley, S., 1974, Purification and properties of 4-hydroxy-2-ketopimelate aldolase from *Acinetobacter*, *J. Bacteriol.*, 120:168.

Liang, C., Li, L., Quin, C., Fan, L., and Yuan, X., 1987, Phenol degradation by *Acinetobacter* in wastewater from Xiangan Chemical Industrial Plant, *Beijing Daxne Xuebao, Ziran Kexueban*, 77.

Ling, Z., Xiong, X., and Qian, C., 1986, Application of microbial degradation in the treatment of petrochemical waste, *Huanjing Huaxne*, 5:7.

Livingstone, A., and Fewson, C.A., 1972, Regulation of the enzymes converting mandelate into benzoate in bacterium NCIB8250, *Biochem. J.*, 130:937.

Livingstone, A., Fewson, C.A., Kennedy, S.I.T., and Zatman, L.J., 1972, Two benzaldehyde dehydrogenases in bacterium NCIB8250. Distinguishing properties and regulation, *Biochem. J.*, 130:927.

Lovrien, R., Jorgenson, G., Ma, M.K., and Sund, W.E., 1980, Microcalorimetry of microorganism metabolism of monosaccharides and simple aromatic compounds, *Biotech. Bioeng.*, 22:1249.

MacKintosh, R.W., and Fewson, C.A., 1987, Microbial aromatic alcohol and aldehyde dehydrogenases, *in* "Enzymology and Molecular Biology of Carbonyl Metabolism," p.259, H. Weiner, and T.G. Flynn, eds., Arthur Liss, New York.

MacKintosh, R.W., and Fewson, C.A., 1988a, Benzyl alcohol dehydrogenase and benzaldehyde dehydrogenase II from *Acinetobacter calcoaceticus.* Purification and preliminary characterization, *Biochem. J.*, 250:743.

MacKintosh, R.W., and Fewson, C.A., 1988b, Benzyl alcohol dehydrogenase and benzaldehyde dehydrogenase II from *Acinetobacter calcoaceticus.* Substrate specificities and inhibition studies, *Biochem. J.*, 255:653.

McCorkle, G.M., Yeh, W.-K., Fletcher, P., and Ornston, L.N., 1980, Repetitions in the NH_2-terminal amino acid sequence of ß-ketoadipate enol-lactone hydrolase from *Pseudomonas putida*, *J. Biol. Chem.*, 255:6335.

Mueller, J.G., Chapman, P.J., and Pritchard, P.H., 1989, Action of a fluoranthene-utilizing bacterial community on polycyclic aromatic hydrocarbon components of creosote, *Appl. Environ. Microbiol.*, 55:3085.

Müller, R., and Lingens, F., 1986, Microbiological degradation of halogenated hydrocarbons: a biological solution to pollution problems?, *Angewandte Chemie, Internat. Edn.*, 25:779.

Neidle, E.L., and Ornston, L.N., 1986, Cloning and expression of *Acinetobacter calcoaceticus* catechol 1,2-dioxygenase structural gene *cat*A in *Escherichia coli, J. Bacteriol.*, 168:815.

Neidle, E.L., and Ornston, L.N., 1987, Benzoate and muconate, structurally dissimilar metabolites, induce expression of *cat*A in *Acinetobacter calcoaceticus, J. Bacteriol.*, 169:414.

Neidle, E.L., Shapiro, M.K., and Ornston, L.N., 1987, Cloning and expression in *Escherichia coli* of *Acinetobacter calcoaceticus* genes for benzoate degradation, *J. Bacteriol.*, 169:5496.

Neidle, E.L., Hartnett, C., Bonitz, S., and Ornston, L.N., 1988, DNA sequence of the *Acinetobacter calcoaceticus* catechol 1,2-dioxygenase I structural gene *cat*A: evidence for evolutionary divergence of intradiol dioxygenases by acquisition of DNA sequence repetition, *J. Bacteriol.*, 170:4874.

Neilson, A.H., Allard, A.S., Reilands, S., Remberger, M., Taernom, A., Victor, T., and Landner, L., 1984, Tri- and tetrachloroveratrole, metabolites produced by bacterial *O*-methylation of tri- and tetrachloroguaiacol: an assessment of their bioconcentration potential and their effects on fish reproduction, *Can. J. Fisher. Aquat. Sci.*, 41:1502.

Neilson, A.H., Lindgren, C., Hynning, P.A., and Remberger, M., 1988, Methylation of halogenated phenols and thiophenols by cell extracts of gram-positive and gram-negative bacteria, *Appl. Environ. Microbiol.*, 54:524.

Ornston, L.N., and Parke, D., 1977, The evolution of induction mechanisms in bacteria: insights derived from the study of the ß-ketoadipate pathway, *Curr. Top. Cell. Reg.*, 12:210.

Ornston, L.N., and Stanier, R.Y., 1964, Mechanism of ß-ketoadipate formation by bacteria, *Nature*, 204:1279.

Ornston, L.N., and Yeh, W.K., 1979, Origin of metabolic diversity: evolutionary divergence by sequence repetition, *Proc. Natl. Acad. Sci. USA*,, 76:3996.

Ornston, L.N., and Yeh, W.K., 1981, Toward molecular natural history, *Microbiology*, 140-143.

Patel, R.N., and Ornston, L.N., 1976, Immunological comparison of enzymes of the ß-ketoadipate pathway, *Arch. Microbiol.*, 110:27.

Patel, R.N., Meagher, R.B., and Ornston, L.N., 1974, Relationships among enzymes of the ß-ketoadipate pathway. IV. Muconolactone isomerase from *Acinetobacter calcoaceticus* and *Pseudomonas putida*, *J. Biol. Chem.*, 249:7410.

Patel, R.N., Mazumdar, S., and Ornston, L.N., 1975, ß-Ketoadipate enol-lactone hydrolases I and II from *Acinetobacter calcoaceticus*, *J. Biol. Chem.*, 250:6567.

Patel, R.N., Hou, C.T., Felix, A., and Lillard, M.O., 1976, Catechol 1,2-dioxygenase from *Acinetobacter calcoaceticus*: purification and properties, *J. Bacteriol.*, 127:536.

Pendrys, J.P., 1989, Biodegradation of asphalt cement-20 by aerobic bacteria, *Appl. Environ. Microbiol.*, 55:1357.

Reber, H.H., 1982, Inducibility of benzoate oxidizing cell activities in *Acinetobacter calcoaceticus* strain Bs5 by chlorobenzoates as influenced by the position of chlorine atoms and the inducer composition, *Eur. J. Appl. Microbiol. Biotech.*, 15:138.

Reber, H.H., and Thierbach, G., 1980, Physiological studies on the oxidation of 3-chlorobenzoate by *Acinetobacter calcoaceticus* strain Bs5, *Eur. J. Appl. Microbiol.*, 10:223.

Reineke, W., and Knackmuss, H.-J., 1988, Microbial degradation of haloaromatics, *Ann. Rev. Microbiol.*, 42:263.

Reiner, A.M., 1971, Metabolism of benzoic acid by bacteria: 3,5-cyclohexadiene-1,2-diol-1-carboxylic acid is an intermediate in the formation of catechol, *J. Bacteriol.*, 108:89.

Rochkind, M.L., Blackburn, J.W., and Sayler, G.S., 1986, "Microbial Decomposition of Aromatic Compounds," United States Environmental Protection Agency, Cincinnati.

Ross, C.M., Kaplan, J.B., and Winkler, M.E., 1990, An evolutionary comparison of *Acinetobacter calcoaceticus trpF* with *trpF* genes of several organisms, *Mol. Biol. Evol.*, 7:74.

Sato, K., Kato, H., and Furusak, C., 1987, Comparative study of soil bacterial flora as influenced by the application of a pesticide, pentachlorophenol (PCP), *Plant Soil*, 100:333.

Sawula, R.V., and Crawford, I.P., 1972, Mapping of the tryptophan genes of *Acinetobacter calcoaceticus* by transformation, *J. Bacteriol.*, 112:797.

Sawula, R.V. and Crawford, I.P., 1973, Anthranilate synthetase of *Acinetobacter calcoaceticus.* Separation and partial characterisation of subunits, *J. Biol. Chem.*, 248:3573.

Seltzer, S., 1973, Purification and properties of maleylacetone cis-trans isomerase from Vibrio 01, *J. Biol. Chem.*, 248:215.

Shanley, M.S., Neidle, E.L., Parales, R.E., and Ornston, L.N., 1986, Cloning and expression of *Acinetobacter calcoaceticus cat* BCDE genes in *Pseudomonas putida* and *Escherichia coli*, *J. Bacteriol.*, 165:557.

Shanley, M.S., Neidle, E.L., Parales, R.E., Harrison, A., and Ornston, L.N., 1989, Organization of catabolic pathway genes in *Acinetobacter calcoaceticus*, *SAAS Bull. Biochem. Biotech.*, 2:1.

Shields, M.S., Hooper, S.W., and Sayler, G.S., 1985, Plasmid-mediated mineralization of 4-chlorobiphenyl, *J. Bacteriol.*, 163:882.

Sparnins, V.L., Chapman, P.J., and Dagley, S., 1974, Bacterial degradation of 4-hydroxyphenylacetic acid and homoprotocatechuic acid, *J. Bacteriol.*, 120:159.

Stanier, R.Y., and Ornston, L.N., 1973, The ß-ketoadipate pathway, *Adv. Microb. Physiol.*, 9:89.

Suzuki, M., and Fujii, T., 1980, Biological treatment of phenolic waste water, *Seisan Kenkyu*, 32:110.

Suzuki, M., and Fujii, T., 1982, Biodegradation rate of aromatic compounds in wastewater treatment, *Seisan Kenkyu*, 34:156.

Suzuki, M., and Fujii, T., 1985, Microbiological treatment of phenol wastewater. Effect of coexisting cyanide and thiocyanate ion on decomposition rate, *Seisan Kenkyu*, 37:312.

Sze, I.S.-Y., and Dagley, S., 1987, Degradation of substituted mandelic acids by meta fission reactions, *J. Bacteriol.*, 169:3833.

Taylor, B.F., Curry, R.W., and Corcoran, E.F., 1981, Potential for biodegradation of phthalic acid esters in marine regions, *Appl. Environ. Microbiol.*, 42:590.

Tempest, D.W., 1978, The biochemical significance of microbial growth yields: a reassessment, *Trends Biochem. Sci.*, 3:180.

Tibbles, B.J., and Baecker, A.A.W., 1989, Effect of pH and inoculum size on phenol degradation by bacteria isolated from landfill waste, *Environ. Poll.*, 59:227.

Tresguerres, M.E.F., de Torrontegui, G., and Canovas, J.L., 1970a, Metabolism of quinate by *Acinetobacter calcoaceticus*, *Arch. Mikrobiol.*, 70:110.

Tresguerres, M.E.F., de Torrontegui, G., Ingledew, W.M., and Canovas, J.L., 1970b, Regulation of the enzymes of the ß-ketoadipate pathway in *Moraxella*. Control of quinate oxidation by protocatechuate, *Eur. J. Biochem.*, 14:445.

Tresguerres, M.E.F., Ingledew, W.M., and Canovas, J.L., 1972, Potential competition for 5-dehydroshikimate between the aromatic biosynthetic route and the catabolic hydroaromatic pathway, *Arch. Mikrobiol.*, 82:111.

Vakeria, D., Vivian, A., and Fewson, C.A., 1984, Isolation, characterization and mapping of mandelate pathway mutants of *Acinetobacter calcoaceticus*, *J. Gen. Microbiol.*, 130:2893.

Vasudevan, N., and Mahadevan, A., 1988, Degradation of *p*-benzyloxyphenol by *Acinetobacter* sp., *FEMS Microbiol. Lett.*, 56:349.

Vasudevan, N., and Mahadevan, A., 1989, Degradation of 3(*o*-methoxyphenoxy)1,2-propanediol (guaiacol glyceryl ester) by *Acinetobacter* sp., *J. Biotech.*, 9:107.

Ventosa, A., Quesada, E., Ferrer, M., Ruiz-Berraquero, F.L., and Ramos-Cormenzana, A., 1984, Degradation of *o*-anisate by soil microorganisms, *Pol. J. Soil Sci.*, 15:37.

Walker, J.D., Colwell, R.R., and Petrakis, L., 1975, Evaluation of petroleum-degrading potential of bacteria from water and sediment, *Appl. Microbiol.*, 30:1036.

Walker, J.D., Colwell, R.R., and Petrakis, L., 1976, Biodegradation of petroleum by Chesapeake Bay sediment bacteria, *Can. J. Microbiol.*, 22:423.

Wheelis, M.L., 1972, Regulation of the synthesis of enzymes of tryptophan dissimilation in *Acinetobacter calcoaceticus*, *Arch. Mikrobiol.*, 87:1.

Williams, R.J., and Evans, W.C., 1975, The metabolism of benzoate by *Moraxella* species through anaerobic nitrate respiration. Evidence for a reductive pathway, *Biochem. J.*, 148:1.

Winstanley, C., Taylor, S.C., and Williams, P.A., 1987, pWW174: a large plasmid from *Acinetobacter calcoaceticus* encoding benzene metabolism by the ß-ketoadipate pathway, *Molec. Microbiol.*, 1:219.

Wyndham, R.C., 1986, Evolved aniline catabolism in *Acinetobacter calcoaceticus* during continuous culture of river water, *Appl. Environ. Microbiol.*, 51:781.

Yall, I., 1979, A petroleum-eating bacterium, *Nat. Hist.*, 88:41.

Yeh, W.-K., and Ornston, L.N., 1980, Origins of metabolic diversity: substitution of homologous sequences into genes for enzymes with different catalytic activities, *Proc. Natl. Acad. Sci. USA*, 77:5365.

Yeh, W.-K., and Ornston, L.N., 1981, Evolutionary homologues $\alpha_2\beta_2$ oligomeric structures in ß-ketoadipate succinyl-CoA transferases from *Acinetobacter calcoaceticus* and *Pseudomonas putida*, *J. Biol. Chem.*, 256:1565.

Yeh, W.-K., and Ornston, L.N., 1984, *p*-Chloromercuribenzoate specifically modifies thiols associated with the active sites of ß-ketoadipate enol-lactone hydrolase and succinyl-CoA:ß-ketoadipate-CoA transferase, *Arch. Microbiol.*, 138:102.

Yeh, W.-K., Davis, G., Fletcher, P., and Ornston, L.N., 1978, Homologous amino acid sequences in enzymes mediating sequential metabolic reactions, *J. Biol. Chem.*, 253:4920.

Yeh, W.-K., Fletcher, P., and Ornston, L.N., 1980a, Evolutionary divergence of co-selected ß-ketoadipate enol-lactone hydrolases in *Acinetobacter calcoaceticus, J. Biol. Chem.*, 255:6342.

Yeh, W.-K., Fletcher, P., and Ornston, L.N., 1980b Homologies in the NH_2-terminal amino acid sequences of ɤ-carboxymuconolactone decarboxylases and muconolactone isomerases, *J. Biol. Chem.*, 255:6347.

Yeh, W.-K., Durham, D.R., Fletcher, P., and Ornston, L.N., 1981, Evolutionary relationships among ɤ-carboxymuconolactone decarboxylases, *J. Bacteriol.*, 146:233.

Yeh, W.-K., Shih, C., and Ornston, L.N., 1982, Overlapping evolutionary affinities revealed by comparison of amino acid compositions, *Proc. Natl. Acad. Sci. USA*, 79:3794.

Zaborsky, O.R., and Schwartz, R.D., 1974, The effect of imidoesters on the protocatechuate 3,4-dioxygenase activity of *Acinetobacter calcoaceticus, FEBS Lett.*, 46:236.

Zaitsev, G.M., 1988, Utilization of 3-chlorobenzoic acid by a mixed culture of microorganisms, *Mikrobiologiya*, 57:550.

Zaitsev, G.M., and Baskunov, B.P., 1985, Utilization of 3-chlorobenzoic acid by *Acinetobacter calcoaceticus, Mikrobiologiya*, 54:203.

METABOLISM OF CYCLOHEXANE AND RELATED

ALICYCLIC COMPOUNDS BY *ACINETOBACTER*

P.W. Trudgill

Department of Biochemistry
University College of Wales
Aberystwyth
Dyfed SY23 3DD
UK

INTRODUCTION

Acinetobacter strains deposited in culture collections have been variously described as species of *Vibrio, Alcaligenes, Achromobacter* and *Bacterium*. One of these (NCIB 8250) deposited in the National Collection of Industrial Bacteria, Aberdeen, Scotland, as *Vibrio* 01 in 1950 by Professor W.C. Evans, has provided those of my colleagues who have been interested in the degradation of aromatic compounds with a topic of interest for decades (Fewson, this volume). The involvement of *Acinetobacter* spp. in the degradation of alicyclic compounds was preceded by a period of gradual development of our understanding of the metabolism of these compounds. Early attempts to isolate organisms capable of growth with simple alicyclic hydrocarbons such as cyclohexane were unsuccessful (Pelz and Rehm, 1971; Beam and Perry, 1973, 1974; De Klerk and van der Linden, 1974). Although reports of organisms capable of growth with cyclohexane have appeared in the literature from time to time (Tausz and Peter, 1919; Johnson et al., 1942; Skarzynski and Czekalowski, 1946; Fredericks, 1966), none has been followed by significant metabolic study.

Commensal and co-oxidative systems have been reported in which the degradation of cycloalkanes has been initiated. Ooyama and Foster (1965) made use of an alkane-degrading strain of *Mycobacterium vaccae* (JOB 5) to convert cyclohexane into cyclohexanol and cyclohexanone. Beam and Perry (1974) and de Klerk and van der Linden (1974) demonstrated that inclusion of cyclohexanol-utilising strains in mixed cultures with organisms capable of the gratuitous hydroxylation of cyclohexane resulted in complete degradation of the cyclic hydrocarbon.

The Biology of Acinetobacter, Edited by K. J. Towner *et al.*
Plenum Press, New York, 1991

More recently, two groups of workers (Stirling et al., 1977; Magor et al., 1986) have clearly established growth of *Nocardia, Pseudomonas* and *Xanthobacter* spp. on cyclohexane as sole source of carbon. Hydroxylation of the compound to form cyclohexanol is the initial step in oxidation (Anderson et al., 1980; Trower et al., 1985). In addition, Kamata et al. (1986) have recently isolated a strain, which they named *Acinetobacter cyclohexanophilus*, that grows rapidly with cyclohexane as sole source of carbon, although no pathway studies have been reported.

Our knowledge of the only clearly established pathway for cyclohexanol oxidation by bacteria was first obtained with *Nocardia* and *Acinetobacter* spp. isolated by enrichment with cyclohexanol and capable of growth on the compound as sole carbon source.

CYCLOHEXANOL METABOLISM BY *ACINETOBACTER* SP. NCIMB 9871

Acinetobacter sp. NCIMB 9871, used in our studies of cyclohexanol metabolism, was a generous gift from Dr. P.J. Chapman who isolated it by elective culture with the alcohol.

It is self-evident that energy and biosynthetic precursor provision from cyclohexanol must involve a ring cleavage step. Cyclohexanol-grown cells oxidised the growth substrate and cyclohexanone rapidly, and extracts of cyclohexanol-grown *Acinetobacter* sp. NCIMB 9871 contained an inducible cyclohexanol dehydrogenase, providing evidence that cyclohexanone is an intermediate in cyclohexanol degradation (Donoghue and Trudgill, 1976). The importance of cyclohexanol dehydrogenase in cyclohexanol degradation has recently been reinforced by the work of Park et al. (1986) in their study of induction of the enzyme in *A. calcoaceticus* strain C10 during growth of the organism on cyclohexanol.

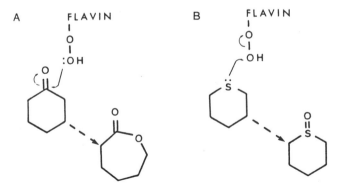

Fig. 1. Mechanism of the chemical Baeyer-Villiger reaction: (a) protonated cyclohexanone; (b) peroxyacid reagent; (c) protonated ε-caprolactone. 1, nucleophilic attack by the peroxyacid; 2, deprotonation; 3, loss of the leaving group; 4, product formation.

The Baeyer-Villiger Reaction

A chemical approach to the problem of ring cleavage of cyclic ketones was first described by Baeyer and Villiger (1899). These authors showed that peroxyacid reagents were able to insert oxygen into cyclic ketones with consequent ring expansion and lactone formation. The accepted mechanism for this reaction is shown in Fig. 1. The reaction is initiated by protonation of the carbonyl group, followed by nucleophilic attack by the peroxy acid. The intermediate resulting from addition of the peroxyacid to the carbonyl has a good leaving group, a carboxylate anion on an oxygen atom, since the oxygen-oxygen bond is weak (approx. 197 kJ/mol) and not difficult to break. As the leaving group departs, a partial positive character develops at the oxygen and a 1-2 alkyl shift from a carbon atom to an adjacent oxygen atom takes place, driven in part by the energy recovery of carbonyl bond reformation.

The Biological Bayer-Villiger Reaction

Ample evidence now exists to show that *Acinetobacter* sp. NCIMB 9871 makes use of a biological parallel to the chemical Baeyer-Villiger reaction in cyclohexanol oxidation. Whole cell oxidation studies, subcellular enzymology and isolation of catabolic intermediates all served to establish the partial catabolic pathway shown in Fig. 2. Manipulation of soluble protein fractions in the presence of the esterase inhibitor paraoxon (diethyl-*p*-nitrophenylphosphate) confirmed that the ϵ-caprolactone arose directly from cyclohexanone and not by ring-closure of 6-hydroxyhexanoate (Donoghue and Trudgill, 1976). All assayed enzymes were induced by growth with cyclohexanol. Subsequent isolation of the enzyme responsible for

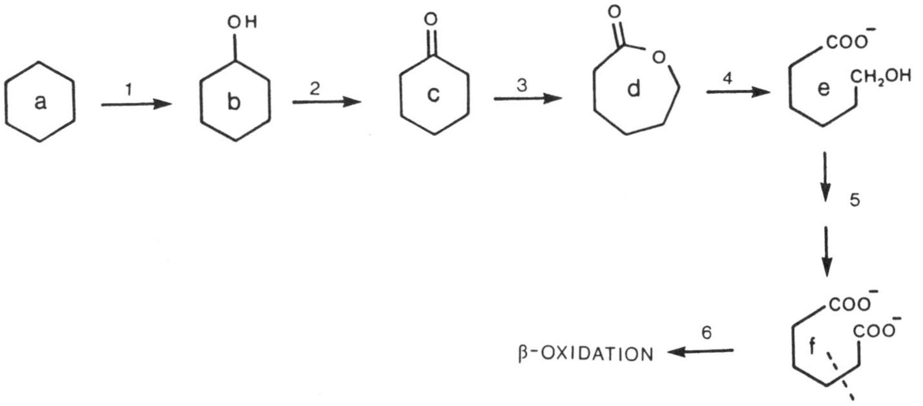

Fig. 2. Oxidation of cyclohexanone by bacteria. Compounds: (a) cyclohexane; (b) cyclohexanol; (c) cyclohexanone; (d) ϵ-caprolactone; (e) 6-hydroxyhexanoate; (f) adipate. Enzymes: 1, cyclohexane monooxygenase; 2, cyclohexanol dehydrogenase; 3, cyclohexanone 1,2-monooxygenase; 4, ϵ-caprolactone hydrolase; 5, 6-hydroxyhexanoate and 6-oxohexanoate dehydrogenases; 6, CoA ester synthetase leading to ß-oxidation.

formation of the ε-caprolactone from cyclohexanone showed it to be a simple flavoprotein (M_r 59,000; 1 FAD) that constituted about 3% of the soluble protein of the induced cell, required NADPH as an electron donor, and molecular oxygen. It had a reaction stoichiometry that was consistent with it being a monooxygenase (Donoghue et al., 1976). The correct systematic name for the enzyme is cyclohexanone NADPH:oxygen oxidoreductase (1,2-lactonising) (EC 1.14.13.-), hereafter called cyclohexanone 1,2-monooxygenase.

Biological Baeyer-Villiger monooxygenases are widely distributed and are also involved in the cleavage of cyclopentanone (Griffin and Trudgill, 1972), both rings of camphor (Conrad et al., 1965a,b; Ougham et al., 1983; Taylor and Trudgill, 1986) and the D-ring of androstene-3,17-dione (Prairie and Talaley, 1963). Removal of the side chains of progesterone (Rahim and Sih, 1966) and acetophenone (Cripps, 1975), and conversion of 2-tridecanone to undecyl acetate by *Pseudomonas cepacia* (Britton and Markovetz, 1977) are also catalysed by this class of enzyme. In all cases where significant characterisation has been undertaken, the enzymes are flavoproteins. Within this group of enzymes the cyclohexanone 1,2-monooxygenase from *Acinetobacter* sp.NCIMB 9871 has a dominant position. Its stability and ease of purification has made it the logical choice for thorough and extensive studies of the reaction mechanism.

Reaction Mechanism of Cyclohexanone 1,2-Monooxygenase

Using an experimental approach that included anaerobic rapid reaction stopped-flow spectrophotometry, Ryerson et al. (1982) were able to establish a *ter-ter* mechanism for the reaction, as shown in Fig. 3, which is compatible with the experimental observation that NADPH and cyclohexanone bind to different forms of the enzyme. This is in accord with the failure of cyclohexanone to influence the rate of enzyme-bound FAD reduction by NADPH. An unusual feature of the enzyme was the very low K_m for oxygen (approx. 10-15 μM) which made it impossible to determine the order of addition of oxygen with any certainty. No stable flavin-oxygen intermediate could be detected in the absence of cyclohexanone and this

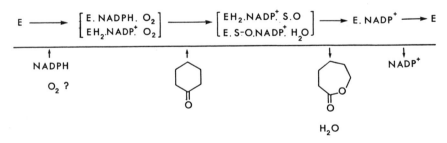

Fig. 3. Reaction sequence for the oxygenation of cyclohexanone by the 1,2-monooxygenase from *Acinetobacter* sp. NCIMB 9871 (after Ryerson et al., 1982).

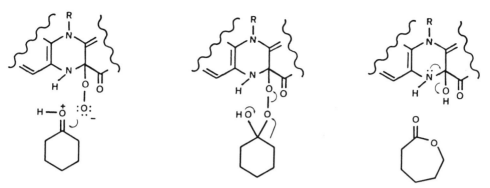

Fig. 4. Proposed mechanism for the biological Baeyer-Villier oxygenation of cyclohexanone by the 1,2-monooxygenase from *Acinetobacter* sp. NCIMB 9871. The reactive oxygen donor is 4a-hydroperoxy FAD (after Ryerson et al., 1982).

observation, plus the fact that the flavin-dependent phenolic monooxygenases all add organic substrate before oxygen, suggests that oxygen adds to the reduced enzyme-NADP complex. Stopped-flow studies, in which the reduced flavoprotein was reacted with oxygen in the presence of NADP, resulted in the formation of an intermediate, proposed from its electronic spectrum to be 4a-hydroperoxyflavin (Entsch et al., 1976), within the dead time (2-3 ms) of the instrument. In the presence of cyclohexanone an additional intermediate, 4a-hydroxyflavin, was also observed.

Reaction of 4a-hydroperoxyflavin with cyclohexanone could theoretically involve cyclohexanone acting as an electrophile at the carbonyl carbon or as a nucleophile through the intermediacy of an enolate tautomer. Studies with deuterated cyclohexanone strongly support an electrophilic role for the cyclohexanone, as shown in Fig. 4, that provides a clear parallel to the peroxy acid mediated Baeyer-Villiger reaction. Schwab (1981), using deuterium NMR, established that the configuration of the migrating C centre was completely retained, further sustaining the analogy between the microbial and chemical versions of the reaction.

The cyclohexanone 1,2-monooxygenase of *Acinetobacter* sp. NCIMB 9871 displays a characteristic unique among flavoprotein monooxygenases in having the ability to deliver an oxygen atom, derived from O_2, to both nucleophilic and electrophilic substrates. The cyclic sulphide thiane yields a stable sulphoxide (thiane-1-oxide), demonstrating that the nature of the peroxide oxygen-oxygen bond cleavage is controlled by electron availability in the compound to be oxygenated (Fig. 5).

The catabolic flexibility of the enzyme is further illustrated by its ability to oxygenate aldehydes (phenylacetone), selenides, boronic acids, a phosphite ester and an iodide ion (Branchaud and Walsh, 1985). Abril et al. (1989) have established a broad ketone specificity for the enzyme using

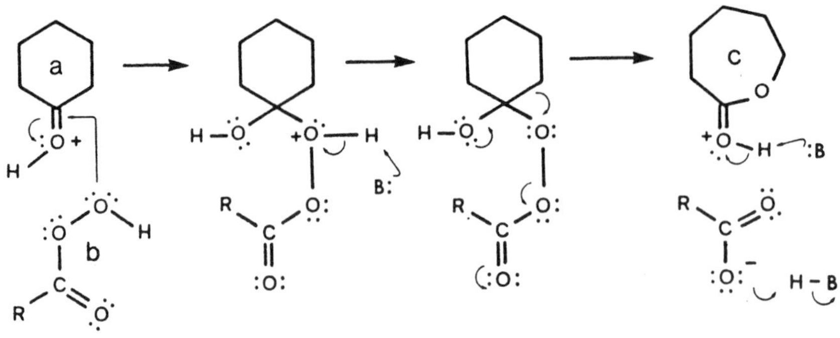

Fig. 5. Electrophilic and nucleophilic oxygenation reactions catalysed by cyclohexanone 1,2-monooxygenase from *Acinetobacter* sp. NCIMB 9871: A. Cyclohexanone acting as an electrophile (carbonyl carbon); B. Thiane acting as a nucleophile with the formation of thiane 1-oxide.

immobilised enzyme and regenerating the NADPH *in situ* with glucose 6-phosphate and glucose-6-phosphate dehydrogenase from *Leuconostoc mesenteroides*.

Gene Sequencing and Structure of Cyclohexanone-1,2-Monooxygenase

The gene coding for cyclohexanone 1,2-monooxygenase has been located, isolated and its nucleotide sequence determined (Chen et al., 1988). The complete sequence of 542 amino acids has been derived by translation of the nucleotide sequence. Analysis of the amino acid sequence showed seven cysteine residues, of which four may be involved in disulphide bridge formation, leaving three available for titration as experimentally determined (Donoghue et al., 1976; Latham and Walsh, 1987). A binding site for FAD has been identified as a ßαß-unit characterised by a pair and a triplet of glycine residues. Typically these regions are near the NH_2 terminus of the polypeptide chain (Wierenga et al., 1983), and cyclohexanone 1,2-monooxygenase is no exception as the region is encompassed by residues 6 to 18. In addition, a potential ADP (NADPH) binding site, based upon a proposed fingerprint of 11 amino acids (Wierenga et al., 1986), has been identified in the region extending from residue 176 to residue 208.

An over-production system for the gene was constructed by using the *tac* promoter vector pKK223-3 in *Escherichia coli*. Although the amount of pure enzyme available from this strain is not significantly greater than that obtained from the same mass of *Acinetobacter* sp. NCIB 9871, the system provides a means for performing genetic manipulations on the plasmid (pKKCO) and creating mutant forms of the enzyme (Chen et al., 1988).

Cyclohexanone 1,2-monooxygenase from *Acinetobacter* sp. NCIMB 9871 is the only enzyme of this class for which the nucleotide and, by inference, the amino acid sequence has been determined. The potential now exists for an interplay of approaches, including X-ray crystallographic methods, that will provide a complete understanding of biological Baeyer-Villiger oxygenation.

A novel exploitation of the biological Baeyer-Villiger reaction has been reported by Alphand et al. (1990). These authors made use of whole cell cultures of two strains of *Acinetobacter* in the conversion of 2-undecylcyclopentanone in the synthesis of (S)-(-)-5-hexadecanolide (a wasp pheromone).

LACTONE HYDROLYSIS

The interest in enzymes that incorporate molecular oxygen into organic molecules has spanned several decades, a variety of biological source material, and a wide range of cyclic and acyclic compounds. In the case of alicyclic ring cleavage, ring oxygen insertion with the consequent formation of a lactone is not, of itself, a ring cleavage mechanism. The ε-caprolactone formed from cyclohexanone, for example, is stable at near neutral pH, and ring cleavage is catalysed by an inducible lactone hydrolase (EC 3.1.1.-). The lactone hydrolase from cyclohexanol-grown *Acinetobacter* sp. NCIMB 9871 has been purified and characterised (Bennett et al., 1988). It constitutes about 1% of the soluble protein of induced cells, has an M_r of about 60,000, and consists of two electrophoretically-identical subunits. Substrate specificity is narrow (5% activity with δ-valerolactone; no activity with γ-lactones) and, of a wide range of putative inhibitors, only paraoxon (diethyl *p*-nitrophenylphosphate) was effective, giving 50% inhibition at 8 μM. In this respect it most closely resembles mammalian acetylcholine esterase (Froede and Wilson, 1971) and carboxyesterase (Heymann et al., 1970), both of which have a functional catalytic centre serine residue, and it contrasts sharply with other bacterial δ-lactone hydrolases that are inhibited by Co^{2+}, Cu^{2+}, Mn^{2+} or Zn^{2+} ions and thiol-reactive reagents, and are thus considered to have catalytically active thiol groups (Kersten et al., 1982; Maruyama, 1983), and with the Ca^{2+}-dependent γ-lactone hydrolases of human blood and rat liver microsomes (Fishbein and Bessman, 1966a,b).

SPONTANEOUS CLEAVAGE OF LACTONES

The formation of stable lactones as a consequence of oxygenation, leading to ring expansion of simple cyclic ketones such as cyclohexanone and cyclopentanone, dictates the requirement for an esterase if ring cleavage is to proceed. Some ketonic substrates, however, yield unstable lactones upon ring oxygenation. *Acinetobacter* sp. NCIMB 9871 and *N. globerula* CL1 are also capable of growth with *trans*-cyclohexan-1,2-diol. This is initially dehydrogenated to 2-hydroxycyclohexanone, with subsequent insertion of a ring oxygen between the two substituent groups forming an unstable lactone that spontaneously ring-opens to form 6-oxohexanoate. Although the first two steps are catalysed by enzymes that are common to cyclohexanol degradation, the induced lactone hydrolase is redundant (Donoghue and Trudgill, 1973). In contrast, *Acinetobacter* sp. strain 6.3, which was isolated by elective culture on *trans*-cyclohexan-1,2-diol, is incapable of growth with cyclohexanol, cyclohexanone or Є-caprolactone, although the induced secondary alcohol dehydrogenase and Baeyer-Villiger oxygenase are of broad specificity and active towards *trans*-cyclohexan-1,2-diol/cyclohexanol and 2-hydroxycyclohexanone/cyclohexanone respectively. One possible explanation of this very restricted growth spectrum is that the organism does not synthesise a lactone hydrolase. Certainly no hydrolase activity towards Є-caprolactone can be detected in extracts of *trans*-cyclohexan-1,2-diol-grown cells (Davey and Trudgill, 1977).

CYCLOHEXANECARBOXYLIC ACID METABOLISM

The presence of a carboxyl group on the cyclohexane ring has a profound effect upon catabolic metabolism and, to date, no organism has been isolated that utilises a biological Baeyer-Villiger reaction in the oxidation of the compound to central metabolites. Organisms that grow with the compound as sole carbon source use one of two catabolic routes. Hydroxylation at the 4-position, followed by oxidation to 4-oxocyclohexanecarboxylic acid and aromatisation to *p*-hydroxybenzoate, leads to protocatechuic acid. Both *ortho* and *meta* fission has been variously demonstrated for strains of *Arthrobacter*, *Corynebacterium* and *Pseudomonas* (Blakley, 1974; Kaneda, 1974; Smith and Callely, 1975; Taylor and Trudgill, 1978). Alternatively, because the carboxyl group is attached to a ring carbon, the molecule is also a potential substrate for ring cleavage by ß-oxidation.

Experimental evidence for this simple direct catabolic route was first obtained by Rho and Evans (1975) who, using a strain of *Acinetobacter anitratum* isolated by enrichment culture on cyclohexane carboxylic acid, demonstrated that pimelate was a metabolite and that ATP, CoA, Mg^{2+} and FAD were required for its formation by cell extracts. Blakley (1978), working with a strain of *Alcaligenes*, has confirmed the aerobic ß-oxidation of cyclohexane carboxylic acid to pimelate and demonstrated inducible enzymes, including a cyclohexane carboxyl-CoA synthetase, in extracts of induced cells.

CONCLUSIONS

This brief survey illustrates the key role played by *Acinetobacter* spp. in the development of our understanding of the pathways and enzymology of the degradation of simple alicyclic compounds. Although important discoveries have also been made with strains of *Arthrobacter*, *Corynebacterium*, *Nocardia* and *Pseudomonas* spp., the unique contribution of *Acinetobacter* spp. to our understanding of biological

Baeyer-Villiger oxygenation is well documented. In particular, *Acinetobacter* sp. NCIMB 9871 was chosen for detailed study because of its rapid growth with cyclohexanol and production of large amounts of a stable oxygenase that is easy to purify.

REFERENCES

Abril, O., Ryerson, C.C., Walsh C.T., and Whitesides, G.M., 1989, Enzymic Baeyer-Villiger type oxidations of ketones catalysed by cyclohexanone oxygenase, *Biorgan. Chem.*, 17:41.

Alphand,V., Archelas, A., and Furstoss, R., 1990, Microbiological transformations. 13. Direct synthesis of both S and R enantiomers of 5-hexadecanolide via an enantioselective microbiological Baeyer-Villiger reaction, *J. Organ. Chem.*, 55:347.

Anderson, M.S., Hall, R.A., and Griffin, M., 1980, Microbial metabolism of alicyclic hydrocarbons: cyclohexane catabolism by a pure strain of *Pseudomonas* sp., *J. Gen. Microbiol.*, 120:89.

Baeyer, A., and Villiger, V., 1899, Einwirkung des caro'schen Reagens auf Ketone, *Berich. Deuts. Chem. Gesellschaft*, 32:3625.

Beam, H.W., and Perry, J.J., 1973, Co-metabolism as a factor in microbial degradation of cycloparaffinic hydrocarbons, *Arch. Microbiol.*, 91:87.

Beam, H.W., and Perry, J.J., 1974, Microbial degradation and assimilation of n-alkyl-substituted cycloparaffins, *J. Bacteriol.*, 118:394.

Bennet, A.P., Strang, E.J., Trudgill, P.W., and Wong, V.T.K., 1988, Purification and properties of ϵ-caprolactone hydrolase from *Acinetobacter* NCIB 9871 and *Nocardia globerula* CL1, *J. Gen. Microbiol.*, 134:161.

Blakley, E.R., 1974, The microbial degradation of cyclohexanecarboxylic acid: a pathway involving aromatization to form *p*-hydroxybenzoic acid, *Can. J. Microbiol.*, 20:1297.

Blakley, E.R., 1978, The microbial degradation of cyclohexanecarboxylic acid by a ß-oxidation pathway with simultaneous induction to the utilization of benzoate, *Can. J. Microbiol.*, 24:847

Branchaud, B.P., and Walsh, C.T., 1985, Functional group diversity in enzymatic oxygenation reactions catalysed by bacterial flavin-containing cyclohexanone oxygenase, *J. Amer. Chem. Soc.*, 107:2153.

Britton, L.N., and Markovetz, A.J., 1977, A novel ketone monooxygenase from *Pseudomonas cepacia*. Purification and properties, *J. Biol. Chem.*, 252:8561.

Chen, J. Y-C., Peoples, O.P., and Walsh, C.T., 1988, *Acinetobacter* cyclohexanone monooxygenase; gene cloning and sequence determination, *J. Bacteriol.*, 170:781.

Conrad, H.E., Dubus, R., Namtvedt, M.J., and Gunsalus, I.C., 1965a, Mixed function oxygenation II. Separation and properties of enzymes catalysing camphor lactonization, *J. Biol. Chem.*, 240:495.

Conrad, H.E., Lieb, K., and Gunsalus, I.C. 1965b, Mixed function oxidation III. An electron transport complex in camphor ketolactonization, *J. Biol. Chem.*, 240:4029.

Cripps, R.E., 1975, The microbial metabolism of acetophenone. Metabolism of acetophenone and some chloroacetophenones by an *Arthrobacter* species, *Biochem. J.*, 152:233.

Davey, J.F., and Trudgill, P.W., 1977, The metabolism of *trans*-cyclohexan-1,2-diol by an *Acinetobacer* species, *Eur. J. Biochem.*, 74:115.

De Klerk, H., and Van Der Linden, A.C., 1974, Bacterial degradation of cyclohexane. Participation of co-oxidation reaction, *Ant. v. Leeuw. J. Microbiol. Serol.*, 40:7.

Donoghue, N.A., and Trudgill, P.W., 1973, The metabolism of cyclohexane-1,2-diols by *Nocardia globerula* CL1, *Trans. Biochem. Soc.*, 1:1287.

Donoghue, N.A., and Trudgill, P.W., 1976, The metabolism of cyclohexanol by *Acinetobacter* NCIB 9871, *Eur. J. Biochem.*, 60:1.

Donoghue, N.A., Norris, D.B., and Trudgill, P.W., 1976, The purification and properties of cyclohexanone oxygenase from *Nocardia globerula* CL1 and *Acinetobacter* NCIB 9871, *Eur. J. Biochem.*, 63:175.

Entsch, B., Ballou, D.P., and Massey, V., 1976, Flavin-oxygen derivatives involved in hydroxylation by *p*-hydroxybenzoate hydroxylase, *J. Biol. Chem.*, 251:2550.

Fishbein, W.N., and Bessman, S.P., 1966a, Purification and properties of an enzyme in human blood and rat liver microsomes catalysing the formation and hydrolysis of Ɣ-lactones. I. Tissue localization, stoichiometry, specificity, distinct from esterase, *J. Biol. Chem.*, 241:4835.

Fishbein, W.N., and Bessman, S.P., 1966b, Purification and properties of an enzyme from human blood and rat liver microsomes catalysing the formation and hydrolysis of Ɣ-lactones. II. Metal ion effects, kinetics and equilibria, *J. Biol. Chem.*, 241:4842.

Fredericks, K.M., 1966, Adaptation of bacteria from one type of hydrocarbon to another, *Nature*, 209:1047.

Froede, H.C., and Wilson, I.B., 1971, Acetylcholine esterase, *in* "The Enzymes, vol. 5," p.87, P.D. Boyer, ed., Academic Press, New York.

Griffin, M., and Trudgill, P.W., 1972, The metabolism of cyclopentanol by *Pseudomonas* NCIB 9872, *Biochem. J.*, 129:595.

Heymann, E., Krisch, K., and Pahlich, E., 1970, Structure of the active centre of microsomal carboxyesterase from pig kidney and liver, *Hoppe-Seyler's Zeits. Physiol. Chem.*, 351:931.

Johnson, F.H., Goodale, W.T., and Turkevich, J., 1942, The bacterial oxidation of hydrocarbons, *J. Cell. Comp. Physiol.*, 19:163.

Kamata, A., Mikawa, T., and Imada, Y., 1986, Bacterial degradation of cyclohexane, *Nippon Kokai Tokkyo Koho*, 61128890.

Kaneda, T., 1974, Enzymatic aromatization of 4-ketocycohexanecarboxylic acid to p-hydroxybenzoic acid, *Biochem. Biophys. Res. Comm.*, 58:140.

Kersten, P.J., Dagley, S., Whittaker, J.W., Arciero, D.M., and Lipscomb, J.D., 1982, 2-Pyrone-4,6-dicarboxylic acid, a catabolite of gallic acid in *Pseudomonas* species, *J. Bacteriol.*, 152:1154.

Latham, J.A., and Walsh, C.T., 1987, Mechanism-based inactivation of the flavoenzyme cyclohexanone oxygenase during oxygenation of cyclic thiol ester substrates, *J. Amer. Chem. Soc.*, 109:3421.

Magor, A.M., Warburton, J., Trower, M.K., and Griffin, M., 1986, Comparative study of the ability of three *Xanthobacter* species to metabolize cycloalkanes, *Appl. Environ. Microbiol.* 52:665.

Maruyama, K.C., 1983, Purification and properties of 2-pyrone-4,6-dicarboxylate hydrolase, *J. Biochem.*, 93:557.

Ooyama, J., and Foster, J.W., 1965, Bacterial oxidation of cycloparaffinic hydrocarbons, *Ant. v. Leeuw. J. Microbiol. Serol.*, 31:45.

Ougham, H.J., Taylor, D.G., and Trudgill, P.W., 1983, Camphor revisited: involvement of a unique monooxygenase in the metabolism of 2-oxo- ³-4,5,5-trimethylcyclopentenylacetic acid by *Pseudomonas putida*, *J. Bacteriol.*, 153:140.

Park, H.D., Choi, S., and Rhee, I.K., 1986, Induction of cyclohexanol dehydrogenase in *Acinetobacter calcoaceticus* C10, *Han'guk Nonghwa Hakhoechi*, 29:304.

Pelz, B.F., and Rehm, H.J., 1971, Isolierung, Substrassimilation und einige produkte alkenabbauender Schimmelpilze, *Arch. Mikrobiol.*, 84:20.

Prairie, R.L., and Talaley, P., 1963, Enzymatic formation of testololactone, *Biochemistry*, 2:203.

Rahim, M.A., and Sih, C.J., 1966, Mechanisms of steroid oxidation by microorganisms. XI Enzymatic cleavage of the pregnone side chain, *J. Biol. Chem.*, 241:3615.

Rho, E.M., and Evans, W.C., 1975, The aerobic metabolism of cyclohexanecarboxylic acid by *Acinetobacter anitratum*, *Biochem. J.*, 148:11.

Ryerson, C.C., Ballou, D.P., and Walsh, C.T., 1982, Mechanistic studies on cyclohexanone oxygenase, *Biochemistry*, 21:2644.

Schwab, J.M., 1981, Stereochemistry of an enzymic Baeyer-Villiger reaction. Application of deuterium NMR, *J. Amer. Chem. Soc.*, 103:1876.

Skarzynski, B., and Czekalowski, J.W., 1946, Utilization of phenols and related compounds by *Achromobacter*, *Nature*, 158:304.

Smith, D.I., and Callely, A.G., 1975, The microbial degradation of cyclohexanecarboxylic acid, *J. Gen. Microbiol.*, 91:210.

Stirling, L.A., Watkinson, R.J., and Higgins, I.J., 1977, Microbial metabolism of alicyclic hydrocarbons: isolation and properties of a cyclohexane degrading bacterium, *J. Gen. Microbiol.*, 99:119.

Tausz, J., and Peter, M., 1919, Neue Methode der Kohlenwasserstoffanalyse mit Hilfe von Bakterien, *Zent. Bact. Parasit. Infekt. Hyg.*, 49:497.

Taylor, D.G., and Trudgill, P.W., 1978, Metabolism of cyclohexanecarboxylic acid by *Alcaligenes strain* W1, *J. Bacteriol.*, 134:401.

Taylor, D.G., and Trudgill, P.W., 1986, Camphor revisited: studies of the 2,5-diketocamphane 1,2-monooxygenase from *Pseudomonas putida* ATCC 17453, *J. Bacteriol.*, 165:489.

Trower, M.K., Buckland, R.M., Higgins, R., and Griffin, M., 1985, Isolation and characterization of a cyclohexane-metabolizing *Xanthobacter* sp., *Appl. Environ. Microbiol.*, 49:1282.

Wierenga, R.K., Drenth, J., and Schulz, G.E., 1983, Comparison of the three-dimensional protein and nucleotide structure of the FAD-binding domain of *p*-hydroxybenzoate hydroxylase with FAD as well as NADPH-binding domains of glutathione reductase, *J. Mol. Biol.*, 167:725.

Wierenga, R.K., Terpstra, P., and Hol, W.G., 1986, Prediction of the occurence of the ADP-binding ßαß-fold in proteins using an amino acid sequence fingerprint, *J. Mol. Biol.*, 187:101.

METABOLISM OF TRIMETHYLAMMONIUM COMPOUNDS BY

ACINETOBACTER

H.-P. Kleber

Sektion Biowissenschaften
Bereich Biochemie
Karl-Marx-Universität Leipzig
Talstrasse 33
7010 Leipzig
Germany

INTRODUCTION

L(-)-Carnitine (R(-)-3-hydroxy-4-trimethylaminobutyrate) is a ubiquitously-occurring substance which is essential for the ß-oxidation of long-chain fatty acids in mitochondria. Bacteria are able to metabolise this trimethylammonium compound in different ways. Species of the genus *Pseudomonas* can utilise L-carnitine as sole source of carbon and nitrogen under aerobic conditions. After growth on L(-)-carnitine, the highly specific L(-)-carnitine dehydrogenase (EC 1.1.1.108) is induced in *P. aeruginosa* (Aurich et al., 1967) and *P. putida* (Kleber et al., 1978). A second group of carnitine-metabolising microorganisms comprising different Enterobacteriaceae, e.g., *Escherichia coli*, *Salmonella typhimurium* and *Proteus vulgaris*, can almost quantitatively convert L(-)-carnitine into ɣ-butyrobetaine under anaerobic conditions (Seim et al., 1980, 1982a,c,d). They do not assimilate the carbon and nitrogen skeleton. The metabolism of L(-)-carnitine by *E. coli* includes at least a two-step reduction of L(-)-carnitine to ɣ-butyrobetaine with crotonobetaine as intermediate (Seim et al., 1982d). The function of this reaction sequence is still unknown. Seim et al. (1982a,c,d) postulated that crotonobetaine serves as an external electron acceptor similar to (e.g.) nitrate (Haddock and Jones, 1977). One of the two enzymes involved, the carnitine dehydratase, was recently purified and characterised (Jung et al., 1989). Conversion of crotonobetaine to L(-)-carnitine also seems to occur in different strains of *Acinetobacter* under aerobic (Yokozeki and Kubota, 1984) and anaerobic (Kawamura et al.,

Table 1. Utilisation of carnitine and structurally-related trimethyl-ammonium compounds $[N^+(CH_3)_3-R]$ as sole carbon source by *A. calcoaceticus* 69-V

Quarternary N-compound (0.3%)	R	Growth	Product formed by cleavage of C-N bond
Glycine betaine	- CH_2-COO^-	-	-
ß-Homobetaine	- CH_2-CH_2-COO^-	-	-
ɣ-Butyrobetaine	- CH_2-CH_2-CH_2-COO^-	+	trimethyl-amine
Crotonobetaine	- CH_2-$CH=CH$-COO^-	-	-
D(+)-Carnitine	- CH_2-$CH(OH)$-CH_2-COO^-	-	-
L(-)-Carnitine	- CH_2-$CH(OH)$-CH_2-COO^-	+	trimethyl-amine
Trimethylamine *N*-oxide	- O^-	-	-
Trimethylammonium	- H	-	-
Choline	- CH_2-CH_2OH	-	-
ɣ-Homocholine	- CH_2-CH_2-CH_2OH	-	-
D,L-ß-Methylcholine	- CH_2-$CH(CH_3)OH$	-	-
D,L-ß-Ethylcholine	- CH_2-$CH(CH_3$-$CH_2)OH$	-	-
Acetonyltrimethylammonium	- CH_2-CO-CH_3	-	-

1986b) conditions. So far as the catabolism of quaternary nitrogen compounds is concerned, under aerobic conditions *Acinetobacter calcoaceticus* 69-V has a different pathway from that which occurs in the other L(-)-carnitine-metabolising bacteria mentioned above.

SPLITTING OF THE C-N BOND

The utilisation of carnitine and structurally-related trimethyl-ammonium compounds (betaines and nitrogen-bases) by *A. calcoaceticus* 69-V is summarised in Table 1 (Kleber et al., 1977). Only L(-)-carnitine and ɣ-butyrobetaine are used as the sole carbon sources. Growth on ɣ-butyrobetaine proceeded with a lag phase considerably longer than that observed with L- or D,L-carnitine as sole sources of carbon (Kleber et al., 1977). L-*O*-Acylcarnitines were also assimilated by this strain of *Acinetobacter*. The utilisation of these compounds and the growth of the organism correlated with the stoichiometric formation of trimethylamine and, presumably, with the cleavage of the C-N bond and the degradation of the carbon backbone (Fig. 1). The bacteria oxidised choline to glycine betaine in the presence of additional carbon sources, but glycine betaine

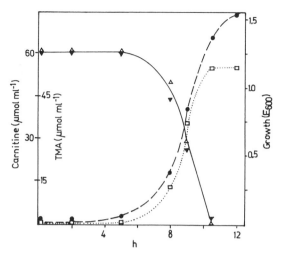

Fig. 1. Degradation of L(-)-carnitine (1 g/l) during growth (●--●) of *A. calcoaceticus* 69-V. Disappearance of L(-)-carnitine (△—△, thin-layer chromatography; ▲—▲, optical activity) and formation of trimethylamine (TMA, □······□) in the culture medium.

itself was not assimilated. D-Carnitine was metabolised by *A. calcoaceticus* 69-V if an additional carbon source, such as L-carnitine, was present in the incubation mixture or if the bacteria were preincubated with L- or D,L-carnitine, but no growth was observed with D-carnitine as the sole carbon source (Kleber et al., 1977).

Miura-Fraboni et al. (1982) investigated the metabolic patterns of utilisation of [1,2,3,4-¹⁴C, methyl-³H]γ-butyrobetaine and D- and L-[1-¹⁴C, methyl-³H]carnitine with resting cell suspensions of *A. calcoaceticus* that had been grown under various conditions. They found that all three betaines were utilised at almost the same rate. Disappearance of each of these quaternary ammonium compounds was accompanied by stoichiometric formation of trimethylamine. The utilisation of the betaines and the corresponding formation of trimethylamine by resting cell suspensions of *A. calcoaceticus* was essentially abolished under conditions of anaerobiosis and was severely impaired in the presence of sodium cyanide, sodium azide, 2,4-dinitrophenol or 2,2'-bipyridine (Miura-Fraboni et al., 1982). The results of these investigations with resting cell suspensions of *A. calcoaceticus* do not support an earlier suggestion (Kleber et al., 1977) that γ-butyrobetaine degradation in this organism proceeds by its prior hydroxylation to L-carnitine. Indeed, cell-free preparations of *A. calcoaceticus* grown on either D,L-carnitine or γ-butyrobetaine show no detectable α-ketoglutarate-linked γ-butyrobetaine hydroxylase activity (Mirua-Fraboni and Englard, 1982; Miura-Fraboni et al., 1982). Furthermore, failure of *A. calcoaceticus* to grow on D-carnitine (Kleber et al., 1977) does not accord with the observations that resting cell suspensions of organisms, grown on either D,L-carnitine or γ-butyrobetaine, vigorously

Table 2. Specifity of the cleavage of the C-N bond of quaternary ammonium compounds by the membrane fraction prepared from *A. calcoaceticus* cells grown on different carbon sources (10 g/l). The protein concentrations in membrane fractions of D,L-carnitine-, acetate- and succinate-grown bacteria were 11.8, 13.6 and 11.2 mg/ml respectively; incubation time 2 h.

Quaternary ammonium compound	Trimethylamine formation after growth on		
	D,L-Carnitine	Acetate	Succinate
D,L-Carnitine	+++	+	+
L(-)-Carnitine	+++	+	+
D(+)-Carnitine	+++	+	+
Acetyl-L(-)-carnitine	++	-	-
४-Butyrobetaine	+++	-	-
Glycine betaine	-	-	-
Acetonyltrimethyl-ammonium	-	-	-
ß-Methylcholine	-	-	-
Choline	-	-	-
Acetylcholine	-	-	-

+++, >10.0 nmol/mg protein; ++, 1.0-10.0 nmol/mg protein;
+, 01-1.0 nmol/mg protein; -, non-detectable (>0.1 nmol/mg protein).

metabolised D-carnitine with stoichiometric formation of trimethylamine and complete degradation of the carbon skeleton (Miura-Fraboni et al., 1982). In addition, Miura-Fraboni and England (1983) showed that D-carnitine effectively supported the growth of *A. calcoaceticus*, and that utilisation of carnitine resulted in stoichiometric formation of trimethylamine and equivalent loss of the [^{14}C]carboxyl-labelled carbon backbone from the growth medium. An explanation for the discrepancy between these results and those reported previously (Kleber et al., 1977) is not obvious.

Cell-free extracts of D,L-carnitine-grown *A. calcoaceticus* are able to split the C-N bond of labelled carnitine and form trimethylamine (Seim et al., 1982b). This activity was found exclusively in the membrane fraction. All quaternary ammonium compounds with four carbon atoms in the chain were split by the membrane fraction of *A. calcoaceticus* cells which had been grown on D,L-carnitine, whereas glycine betaine and the closely related trimethylammonium bases were not (Table 2). Thus, the four carbon atom chain and the negative charge of the C_1-atom, but not the hydroxyl group of the C_3-atom, are necessary for the enzyme action to occur. Cleavage of the C-N bond of 4-*N*-trimethylaminocrotonic acid by *A. calcoaceticus*, and lack of stereochemical specificity with respect to the configuration of C-3 for other betaines undergoing a similar degradation with formation of trimethylamine, are inconsistent with a mechanism involving a simple Hofmann-type of elimination reaction (Englard et al., 1983).

The carnitine-splitting activity of acetate- and succinate-grown bacteria was very low, but was increased by more than two orders of magnitude when the enzyme was induced by growth of the bacteria on D,L-carnitine. The formation of trimethylamine from ɣ-butyrobetaine could not be detected after growth on succinate or acetate. However, in these experiments the specific radioactivity of ɣ-butyrobetaine was less than one-tenth that of carnitine, so a degradation comparable to that of carnitine cannot be ruled out (Seim et al., 1982b).

Although the remaining unlabelled carbon skeleton of the carnitine has not yet been identified, the results show that *A. calcoaceticus* 69-V possesses a carnitine degradation pathway which is different from that in *Pseudomonas* spp. (Aurich et al., 1967; Kleber et al., 1978), Enterobacteriaceae (Seim et al., 1980, 1982a,c,d), and other *Acinetobacter* spp. (see below).

SYNTHESIS OF L(-)-CARNITINE

At present there is an increasing demand for L(-)-carnitine in medicine since it is a registered drug for the treatment of carnitine deficiency syndromes. It is used in replacement therapy for haemodialysis patients, for therapy and prevention of heart diseases linked to disturbed fat metabolism and related indications, e.g. hyperlipoproteinaemia (Gitzelmann et al., 1987).

Cheap racemic carnitine can no longer be applied in clinics because the D-enantiomer is not as harmless as was assumed in the past (Gitzelmann et al., 1987; Seim and Kleber, 1988). Therefore, many chemical, enzymic and microbiological procedures have been developed in the last few years for the production of enantiomerically pure L(-)-carnitine (Seim and Kleber, 1988).

Different strains of *Acinetobacter* have been used for the resolution of racemic D,L-carnitine or its derivatives. Besides *Achromobacter parafinoclastus, Aeromonas liquefaciens, E. coli, P. aeruginosa, Rhodotorula rubra*, etc., the stereoselective hydrolysis of *O*-acyl-D,L-carnitines was also effected by *A. calcoaceticus* (Nakamura et al., 1985). Another process for resolving a racemic mixture of D,L-carnitine, described by Sih (1985, 1988), involves the growth of organisms of the genus *Acinetobacter (A. calcoaceticus* ATCC 39648, *A. lwoffii* ATCC 39770), or mutants of these strains (*A. calcoaceticus* ATCC 39647, *A. lwoffii* 39769), which are characterised by the unique ability to preferentially degrade D(+)-carnitine in the racemic mixture under aerobic conditions, followed by recovery of the L(-)-carnitine. It should be mentioned that the enzymes for the stereoselective hydrolysis of *O*-acyl-D,L-carnitines and the degradation of D(+)-carnitine have not yet been described.

In contrast to our results (Jung and Jung, 1988; Seim and Kleber, 1988), various strains of bacteria and yeasts, among them *A. calcoaceticus* IFO 13006 (Kawamura et al., 1986a) and *A. lwoffii* ATCC 9036 (Yokozeki and Kubota, 1984), have been described as producing L(-)-carnitine from crotonobetaine under aerobic growth conditions. Culture media used for growth of these microorganisms contained the usual carbon and nitrogen sources and inorganic ions plus crotonobetaine. In addition, *A. calcoaceticus* IFO 13006 also forms L(-)-carnitine from crotonobetaine under anaerobic conditions (Kawamura et al., 1986b). In our experiments, however, a variety of *Acinetobacter* strains (*A. calcoaceticus* 69-V, *A. lwoffii* CCM 2376, *A. lwoffii* ATCC 9036) were unable to synthesise L(-)-carnitine from crotonobetaine under both aerobic and anaerobic growth conditions (Jung and Jung, 1988). We have always found that aerobic growth conditions, as well as the addition of nitrate to anaerobically growing cultures, diminished or inhibited completely the formation of L(-)-carnitine (Seim and Kleber, 1988). The ability to synthesise L(-)-carnitine from crotonobetaine, as well as the activity of carnitine dehydratase, which catalyses the reversible dehydration of L(-)-carnitine:

$$
\begin{array}{cc}
\underset{\substack{|\\CH_3\text{-}N^+\text{-}CH_2\text{-}CH\text{-}CH_2\text{-}COO^-\\|\\CH_3}}{\overset{CH_3 \qquad OH}{}} \rightleftharpoons
\underset{\substack{|\\CH_3\text{-}N^+\text{-}CH_2\text{-}CH=CH\text{-}COO^-\\|\\CH_3}}{\overset{CH_3}{}}
\end{array}
$$

L(-)-carnitine   crotonobetaine

was detected only in cells of *E. coli* which had been grown under anaerobic conditions in complex media supplemented with L(-)-carnitine or crotonobetaine as inducers (Jung and Jung, 1988; Jung et al., 1989). Only if the carnitine dehydratase and the transport system had been induced previously could aerobic carnitine synthesis be achieved (Seim and Kleber, 1988).

On the basis of the facts known so far, and even taking into account the different cultivation conditions that have been used, it appears that different acinetobacters have different pathways for the degradation/metabolism of carnitine and its derivatives.

REFERENCES

Aurich,H., Kleber, H.-P., and Schopp, W.-D., 1967, An inducible carnitine dehydrogenase from *Pseudomonas aeruginosa*, *Biochim. Biophys. Acta*, 139:505.

Englard, S., Blanchard, J.S., and Miura-Fraboni, J., 1983, Production of trimethylamine from structurally related trimethylammonium compounds by resting cell suspensions of ɣ-butyrobetaine- and D,L-carnitine- grown *Acinetobacter calcoaceticus* and *Pseudomonas putida*, *Arch. Microbiol.*, 135:305.

Gitzelmann, G., Baerlocher, K., and Steinmann, B., eds., 1987, "Carnitin in der Medizin," Schattauer, Stuttgart.

Haddock, B.A., and Jones, C.W., 1977, Bacterial respiration, *Bacteriol. Rev.*, 41:47.

Jung, K., and Jung, H., 1988, Zur Charakterisierung und Regulation der L(-)-Carnitinmetabolisierung in *Escherichia coli, Dissertation A, Karl-Marx Universität, Leipzig.*

Jung, H., Jung, K., and Kleber, H.-P., 1989, Purification and properties of carnitine dehydratase from *Escherichia coli* - a new enzyme of carnitine metabolization, *Biochim. Biophys. Acta*, 1003:270.

Kawamura, M., Akutsu, S., Fukuda, H., Hata, H., Morishita, T., Kano, K., and Nishimori, H., 1986a, Microbial production of L-carnitine from crotonobetaine, *J P*, 61,271,995.

Kawamura, M., Akutsu, S., Fukuda, H., Hata, H., Morishita, T., Kano, K., and Nishimori, H., 1986b, Production of L-carnitine, *J P*, 61,234,794.

Kleber, H.-P., Seim, H., Aurich, H., and Strack, E., 1977, Verwertung von Trimethylammoniumverbindungen durch *Acinetobacter calcoaceticus, Arch. Microbiol.*, 112:201.

Kleber, H.-P., Seim, H., Aurich, H., and Strack, E., 1978, Beziehungen zwischen Carnitinstoffwechsel und Fettsäureassimilation bei *Pseudomonas putida, Arch. Microbiol.*, 116:213.

Miura-Fraboni, J., and Englard, S., 1982, Utilization of ɣ-butyrobetaine and D- and L-carnitine by *Acinetobacter calcoaceticus* and *Pseudomonas putida, Fed. Proc.*, 41:1166.

Miura-Fraboni, J., and Englard, S., 1983, Quantitative aspects of ɣ-butyrobetaine and D- and L-carnitine utilisation by growing cell cultures of *Acinetobacter calcoaceticus* and *Pseudomonas putida, FEMS Microbiol. Lett.*, 18:113.

Miura-Fraboni, J., Kleber, H.-P., and Englard, S., 1982, Assimilation of ɣ-butyrobetaine, and D- and L-carnitine by resting cell suspensions of *Acinetobacter calcoaceticus* and *Pseudomonas putida, Arch. Microbiol.*, 133:217.

Nakamura, T., Ito, M., and Ueno, M., 1985, Optically active carnitines, *J P*, 60,214,898.

Seim, H., and Kleber, H.-P., 1988, Synthesis of L(-)-carnitine by hydration of crotonobetaine by enterobacteria, *Appl. Microbiol. Biotech.*, 27:538.

Seim, H., Ezold, R., Kleber, H.-P., and Strack, E., 1980, Stoffwechsel des L-Carnitins bei Enterobakterien, *Zeits. Allg. Mikrobiol.*, 20:591.

Seim, H., Löster, H., Claus, R., Kleber, H.-P., and Strack, E., 1982a, Formation of ɤ-butyrobetaine and trimethylamine from quaternary ammonium compounds structure-related to L-carnitine and choline by *Proteus vulgaris, FEMS Microbiol. Lett.,* 13:201.

Seim, H., Löster, H., Claus, R., Kleber, H.-P., and Strack, E., 1982b, Splitting of the C-N bond in carnitine by an enzyme (trimethylamine forming) from membranes of *Acinetobacter calcoaceticus, FEMS Microbiol. Lett.,* 15:165.

Seim, H., Löster, H., Claus, R., Kleber, H.-P., and Strack, E., 1982c, Stimulation of the anaerobic growth of *Salmonella typhimurium* by reduction of L-carnitine, carnitine derivatives and structure-related trimethylammonium compounds, *Arch. Microbiol.,* 132:91.

Seim, H., Loster, H., and Kleber, H.-P., 1982d, Reduktiver Stoffwechsel des L-Carnitins und strukturverwandter Trimethylammonium-verbindungen in *Escherichia coli, Acta Biol. Med. German.,* 41:1009.

Sih, C.J., 1985, L-Carnitine, *W O*, 85/04,900.

Sih, C.J., 1988, L-Carnitine and its preparation from D,L-carnitine with D-carnitine-metabolizing microorganisms, *US*, 475,1182.

Yokozeki, K., and Kubota, K., 1984, Method for producing L-carnitine, *EP*, 0,122,794.

APPLICATIONS OF *ACINETOBACTER* AS AN INDUSTRIAL

MICROORGANISM

D.L. Gutnick, R. Allon, C. Levy, R. Petter and W. Minas

Department of Microbiology
George S. Wise Faculty of Life Sciences
Tel Aviv University
Ramat Aviv, 69978
Israel

INTRODUCTION

Bacteria have a unique role to play in the development of the biotechnology industry, not only because of their importance in modern recombinant DNA technology and molecular biology, but more importantly because of the incredible diversity of microbial processes and products which are of potential commercial and industrial interest. In many cases, these systems are readily obtained from organisms whose isolation depends simply on suitable enrichment procedures, sometimes coupled with specific screening techniques.

In the case of *Acinetobacter*, several strains of industrial importance have been isolated by either specific enrichment or simple screening (Gutnick, 1987). Two major activities of applied interest associated with various *Acinetobacter* isolates which have received a great deal of attention in recent years are (i) the production of surface polysaccharide-containing biopolymers of potential industrial importance (Gutnick and Minas, 1987; Gutnick and Shabtai, 1987), and (ii) the ability of acinetobacters to sequester large quantities of inorganic phosphate in the form of intracellular polyphosphate, suggesting an application in phosphate removal from waste-water environments (Fuhs and Chen, 1975).

Hospitals are also a "natural environment" for *Acinetobacter* in its role as an opportunistic pathogen (Ramphal and Kluge, 1979; Hinchliffe and Vivian, 1980; Noble, this volume). Large collections of hospital isolates represent potentially useful banks of *Acinetobacter* strains. A study of hospital isolates showed that a significant percentage of strains grew on hexadecane and produced extracellular bioemulsifier activity (Sar and Rosenberg, 1983; Gutnick et al., US Patent No. 4,883,757; see below).

Other activities of applied interest include the potential for hydrocarbon and xenobiotic degradation (Gutnick and Rosenberg, 1977; Gutnick, 1984; Cain, 1987; Asperger and Kleber, this volume; Fewson, this volume), coal desulphurisation (Stevens and Daniels, European Patent No. EP126443; Isbister, US Patent No. 4808535) and manganese leaching (Karavaiko et al., 1986a,b; Yurchenko et al., 1987).

This article examines *Acinetobacter* as an industrial microorganism. The identification, chemical and physical properties, and mode of action of the products themselves are discussed wherever possible. We have attempted to include relevant information on the physiology of production which is essential for the design of an efficient fermentation process. In some cases it has been possible to describe simple approaches to select improved mutant strains which either (a) exhibit enhanced productivity, or (b) produce modified products with a broader range of activities. In this regard we present some of our own results dealing with the cloning and expression of relevant *Acinetobacter* genes in either *Escherichia coli* or heterologous strains of *A. calcoaceticus*.

Some of the work from our own laboratory has been partially reviewed elsewhere (Gutnick and Rosenberg, 1977; Gutnick, 1987; Gutnick and Minas, 1987; Gutnick and Shabtai, 1987; Gutnick et al., 1987).

BIOPOLYMER PRODUCTION BY *ACINETOBACTER*

Emulsan - Acinetobacter strain RAG-1

Oil degradation and bioemulsification. Over 200 different species of eukaryotic and prokaryotic hydrocarbon-utilising microorganisms have been described (Beerstecher, 1954; ZoBell, 1973; Gutnick and Rosenberg, 1977; Rosenberg and Gutnick, 1981; Atlas, 1984). *Acinetobacter* spp. are generally among the most prevalent natural isolates. This is not a surprising observation since, in addition to possessing the catalytic and regulatory genetic potential for hydrocarbon oxidation, *Acinetobacter* spp. also exhibit a major requirement for efficient utilisation of this potential in the natural environment: i.e. the ability to interact physically with the hydrocarbon substrate (Gutnick and Rosenberg, 1977). Two major forms of physical interaction are (i) direct contact between the cells and the hydrophobic growth substrate, which is apparently mediated by cell surface fimbrae that are similar in function to those of other organisms which interact with hydrophobic surfaces (Rosenberg and Kjelleberg, 1986), and which are responsible for the avid adherence of the cells to the oil; (ii) the production of biopolymers which bring about the emulsification of the hydrocarbon growth substrate in water.

The possibility of exploiting the in-situ biodegradation of oil pollutants by microbes led to the isolation and characterisation of *A. calcoaceticus* RAG-1 (ATCC 31012) from a polluted beach in Tel Aviv (Reisfeld et al., 1972). Four basic observations characterised the degradation of crude oil by RAG-1: (i) microscopic - the organism initially adhered to and appeared to grow at the oil/water interface; (ii) nutritional - there was an absolute requirement for nitrogen and phosphorus; (iii) physiological - only the alkane components of the oil were degraded (Horowitz et al., 1975); (iv) physical - the residual oil was emulsified. These factors were incorporated into a large-scale field experiment aboard an oil tanker (Gutnick and Rosenberg, 1977). The experiment had three basic objectives: (a) the in-situ growth of RAG-1 on crude oil residues in an oil storage tank; (b) the dispersion of these polluting residues as a stable oil/water emulsion capable of discharge without the concomitant oil slick formation; (c) the cleaning of the dirty oil storage tank as a result of the efficient emulsification of the oil waste pollutants. All these objectives were achieved (Gutnick and Rosenberg, 1977; US Patent No. 3,941,692) and some of the potential advantages and disadvantages of this approach have been discussed previously (Gutnick, 1987; Gutnick and Minas, 1987). One of the major consequences of this type of large-scale experiment (Gutnick and Shabtai, 1987) was the development of a programme to isolate, characterise and examine the potential applications of the major factor(s) from *Acinetobacter* and other oil-degrading isolates responsible for the formation of stable emulsions. In the case of RAG-1, emulsification was found to be due to the action of an extracellular non-dialysable polymer which was termed emulsan.

Characteristics and interaction with hydrocarbons. The emulsan produced by strain RAG-1 is an extracellular anionic heteropolysaccharide bioemulsifier (Rosenberg et al., 1979a,b; Zuckerberg et al., 1979; Gutnick, 1987). Unlike many surfactants produced by oil-degrading microorganisms, emulsan is also produced on a water-soluble growth substrate, ethanol. Purified emulsan (M_r 10^6) consists of D-galactosamine, D-galactosamine uronic acid (pKa 3.05) and an unidentified hexosamine. The amphipathic properties of emulsan are due in part to the presence of fatty acids linked to the polysaccharide backbone in both ester and amide linkages (Belsky et al., 1979). In addition, when recovered from the growth medium, emulsan is complexed with about 15-25% protein. Removal of the emulsan-associated protein (ranging in M_r from 10-90,000) yields a product, termed apoemulsan, which retains most of the original properties of emulsan itself (Zuckerberg et al., 1979; Zosim et al., 1982).

Emulsan exhibits several activities which are of applied and industrial interest: (i) stabilisation of preformed oil-in-water emulsions; (ii) formation of oil-in-water emulsions under conditions of low energy input; (iii) removal of adherent microorganisms from hydrophobic surfaces. The last activity, while related to the high affinity of emulsan for hydrophobic surfaces, will be discussed separately.

Under conditions of low energy input, emulsan requires a mixture of both aliphatic and aromatic hydrocarbons for optimum emulsification (Rosenberg et al., 1979b). However, we have recently found that this specificity can be modified, either by the addition of small amounts (μg) of aliphatic alcohols, or by the addition of RAG-1 proteins normally associated with crude emulsan which are inactive by themselves, but which enhance the activity of apoemulsan towards pure aliphatics (Zosim et al., 1986, 1987, 1989).

Under conditions of high energy input, such as rapid mixing or sonic oscillation, both emulsan and apoemulsan bind tightly at the interface between the droplet and the bulk aqueous phase. Both activities (at either high or low energy input) require relatively low ratios of emulsan to oil (between 1:500 and 1:1000) because of the very high affinity of the polymer for the oil/water interface (Zosim et al., 1982). Emulsan forms a film of about 20 nm on the surface of the oil droplet with the hydrophilic groups orientated towards the aqueous phase, thus preventing coalescence of the droplets (stabilisation). The affinity is sufficiently high that, even under conditions of strong centrifugation, the emulsion is not broken, but separates into an upper cream layer and a lower aqueous phase. The cream, termed an emulsanosol, is itself an oil/water emulsion consisting of about 70% by weight oil (Zosim et al., 1982). Because of the orientation of the polymer on the surface of the droplet, both organic and inorganic cations can be partitioned into the oil layer as a result of electrostatic interactions with the negatively charged uronic acid residues in the polymer (Zosim et al., 1983). In several instances the binding of the cations to the emulsanosol was even greater than equimolar, probably because of a conformational change in the polymer associated with its binding to the oil/water interface (Zosim et al., 1983).

The stability of the emulsanosols arises from the ability of the polymer to form a stable film around the surface of the oil droplet (Zosim et al., 1983), which is probably stabilised by hundreds of points of contact between the large amphipathic polymer and the oil droplet. One prediction of this model for the rigid interaction of the biopolymer with the interface is that cleavage of the polymer to lower M_r subunits should severely impair the stabilising ability of emulsan. An emulsan depolymerase, isolated from the organism YUV-1 (Shoham and Rosenberg, 1983; Shoham et al., 1983), has been shown to degrade emulsan via a *trans* elimination reaction and to produce oligosaccharides with an average M_r of about 3,000 after exhaustive digestion. Depolymerisation of the emulsan polymer required the presence of ester linkages. Breakage of only 5% of the glycosidic linkages (as assayed by release of reducing groups) was sufficient to reduce the emulsifying activity of emulsan by over 75%, thereby indicating a strong dependence for activity on polymer size. Furthermore, in addition to its ability to degrade emulsan in solution, the enzyme was able to degrade the biopolymer at the oil/water interface (Shoham and Rosenberg, 1983) or on the surface of the producing cell (Pines et al., 1988; see below).

Physiology of emulsan production. The successful applications and technologies for emulsan utilisation depend to a great extent on the availability of the bioemulsifier and the economics of its manufacture (Gutnick, 1984). A cell-associated form of emulsan constitutes a minicapsular layer on the surface of RAG-1 cells growing exponentially on minimal medium, and is released into the medium as the cells approach stationary phase (Goldman et al., 1982; Pines et al., 1983). This capsule is orientated on the cell surface such that (a) emulsan is a receptor for a specific bacteriophage, and (b) the cells are partially protected from the effects of toxic cationic surfactants such as cetyltrimethylammonium bromide (Shabtai and Gutnick, 1985b). Evidence for a natural role for the emulsan capsule comes from the finding that mutants of strain RAG-1 defective in emulsan do not grow on crude oil (Pines and Gutnick, 1986). This deficiency was not corrected by either the addition of emulsan to the culture or by growth in mixed culture with the wild-type RAG-1 which itself produces extracellular emulsan. The results indicate that the form of emulsan required for growth on crude oil is the cell-bound biopolymer (Pines and Gutnick, 1986). Emulsan-producing revertants of the emulsan-defective mutants could be isolated on the basis of their ability to grow on crude oil.

The emulsan minicapsule was easily detected using a crossed immunoelectrophoretic technique and has been shown to be identical to purified apoemulsan (Bayer et al., 1983). The cell-associated biopolymer, which can comprise as much as 30% of the dry cell weight, also serves as a receptor for a specific RAG-1 phage, ap3 (Pines and Gutnick, 1981, 1984a). Thus, mutants of RAG-1 defective in emulsan production are readily isolated on the basis of ap3 resistance (Pines and Gutnick, 1981). Emulsan-deficient strains exhibit a characteristic pale translucent colony morphology which is also observed when the cells are treated with emulsan depolymerase (Shoham et al., 1983; Pines et al., 1988). These enzyme-treated RAG-1 cells become resistant to phage ap3 and simultaneously become sensitive to a second *Acinetobacter* phage, nφ (Pines and Gutnick, 1984b; Pines et al., 1988). Evidence that emulsan undergoes a conformational change once released from the cell surface comes from the finding that phage receptor activity is not normally observed with cell-free emulsan, but can be reconstituted in an emulsanosol. The results are consistent with the cation-binding experiments described above (Zosim et al., 1983) and suggest that the conformation of emulsan at the oil/water interface resembles its conformation at the cell surface (Pines and Gutnick, 1984a).

Emulsan release from the microbial cell surface is mediated by the action of an exocellular esterase which appears in the growth medium just prior to the appearance of cell-free emulsifying activity (Shabtai and Gutnick, 1985a). A role for esterase in capsule release was indicated by studies involving emulsan production in the presence of the protein synthesis inhibitor chloramphenicol (Rubinovitz et al., 1982; Shabtai and Gutnick, 1985a; Shabtai et al., 1985). Both emulsan and esterase release were stimulated in the presence of the antibiotic provided that the medium contained both a carbon and a nitrogen source. In the absence of a carbon source only the esterase was released. Subsequent addition of a carbon source in the presence of chloramphenicol did not bring about release, presumably because the esterase had already been released from the cell surface and could not be made when protein synthesis was inhibited. Mutants defective in esterase were also found to be defective in emulsan production (Shabtai, 1982). In addition to weakening the association of the emulsan minicapsule with the cell surface during fermentation, the esterase allows the cells to utilise simple triglycerides as carbon sources.

Studies using a metabolic control strategy, combined with a computer-aided system for regulating the rate of oxygen uptake by the bacteria during their growth on triglycerides, revealed a complex and highly organised network of cells interconnected through cell surface appendages (Gutnick and Shabtai, 1987; Shabtai and Wang, in preparation). Under conditions of oxygen limitation, these aggregates become destabilised as the emulsan concentration increases significantly. The results of the fermentation study support the hypothesis that restricting the bacterial oxidative metabolism enhances the destabilisation of the minicapsule, resulting in a faster production rate and easier separation of biopolymer from the cells. In this regard it should be noted that emulsan is not the only molecule released from the cell surface during the fermentation. About 25-30% of the released extracellular material consists of a number of proteins ranging in M_r from 20,000 to 90,000 (Shabtai, 1982).

Genes involved in emulsan production. An esterase-coding gene from RAG-1 (*est*) was cloned into *E. coli* (Gutnick et al., 1987; Reddy et al., 1989; Fig. 1). A simple screen for esterase-positive clones exploited the fact that strains of *E. coli* normally do not utilise triglycerides such as triacetin as carbon sources, even though the same strains grow on the

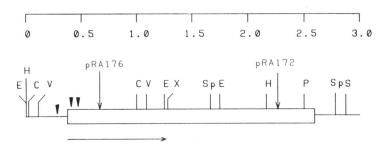

Fig. 1. Physical map of pRA17. The open box indicates the insert of RAG-1 DNA. The arrow at the bottom shows the direction and extent of the open reading frame of the *est* gene. pRA176 (esterase-negative) and pRA172 (esterase-positive) carry Tn5 insertions at the positions indicated by the long downward pointing arrows. Tn*phoA* insertions into *est* are identified by the dark arrow heads. Restriction enzyme sites are as follows: C,*Cla*I; E,*Eco*RI; V,*Eco*RV; H,*Hind*III; P,*Pvu*II; S,*Sal*I; Sp,*Sph*I; X,*Sba*I.

```
5'   ATG AAA TTT GGT ACT GTT TGG AAA TAT TAT TTT ACT GAA TCG TTA
      M   K   F   G   T   V   W   K   Y   Y   F   T   E   S   L

     CTA AAG GCG   3'
      L   K   A
```

Fig. 2. Putative leader sequence of the *est* gene from *A. lwoffii* RAG-1. The translation product of the first 54 bp of the *N*-terminal sequence of the esterase (*est*) gene from RAG-1 (from Reddy, et al., 1989).

hydrolysis products, glycerol and acetate. Esterase-negative mutants of RAG-1 grow on acetate, but not on triacetin (RAG-1 itself cannot grow on glycerol). The same insert was obtained in four independent cloning experiments. Esterase-positive clones exhibited high levels of esterase activity, even when assayed with intact cells. The original esterase-positive plasmid pRA17 carried a 2.2 kb insert which gave a single band following Southern hybridisation with RAG-1 DNA. It was, however, interesting to note that other *Acinetobacter* strains which exhibit cell surface esterase activity and grow on triglycerides, such as BD413 or RA57 (Rusansky et al., 1987), show no homology with the pRA17 *est* probe. After subcloning, the *est* gene was sequenced. An open reading frame of 909 bp suggested that *est* encoded a peptide of M_r 32,700, which correlated well with an in-vitro translation product of 32,500. This polypeptide was not found in clones in which Tn5 was inserted in the gene. In contrast to RAG-1, no esterase activity was found in the cell-free broth. In addition, Tn*phoA* insertions were obtained within the first 300 bp of the insert. Localisation of these fusions by sequencing and restriction analysis indicate that the enzyme is exported out of the cytoplasm in *E. coli*. A putative signal peptide for *est* is shown in Fig. 2.

As described above, expression of the RAG-1 *est* gene in *E. coli* conferred a new physiological function on the *E. coli* host, i.e. the ability to grow on triacetin. Similar results were obtained with esterase-positive clones of BD413 (Gutnick et al., 1987). A second *Acinetobacter* system involves the ability to utilise ethanol as a sole source of carbon and energy. This system is important for emulsan fermentation since polysaccharide yields with RAG-1 are highest when the cells are grown on ethanol (Shabtai et al., 1985). There have been a number of reports describing multiple alcohol and aldehyde dehydrogenases from various strains of *Acinetobacter* (Duine and Frank, 1981; Beardmore-Gray and Anthony, 1983; Fixter and Nagi, 1984; Singer and Finnerty, 1984, 1985).

Several clones were isolated when the RAG-1 genomic library was transformed into *E. coli* and plated on minimal plates containing ethanol as a sole source of carbon and energy. These strains all possessed a plasmid, pRET1, which carried a 5.3 kb insert (Fig. 3). When tested for alcohol dehydrogenase activity (ADH), crude extracts of pRET1-containing cells

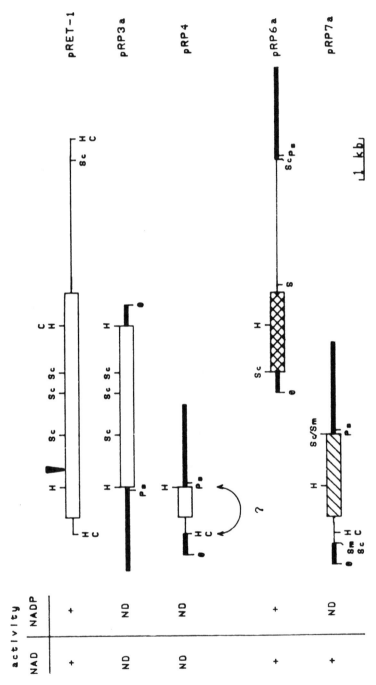

Fig. 3. Subclones of the alcohol dehydrogenase (ADH) complex from RAG-1. Plasmid pRET1 was digested with various restriction endonucleases and cloned into suitably digested pUC18 to generate plasmids pRP3a, 4, 6a and 7a respectively. Cells carrying each of the plasmids were analysed in crude extracts (as shown on the left) or by zymogram analysis for either NAD- or NADP-dependent ADH activities (ND: no detectable ADH activity). The site of Tn5 insertion in pRET1 leading to the elimination of either growth on or oxidation of ethanol is shown by the dark arrow head. Restriction enzyme sites are as follows: C, *ClaI*; H, *HindIII*; Ps, *PstI*; S, *SalI*; Sc, *SacI*; Sm, *SmaI*.

showed both NAD-dependent as well as NADP-dependent activities. In addition, using in-situ specific staining for zymogram analysis on native gels, three distinct NADP-dependent ADH-positive bands were detected, one of which was also NAD-dependent. Expression of the insert in mini-cells showed three unique polypeptides of apparent M_r 60, 49, and 48 kDa respectively. Insertion of Tn5 into one end of the insert in pRET1 eliminated growth on ethanol, all ADH activities, and appearance of the three peptides in a minicell expression system. The results suggested that the genes encoding these activities might be organised or associated within an "ethanol regulon". When pRET1 was subcloned into pUC18, two plasmids, pRP6 and pRP7 were obtained. Clones containing pRP6 exhibited both NAD- and NADP-dependent activities, while the ADH from pRP7 was dependent on NADP alone. Experiments are currently in progress in our laboratory to identify which of the cloned ADH genes are involved in capsule production and to study their regulation.

Mutants over-producing emulsan can be isolated on the basis of resistance to a cationic surfactant such as cetyltrimethylammonium bromide (CTAB) which is highly toxic to RAG-1 (Shabtai and Gutnick, 1986). Emulsan-overproducers generate elevated levels of the polyanionic bioemulsifier which neutralises the toxic effect of cetyltrimethylammonium bromide (Shabtai and Gutnick, 1985b, 1986). This protection is mediated by either cell-associated or cell-free emulsan (Shabtai and Gutnick, 1985b). Over-producers isolated in this fashion have yielded between two and three times more emulsan /g cells than the wild-type strain. Our most recent experiments indicate that this approach may be applicable to the isolation of over-producers which produce a variety of different polyanionic bioemulsifiers.

A second approach to strain improvement involves the use of flow cytometry, in conjunction with immunocytochemical labelling and cell sorting, to isolate cells which exhibit larger amounts of a specific antigen on the cell surface. In addition, this approach may be used to screen genomic libraries for the expression of new specific cell-surface antigens in *E. coli*. The feasibility of this approach has been demonstrated with a population of *E. coli* clones containing a genomic library from RAG-1 (Minas et al., 1988).

Adhesion to hydrophobic surfaces. Strains of *Acinetobacter* grow at an oil/water interface as a result of their ability to adhere to the substrate (Rosenberg and Rosenberg, 1985). The role of hydrophobic interactions in microbial adherence has been reviewed extensively by Rosenberg and Kjelleberg (1986). A simple and semi-quantitative assay for adherence was initially developed to study the interaction of strain RAG-1 with oil (Rosenberg et al., 1980). In this method washed bacterial suspensions are mixed with liquid hydrocarbons, an emulsion is formed and, upon settling, the adherent organisms rise to the surface with the hydrocarbon droplets, resulting in a concomitant measurable decrease in optical density of the aqueous phase. This decrease was taken as an approximation of the cell surface hydrophobicity of the bacterial cells. Adherent organisms could be removed from the oil droplets with alcohols or with emulsan (Pines and Gutnick, 1986). Using this approach it was possible to describe a number of common features related to the cell surface hydrophobicity of RAG-1 and other acinetobacters: (i) The adherence to the hydrocarbon (whether liquid or solid) was not dependent on the ability of the organism to degrade it. (ii) Cells grown under conditions which were optimal for emulsan production were actually less hydrophobic than when grown on rich medium in which emulsan is not produced. (iii) Mutants defective in emulsan production were more hydrophobic than the wild-type (Pines and Gutnick, 1984b). Similar results were obtained when emulsan was removed from the RAG-1 cell surface with emulsan depolymerase, or from the cell surface of *A. calcoaceticus* BD4 with a similar enzyme active against the BD4 capsule (Rosenberg et al., 1983c; Pines et al., 1988). (iv) RAG-1 mutants lacking thin fimbrae on the cell surface were found to be defective in adherence to hydrocarbons (Rosenberg et al., 1982). (v) Mutants defective in adherence to hydrocarbon could still produce emulsan (Pines and Gutnick, 1984b). Selection for emulsan-defective derivatives of these non-adherent mutants restored the hydrophobicity even though the revertant cells still lacked the thin fimbrae. The results suggest that alternative sites on the surface of the RAG-1 cell can mediate adhesion to hydrocarbons (Pines and Gutnick, 1984b). (vi) Under conditions of poor mixing and with a small inoculum, mutants defective in adherence lost the ability to grow on hydrocarbons, suggesting a role for adherence in the early stages of growth on these substrates (Rosenberg and Rosenberg, 1985). Growth was restored by the addition of emulsan to the medium to disperse the growth substrate. It thus appears that, in the case of RAG-1, early growth on hydrocarbons occurs primarily via direct contact at the oil/water interface. During subsequent growth, emulsan which accumulates in the culture broth: (a) brings about a detachment of the cells from the hydrocarbon droplets, and (b) enhances emulsion formation which further increases the interfacial area.

According to the adherence assay described above, many other bacteria such as pathogenic streptococci (Rosenberg et al., 1983b), *Yersinia* spp. (Lachica and Zink, 1984), *Serratia* (Rosenberg and Kjellenberg, 1986), cyanobacteria (Fattom and Shiloh, 1985), myxobacteria (Kupfer and Zusman, 1984), group A streptococci (Ofek et al., 1983) and gramicidin-producing bacilli (Rosenberg, 1986) present hydrophobic cell surfaces. Moreover, the adherence of many of these strains to hydrocarbons was found to be inhibited by emulsan. It was of interest that RAG-1 itself could bind to other biological (buccal epithelial cells) or non-biological (polystyrene) hydrophobic surfaces (Rosenberg et al., 1981). Finally, in addition to removing the bound cells, emulsan prevented adherence of washed suspensions to the surface (Rosenberg et al., 1983a,d). The potential industrial applications of these "anti-adherence" properties of *Acinetobacter* biopolymers will be discussed in a separate section.

Strains BD4 and BD413.   Taylor and Juni (1961a,b,c) originally characterised the capsular polysaccharide of the heavily encapsulated *A. calcoaceticus* strain BD4.   Kaplan and Rosenberg (1982) studied the protein-dependent emulsifying factors from BD4 and a minicapsulated derivative mutant, BD413, originally characterised by Juni and Janick (1969).   Both these strains are highly competent in specific transformation assays with DNA from virtually all known isolates of *Acinetobacter*.   It was found that these strains grow on hydrocarbons and produce extracellular emulsifiers.   Both strains produce an extracellular polysaccharide, with a 3:1 molar ratio of rhamnose to glucose, which was similar in composition to the cell-bound capsular material.   The structure of this material was shown to consist of a repeating heptasaccharide branched subunit (Kaplan et al., 1985).   While strain BD4 produced about four times more total polysaccharide (cell bound and extracellular) than the mutant BD413, the latter strain released up to half of the polysaccharide into the medium.   The parent strain BD4 released only about 10% of the rhamnose-containing polymer into the medium.   Unlike RAG-1, the BD strains grow and produce emulsifier on glucose.   This may be of importance in any scale-up process for emulsifier production.   The emulsifying factor from BD413 was precipitated in the presence of 50% ammonium sulphate and shown to resemble emulsan in its requirement for mixtures of aliphatic and aromatic hydrocarbons as substrates for emulsification.   It is of interest that, in contrast to emulsan, the BD413 product is relatively inactive with crude oil.   In this regard it should be noted that, unlike RAG-1, *A. calcoaceticus* BD413 does not grow on crude oil (Rusansky, 1985).   When the capsular polysaccharide of BD4 is released as an extracellular polymer during growth, the product is active as an emulsifier.   However, when the capsule is mechanically removed by shearing from the cell surface, the polysaccharide is not active as an emulsifier.   Extracellular protein produced by a mutant which does not make the polysaccharide has been shown to reconstitute the emulsifying activity of the rhamnose polymer.   The amphipilic properties of the BD413 polymer are thus provided through the interaction of protein with the polysaccharide.   It remains to be determined whether the hydrophobic portion of the emulsifier complex is due to the presence of hydrophobic proteins, or whether the amphilicity is due to conformational changes in the complex.   Recently the lipoprotein from *E. coli* has been shown to be active in restoring emulsifying activity to the BD413 polysaccharide (Z. Zosim, unpublished results).

Biodispersan - *A. calcoaceticus* strain A2. The *Acinetobacter* emulsifiers, emulsan or the polysaccharide-protein complex of BD4, exert their effects by adhering to a hydrophobic organic surface.   Rosenberg et al., (1988a,b) described the production by *A. calcoaceticus* strain A2 of an extracellular polysaccharide, termed Biodispersan, which binds to inorganic materials such as calcium carbonate.   The organism was first isolated from soil by an enrichment procedure for *Acinetobacter* (Baumann, 1968), followed by a screening of the extracellular culture broth for its ability to disperse limestone powders in water.   Strain A2 was identified as a member of the genus *Acinetobacter* on the basis of its ability to transform BD413. However, since the organism was unable to grow on glucose (Rosenberg et al., 1988a), it may be more appropriate to assign it to the species *A. lwoffii*. The biodispersan polymer accumulated in the culture broth of an ethanol-grown culture during late exponential and stationary growth (Rosenberg et al., 1988a). The material was precipitated from the culture broth with ammonium sulphate and subsequently purified by hot phenol treatment to produce a polysaccharide of M_r 51,400; yields of about 30% by weight and 84% by activity were obtained. Biodispersan appeared to consist of four different amino sugars, including glucosamine, rhamnosamine or fucosamine, galactosamine uronic acid and an unidentified amino sugar. Potentiometric titration and ^{13}C NMR analyses suggested that at least 90% of the amino groups were acetylated. The polymer showed no emulsifying activity, but was effective in dispersing the inorganic limestone powders (Rosenberg et al., 1988b). In addition to its dispersing activity, Biodispersan was shown to dramatically reduce the time necessary to grind calcium carbonate to fine particles, apparently as a result of binding of the polymer (Rosenberg et al., 1989).

***A. calcoaceticus* strain *HO1-N* - *Particulate surfactant*.**
Kappeli and Finnerty (1979, 1980) observed the accumulation of extracellular membrane vesicles in cell-free broth from a culture of *A. calcoaceticus* strain HO1-N growing on hexadecane. These vesicles were composed of proteins, phospholipids and lipopolysaccharides, suggesting that they might be derived from the outer membrane of the cells. However, the phospholipid to protein ratio was five times greater than in the outer membrane, while the lipopolysaccharide content (determined by assaying for keto-deoxyoctulosonic acid) appeared to be enriched some 360-fold. The vesicles were found only in the culture broths of cells growing on alkanes. The authors suggest that these extracellular vesicles play an important role in alkane uptake since: (i) they are only found in cultures growing on hydrocarbons; (ii) radioactive hexadecane bound to the vesicles was taken up by intact cells; (iii) the vesicle-solubilised hexadecane was transferred to hydrocarbon inclusion bodies normally observed in alkane-growing cells of this strain (Singer and Finnerty, 1984).

Molecular Analysis of Bioemulsifier Production by Acinetobacter

Apart from RAG-1, BD4 and BD413, a number of other strains of *Acinetobacter* have been isolated from natural environments and shown to grow on hydrocarbons and to produce extracellular emulsifying activity (Sar and Rosenberg, 1983). Studies with specific phages (Pines and Gutnick, 1981) indicated that these emulsifier complexes exhibit widely different structures. In most cases the emulsifiers required both a capsular polysaccharide and a mixture of extracellular proteins for activity.

Both BD4 and BD413 are naturally competent for transformation with genomic DNA from all known *Acinetobacter* strains (Juni, 1972), suggesting a high degree of homology in both sequence and function.

Moreover, it has been found (W.Minas and D.Gutnick, in preparation) that while plasmids derived from ColE1 are not stably maintained in BD413, they do integrate into the chromosome provided a fragment of homologous *Acinetobacter* DNA is present in the plasmid. It was postulated that if RAG-1 DNA contains homologous sequences to emulsifier-related genes in BD4 or BD413, it might be possible to detect such sequences by virtue of their ability to integrate into the BD4 chromosome, thereby giving rise to emulsifier-defective mutants (Levy, 1990; W.Minas and D.Gutnick, in preparation). Genomic libraries of RAG-1 DNA cloned into a derivative of pBR322 were used to transform either BD413 or BD4 to kanamycin resistance. Among the BD4 transformants, six mutants defective in capsular polysaccharide production were isolated, initially on the basis of translucent (*tlu*) colony morphology. Further evidence that these mutants were defective in exopolysaccharide (EPS) production included: (i) India ink staining; (ii) resistance to exopolysaccharide-specific phages; (iii) lack of extracellular emulsifying activity; (iv) enhanced adherence to hydrocarbon. Chromosomal DNA isolated from the kanamycin-resistant transformants could be used to transform BD4 to *tlu*. Southern blot analysis demonstrated

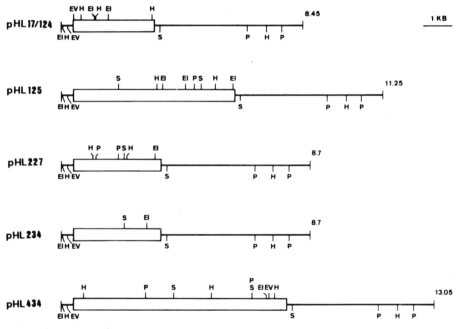

Fig. 4. Physical maps of rescued plasmids from exopolysaccharide mutants of strain BD4. Following insertional mutagenesis by homologous recombination, the integrated plasmids were rescued from the chromosome by digestion with *Sac*I followed by self-ligation and transformation into *E. coli* MM294 as described in the text. The inserts are represented by the blank boxes. The total size (kb) of each plasmid is shown at the right hand end; the vector itself, pWM3, is 6.5 kb. Restriction enzyme sites are as follows: EI, *Eco*RI; EV, *Eco*RV; H, *Hind*III; P, *Pst*I; S, *Sal*I.

that the entire plasmid was integrated into the chromosome. The flanking sequences adjacent to the integrated vectors were rescued from the chromosomal DNA of the six different mutants by cleavage with *SacI* (which does not cut the vector), and transformed following self-ligation into *E. coli*. As evidenced by restriction analysis (Fig. 4), five unique sequences were isolated from six mutants. These were used as probes in subsequent Southern blot analysis of DNA from either RAG-1 or BD4.

The functions encoded by these sequences are not known. However, it was of interest to note that three of the six BD4 mutants were unable to grow on minimal medium with glucose as carbon source (Levy, 1990). This was at first surprising in view of the finding that RAG-1 cannot grow on glucose (Reisfeld et al., 1972). One possible explanation emerged from the finding that the mutants unable to grow on glucose are also defective in glucose oxidation. In certain strains of *A. calcoaceticus*, such as BD4, glucose oxidation is generally catalysed by glucose dehydrogenase (GDH), whose activity is mediated by the co-factor pyrroloquinoline-quinone (PQQ) (Duine et al., 1979; van Schie et al., 1984, 1985). In addition, GDH activity can be restored by the addition of PQQ to many of the strains of *A. lwoffii* (e.g. RAG-1) which cannot grow on or oxidise glucose. Such strains possess an apo-GDH protein, but lack the genes necessary for the biosynthesis of PQQ (Duine, this volume). In some cases it is, therefore, likely, that the mutants generated by homologous recombination between RAG-1 and BD4 are the result of insertion of the gene for apo-GDH. This cannot, however, explain the behaviour of all of the mutants since the restriction maps are completely different (Fig. 4). The genes from BD4 that encode the synthesis of PQQ have recently been cloned and sequenced (Goosen et al., 1987, 1989). Four gene products, encoded on a 5.1 kb fragment, are required for production of PQQ in *A. calcoaceticus*. When a plasmid carrying all four genes was introduced into a strain of *A. lwoffii* which produced apo-GDH, the resulting transconjugants could produce acid from a variety of sugars. All four genes were required to restore holo-GDH activity in the *A. lwoffii* system. The ability to synthesise PQQ could also be introduced into a phosphotransferase-defective mutant of *E. coli* K12 that was unable to grow on glucose, produced apo-GDH, but lacked the PQQ co-factor. Transformants were able to grow slowly on glucose provided the PQQ genes were placed under the control of the *lac* promoter. Restoration of the PQQ synthesising system was, however, less than complete since exogenous addition of the cofactor enhanced the growth of *E. coli* transformants even further. The results with the PQQ system suggest the possibility of manipulating RAG-1 genetically so that growth and/or production of emulsan or other biopolymers could be achieved using glucose or other sugars as carbon substrates.

Cryptic plasmids in BD413 and the construction of shuttle vectors. In one of the BD413 mutants generated by the homologous recombination method mentioned above, extrachromosomal DNA was detected which was not observed before the transformation. However, using Southern blot analysis, no homology was detected between this DNA and the

424

original plasmid vector used for transformation. Further analysis of the extrachromosomal DNA revealed that BD413 itself (even before transformation) contains three cryptic plasmids of 2.3, 2.6 and 7.4 kb respectively. The three plasmids were observed only in preparations of total genomic DNA. Two plasmids were detected with a similar extraction procedure in BD4. Southern blot analysis revealed homology between all three plasmids of BD413 and the two plasmids of BD4. Two of the BD413 plasmids were sub-cloned into pUC19, and restriction maps established (Fig. 5). Identical restriction sites were found in a 2.1 kb fragment. The recombinant plasmids were stably maintained as independent replicons in *E. coli* MM294 and *A. calcoaceticus* BD413 or BD4. No apparent modification of the plasmids was detected, even after several reciprocal transfers between *E. coli* and *A. calcoaceticus*. When sub-clones carrying partial deletions of the cloned 2.3 kb *Acinetobacter* plasmid were examined for transformation efficiency (Fig. 6), it was found that a minimum size fragment of about 190 bp (Figs. 6 and 7) was sufficient to support

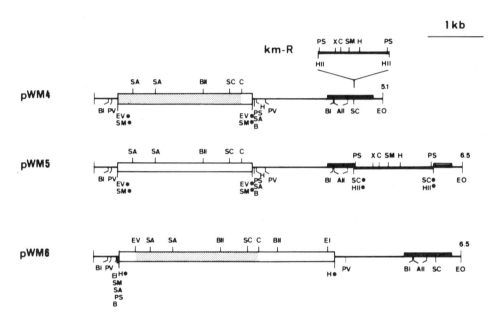

Fig. 5. Partial restriction maps of the cloned cryptic plasmids from strain BD413. pWM4 carries the cloned 2.3 kb plasmid from BD413 (shown as a box), linearised with *Eco*RV and cloned into the *Sma*I site of pUC19. Plasmid pWM6 carries a 3.9 kb *Hin*dIII fragment of the 7.4 kb cryptic plasmid from BD413. pWM5 is derived from pWM4 and carries the kanamycin resistance gene of Tn903 (heavy line) inserted into the *bla* gene (striped box) of the pUC19 part (line) of the shuttle vector. Shaded areas of the inserts indicate regions with identical restriction sites in pWM4 and pWM6. Restriction enzyme sites are as follows: AII,*Ava*II; B,*Bam*HI; BI,*Bgl*I; BII,*Bgl*II; BS,*Bst*EII; C,*Cla*I; EO,*Eco*0109; EI,*Eco*RI; EV,*Eco*RV; H,*Hin*dIII; HII,*Hin*cII; PS,*Pst*I; PV,*Pvu*II; SA,*Sal*I; SM,*Sma*I; X,*Xho*I.

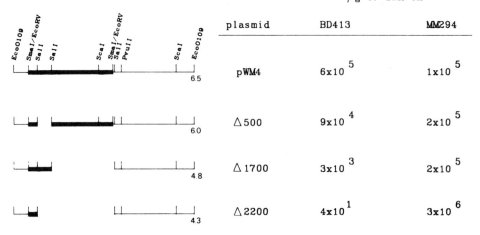

Transformants per

μg of DNA in

plasmid	BD413	MM294
pWM4	6×10^5	1×10^5
△500	9×10^4	2×10^5
△1700	3×10^3	2×10^5
△2200	4×10^1	3×10^6

Fig. 6. Efficiencies of transformation in *A. calcoaceticus* BD413 and *E. coli* MM294 using shuttle plasmids pWM4 and deletions. Deletions in pWM4 were generated by partial digestion with *Sal*I followed by self-ligation. Sites of the deletions are indicated as gaps in the restriction maps shown on the left side of the figure. The size of the deletion is expressed in bp. The remaining portion of the insert is indicated by the filled boxes. Transformation efficiency is expressed as the number of ampicillin-resistant transformants of either BD413 or *E. coli* MM294 per μg plasmid DNA.

```
          10          20          30          40          50          60
          *           *           *           *           *           *

      CCCATCCGCT TCAATTGGAA AATCAGACT GTTTGAACTG GCGAAGCAAA GCTATATTTG


          70          80          90         100         110         120
          *           *           *           *           *           *

      ATGAATCCTG CCAATTGAAA GTCTCACAA CCATTGCCGA TTGCATTACC TTGAGATAAA


         130         140         150         160         170         180
          *           *           *           *           *           *

      TCCATTGGGA TTTCACAATA CAAAGCAAC ATGGTCATAG TGACGGTTTG AACCGTCAGA


         190
          *

      CGGTTTAAAG
```

Fig. 7. Sequence of the 190 bp fragment which still permits independent replication of pUC19 in *A. calcoaceticus* BD413.

replication in the *Acinetobacter* strains. However, the efficiency of transformation with the various sub-clones was reduced by up to four orders of magnitude, depending on the size of the deletion (Fig. 6). No effects resulting from the deletions were observed with respect to replication in *E. coli*. Although we were not able to introduce DNA into RAG-1 via conventional methods of transduction, transformation or conjugation, the shuttle vector could be transferred into the emulsan-producing strain using electroporation. Although the copy number of the vector in RAG-1 was too low for it to be observed using normal extraction procedures, it could be readily isolated as an unmodified shuttle vector following extraction of total DNA from the electroporants and transformation into either *E. coli* or BD413. Both restriction and Southern blot analysis demonstrated that the plasmid was not integrated into the RAG-1 chromosome. Finally, the shuttle vector was used to clone alcohol dehydrogenase genes from RAG-1, which were shown subsequently to positively complement mutants unable to grow on ethanol (selected on the basis of their resistance to allyl alcohol). In further experiments the 1.2 kb gene for the extracellular esterase from RAG-1 has been inserted. Expression of this gene in *E. coli* allows growth on simple triglycerides such as triacetin. Subsequent insertion of foreign sequences within the esterase gene abolishes growth on these substrates (R. Alon, R.Petter and D.Gutnick, unpublished results).

Hunger et al. (1990) described the construction of a shuttle vector by fusing a cryptic plasmid from *A. lwoffii* with pBR322. This vector was found to be functional in both *E. coli* and *A. calcoaceticus*. The plasmid was sub-cloned and shown to contain a 1350 bp fragment which apparently contained the origin of replication. This *ori* consisted of an AT-rich region, an 18 bp element with palindromic symmetry, and 11 repeats of a consensus sequence, AAAAATAT, eight of which were clustered within about 360 bp. Interestingly, this sequence was not present in the 190 bp segment minimally required for stability, maintenance and replication of pWM4 in *Acinetobacter* (Figs. 6 and 7). The 9 bp repeat was, however, found in several locations throughout the 2.1 kb insert.

Kaplan and Silverman (1989) have begun a study of the genetic and regulatory aspects of exopolysaccharide biosynthesis (EPS) in BD4. A recombinant cosmid genomic library bank of BD4 constructed in *E. coli* was used to complement spontaneous EPS⁻ mutants. All of the EPS⁻ mutants mapped to one of two separate loci, *epsA* or *epsB*. Most of the EPS⁻ mutants carried deletions in *epsA*, and transposon mutagenesis demonstrated that a region of about 10 kb in *epsA* was essential for EPS production. A strange result was obtained when an EPS⁻ mutant, termed 4011, carrying a deletion of over 30 kb, of which 20 kb were missing from *epsA* and 10 kb were missing from *epsB*, was transformed with purified *epsB* DNA. The transformants were found to contain both "missing" fragments, the 20 kb of *epsA* and the 10 kb of *epsB*. Moreover, chromosomal DNA from mutant 4011 could transform other *epsA* mutants to wild-type despite the fact that the deletion in 4011 was larger than the deletion in the recipient *epsA* mutants. Southern blot analysis indicated that the transformants contained a wild-type *epsA* locus. Kaplan and Silverman (1989) suggested that the *eps* DNA in deletion 4011 is maintained in some sort of "archived" or "invisible" form which escapes detection by classical means of gene expression or hybridisation. In addition to elucidating the mechanisms associated with DNA "archiving", it will be of interest to determine whether the phenomenon is general, or specific to the EPS system in *Acinetobacter*, and whether the process plays some role in the regulation of capsule production.

APPLICATIONS OF *ACINETOBACTER* SURFACE POLYMERS

The chemical industry has produced very large numbers of synthetic materials which carry out many of the activities described above. This vast myriad of products, which ostensibly do the same thing, stems largely from industrial experience which has shown that there is no universally accepted surfactant which fits the needs of every application. Moreover, the ability to formulate such materials in combination with others to generate surfactant packages for specific applications has added to the demand for surfactants in a number of industries. Over 2.5 million tons of surfactants were sold in the USA alone in 1983 (Gutnick, 1984). If chemically-enhanced oil recovery technology becomes applicable, surfactant consumption is likely to double by 1995. These optimistic projections for growth of the surfactant industry have stimulated a great deal of interest in the potential application of biotechnology as a new dimension for commercial development in this area. The advantages of the approach include the ability to: (i) exploit the vast amount of knowledge and experience from the fermentation industry; (ii) design simple yet powerful selection techniques for the isolation of suitable surfactant-producing organisms from nature; (iii) utilise classical techniques of mutation and selection, as well as recombinant DNA technology, in strain improvement programmes; (iv) exploit unique biological specificity for producing specific formulations; (v) produce materials which are non-toxic. Along with the advantages come a number of problems which must be overcome if biosurfactant production and technology is to become a reality. These may include: (a) the high cost of fermentation products in comparison to those which are chemically synthesised; (b) the difficulty of introducing new products to compete with materials already on the market; (c) the high cost of health risk assessments for new products introduced into certain industries.

Kosaric et al., (1987) and Kachholz and Schlingmann (1987) have considered many of the properties of potential biosurfactants as well as the industries where they may be applied. These properties include emulsification, de-emulsification, wetting, spreading and penetration, solubilisation, dispersal of solids, air-entrapment, foaming, de-foaming, detergency, antistatic and corrosion inhibition. The industries include agriculture, building and construction, elastomer and plastics, foods and beverages, industrial cleaning, leather, metals, paper, paint and protective coating, petroleum and petrochemical products, and textiles. However, we have found only a few published reports summarising the results of large-scale field experiments with biosurfactants. Such experiments require that: (i) the fermentation and production technology be developed to the point where sufficient product can be produced for testing; (ii) sufficient information is available so that the biosurfactant product can be suitably formulated for the particular application to be tested; (iii) enough is known about the particular application so that it can be tested on a large scale using suitable equipment, storage conditions, means of monitoring results etc.; (iv) prior governmental approval for certain tests is obtained. Developmental programmes of this magnitude are usually sponsored within the industry itself and, as such, the results are not always available.

The Use of Emulsan in Stabilisation of Heavy-Oil-in-Water Emulsions

Although high molecular mass extracellular bioemulsifiers such as emulsan are not particularly effective in reducing interfacial tension, they do have the property of binding tightly to an interface. The practical outcome of such a property is the ability to utilise relatively small amounts of biopolymer to stabilise oil-in-water emulsions. This section will describe two field experiments illustrating potential applications of emulsion stabilisation which represented part of a programme by Petroferm USA to explore two technologies (Hayes et al., 1986): (a) pipeline transportation of oil-in-water emulsions of heavy viscous crude oils, and (b) direct combustion (without de-watering) of oil-in-water emulsions containing viscous hydrocarbons.

Heavy oil transportation. A pipeline test was carried out in which about 40 barrels of Venezuelan crude oil (viscosity of about 200,000 cp) were emulsified to form a 70% oil-in-water emulsion with a commercial surfactant-emulsan formulation at a ratio of one part surfactant package to 500 parts oil. The emulsion was pumped a physical distance of approximately 380 miles at ambient temperature through a 3.125 in. half-mile pipeline loop. The apparent viscosity of the emulsion was reduced to 70 cp. During the test the flow was stopped for threedays to simulate a pump failure. When pumping was resumed there were no effects observed on pressure drop, flow rates, or emulsion characteristics. It was calculated that when pipe diameters, pump diameters, flow rates, frequency of pump transits, etc., were considered, the emulsion was subjected to stresses and shear forces comparable to having been pumped a distance of 26,000 miles in a commercial pipeline. Conditions of high stress during pump transit generally give rise to inversions of oil-in-water emulsions stabilised by classical surfactants.

Direct combustion of heavy fuel emulsions. Injection of small amounts of water into fuel to form water-in-oil emulsions improve combustion characteristics of the fuel. It was of interest, therefore, to determine whether emulsan-stabilised oil-in-water emulsions could be combusted directly. In this test a high-asphaltene no.6 fuel oil-in-water emulsion (70:30 w/w) was prepared one week prior to the combustion test.

The results of the test can be summarised as follows: (i) ignitability and stability of the biopolymer-stabilised emulsion were comparable to the unemulsified no.6 oil; (ii) minimum excess air levels of less than 2% were obtained without any visible smoke or CO in the flue gas; (iii) flame appearance was the same as that of the no.6 fuel oil itself; (iv) carbon burnout and light-off were also similar to the parent oil; (v) axial flame temperatures were lower by 100-150°C along the length of the combustion chamber; (vi) nitrogen oxide gases (Nox emissions) were significantly reduced during combustion of the emulsions. The direct combustion test of emulsan-stabilised emulsion fuels suggested several advantages including: (a) ease of transportation; (b) uniform flow behaviour for various fuels - the hydrocarbon fuel itself does not significantly influence the emulsion viscosity; (c) ease of handling cumbersome heavy fuels such as high-asphaltene vacuum residuals; (d) upgrading of low cost fuels - cutter stock need not be added in order to burn the lower quality fuel. This technology has been tested on a large scale and shown to be feasible. Its commercial viability will depend on the relative cost of crude oil compared to the savings associated with upgrading a heavy fuel by emusifying it in water.

The Use of Emulsan in the Treatment of Dental Plaque

A patent filed by Eigen et al. (US Patent No. 4,619,825), on behalf of the Colgate-Palmolive Company, describes the application of emulsan to improve oral hygiene by removing dental plaque. Dental plaque is a deposit of a mucilaginous gelatinoid-type material which forms on the surface of teeth and which is readily produced by and/or invaded by bacteria. Different organic acids produced by the plaque-forming bacteria can cause decalcification of the dental enamel, leading to dental caries. Other microbial products are thought to promote gingivitis by attacking soft gum tissue. *Streptococcus mutans* is generally recognised as the major aetiological source of dental caries. Emulsan solutions (0.1%) were shown to reduce dental plaque formation by *S. mutans in vitro* by about 81%. Similar results were obtained with *S. salivarius* SS2 and *S. sanguis* FC4. In an animal model employing hamsters whose mouths were infected with *S. mutans*, it was found that: (a) pre-treatment of the infecting organisms with apoemulsan resulted in about a 61% decrease in the recoverable *S. mutans* after feeding the hamsters for five weeks on a diet designed to induce dental caries; (b) when the infection was carried out with *S. mutans* which were not pre-treated, but with emulsan (1%) included in the drinking water, the reduction was 77%; (c) a large decrease in dental caries was observed following either pre-treatment with emulsan solutions (47%), or inclusion in drinking water (63%). Finally, after infection, hamster teeth were swabbed twice daily with solutions containing either sodium fluoride or emulsan (0.25 or 1% respectively). At the end of a 12-week period the hamster teeth were examined for *S. mutans* infections and dental caries. In the case of fluoride, the *S. mutans* population actually increased (+21%), while the emulsan treatment resulted in a 53% reduction compared with a control containing only water. However, fluoride treatment did bring about a 90% reduction in dental caries compared with 61% for the emulsan treatment. The authors described emulsan-containing formulations for either mouthwash or toothpaste which combine the anti-caries activity of fluoride and the activity of emulsan in reducing plaque formation. The results with emulsan were somewhat surprising in view of the finding that emulsan interactions with microbes are generally hydrophobic in nature (Rosenberg et al., 1981; Rosenberg and Kjelleberg, 1986), while the adhesion of *S. mutans* is thought to be primarily via lectin-like interactions (Gibbons, 1980; Gibbons and Etherden, 1983). The patent authors explain the strong effect of emulsan in the reduction of dental caries and plaque formation as an interaction of the galactosamine residues in the apoemulsan polysaccharide backbone with lectins on the surface of the *S. mutans* cells, thereby preventing their adherence. If this is true, than de-esterified emulsan, which is no longer amphipathic, should also be effective.

The Use of Biodispersan in Grinding Limestone and Making Paper

Limestone (calcium carbonate) is a component of paper, the quality of which is determined to some extent by the size, shape and homogeneity of the limestone particles. The major cost in the processing of limestone powder is grinding, since chemical dispersants are frequently not used because of their adverse effects on the quality of the paper, as well as the difficulties of disposal. When Biodispersan (0.1%) was tested in this system it was found to reduce the time necessary for grinding from 6 h to 3 h. After 9 h about 70% of the particles were about 2μm in diameter. Production of Biodispersan was scaled-up and several trials showed that paper blended with Biodispersan-ground limestone was of high quality. It was concluded that, in addition to its properties as a flocculating surfactant, Biodispersan is also a "*surfractant*". The mechanism of action of this interesting biopolymer is currently unknown. Biodispersan has been patented in a number of countries (e.g. Rosenberg and Ron, Swiss Patent No. SW122 672500).

Drag Reduction

The turbulent flow of fluids can be significantly impaired by the friction of the fluids. This phenomenon, known as drag, has been shown to be reduced by viscous polymers. Both synthetic and biopolymers have been shown to be effective drag reducers (Hoyt, 1985). Sar and Rosenberg (1987) reported that the isolated capsular polysaccharide from BD4 was an effective drag-reducing polymer. A three-fold higher drag-reducing activity was observed when the capsule was tightly bound to the cell surface (3.6×10^9 cells /ml reduced the friction of turbulent water by 55%). Cells from a capsule-deficient mutant did not have any significant drag-reducing activity. The cells apparently provide a natural system for immobilisation of the polymers.

PHOSPHATE REMOVAL FROM WASTE-WATER
BY *ACINETOBACTER*

Phosphate removal from waste-water is an essential feature of any sewage treatment facility because of the threat of eutrophication. Although the addition of various agents, such as salts of calcium, iron or aluminium, to activated sludge can be used to remove phosphates from the waste-water by settling, the treatment causes the accumulation of large quantities of chemical sludge. In addition, the initial cost of the materials required for chemical treatment has induced investigators to seek alternative biological methods for phosphate removeal (e.g. Levin and Shapiro, 1965; Fuhs and Chen, 1975; Buchan, 1983; Marais et al., 1983; Hao and Chang, 1987). The unique role of *Acinetobacter* in this process of phosphate removal will be discussed briefly below. Fixter and Sherwani (this volume) provide a critical assessment of the evidence for the function of polyphosphate as an energy reserve and a phosphate reserve in acinetobacters.

Accumulation of Polyphosphate by Microorganisms

Any practical application of metabolic uptake to remove dissolved orthophosphate (Pi) from sewage effluent depends on the ability of the sludge organisms to take up the phosphate in quantities exceeding those required for normal metabolic activities. Polyphosphate (Pn), the form in which excess phosphate is frequently stored, has been detected in a wide variety of microorganisms (Kulaev, 1979; Kulaev and Vagabov, 1983) and is usually stored either in the periplasm (Umnov et al., 1975; Tijssen and van Steveninck, 1984) or in the form of large cytoplasmic volutin granules (Harold, 1966; Kulaev, 1979; Miller, 1984).

There are generally thought to be two distinct physiological conditions leading to excess Pi accumulation by microorganisms. The so-called "phosphate overplus" phenomenon (Harold, 1966) involves the uptake of excess Pi which occurs when phosphate-starved microorganisms are transferred to a phosphate-rich growth medium. Sewage treatment plants are unlikely to be limited for phosphate and so this form of accumulation is probably not applicable. The second mode of phosphate accumulation, which is often referred to as "luxury uptake" (Harold, 1966; Deinema et al., 1980, 1985), involves the uptake of Pi by certain microorganisms under conditions of either balanced growth or, more frequently, under conditions of nutrient limitation.

A number of reports have demonstrated that effective phosphate removal from activated sludge effluents requires alternating aerobic and anaerobic cycles (Fuhs and Chen, 1975; Timmerman, 1979; Deinema et al., 1985; Fukase et al., 1985; Suresh et al., 1985). Examination of activated sludge samples from these sewage treatment plants for prevalent strains demonstrated that the dominant organism is *Acinetobacter*, either *calcoaceticus* or *lwoffii* (Fuhs and Chen, 1975; Buchan, 1983; Marais et al., 1983; Lotter, 1985; Lotter and Murphy, 1985; Murphy and Lotter, 1986;

Hao and Chang, 1987). Under aerobic conditions the *Acinetobacter* strains store the accumulated phosphate as Pn, which is subsequently broken down and released as Pi in response to the anaerobic treatment. Suresh et al. (1985) studied a strain of *A. lwoffii* isolated from a sewage treatment plant, using ^{31}P-NMR spectroscopy, and demonstrated that both the phosphate released during the anaerobic phase and that originally present in the feed were accumulated when the cells were subsequently exposed to aerobic conditions, confirming the previous observations (Fuhs and Chen, 1975; Buchan, 1983; Hao and Chang, 1987). However, NMR analysis (Suresh et al., 1985) monitors only the surface (periplasmic) water-soluble Pn granules which are dispersed by membrane impermeable chelators such as EDTA (Halvorson et al., 1987) and cannot detect the large cytoplasmic volutin granules. It was hypothesised (Suresh et al., 1985), and subsequently confirmed (Halvorson et al., 1987), that the Pn is maintained in two distinct pools, and that only the periplasmic polyphosphate is mobilised and released during the aerobic-anaerobic transition. The cytoplasmic volutin granules remained relatively inactive metabolically during this period. Surface Pn breakdown occurs under metabolic stress, e.g. ATP limitation, which would be expected to occur under anaerobic conditions (*Acinetobacter* strains are strictly aerobic). According to Fuhs and Chen (1975), during the aerobic period a natural enrichment for *Acinetobacter* occurs because of their ability to grow on fermentation products such as ethanol, organic acids and alcohols, which accumulate as a result of glucose metabolism during the anaerobic cycle. Recently, van Groenestijn et al. (1989) determined the optimum conditions for Pn accumulation in various *Acinetobacter* spp. isolated from sewage treatment plants. Pn accumulation was stimulated by streptomycin and inhibited by energy poisons. Maximum accumulation occurred when the energy source was in excess and the cells were limited for sulphur at low growth rates, or when nitrogen or sulphur was depleted during the stationary phase. Moreover, Pn accumulation appeared to be relatively insensitive to changes in pH over a range of 6.5-9.

CONCLUDING REMARKS

In this review we have presented several lines of evidence pointing to the advantages (both potential and actual) of *Acinetobacter* as an industrial microorganism. The unique cell surface properties of many acinetobacters, together with their ability to compete in Nature once a selective advantage is presented (e.g. sewage treatment plants), offers the possibility of isolating surface active polymers, many of which exhibit specific interactions at interfaces. Moreover, experience with large-scale production at both pilot and industrial levels has demonstrated that *Acinetobacter* is easily handled in the factory, quite stable in its growth and production characteristics, and lends itself to strain improvement. Finally, the development of suitable recombinant techniques for use with strains of industrial importance is likely to lead to an understanding of the underlying mechanisms of product synthesis and regulation, and offers an opportunity for producing second and third generation products.

ACKNOWLEDGEMENTS

The authors are indebted to Rina Avigad for excellent technical and administrative assistance for a large portion of the work carried out in the laboratory of DLG. The work done in Tel Aviv was supported in part by a grant from Petroleum Fermentations, Inc.

REFERENCES

Atlas, R.M., ed., 1984, "Petroleum Microbiology," Macmillan, New York.

Baumann, P., 1968, Isolation of *Acinetobacter* from soil and water, *J. Bacteriol.*, 96:34.

Bayer, E.A., Skutelsky, E., Goldman, S., Rosenberg, E., and Gutnick, D.L., 1983, Immunochemical identification of the major cell surface agglutinogen of *Acinetobacter calcoaceticus* RAG-92, *J. Gen. Microbiol.*, 129:1109.

Beardmore-Gray, M., and Anthony, C., 1983, The absence of quinoprotein alcohol dehydrogenase in *Acinetobacter calcoaceticus*, *J. Gen. Microbiol.*, 129:2979.

Beerstecher, E., 1954, "Petroleum Microbiology," Elsevier, New York.

Belsky, I., Gutnick, D.L., and Rosenberg, E., 1979, Emulsifier of *Arthrobacter* RAG-1, determination of emulsifier-bound fatty acids, *FEBS Lett.*, 101:175.

Buchan, L., 1983, Possible biological mechanism of phosphorous removal, *Water Sci. Tech.*, 15:87.

Cain, R., 1987, Biodegradation of anionic surfactants, *Biochem. Soc. Trans.*, 15S:7S.

Deinema, M.H., Habets, L.H.A., Scholten, J., Turkstra, E., and Webers, H.A.A.M., 1980, The accumulation of polyphosphate in *Acinetobacter* spp., *FEMS Microbiol. Lett.*, 9:275.

Deinema, M.H., van Loosdrecht, M., and Scholten, A., 1985, Some physiological characteristics of *Acinetobacter* spp. accumulating large amounts of phosphate, *Water Sci.Tech.*, 17:119.

Duine, J.A., and Frank, J., 1981, Quinoprotein alcohol dehydrogenase from a non-methylotroph *Acinetobacter calcoaceticus*, *J. Gen. Microbiol.*, 122:201.

Duine, J.A., Frank, J., and van Zeeland, J.K., 1979, Glucose dehydrogenase from *Acinetobacter calcoaceticus*: a quinoprotein, *FEBS Lett.*, 108:443.

Fattom, A., and Shilo, M., 1984, Hydrophobicity as an adhesion mechanism of benthic Cyanobacteria, *Appl. Environ. Microbiol.*, 47:135.

Fixter, L.M., and Nagi, M.N., 1984, The presence of an NADP-dependent alcohol dehydrogenase activity in *Acinetobacter calcoaceticus*, *FEMS Microbiol. Lett.*, 22:297.

Fuhs, G.W., and Chen, M., 1975, Microbiological basis of phosphate removal in the activated sludge process for the treatment of wastewater, *Microb. Ecol.*, 2:119.

Fukase, T., Shibata, M., and Miyaji, Y., 1985, The role of an anaerobic stage on biological phosphorous removal, *Water Sci. Tech.*, 17:69.

Gibbons, R.J., 1980, Adhesion of bacteria to the surfaces of the mouth, *in* "Adsorption of Microorganisms to Surfaces," p.351, R.C.W. Berkeley, J.M. Lynch, J. Melling, P.R. Rutter and B. Vincent, eds., Ellis Horwood, Chichester.

Gibbons, R.J., and Etherden, I., 1983, Comparative hydrophobicities of oral bacteria and their adherence to salivary pellicles, *Infect. Immun.*, 41:1190.

Goldman, S., Shabtai, Y., Rubinovitz, C., Rosenberg, E., and Gutnick, D.L., 1982, Emulsan in *Acinetobacter calcoaceticus* RAG-1, distribution of cell-free and cell-associated cross reacting material, *Appl. Environ. Microbiol.*, 44:165.

Goosen, N., Vermaas, A.M., and van de Putte, P., 1987, Cloning of the genes involved in synthesis of coenzyme pyrroloquinoline-quinone from *Acinetobacter calcoaceticus*, *J. Bacteriol.*, 169:303.

Goosen, N., Horsman, H.P.A., Huinen, R.G.M., and van de Putte, P., 1989, *Acinetobacter calcoaceticus* genes involved in biosynthesis of the coenzyme pyrrolo-quinoline-quinone: nucleotide sequence and expression in *Escherichia coli* K-12, *J. Bacteriol.*, 171:447.

Gutnick, D., 1984, Biosurfactants in the oil industry, *World Biotech Report*, 2:645.

Gutnick, D.L., 1987, The emulsan polymer: Perspectives on a microbial capsule as an industrial product, *Biopolymers*, 26S:223.

Gutnick, D.L., and Minas, W., 1987, Perspectives on microbial surfactants, *Biochem. Soc. Trans.*, 15S:22.

Gutnick, D.L., and Rosenberg, E., 1977, Oil tankers and pollution, a microbiological approach, *Ann. Rev. Microbiol.*, 31:379.

Gutnick, D.L., and Shabtai, Y., 1987, Exopolysaccharide emulsifiers, *in* "Biosurfactants and Biotechnology," p.211, N. Kosaric, W.L. Cairns and N.C.C. Gray, eds., Marcel Dekker, New York.

Gutnick, D.L., Reddy, P.G., Allon, R., Rusansky, S., and Avigad, R., 1987, Cloning and expression in *E. coli* of Emulsan related genes from the oil-degrading bacterium *Acinetobacter calcoaceticus* RAG-1, *in* "Genetics of Industrial Microorganisms. Proceedings of the Fifth International Symposium on the Genetics of Industrial Microorganisms," p.313, M.Alacevic, D. Hranueli and Z. Toman, eds., Ognjen Prica, Karlovac.

Halvorson, H.O., Suresh, N., Roberts, M.F., Coccia M., and Chikarmane, H.M., 1987, Metabolically active surface polyphosphate pool in *Acinetobacter lwoffi*, *in* "Phosphate Metabolism and Cellular Regulation in Microorganisms," p.220, A. Torriani-Gorini, F.G. Rothman, S. Silver, A. Wright and E. Yagil, eds., American Society for Microbiology, Washington, D.C.

Hao, O.J., and Chang, C.H., 1987, Kinetics of growth and phosphate uptake in pure culture studies of *Acinetobacter Species*, *Biotech. Bioeng.*, 24:819.

Harold, F.M., 1966, Inorganic polyphosphates in biology: structure, metabolism and function, *Bacteriol. Rev.*, 30:772.

Hayes, M.E., Hrebenar, K.R., Murphy, P.L., Futch, L.E., and Deal, J.F., 1986, Combustion of viscous hydrocarbons, *US Patent 4,618,348.*

Hinchliffe, E., and Vivian, A., 1980, Naturally occurring plasmids in *Acinetobacter calcoaceticus*: pAV1, a P class R factor of restricted host range, *J. Gen. Microbiol.,* 116:75.

Horowitz, A., Gutnick, D.L., and Rosenberg, E., 1975, Sequential growth of bacteria on crude oil, *Appl. Environ. Microbiol.,* 30;10.

Hoyt, J.W., 1985, Drag reduction in polysaccharide solutions, *Trends Biotech.,* 3:17.

Hunger, M., Schmucker, R., Kishan, V., and Hillen, W., 1990, Analysis and nucleotide sequence of an origin of DNA replication in *Acinetobacter calcoaceticus* and its use for *Escherichia coli* shuttle plasmids, *Gene,* 87:45.

Juni, E., 1972, Interspecies transformation of Acinetobacter: genetic evidence for a ubiquitous genus, *J. Bacteriol.,* 112:917.

Juni, E., and Janick, A., 1969, Transformation of *Acinetobacter calcoaceticus (Bacterium anitratum),* J. Bacteriol., 98:281.

Kacholz, T., and Schlingmann, M., 1987, Possible food and agricultural application of microbial surfactants, an assessment, *in* "Biosurfactants and Biotechnology," p.183, N. Kosaric, W.L. Cairns and N.C.C. Gray, eds., Marcel Dekker, New York.

Kaplan, N., and Rosenberg, E., 1982, Exopolysaccharide distribution of and bioemulsifier production by *Acinetobacter calcoaceticus* BD4 and BD413, *Appl. Environ. Microbiol.,* 44:1335.

Kaplan, N., and Silverman, M., 1989, Analysis of cloned exopolysaccharide genes and of "archived" DNA from *Acinetobacter calcoaceticus* BD4, *Abst. Ann. Meet. Amer. Soc. Microbiol.,* H130.

Kaplan, N., Rosenberg, E., Jann, B., and Jann, K., 1985, Structural studies of the capsular polysaccharide of *Acinetobacter calcoaceticus* BD4, *Eur. J. Biochem.,* 152:453.

Kappeli, O., and Finnerty, W.R., 1979, Partition of alkane by an extracellular vesicle derived from hexadecane-grown *Acinetobacter, J. Bacteriol.,* 140:707.

Kappeli, O., and Finnerty, W.R., 1980, Characteristics of hexadecane partition by the growth medium of *Acinetobacter, Biotech. Bioeng.,* 22:495.

Karavaiko, G.I., Yurchenko, V.A., Remizov, V.I., and Klyushnikova, T.M., 1986a, Release of hydrogen peroxide by the *Acinetobacter calcoaceticus* culture for manganese leaching from oxides, *Microbiol. Zh.,* 48:41.

Karavaiko, G.I., Yurchenko, V.A., Remizov, V.I., and Klyushnikova, T.M., 1986b, Manganese dioxide reduction by *Acinetobacter calcoaceticus* cell-free extracts, *Microbiologiya,* 55:709.

Kosaric, N., Gray, N.C.C., and Cairns, W.L., 1987, Introduction: Biotechnology and the surfactant industry, *in* "Biosurfactants and Biotechnology," p.1, N. Kosaric, W.L. Cairns and N.C.C. Gray, eds., Marcel Dekker, New York.

Kulaev, I.S., 1979, "Biochemistry of Inorganic Polyphosphates," John Wiley, New York.

Kulaev, I.S., and Vagabov, V.M., 1983, Polyphosphate metabolism in microorganisms, *Adv. Microb. Physiol.*, 24:83.

Kupfer, D., and Zusman, D.R., 1984, Changes in cell surface hydrophobicity of *Myxococcus xanthus* are correlated with sporulation-related events in the developmental program, *J. Bacteriol.*, 159:776.

Lachica, R.V., and Zink, D.L., 1984, Plasmid-associated cell surface charge and hydrophobicity of *Yersinia enterocolitica, Infect. Immun.*, 44:540.

Levin, G.V., and Shapiro, J., 1965, Metabolic uptake of phosphorous by waste-water organisms, *J. Water Pollut. Control Fed.*, 37:800.

Levy, C., 1990, Using insertional mutagenesis to study genes involved in polysaccharide synthesis and glucose utilization in *Acinetobacter calcoaceticus* RAG-1 and BD4, *MSc Thesis, Tel Aviv University, Israel.*

Lotter, L.H., 1985, The role of polyphosphate metabolism in enhanced phosphorous removal from the activated sludge process, *Water Sci. Tech.*, 17:127.

Lotter, L.H., and Murphy, M., 1985, The identification of heterotrophic bacteria in an activated sludge plant with particular reference to polyphosphate accumulation, *Water SA*, 11:179.

Marais, G.V.R., Loewenthal, R.E., and Siebritz, I.P., 1983, Observations supporting phosphate removal by biological excess uptake - a review, *Water Sci. Tech.*, 115:15.

Miller, J.J., 1984, *In vitro* experiments concerning the state of polyphosphate in the yeast vacuole, *Can. J. Microbiol.*, 30:236.

Minas, W., Sahar, E., and Gutnick, D., 1988, Flow cytometric screening and isolation of *Escherichia coli* clones which express surface antigens of the oil-degrading microorganism *Acinetobacter calcoaceticus* RAG-1, *Arch. Microbiol.*, 150:432.

Murphy, M., and Lotter, L.H., 1986, The effect of acetate and succinate on polyphosphate formation and degradation in activated sludge, with particular reference to *Acinetobacter calcoaceticus, Appl. Microbiol. Biotech.*, 24:512.

Ofek, I., Whitnack, E., and Beachey, E.H., 1983, Hydrophobic interactions of group A streptococci with hexadecane droplets, *J. Bacteriol.*, 154:139.

Pines, O., and Gutnick, D.L., 1981, Relationship between phage resistance and emulsan production; interaction of phages with the cell surface of *Acinetobacter calcoaceticus* RAG-1, *Arch. Microbiol.*, 130:129.

Pines, O., and Gutnick, D.L., 1984a, Specific binding of a bacteriophage at a hydrocarbon-water interface, *J. Bacteriol.*, 157:179.

Pines, O., and Gutnick, D.L., 1984b, Alternate hydrophobic sites on the cell surface of *Acinetobacter calcoaceticus* RAG-1, *FEMS Microbiol. Lett.*, 22:307.

Pines, O., and Gutnick, D.L., 1986, A role for emulsan for growth of *Acinetobacter calcoaceticus* on crude oil, *Appl. Environ. Microbiol.*, 51:661.

Pines, O., Bayer, E.A., and Gutnick, D.L., 1983, Localisation of emulsan-like polymers associated with the cell surface of *Acinetobacter calcoaceticus, J. Bacteriol.*, 154:893.

Pines, O., Shoham, Y., Rosenberg, E., and Gutnick, D.L., 1988, Unmasking of surface components by removal of cell-associated emulsan from *Acinetobacter* sp. RAG-1, *Appl. Microbiol. Biotech.*, 28:93.

Ramphal, R., and Kluge, R.M., 1979, *Acinetobacter calcoaceticus* variety *anitratus*: an increasing nosocomial problem, *Amer. J. Med. Sci.*, 277:57.

Reddy, P.G., Allon, R., Mevarech, M., Mendelovitz, S., Sato, Y., and Gutnick, D.L., 1989, Cloning and expression in *Escherichia coli* of an esterase-coding gene from the oil-degrading bacterium *Acinetobacter calcoaceticus* RAG-1, *Gene*, 76:145.

Reisfeld, A., Rosenberg, E., and Gutnick, D.L., 1972, Microbial degradation of crude oil: factors affecting the dispersion in sea water by mixed and pure cultures, *Appl. Microbiol.*, 24:363.

Rosenberg, E., 1986, Microbial surfactants, *CRC Crit. Rev. Biotech.*, 3:109.

Rosenberg, E., and Gutnick, D.L., 1981, The hydrocarbon-oxidising bacteria, *in* "The Procaryotes: A Handbook on Habitats, Isolation & Identification of Bacteria," p.903, M.P. Starr, H. Stolp, H.G. Trueper, A. Balows and H.G. Schlegel, eds., Springer Verlag, Heidelberg.

Rosenberg, M., and Kjelleberg, S., 1986, Hydrophobic interactions: role in bacterial adhesion, *Adv. Microb. Ecol.* 9:353.

Rosenberg, M., and Rosenberg, E., 1985, Bacterial adherence at the hydrocarbon-water interface, *Oil Petrochem. Pollut.*, 2:155.

Rosenberg, E., Perry, A., Gibson, D., and Gutnick, D.L., 1979a, Emulsifier of *Arthrobacter* RAG-1: specificity of hydrocarbon substrate, *Appl. Environ. Microbiol.*, 37:409.

Rosenberg, E., Zuckerberg, A., Rubinovitz, C., and Gutnick, D.L., 1979b, Emulsifier of *Arthrobacter* RAG-1: isolation and emulsifying properties, *Appl. Environ. Microbiol.*, 37:402.

Rosenberg, M., Gutnick, D., and Rosenberg, E., 1980, Adherence of bacteria to hydrocarbons: a simple method for measuring cell-surface hydrophobicity. *FEMS Microbiol. Lett.*, 9:29.

Rosenberg, M., Perry, A., Bayer, E.A., Gutnick, D.L., Rosenberg, E., and Ofek, I., 1981, Adherence of *Acinetobacter calcoaceticus* RAG-1 to human epithelial cells and to hexadecane, *Infect. Immun.*, 33:29.

Rosenberg, M., Bayer, E.A., Delarea, J., and Rosenberg, E., 1982, Role of thin fimbriae in adherence and growth of *Acinetobacter calcoaceticus* on hexadecane, *Appl. Environ. Microbiol.*, 44:929.

Rosenberg, E., Gottlieb, A., and Rosenberg, M., 1983a, Inhibition of bacterial adherence to epithelial cells and hydrocarbons by emulsan, *Infect. Immun.*, 39:1024.

Rosenberg, M., Judes, H., and Weiss, E., 1983b, Cell surface hydrophobicity of dental plaque microorganisms in situ, *Infect. Immun.*, 42:831.

Rosenberg, E., Kaplan, N., Pines, O., Rosenberg, M., and Gutnick, D., 1983c, Capsular polysaccharides interfere with adherence of *Acinetobacter calcoaceticus* to hydrocarbons, *FEMS Microbiol. Lett.*, 17:157.

Rosenberg, M., Rosenberg, E., Judes, H., and Weiss, E., 1983d, Bacterial adherence to hydrocarbons and to surfaces in the oral cavity, *FEMS Microbiol. Lett.*, 20:1.

Rosenberg, E., Rubinovitz, C., Gottlieb, A., Rosenhak, S., and Ron, E.Z., 1988a, Production of biodispersan by *Acinetobacter calcoaceticus* A2, *Appl. Environ. Microbiol.*, 54:317.

Rosenberg, E., Rubinovitz, C., Legmann, R., and Ron, E.Z., 1988b, Purification and chemical properties of *Acinetobacter calcoaceticus* A2 biodispersan, *Appl. Environ. Microbiol.*, 54:323.

Rosenberg, E., Schwartz, Z., Tenenbaum, A., Rubinovitz, C., Legmann, R., and Ron, E.Z., 1989, A microbial polymer that changes the surface properties of limestone: effect of biodispersan in grinding limestone and making paper, *J. Disp. Sci. Tech.*, 10:241.

Rubinovitz, C., Gutnick, D.L., and Rosenberg, E., 1982, Emulsan production by *Acinetobacter calcoaceticus* in the presence of chloramphenicol, *J. Bacteriol.*, 152:126.

Rusansky, S., 1985, Growth and emulsification of crude oil by R57, a multiplasmid-containing strain of *Acinetobacter*, *MSc Thesis, Tel Aviv University, Israel.*

Rusansky, S., Avigad, R., Michaeli, S., and Gutnick, D.L., 1987, Involvement of a plasmid in growth on and dispersion of crude oil by *Acinetobacter calcoaceticus* R57, *Appl. Environ. Microbiol.*, 53:1918.

Sar, N., and Rosenberg, E., 1983, Emulsifier production by *Acinetobacter calcoaceticus* strains, *Curr. Microbiol.*, 9:309.

Sar, N., and Rosenberg, E., 1987, Drag reduction by *Acinetobacter calcoaceticus* BD4, *Appl. Environ. Microbiol.*, 53:2269.

Shabtai, Y., 1982, Enhancement of release of an amphipathic exopolysaccharide from *Acinetobacter calcoaceticus* RAG-1, *PhD Thesis, Tel Aviv University, Israel.*

Shabtai, Y., and Gutnick, D.L., 1985a, Exocellular esterase and emulsan release from the cell surface of *Acinetobacter calcoaceticus* RAG-1, *J. Bacteriol.*, 161:1176.

Shabtai, Y., and Gutnick, D.L., 1985b, Tolerance of *Acinetobacter calcoaceticus* RAG-1 to the cationic surfactant cetyltrimethylammonium bromide (CTAB) - role of the bioemulsifier emulsan, *Appl. Environ. Microbiol.*, 49:192.

Shabtai, Y., and Gutnick, D.L., 1986, Enhanced emulsan production in mutants of *Acinetobacter calcoaceticus* RAG-1 selected for resistance to cetyltrimethylammonium bromide, *Appl. Environ. Microbiol.*, 52:146.

Shabtai, Y., Pines, O., and Gutnick, D.L., 1985, Emulsan: a case study of microbial capsules as industrial products, *Develop.Indust.Microbiol.*, 26:291.

Shoham, Y., and Rosenberg, E., 1983, Enzymatic depolymerization of emulsan, *J. Bacteriol.*, 156:161.

Shoham, Y., Rosenberg, M., and Rosenberg, E., 1983, Bacterial degradation of emulsan, *Appl. Environ. Microbiol.*, 46:573.

Singer, M.E., and Finnerty, W.R., 1984, Microbial metabolism of straight-chain and branched alkanes, *in* "Petroleum Microbiology," p.1, R.M. Atlas, ed., Macmillan, New York.

Singer, M.E., and Finnerty, W.R., 1985, Alcohol dehydrogenase in *Acinetobacter* sp. strain HO1-N: Role in hexadecane and hexadecanol metabolism, *J. Bacteriol.*, 164:1017.

Suresh, N., Warburg, R., Timmerman, M., Wells, J., Coccia, M., Roberts, M.F., and Halvorson, H.O., 1985, New strategies for the isolation of microorganisms responsible for phosphate accumulation, *Water Sci.Tech.*, 17:99.

Taylor, W.H., and Juni, E., 1961a, Pathways for biosynthesis of a bacterial capsular polysaccharide. I. Characterisation of the organism and polysaccharide, *J. Bacteriol.*, 81:688.

Taylor, W.H., and Juni, E., 1961b, Pathways for biosynthesis of bacterial capsular polysaccharide. II. Carbohydrate metabolism and terminal oxidation mechanisms of a capsule producing coccus, *J. Bacteriol.*, 81:694.

Taylor, W.H., and Juni, E., 1961c, Pathways for biosynthesis of bacterial capsular polysaccharide. III. Synthesis from radioactive substrates, *J. Biol. Chem.*, 236:1231.

Tijssen, J.P.F., and van Steveninck, J., 1984, Detection of a yeast polyphosphate fraction localized outside the plasma membrane by the method of phosphorous-31 nuclear magnetic resonance, *Biochem. Biophys. Res. Commun.*, 119:447.

Timmerman, W.M., 1979, Biological phosphate removal from domestic waste water using anaerobic/aerobic treatment, *Devel. Indust. Microbiol.*, 20:285.

Umnov, A.M., Steblyak, A.G., Umnova, S., Mansurova, S.E., and Kulaev, I.S., 1975, Possible physiological role of the high molecular weight polyphosphate phosphohydrolase system in *Neurospora crassa*, *Mikrobiologiya*, 44:414.

van Groenestijn, J.W., Zuidema, M., Van de Worp, J.J.M., Deinema, M.H., and Zehnder, A.J.B., 1989, Influence of environmental parameters on polyphosphate accumulation in *Acinetobacter* sp., *Ant. v. Leeuw.*, 55:67.

van Schie, B.J., van Dijken, J.P., and Kuenen, J.G., 1984, Non-coordinated synthesis of glucose dehydrogenase and its prosthetic group in *Acinetobacter* and *Pseudomonas* species, *FEMS Microbiol. Lett.*, 24:133.

van Schie, B.J., Hellingwerf, K.J., van Dijken, J.P., Elferink, M.G.L., van Dijl, J.M., Kuenen, J.G., and Konings, W.N., 1985, Energy transduction by electron transfer via a pyrroloquinoline-quinone-dependent glucose dehydrogenase in *Escherichia coli, Pseudomonas aeruginosa*, and *Acinetobacter calcoaceticus, J. Bacteriol.*, 163:493.

Yurchenko, V.A., Karavaiko, G.I., Remizov, V.I., Klyushnikova, T.M., and Tarusin, A.D., 1987, Role of the organic acids produced by *Acinetobacter calcoaceticus* in manganese leaching, *Prikl. Biochim. Mikrobiol.*, 23:404.

ZoBell, C.E., 1973, Microbial degradation of oil: present status, problem and perspectives, *in* "The Microbial Degradation of Oil Pollutants," p.3, Georgia State University, Atlanta.

Zosim, Z., Gutnick, D.L., and Rosenberg, E., 1982, Properties of hydrocarbon-in-water emulsions stabilized by *Acinetobacter* RAG-1 emulsan, *Biotech. Bioeng.*, 24:281.

Zosim, Z., Gutnick, D.L., and Rosenberg, E., 1983, Uranium binding by emulsan and emulsanosols, *Biotech.Bioeng.*, 25:1725.

Zosim, Z., Rosenberg, E., and Gutnick, D.L., 1986, Changes in hydrocarbon emulsification specifity of the polymeric bioemulsifier emulsan: effects of alkanols, *Colloid Polymer Sci.*, 264:213.

Zosim, Z., Gutnick, D.L., and Rosenberg, E., 1987, Effect of protein content on the surface activity and viscosity of emulsan, *Colloid Polymer Sci.*, 265:442.

Zosim, Z., Fleminger, G., Gutnick, D., and Rosenberg, E., 1989, Effect of protein on the emulsifying activity of emulsan, *J. Disp. Sci.Tech.*, 10:307.

Zuckerberg, A., Diver, A., Perri, Z., Gutnick, D.L., and Rosenberg, E., 1979, Emulsifier of *Arthrobacter* RAG-1, chemical and physical properties, *Appl. Environ. Microbiol.*, 37:414.

INDEX

Cadmium ion transport, 284
Caprolactone, 392-4, 397-8
Capsular polysaccharide, 6, 78,
    259, 415, 421-4
CARB enzymes, 89, 91
Carbon monoxide dehydrogenase, 11
L-Carnitine, 13
    metabolism of, 13, 403-10
    synthesis of, 407-8
CAT, *see* Chloramphenicol resistance
Cat genes, 203-30
Catechol 1,2-dioxygenase, 202,
    224-8, 356, 366
Catechol metabolism, 203, 225,
    334-7, 360-2, 366-7,
    371-2
CAZ enzymes, 95
Cell envelope protein profiles,
    29, 43
Cephalosporinase, *see* ß-Lactamase
Chemotaxis, 353
Chloramphenicol resistance, 137-48
4-Chlorobiphenyl, 161, 373
Chlorocatechol, 225
Chlorocatechol oxygenase, 225-6
1-Chlorohexadecane, 330
Chromosome mobilisation, 152, 193
Chromosome organisation, 7, 194
Cinnamate, 370
Citrate synthase, 313
Citric acid cycle, 313-31
    biosynthetic branch points, 319
    energy control of, 317
    multipoint control of, 320
Classification, *see* Taxonomy
Clinical importance, 4-5, 54-9
Cloning, (*see also* Genes)
    of antibiotic resistance genes,
        136-8
    of benzene utilisation genes, 163
    of *est* genes, 416-9
    of glucose dehydrogenase genes,
        300-2
    of *pca* genes, 162
Cloning vectors, *see* Vector plasmids
Co-factor, *see* PQQ
Codon usage, 206-8, 217-20, 241,
    251-8
Collagenase, 263

Colonisation, *see* Occurrence
Competence, development of, 185
Composition of outer membrane,
    264-6
Conjugative transfer,
    of chromosomal DNA, 152, 193-6
    of plasmids, 150, 193
Continuous culture, 8, 274, 283, 303
Control of transcription, 145
Crotonobetaine, 403
Crude oil degradation,
        see Oil degradation
CTX enzymes, 95
Cycloalkanes, 391
Cyclohexane metabolism, 391-402
Cyclohexanecarboxylic acid, 398
Cyclohexanol, 391-3, 398
Cyclohexanone, 391-5, 398
Cyclohexanone 1,2-monooxygenase,
    394-6
    gene coding for, 396
Cycloisomerase, 202
Cyclopentanone, 397
Cytochromes, 9, 298, 325-6, 335-7,
    341-2
    b_{562}, 298
    P-450, 325-6, 334-7
Cytoplasmic membranes, 340-1

Decarboxylase, 202
Dental plaque, treatment of, 430-1
Detergents, sensitivity to, 262
2,4-Dichlorophenoxyacetate, 373
DNA probes,
    for AAD(2"), 135
    for CATI, 134
DNA sequence exchange, 204, 219-28
DNA slippage structures, 204, 222-8
n-Dodecane, 324
Drag reduction, 431

Electron transport, 8, 334, 341
Emulsan, 331, 340, 412-20, 429-30
    genes for production of, 416-9
Emulsification,
        see Bioemulsifying agents
Endemic infection, 54
Endotoxin, 80
Energy conservation, 8, 303